Proteomics: Technology, Analysis and Applications

Proteomics: Technology, Analysis and Applications

Edited by Belinda Pitman

SYRAWOOD
PUBLISHING HOUSE
New York

Published by Syrawood Publishing House,
750 Third Avenue, 9th Floor,
New York, NY 10017, USA
www.syrawoodpublishinghouse.com

Proteomics: Technology, Analysis and Applications
Edited by Belinda Pitman

International Standard Book Number: 978-1-68286-753-2 (Hardback)

Cataloging-in-Publication Data

Proteomics : technology, analysis and applications / edited by Belinda Pitman.
    p. cm.
Includes bibliographical references and index.
ISBN 978-1-68286-753-2
1. Proteomics. 2. Proteins. 3. Proteins--Biotechnology. 4. Molecular biology. I. Pitman, Belinda.
TP248.65.P76 P76 2019
572.6--dc23

TABLE OF CONTENTS

PREFACE

This book aims to highlight the current researches and provides a platform to further the scope of innovations in this area. This book is a product of the combined efforts of many researchers and scientists, after going through thorough studies and analysis from different parts of the world. The objective of this book is to provide the readers with the latest information of the field.

Proteins are large biomolecules or macromolecules that are made of long chains of amino acid residues. The entire set of proteins that are produced by an organism is called the proteome. It differs from cell to cell and also varies with time. A cell makes different proteins at different times and conditions, such as during cellular differentiation, carcinogenesis, cell cycle and development. The study of proteins is covered by the discipline of proteomics. This field has evolved tremendously due to the development of many innovative approaches. The common techniques for the study of proteins are mass spectrometry and protein profiling, reverse-phased protein microarrays and protein chips. This book brings forth some of the most innovative concepts and elucidates the unexplored aspects of proteomics. Different approaches, evaluations, methodologies and advanced studies have been included herein. The extensive content of this book provides the readers with a thorough understanding of the subject.

I would like to express my sincere thanks to the authors for their dedicated efforts in the completion of this book. I acknowledge the efforts of the publisher for providing constant support. Lastly, I would like to thank my family for their support in all academic endeavors.

Editor

Proteome changes in the small intestinal mucosa of broilers (*Gallus gallus*) induced by high concentrations of atmospheric ammonia

Jize Zhang, Cong Li, Xiangfang Tang, Qingping Lu, Renna Sa and Hongfu Zhang[*]

Abstract

Background: Ammonia is a well-known toxicant both existing in atmospheric and aquatic system. So far, most studies of ammonia toxicity focused on mammals or aquatic animals. With the development of poultry industry, ammonia as a main source of contaminant in the air is causing more and more problems on broiler production, especially lower growth rate. The molecular mechanisms that underlie the negative effects of ammonia on the growth and intestine of broilers are yet unclear. We investigated the growth, gut morphology, and mucosal proteome of Arbor Acres broilers (*Gallus gallus*) exposed to high concentrations of atmospheric ammonia by performing a proteomics approach integrated with traditional methods.

Results: Exposure to ammonia interfered with the development of immune organ and gut villi. Meanwhile, it greatly reduced daily weight gain and feed intake, and enhanced feed conversion ratio. A total of 43 intestinal mucosal proteins were found to be differentially abundant. Up-regulated proteins are related to oxidative phosphorylation and apoptosis. Down-regulated proteins are related to cell structure and growth, transcriptional and translational regulation, immune response, oxidative stress and nutrient metabolism. These results indicated that exposure to ammonia triggered oxidative stress, and interfered with nutrient absorption and immune function in the small intestinal mucosa of broilers.

Conclusions: These findings have important implications for understanding the toxic mechanisms of ammonia on intestine of broilers, which provides new information that can be used for intervention using nutritional strategies in the future.

Keywords: Proteome, Ammonia, Small intestinal mucosa, Broilers

Background

Ammonia is a colorless and highly water-soluble gas, which is a well-known toxicant both in aquatic and atmospheric system. In animal houses, ammonia may be formed mainly from animal manure by hydrolysis, mineralization, and volatilization [1]. Animal produced ammonia accounts for almost 50% of the total annual anthropogenic emission of ammonia, in which poultry operations produced the highest ammonia emission as compared with other animal groups [2,3]. The limiting level of ammonia for poultry is under 25 μL/L. But in practice, birds are often exposed to higher concentrations of ammonia (50–200 μL/L) in some poorly ventilated facilities. High level of atmospheric ammonia induces several problems in broiler production, such as decreased growth rate, body weight, and increased feed conversion [4,5]. Longtime exposure can create many health issues in broilers and severely interfere with broiler welfare [6,7].

In previous research, degenerative vacuole and necrosis of renal tubulae were observed in livers and kidneys of ammonia-exposed broilers, respectively [8]. Apoptosis of epithelium cells of tracheal mucosa has been demonstrated in ammonia-exposed broilers in our study (unpublished data). The neurotoxicity of ammonia induces an increase in expression of tumor necrosis factor α (TNF-α) and interleukin 1 β (IL-1β), which can be associated with the production of reactive oxygen species (ROS), nitric oxide (NO) involved with protein kinase A

* Correspondence: zhanghf6565@vip.sina.com
State Key Laboratory of Animal Nutrition, Institute of Animal Sciences, Chinese Academy of Agricultural Sciences, Beijing 100193, People's Republic of China

(PKA), extracellular signal regulated kinase (ERK) pathway and nuclear factor-κB (NF-κB) activation in astrocytes in rats [9].

Negative effects of ammonia on gastrointestinal (GI) tract were also reported in previous studies that may be related to nutrient metabolism and energy production. In rat colonocytes, it showed that ammonia suppressed short-chain fatty acid (SCFA) oxidation [10]. Inhibition of oxygen consumption induced lower energetic efficiency and decreased cellular energy production were also observed in the similar animal model due to elevated concentration of ammonia by the ingestion of high protein diet [11]. Tsujii et al. [12] reported that ammonia impaired mitochondrial and cellular respiration, and energy metabolism of gastric mucosa, which triggered a decrease of mucosal cell viability leading to mucosal damage subsequently. Moreover, Igarashi et al. [13] demonstrated that ammonia accelerated cytokine-induced apoptosis in human gastric epithelial cell lines.

Gastrointestinal (GI) tract is regarded as the essential sensory organ for nutrition absorption, immune response, and pathogen prevention [14]. Previous research have demonstrated that changes of animal growth performance are closely related to alterations of protein expression in the small intestinal mucosa [15-17]. There are numerous enzymes in the small intestinal mucosa involved in different physiological functions, such as protein metabolism, lipid metabolism, carbohydrates metabolism, energy production, mucosal integrity and so on [18-21]. However, it is almost impossible to detect a huge number of proteins in the GI mucosa at the same time using traditional methods, for example western blots, immunohistochemical staining or ELISAs. Currently, most studies relevant to toxic mechanisms of ammonia are on mammals and aquatic animals. Little is known about the alteration of proteins in the small intestinal mucosa of broilers that have been exposed to high concentrations of atmospheric ammonia.

Based on previous research, we hypothesized that high concentrations of atmospheric ammonia exposure can confer negative effects on growth via changes of proteins involved in different physiological processes in the small intestinal mucosa of broilers, which requires further study to elucidate. Therefore, in this study, we utilized a label-based iTRAQ procedure (isobaric tags for relative and absolute quantitation), followed by LC-MS/MS to quantitate altered proteins that are induced differentially in the small intestinal mucosa of broilers exposed to high concentrations of atmospheric ammonia.

Materials and methods
Animals and exposure conditions
A total of 60 1-day-old Arbor Acers (AA) male broilers were obtained from a commercial hatchery in Beijing (Beijing Arbor Acers Broiler Co., Beijing, China). All birds were housed in individual wire-bottom cages in an environmentally controlled room under standard brooding practices, and given *ad libitum* access to water and a maize-soybean basal diet during the first 21 days. Then, broilers were transferred to environmentally controlled exposure chambers. The diet during the experiment was formulated to achieve the National Research Council (NRC, 1994) recommended requirements for all nutrients containing ME, 12.76 MJ kg^{-1}; and crude protein 19.94% (Additional file 1: Table S1). The concentrated ammonia was delivered in a whole-body animal exposure chamber [7] from days 22 to 42. Each exposure chamber was a $4500 \times 3000 \times 2500$ mm (length × width × height) sealed unit, sectioned for housing 30 birds per chamber. Temperature and airflow were controlled during the exposures to ensure adequate ventilation, minimize buildup of animal-generated contaminants (dander, H_2S, CO_2) and to avoid thermal stress [22].

The setting of the concentration of ammonia in the present study was according to previous studies that the growth performance of broilers was severely interfered with ammonia level over 70 μL/L [4,7,8]. Treatment (TRET) group of broilers were exposed to 75 ± 3 μL/L ammonia during the experimental period. Control (CTRL) broilers were raised in a separated chamber without ammonia for the same period, and the concentration of ambient ammonia was kept at 3 ± 3 μL/L. The concentration of ammonia in both chambers was monitored with a LumaSense Photoacoustic Field Gas-Monitor Innova-1412 (Santa Clara, CA, USA) during the entire experimental period. Body weight (BW) and feed consumption were recorded weekly for feed-conversion ratio evaluation. This study was carried out in strict accordance with the Regulations for the Administration of Affairs Concerning Experimental Animals of the State Council of the People's Republic of China. The protocol was approved by the Committee on Experimental Animal Management of Chinese Academy of Agricultural Sciences.

Sample collection
At day 42, all birds were weighed after a 12 h-fasting (12 h food withdrawal) period. The growth parameters (n = 30) including body weight gain, feed intake and feed-conversion ratio were determined. Twelve birds (6 per each group) were randomly selected for blood and small intestine sample collection. Each blood sample was obtained from a wing vein using a sterilized syringe within 30 s. Blood was incubated in a water bath for 1 h at 37°C then centrifuged at 400 × g for 10 min at 4°C, and the sera obtained were stored at −80°C for further analysis [23]. After blood sampling, the chickens were sacrificed by cervical dislocation and then exsanguinated. Immediately

after death, the intestinal mucosa was scraped from the intestinum tenue with the back of a surgery knife as described by Luo et al. [24], frozen in liquid nitrogen, and stored at −80°C for further proteome and qPCR analyses. Samples of about 1 cm of medial duodenum (apex of the duodenum), medial jejunum (midway between the point of entry of the bile duct and Meckel's diverticulum) and medial ileum (midway between Meckel's diverticulum to the ileocecal junction) were taken and fixed in buffered 4% formal-saline solution before processing for embedding in paraffin. To calculate the indices of immune organs, another twelve birds (6 per each group) were killed as described above, and the bursa of Fabricius, spleen, thymus and intestine of were excised and weighted, respectively.

Biochemical and histological analyses

For biochemical analysis, the activities of creatine kinase (CK) and total superoxide dismutase (T-SOD) in the serum were measured using a corresponding diagnostic kit (Nanjing Jiancheng Bioengineering Institute, Nanjing, China) according to the instructions of the manufacturer. Histological examination was carried out according to the method described by [25]. Briefly, villus height was determined from the tip of the villus to the villus crypt junction and crypt depth was defined as the depth of invaginations between adjacent villi.

Small intestinal mucosa preparation and protein extraction

Sample pooling is a commonly used strategy to reduce the influence of individual variation on candidate target selection in proteomic studies [24,26,27]. To avoid erroneous conclusions due to individual variations, the same amount of the intestinal mucosa (weight: weight as 1: 1 ratio) from two chickens in the same group was pooled as a biological replicate, and three biological replicates were acquired for each group.

Each pooled small intestinal mucosal sample (~0.5 g) was ground in a Dounce glass grinder using liquid nitrogen. Ground samples were precipitated with 10% trichloroacetic acid (TCA) (w/v), 90% ice-cold acetone at −20°C for 2 h. The samples were then centrifuged at $20,000 \times g$ for 30 min at 4°C. The supernatants were decanted and the pellets washed with ice-cold acetone. The pellets were lysed in lysis buffer consisting of 8 M urea, 30 mM 4-(2-hydroxyethyl)-1-piperazineethanesulfonic acid (HEPES), 1 mM phenylmethanesulfonyl fluoride (PMSF), 2 mM ethylene diamine tetraacetic acid (EDTA), and 10 mM dithiothreitol (DTT). The crude tissue extracts were centrifuged for 30 min at $20,000 \times g$ to remove the undissolved pellets. The tissue lysates were reduced for 1 h at 36°C in water bath by addition of 1 M DTT to a final concentration of 10 mM DTT and then alkylated for 1 h by addition of 1 M iodoacetamide (IAM) to a final concentration of 55 mM in the dark. After reduction and alkylation, proteins were precipitated by adding 4 volumes of ice-cold acetone. The pellets were then washed three times with ice-cold pure acetone and resuspended in buffer consisting of 50% tetraethyl ammonium bromide (TEAB) and 0.1% sodium dodecyl sulfonate (SDS). The samples were then centrifuged for 30 min at $20,000 \times g$ and the undissolved pellets were removed and protein quantitation performed using a Bio-Rad Bradford Protein Assay Kit (Hercules, CA, USA).

Trypsin digestion, iTRAQ labeling and strong cation exchange chromatography

Modified sequence grade trypsin (Promega Corporation, Madison, WI) was added to each sample at a 1:30 ratio (3.3 μg trypsin : 100 μg target) and digested overnight at 37°C.

Each isobaric tag was solubilized in 70 μL isopropanol. Tags (113, 114, 115, 116, 117 and 121) were added to respective pooled samples (3 pooled replicates in each group) individually and incubated at room temperature for 2 h. Additional isopropanol was added to samples to ensure an organic composition > 60% prior to incubation.

The strong cation exchange fractionation protocol followed a previous report [28] with slight modification. Briefly, the samples were loaded onto a strong cation exchange column (Phenomenex Luna SCX 100A) equilibrated with buffer A (10 mM KH_2PO_4 in 25% acetonitrile, pH 3.0) using an Agilent 1100 (Santa Clara, CA) system. The peptides were separated using a linear gradient of buffer B (10 mM KH_2PO_4 and 2 M KCl in 25% acetonitrile, pH 3.0) increasing to 5% after 36 min, 50% after 66 min and 100% after 71 min, at a flow rate of 1 ml/min. Elution was monitored by setting the absorbance at 214 nm. The eluted peptides were pooled into 10 fractions, desalted with a Strata X C18 column (Phenomenex) and vacuum-dried.

Mass spectrometry

Each fraction was resuspended in buffer A (2% acetonitrile, 0.1% formic acid) and centrifuged at $20,000 \times g$ for 10 min. In each fraction, the final concentration of peptides was approximately 0.25 μg/μl. Using an autosampler, 20 μl of supernatant was loaded onto a 2 cm C18 trap column (inner diameter 200 μm) on an UltiMate® 3000 Nano LC system (Bannockburn, IL). Peptides were eluted onto a resolving 100 mm × 75 μM analytical C18 column containing 5-μm particles that was assembled in-house. Samples were loaded at 15 μl/min for 4 min and eluted with a 45-min gradient at 400 nl/min from 5 to 60% buffer B (98% acetonitrile, 0.1% formic acid), separated with a 3 min linear gradient to 80% B, maintained at 80% B for 7 min, returned to 5% B over

3 min, and finally combined with a Q-Exactive mass spectrometer (Thermo Scientific, MA, USA). The mass spectrometer was operated in data dependent acquisition mode, with MS performed in the Q-Exactive at a resolution of 70,000 full width at half maximum (FWHM). MS/MS was performed in high-energy collision dissociation (HCD) operating mode and product ions were detected in the Q-Exactive at 17,500 FWHM resolution. Data were acquired using a data-dependent data acquisition mode in which, for each cycle, the ten most abundant multiply charged peptides (2^+ to 4^+) with an m/z between 350 and 2000 were selected for MS/MS with a 15-s dynamic exclusion setting.

Data processing and analyses

For iTRAQ protein identification, the raw mass data were processed with Proteome Discover 1.3 (Thermo Fisher Scientific) and searched with in-house MASCOT software (Matrix Science, London, U.K.; version 2.3.0) against the database Uniprot_Gallus gallus_9031 (Apr 11th, 2014) the following parameters: enzyme: trypsin; fixed modification: carbamidomethyl (C); variable modifications: oxidation (M), gln-pyro-glu (N-term Q), iTRAQ 8-plex (N-term, K, Y); peptide tolerance:15 ppm; MS/MS tolerance: 20 mmu; maximum missed cleavages: 1. All identified peptides had an ion score above the Mascot peptide identity threshold, and a protein was considered identified if at least one such unique peptide match was apparent for the protein. For iTRAQ quantitation, MASCOT software was also used. Protein quantitative values were derived only from uniquely assigned peptides. Intra-sample channels were normalized based on the median ratio for each channel across all proteins. Ratios for each iTRAQ label were obtained using a pooled sample in the control group (sample tagged with 113) as the denominator. Inter-sample, protein reference, and spectrum normalizations were performed. Differential expression in the TRET samples was then presented as a $\log_{2\text{-fold}}$ change relative to the CTRL. Thus, the fold change for each individual reporter ion is based on referencing a reporter channel which is then log transformed to base 2. Proteins were deemed to be differentially expressed using Student's t-test corrected for multiple testing using the Benjamini and Hochberg correction [29]. Proteins with a 1.2-fold change or greater were considered to be differentially expressed.

Bioinformatics analysis of proteins differential abundance

Gene Ontology (GO) distribution for all of the proteins that were significantly altered in the small intestinal samples of ammonia exposed chickens were classified using Blast2GO software (http://www.blast2go.com/) and WEGO (http://wego.genomics.org.cn) that were provided by the Institute for Genomic Research [30,31].

Validation of proteins of differential abundance

Real-time quantitative PCR (qPCR) was used to verify seven intestinal mucosal proteins of differential abundance at the mRNA level.

Total RNA from intestinal mucosal samples was isolated using a Qiagen RNeasy Plus Mini Kit (Valencia, CA). The quality of the RNA was evaluated by electrophoresis on an agarose gel, and the quantity of the RNA was measured with a spectrophotometer (Nanodrop 2000, Thermo Scientific, Waltham, MA).

Reverse transcription was performed immediately following total RNA isolation using PrimeScript™ Reverse Transcriptase, D2680A (Takara, Dalian, China). RT-qPCR was performed using an Applied Biosystems 7500 Fast Real-Time PCR System (Foster City, CA). RT-qPCRs were performed at 95°C for 30 s, followed by 40 cycles of 95°C for 10 s and 60°C for 30 s. SYBR green fluorescence was detected at the end of each cycle to monitor the amount of PCR product. A standard curve was constructed using a 10-fold dilution series, and its slope was used to calculate the efficiency of the qPCR primers. Primer sequences are listed in Additional file 2: Table S2.

The relative amount of a target gene mRNA was calculated as previously described [23]. The expression level of a target gene mRNA was normalized to the mRNA level of β-actin. The $\Delta\Delta C_T$ was calibrated against an average from the control group. The linear amount of the target gene expression to the calibrator was calculated by $2^{-\Delta\Delta CT}$. Therefore, all gene expression results are reported as the fold difference between treated and control groups. The specificity of the real-time PCR product was verified using a melting curve and DNA sequencing.

Statistical analysis

Data on growth parameters, immune organ indices, serum parameters, gut morphological structure and gene expressions were analyzed by one-way ANOVA (SAS Version 9.2, SAS institute Inc., Cary, NC). A group difference was assumed to be statistically significant when $P < 0.05$. All results were expressed as means ± S.D.

Results

Body weight gain, feed intake, feed-conversion ratio and immune organ indices

The body weight gain and feed intake are key parameters to assess the growth of animal. In this study, all birds (CTRL and TRET) started at the same age (d 22). During the entire experimental period (20 days), TRET birds had 15.4% less ($P < 0.05$) body weight gain and 9.6% less ($P < 0.05$) feed intake. On the contrary, feed-conversion ratio (FCR) in TRET group was greatly increased ($P < 0.05$) compared with CTRL group (Table 1).

Table 1 Effect of atmospheric ammonia on the body weight gain, feed intake, feed-conversion ratio and immune organ indices of broilers

	Groups	
	Control	Treatment
Body weight gain (g/day)	91 ± 3.6^a	77 ± 2.5^b
Feed intake (g/day)	150 ± 1.9^a	135 ± 2.8^b
Feed-conversion ratio[c]	1.64 ± 0.09^b	1.75 ± 0.05^a
Index of bursal (%)[d]	0.73 ± 0.06	0.61 ± 0.05
Index of spleen (%)[d]	1.07 ± 0.04^a	0.75 ± 0.02^b
Index of thymus (%)[d]	2.35 ± 0.50	2.19 ± 0.44
Index of intestine (%)[d]	3.70 ± 0.15^a	2.67 ± 0.11^b

[a, b]Values within a row not sharing a common superscript letter indicate significant difference at $P < 0.05$. Numbers are means $\pm$ S.D. (n = 30 for growth parameters; n = 6 for indices of immune organs).
[c]Feed-conversion ratio = the ratio of feed intake to body weight gain.
[d]The ratio of organ weight to body weight.

Of four tested immune organs, indices of spleen and intestine of chickens in TRET group were lighter than those of CTRL group ($P < 0.05$). Thus, exposure to high concentrations of atmospheric ammonia interfered with immune organ development of AA broilers and resulted in a reduction of feed conversion efficiency [24].

Serum parameters and gut morphological structure

Activity of serum CK is an important indicator under stress in the body [25,32]. T-SOD represents the oxidation resistance in the animal [33]. In TRET broilers, activity of serum CK was significantly elevated ($P < 0.05$) compared with the control group indicating extensive organ injury (Table 2). Antioxidase T-SOD was decreased significantly compared with the control group ($P < 0.05$), illustrating lower oxidation resistance (Table 2). As shown in Figure 1A, B, C, D, E and F the VH and CD in all small intestinal segments of birds in CTRL group were significantly higher than those in TRET group, which implicates the absorptive area of small intestine was greatly reduced in broilers under high level of ambient ammonia.

Identification and comparison of proteins of differential abundance

Using iTRAQ analysis, a total of 2726 proteins were identified within the FDR of 1% (Additional file 3: Table S3). Following statistical analysis, 70 proteins were found to be differentially expressed in the small intestinal mucosa between CTRL and TRET broilers, with 26 being up-regulated and 44 down-regulated (Additional file 4: Table S4).

A total of 43 proteins of differential abundance were grouped into eleven classes based on putative functions: transcriptional and translational regulation (23.3%), immune response (18.6%), energy metabolism (16.3%), cell

Table 2 Effect of atmospheric ammonia on the serum biochemical parameters of broilers

	Groups	
	Control	Treatment
CK (U/L)[c]	6224.50 ± 172.26^b	7173.63 ± 309.05^a
T-SOD (U/mL)[d]	77.81 ± 6.55^a	61.12 ± 2.11^b

[a, b]Values within a column not sharing a common superscript letter indicate significant difference at $P < 0.05$. Numbers are means $\pm$ S.D. (n = 6).
[c]CK = creatine kinase.
[d]T-SOD = total superoxide dismutase.

growth and proliferation (9.3%), oxidative stress (7.0%), apoptosis (7.0%), cell cytoskeleton (4.7%), lipid metabolism (4.7%), amino acid metabolism (4.7%), vitamin metabolism (2.3%) and neurotoxicity (2.3%) (Figure 2). Those related to transcriptional and translational regulation, immune response and energy metabolism were predominant and accounted for approximately 60% of the differentially-expressed proteins. A comparison of proteins of differential abundance with functional grouping between the groups indicated that fewer protein species were up-regulated in ammonia-exposed chickens (11 versus 32, respectively) (Table 3). These 11 up-regulated protein species were distributed in five categories: four in energy metabolism, three in apoptosis, two in transcriptional and translational regulation, one in oxidative, and one in neurotoxicity. The 32 down-regulated protein species were distributed in nine categories: eight in transcriptional and translational regulation, eight in immune response, four in cell growth and proliferation, three in energy metabolism, two in oxidative stress, two in lipid metabolism, two in amino acid metabolism, two in cell cytoskeleton and one in vitamin metabolism.

GO annotations of proteins of differential abundance

In the cellular component group, the differentially expressed proteins are concentrated in intracellular organelles (mitochondrion, cytoskeletal part and nuclear part) and the cytoplasm part (Figure 3). In the molecular functional group, the differentially expressed proteins that are metabolic enzymes (oxidoreductase activity and hydrolase activity), binding proteins (protein binding and nucleotide binding) and enzyme regulator were ranked at the top of the category occupancy, suggesting that the relevant functions were important in the small intestinal mucosa of broilers (Figure 3). In the biological process category, the proteins that participate in cellular processes, metabolism and biological regulation were at the top ratio in the differentially expressed proteins (Figure 3), suggesting that exposure to ammonia changes the cellular metabolic process, like cellular biosynthetic process, nucleotide and nucleic acid metabolic process, alters metabolism in the intestine, such as

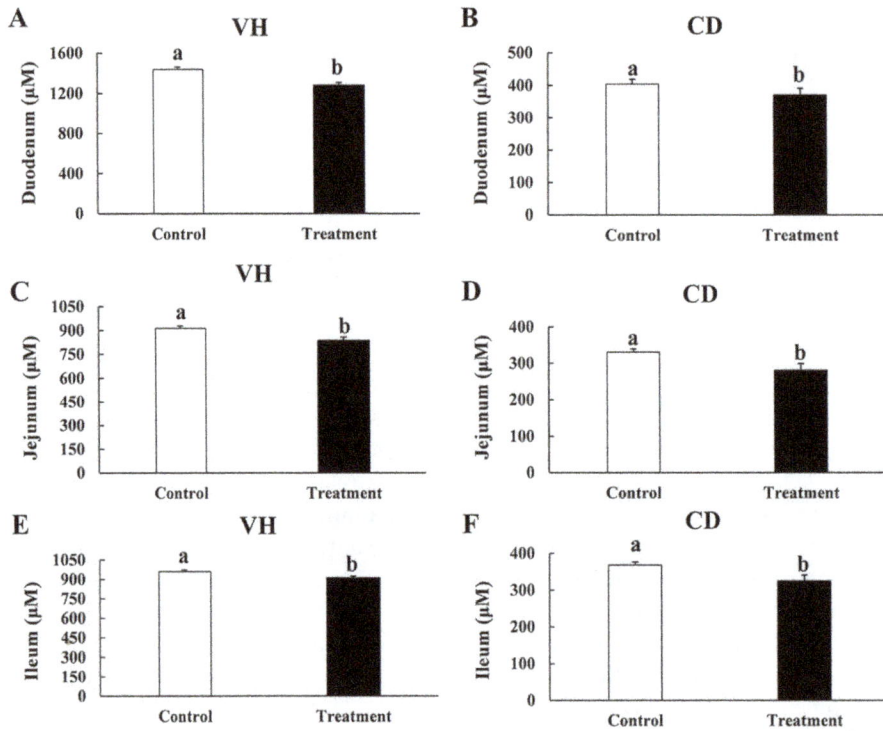

Figure 1 Effects of atmospheric ammonia on villus height (VH) and crypt depth (CD) of duodenum (A and B), jejunum (C and D) and ileum (E and F) in control and treatment groups. Vertical lines represent ± S.D, and different letters denote significant difference at $P < 0.05$ (n = 6).

nutrient (carbohydrate, amino acid and lipid) metabolism, and have various effects on biological processes, for example transcriptional and translational regulation, cell growth and proliferation, oxidative stress and so on.

Validation of proteins of differential abundance

Seven differentially expressed proteins (GLUD1 involved in amino acid metabolism; fatty acid translocase (CD36) involved in lipid metabolism; IRF3 involved in immune response; FTH involved in oxidative stress; SDHA

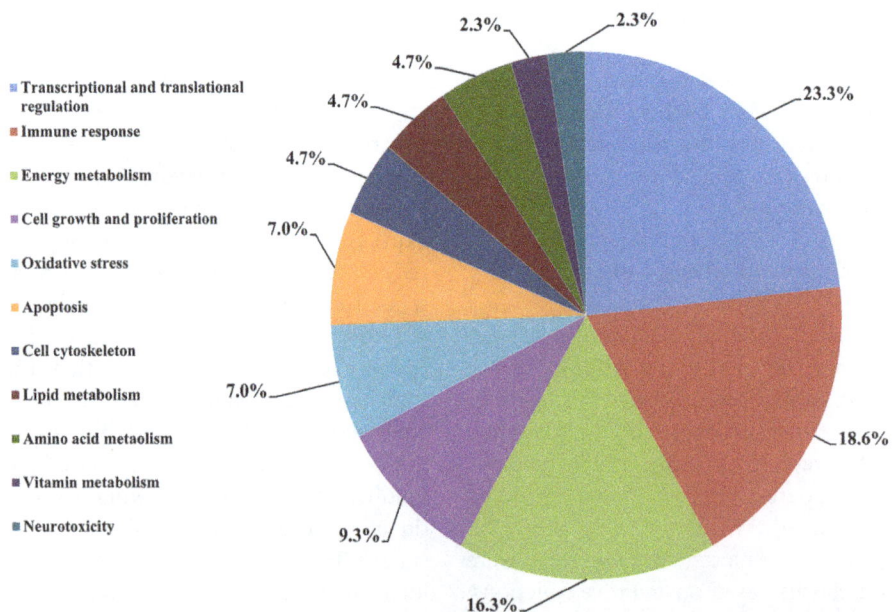

Figure 2 Functional classification of the proteins of differential abundance identified from the small intestinal mucosa of 42-day-old broilers.

Table 3 List of differentially expressed proteins in small intestinal mucosal samples from treatment group and control group

Accession[a]	Description[b]	Gene symbol	Theoretical MW[c]/pI[d]	Score[e]	Pep. no.[f]	Log$_2$ fold change[g]	P-value	Biological process GO term
Energy metabolism								
F1NTZ0	Uncharacterized protein (Fragment) OS = Gallus gallus GN = ADH6 PE = 3 SV = 2 - [F1NTZ0_CHICK]	ADH6	39.5/7.49	565.75	14	-0.60	0.0162	Oxidoreductase activity
R4GI36	Phosphoenolpyruvate carboxykinase, cytosolic [GTP] OS = Gallus gallus GN = PCK1 PE = 3 SV = 1 - [R4GI36_CHICK]	PCK1	61.3/7.75	64.06	2	-0.35	0.0038	Gluconeogenesis
P07322	Beta-enolase OS = Gallus gallus GN = ENO3 PE = 1 SV = 3 - [ENOB_CHICK]	ENO3	47.2/7.61	259.39	5	-0.29	0.0497	Gluconeogenesis
E1C4U7	Uncharacterized protein OS = Gallus gallus GN = NDUFB3 PE = 4 SV = 1 - [E1C4U7_CHICK]	NDUFB3	11.1/9.82	35.42	1	0.39	0.0179	Electron transport chain
E1BQJ6	Uncharacterized protein OS = Gallus gallus PE = 3 SV = 2 - [E1BQJ6_CHICK]		257.7/5.17	23.38	1	0.87	0.0490	ATP binding
Q9YHT1	Succinate dehydrogenase [ubiquinone] flavoprotein subunit, mitochondrial OS = Gallus gallus GN = SDHA PE = 1 SV = 2 - [SDHA_CHICK]	SDHA	72.9/7.08	773.5	18	1.02	0.0211	Succinate dehydrogenase (ubiquinone) activity
F1NZI4	Uncharacterized protein OS = Gallus gallus GN = ATHL1 PE = 4 SV = 2 - [F1NZI4_CHICK]	ATHL1	77/5.73	30.07	1	1.15	0.0239	Catalytic activity
Lipid metabolism								
Q2MJT5	Fatty acid translocase OS = Gallus gallus PE = 2 SV = 1 - [Q2MJT5_CHICK]	CD36	52.6/8.37	110.81	4	-0.45	0.0471	Lipid uptake
E1BS15	Uncharacterized protein OS = Gallus gallus GN = ACSF2 PE = 4 SV = 2 - [E1BS15_CHICK]	ACSF2	68.7/8.7	441.08	15	-0.33	0.0192	Fatty acids catalytic activity
Amino acid metabolism								
P00368	Glutamate dehydrogenase 1, mitochondrial OS = Gallus gallus GN = GLUD1 PE = 1 SV = 1 - [DHE3_CHICK]	GLUD1	55.7/8.28	641.29	19	-0.50	0.0155	Glutamate dehydrogenase [NAD(P)+] activity
F1P3F9	Glutamate dehydrogenase OS = Gallus gallus GN = GLUD1 PE = 3 SV = 2 - [F1P3F9_CHICK]	GLUD1	47.6/8.18	695.44	19	-0.39	0.0288	Glutamate dehydrogenase [NAD(P)+] activity
Vitamin metabolism								
F1P4K4	Uncharacterized protein OS = Gallus gallus GN = ALDH8A1 PE = 3 SV = 2 - [F1P4K4_CHICK]	ALDH8A1	53.2/7.83	31.26	1	-0.54	0.0080	9-cis-retinoic acid biosynthetic process
Cell cytoskeleton								
Q90WF1	Filamin OS = Gallus gallus PE = 2 SV = 1 - [Q90WF1_CHICK]	FLNA	272.8/6.35	103.29	2	-0.72	0.0018	Actin-binding
D2Z1L9	LIM and SH3 protein 1 OS = Gallus gallus GN = LASP1 PE = 2 SV = 1 - [D2Z1L9_CHICK]	LASP1	29.6/7.4	417.26	11	-0.43	0.0284	Zinc ion binding
Cell growth and proliferation								
R4GF89	Uncharacterized protein OS = Gallus gallus GN = SFN PE = 3 SV = 1 - [R4GF89_CHICK]	SFN	27.7/5.01	114.99	3	-0.54	0.0239	Positive regulation of cell growth
E1BY89	Uncharacterized protein OS = Gallus gallus GN = RPL23 PE = 3 SV = 2 - [E1BY89_CHICK]	RPL23	14.9/10.51	286.2	8	-0.46	0.0019	Structural constituent of ribosome
F1NQS9	Uncharacterized protein OS = Gallus gallus GN = ZNF598 PE = 4 SV = 2 - [F1NQS9_CHICK]	ZNF598	99.6/8.57	28.5	1	-0.42	0.0041	Zinc ion binding

Table 3 List of differentially expressed proteins in small intestinal mucosal samples from treatment group and control group *(Continued)*

Accession	Description	Gene	MW/pI	Score	Count	Ratio	p-value	Function
E1BTA6	Uncharacterized protein OS = Gallus gallus GN = SEPT12 PE = 3 SV = 2 - [E1BTA6_CHICK]	SEPT12	47.1/8.68	52.61	2	-0.33	0.0202	GTP binding
Transcriptional and translational regulation								
F1P0C0	Uncharacterized protein OS = Gallus gallus GN = HMGN1 PE = 4 SV = 1 - [F1P0C0_CHICK]	HMGN1	11.1/9.26	68.36	1	-0.55	0.0223	Chromatin binding
E1C9F0	Uncharacterized protein OS = Gallus gallus GN = DYNC2H1 PE = 4 SV = 2 - [E1C9F0_CHICK]	DYNC2H1	491.7/6.43	22.2	1	-0.50	0.0062	Protein processing
E1BT82	Uncharacterized protein OS = Gallus gallus GN = EIF2S2 PE = 2 SV = 1 - [E1BT82_CHICK]	EIF2S2	37.9/6.13	71.23	4	-0.44	0.0040	Translation initiation factor activity
Q5ZJ39	Density-regulated protein OS = Gallus gallus GN = DENR PE = 2 SV = 1 - [DENR_CHICK]	DENR	22.1/5.21	60.49	1	-0.41	0.0080	Translation initiation factor activity
F1NS60	Uncharacterized protein (Fragment) OS = Gallus gallus GN = MMS19 PE = 4 SV = 2 - [F1NS60_CHICK]	MMS19	111.8/5.99	55.46	1	-0.38	0.0038	Transcription coactivator activity
E1C4N0	Uncharacterized protein OS = Gallus gallus GN = RPS10 PE = 4 SV = 2 - [E1C4N0_CHICK]	RPS10	18.9/10.15	183.89	6	-0.36	0.0068	Structural constituent of ribosome
R4GL23	Uncharacterized protein OS = Gallus gallus GN = CHTOP PE = 4 SV = 1 - [R4GL23_CHICK]	CHTOP	26.3/12.23	83.09	2	-0.30	0.0036	Transcription export complex
F1NLT8	Uncharacterized protein (Fragment) OS = Gallus gallus GN = ARHGDIB PE = 4 SV = 1 - [F1NLT8_CHICK]	ARHGDIB	23.2/5.2	235.16	7	-0.29	0.0435	Rho GDP-dissociation inhibitor activity
F1NA55	Eukaryotic translation initiation factor 2A OS = Gallus gallus GN = EIF2A PE = 2 SV = 2 - [F1NA55_CHICK]	EIF2A	62.7/8.79	62.63	1	0.27	0.0449	Translation initiation factor activity
Q800W4	TIA-1 OS = Gallus gallus GN = TIA1 PE = 2 SV = 1 - [Q800W4_CHICK]	TIA1	41.3/7.72	62.06	1	0.29	0.0074	Nucleotide binding
Immune response								
P40618	High mobility group protein B3 OS = Gallus gallus GN = HMGB3 PE = 2 SV = 3 - [HMGB3_CHICK]	HMGB3	23/8.12	126.17	4	-0.59	0.0009	DNA binding
Q9YH06	High mobility group protein B1 OS = Gallus gallus GN = HMGB1 PE = 1 SV = 1 - [HMGB1_CHICK]	HMGB1	24.9/5.74	165.42	6	-0.44	0.0084	DNA binding
E1C9I0	Unconventional myosin-Ig OS = Gallus gallus GN = MYO1G PE = 4 SV = 2 - [E1C9I0_CHICK]	MYO1G	115/8.78	203.91	7	-0.47	0.0410	Nucleotide-binding(motor activity)
E1BTE2	Uncharacterized protein OS = Gallus gallus GN = SERPINB5 PE = 3 SV = 2 - [E1BTE2_CHICK]	SERPINB5	42.6/5.96	218.9	7	-0.40	0.0017	Serine-type endopeptidase inhibitor activity
Q90643	Interferon regulatory factor 3 OS = Gallus gallus GN = IRF3 PE = 2 SV = 1 - [IRF3_CHICK]	IRF3	54.4/5.21	45.52	2	-0.33	0.0460	Activation of innate immune response
E1BWS0	Uncharacterized protein OS = Gallus gallus GN = GIT2 PE = 4 SV = 2 - [E1BWS0_CHICK]	GIT2	84.5/6.98	29.12	1	-0.32	0.0461	ARF GTPase activator activity
E1BUY6	Uncharacterized protein OS = Gallus gallus GN = HMHA1 PE = 4 SV = 2 - [E1BUY6_CHICK]	HMHA1	103.3/7.97	29.79	1	-0.31	0.0080	GTPase activator activity
E1BVP2	Uncharacterized protein OS = Gallus gallus GN = PLD1 PE = 4 SV = 2 - [E1BVP2_CHICK]	PLD1	118.9/8.98	50.12	2	-0.27	0.0397	Defense response to Gram-positive bacterium

Table 3 List of differentially expressed proteins in small intestinal mucosal samples from treatment group and control group (*Continued*)

Accession	Description	Gene	MW/pI	Score	Peptides	log2 fold change	p-value	GO term
Apoptosis								
I3VQH4	Interleukin enhancer binding factor 3-like protein OS = Gallus gallus GN = ILF3 PE = 2 SV = 1 - [I3VQH4_CHICK]	ILF3	95/8.81	68.88	3	0.30	0.0025	Participate in the apoptosis
R4GLP0	Uncharacterized protein OS = Gallus gallus GN = COX7C PE = 4 SV = 1 - [R4GLP0_CHICK]	COX7C	72/10.96	27.81	1	0.55	0.0380	Cytochrome-c oxidase activity
F1NIC5	Uncharacterized protein (Fragment) OS = Gallus gallus GN = TIMM8A PE = 4 SV = 2 - [F1NIC5_CHICK]	TIMM8A	122/7.78	135.92	4	0.57	0.0346	Protein transport (metal ion binding)
Oxidative stress								
H9L201	Uncharacterized protein OS = Gallus gallus GN = PNKD PE = 3 SV = 2 - [H9L201_CHICK]	PNKD	48/9.35	31.96	1	-0.46	0.0095	Glutathione biosynthetic process
F1NQC3	Glutamine synthetase OS = Gallus gallus GN = GLUL PE = 3 SV = 2 - [F1NQC3_CHICK]	GLUL	46.5/7.72	479.53	13	-0.40	0.0483	Glutamate-ammonia ligase activity
P08267	Ferritin heavy chain OS = Gallus gallus GN = FTH PE = 2 SV = 2 - [FRIH_CHICK]	FTH	21.1/6.21	311.58	8	0.34	0.0248	Ferroxidase activity
Neurotoxicity								
F1P4B2	Protein piccolo (Fragment) OS = Gallus gallus GN = PCLO PE = 4 SV = 2 - [F1P4B2_CHICK]	LOC768552	560.2/6.77	33.63	1	0.62	0.0247	cAMP-mediated signaling

[a]Uniprot_Gallus gallus_9031 database accession number.
[b]The name of the protein exclusive of the identifier that appears in the database.
[c]Theoretical molecular mass (kDa).
[d]Theoretical pI.
[e]The sum of the scores of the individual peptides.
[f]The number of distinct peptide sequences in the protein group.
[g]Differential protein expression in the treatment group was presented as a $\log_2$ fold change relative to the control group.

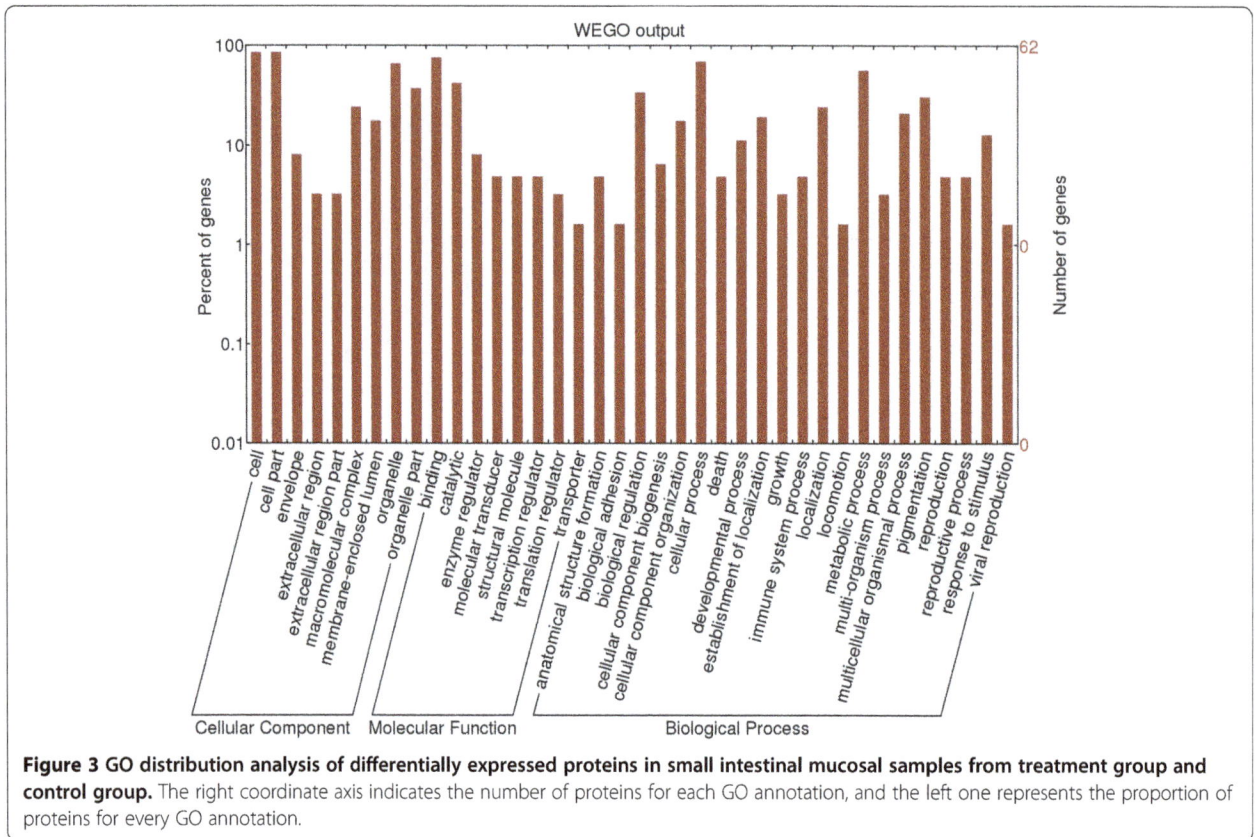

Figure 3 GO distribution analysis of differentially expressed proteins in small intestinal mucosal samples from treatment group and control group. The right coordinate axis indicates the number of proteins for each GO annotation, and the left one represents the proportion of proteins for every GO annotation.

involved in energy metabolism; SFN involved in cell growth and proliferation; and EIF2A involved in transcriptional and translational regulation) were selected for functional validation at the mRNA level using qPCR (Figure 4). The protein levels of GLUD1, CD36, SDHA and EIF2A were consistent with their mRNA expression levels. The results for the remaining three proteins were inconsistent between the mRNA levels and the protein levels. Possible reasons for these inconsistent results include the following: 1) the relationships between the mRNA levels and the protein levels were indirect, 2) there were some post-translational effects and/or the function of other regulatory mechanisms, and 3) there was a time delay between responses on the mRNA and protein levels [24].

Discussion

Ammonia influences different organs and physiological functions in animals due to oxidative stress and inflammation; therefore, excess concentration of ammonia lead to plenty of health problems in the body. More and more evidence demonstrates that high concentrations of ammonia impairs energy metabolism, and induces cell apoptosis and mitochondrial damage in the mucosa of GI tract [10,12,13]. To identify the molecular mechanisms related to the exposure to high concentration of

atmospheric ammonia in broilers, we compared the growth parameters, immune organ development, gut morphology, serum parameters and small intestinal mucosa proteome of control (ammonia concentration, 3 ± 3 µL/L) with those exposed to high level of ambient ammonia (ammonia concentration, 75 ± 3 µL/L). On the whole, exposure to high concentrations of ambient ammonia (75 ± 3 µL/L) greatly reduced the growth of broilers.

In addition to the reduction of growth performance, exposure to high concentrations of ammonia also resulted in interference with multiple physiological functions in broilers. As two of the most important immune organs, indices of intestine and spleen were reduced in TRET group compared to CTRL group. Moreover, ammonia-exposed chickens had much lower villus height and crypt depth among different segments of small intestine. These results indicated that exposure to high concentrations of atmospheric ammonia mainly exerts negative impacts on intestine mucosal structure and immune organ development of chickens [25], which may cause huge damages to nutrients absorption and immune system. Increased activity of serum CK and decreased activity of serum T-SOD indicated oxidative stress in ammonia-exposed broilers. Previous study also reports that obvious pathomorphological changes were observed in kidneys and livers in broiler chickens

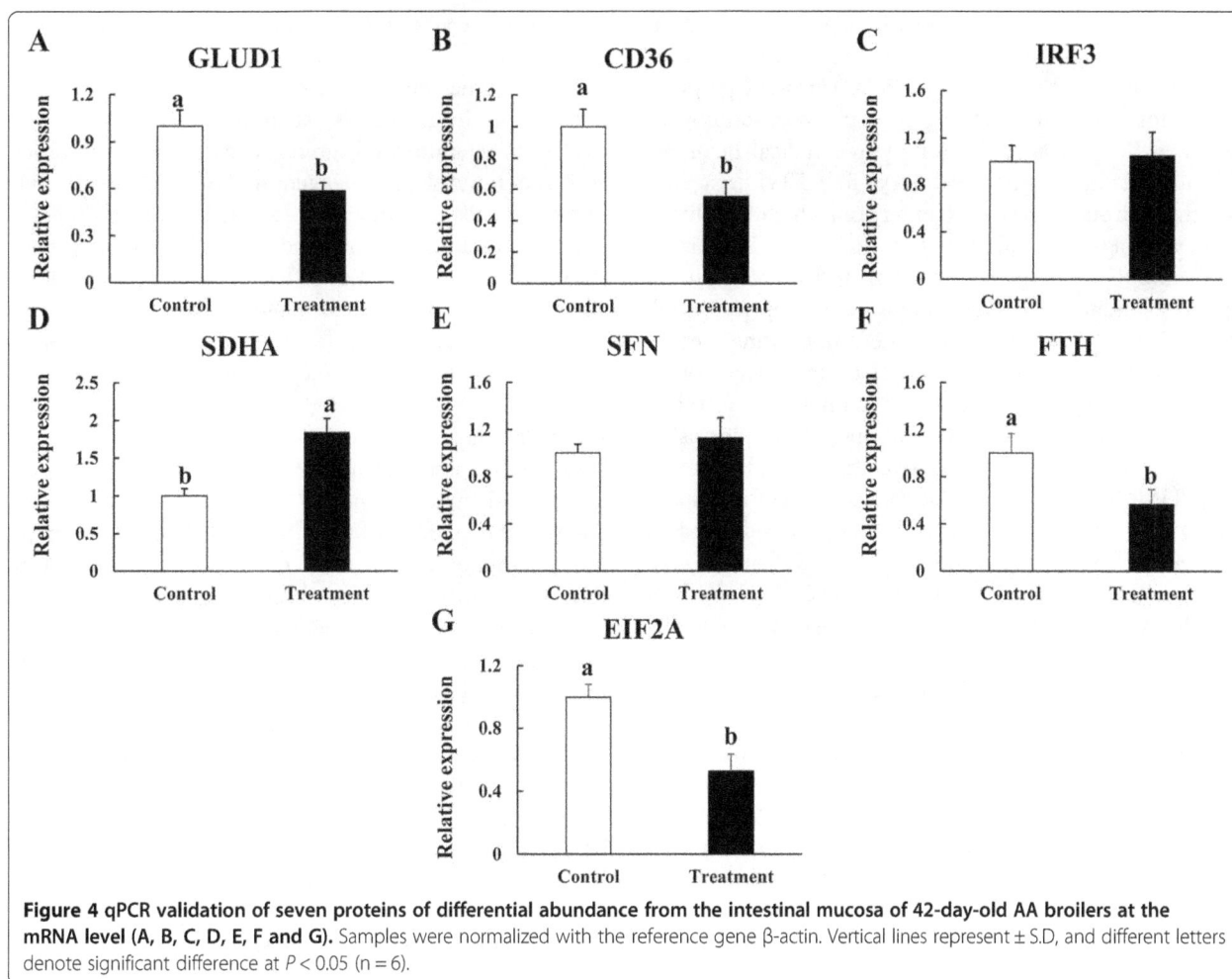

Figure 4 qPCR validation of seven proteins of differential abundance from the intestinal mucosa of 42-day-old AA broilers at the mRNA level (A, B, C, D, E, F and G). Samples were normalized with the reference gene β-actin. Vertical lines represent ± S.D, and different letters denote significant difference at $P < 0.05$ (n = 6).

under the dynamic range of atmospheric ammonia (31–95 ppm) [8].

A total of 43 proteins related to nutrient metabolism, apoptosis, immune and oxidative response, transcriptional and translational regulation, and cell cytoskeleton and growth altered in abundances corresponding to the change in intestinal histomorphology of ammonia-exposed broilers. Of these, up-regulation proteins involved in energy metabolism and apoptosis may induce the mitochondrial apoptosis resulting in an increase rate of oxidative phosphorylation under stress, whereas down-regulation of immune and nutrient metabolic proteins may decrease the anti-microbial ability and nutrient absorption in the intestine itself.

Cytoskeletal proteins have crucial roles in the maturation, migration, and renewal of epithelial cells along the crypt-villus axis [24,34,35]. In this study, two differential protein species related to cytoskeleton were down-regulated in the small intestinal mucosa of ammonia exposed broilers. FLNA is an actin cross-linking protein that is crucial for actin cytoskeleton organization participating in cellular architectural and signaling functions

[36-38]. LASP1 is a cytoskeletal adaptor protein, which has been reported as a signal molecule playing role in the differentiation of parietal cells [39,40]. This is consistent with the finding in this study that lower villus height and crypt depth among different segments of small intestine was observed in TRET group. As a result, the surface area of the intestine was decreased, and finally resulted in impairment of digestion and absorption efficiency in the gut. Other proteins involved in cell growth and proliferation, including SFN, RPL23, ZNF598 and SEPT12, are also down-regulated, and may harm the mucosal regeneration in GI tract due to ammonia exposure related injury [41-43]. Furthermore, the reduced abundance of proteins relevant to transcriptional and translational regulation, including HMGN1, DYNC2H1, EIF2S2, DENR, MMS19, RPS10, CHTOP and ARHGDIB, is observed in TRET group, which indicates a decreased capacity for protein synthesis to impair overall gut function and integrity [44-50].

Excess concentration of ammonia induces oxidative stress in various tissues, which can trigger inflammation and subsequent apoptosis [13]. In the present study, three and three differential protein species were identified in the categories of oxidative stress and apoptosis, respectively. Of these proteins, PNKD protein plays an important role in maintaining cellular redox status [51]; GLUL activity is an indicator of free radical-mediated oxidative damage in tissue injury [52]; FTH is a core subunit of iron-binding protein ferritin, which is induced to protect against oxidative stress [53,54]; ILF3 participates in apoptosis, its expression is up-regulated during apoptosis induced by H_2O_2 in murine macrophages [55]; COX7C is shown to represent the rate-limiting step of mitochondrial electron transport chain in normal condition [56], however, its expression cannot be controlled under oxidative stress induced apoptosis, increasing thermogenesis and the rate of oxidative phosphorylation [57]; TIMM8A is a mitochondrial intermembrane space (IMS) protein that is involved in caspase-independent cell death [58]. Down-regulation of PNKD and GLUL along with up-regulation of FTH, ILF3, COX7C and TIMM8A in the treatment group suggests that, the small intestinal mucosa of ammonia-exposed broilers are under oxidative stress, which triggers the elevation of apoptosis. Moreover, previous research has proven that oxidative stress is related to the impairment of energy metabolism [59]. In this study, proteins involved in oxidative phosphorylation, including NDUFB3, SDHA and ATHL1 [60,61], are up-regulated, and indicate that ATP production and oxidative phosphorylation are uncoupled due to oxidative stress induced by ammonia, which may explain why the feed-conversion efficiency is reduced in ammonia-exposed broilers.

As the biggest immune organ in the body, intestine plays very important roles in defense of invasion of harmful bacteria and xenobiotics [14]. In the present study, eight differential protein species related to immune response were down-regulated in the treatment group. HMGB1 and HMGB3 serve as immunogenic nucleic acids binding proteins that are generally involved in the nucleic acid receptor-mediated activation of innate immune responses [62,63]; MYO1G is a plasma membrane-associated class I myosin contributing to T-cell activation [64]; SERPINB5 is a tumor suppressor that plays a role in protein binding [65]; IRF3 is a transcription factor that plays distinct role in innate antiviral response [66]; GIT2 is one of regulators of G protein-coupled receptor (GPCR), and loss of GIT2 *in vivo* leads to an immunodeficient state [67]; HMHA1 is a major target of immune responses also playing a role in T-cell activation [68]; and PLD1 contributes to the essential function of macrophages for protecting against a wide variety of invading microorganisms [69]. Down-regulation of these proteins in the treatment group suggests that, the immunity of gut is under low condition in ammonia-exposed broilers, which increases possibilities

of bacterial or viral infection and probably leads to lower growth rate.

Ammonia has been reported to interfere with nutrient metabolism in mammals, such as reduced fatty acid oxidation, vitamins and amino acids synthesis disorder, and inhibitory of gluconeogenesis [10,70,71]. In the intestinal mucosa of ammonia-exposed broilers, differential proteins involved in carbohydrate/amino acid/lipid/vitamin metabolism indicate that impairment of nutrient absorption and digestion is related to metabolic changes in the intestine, which affects gluconeogenesis, vitamin A synthesis and fatty acid metabolism.

Conclusions

This study integrates traditional nutritional, morphological and state of the art proteomic approaches to identify the impact of high concentrations of atmospheric ammonia exposure on intestine of broilers. Reduced growth rate was observed in broilers exposed to high level of environmental ammonia. Possible reasons for exposure to ammonia derived influence on broilers are related to intestinal immune and histomorphology. Integrative data analysis indicates that exposure to high amount environmental ammonia resulted in significant changes in the development of immune organs and intestinal villi, and mucosal proteome of AA broilers. These changes might be resulting from oxidative stress induced by ammonia. Several proteins are identified to related to immune response, oxidative stress, apoptosis and mucosal structure, and thus play key roles in nutrient consumption and absorption. This study identifies the potential molecular mechanisms of high concentrations of atmospheric ammonia exposure to broilers and provides new knowledge that can be used for possible intervention using nutritional strategies in the future.

Additional files

Additional file 1: Table S1. Composition of the experimental diet and calculated proximate composition of the diet.

Additional file 2: Table S2. The qPCR primers used for verification of the differentially expressed genes of the AA broiler small intestinal mucosa.

Additional file 3: Table S3. List of all proteins (n = 2726) identified in the study.

Additional file 4: Table S4. List of all differently expressed proteins (n = 70) identified in the study.

Abbreviations

CD36: Cluster of differentiation 36; SFN: Stratifin; RPL23: Ribosomal protein L23; ZNF598: Zinc finger protein 598; SEPT12: Septin 12; HMGN1: High mobility group nucleosome binding domain 1; DYNC2H1: Dynein, cytoplasmic 2, heavy chain 1; EIF2S2: Eukaryotic translation initiation factor 2, subunit 2 beta; MMS19: MMS19 nucleotide excision repair homolog; RPS10: Ribosomal protein S10; CHTOP: Chromatin target of PRMT1; ARHGDIB: Rho GDP dissociation inhibitor (GDI) beta; PNKD: Paroxysmal nonkinesigenic dyskinesia; COX7C: Cytochrome c oxidase subunit VIIc; TIMM8A: Translocase of inner mitochondrial membrane 8 homolog A; NDUFB3: NADH dehydrogenase (ubiquinone) 1 beta subcomplex, 3; ATHL1: Acid trehalase-like 1; SERPINB5: Serpin peptidase inhibitor, clade B (Ovalbumin), member 5; GIT2: G protein-coupled receptor kinase interacting ArfGAP 2; HMHA1: Histocompatibility (Minor) HA-1; PLD1: Phospholipase D1.

Competing interests
The authors declare that they have no competing interests.

Authors' contributions
JZ and HZ designed the study. JZ and CL performed the experiments and analyzed the data. JZ, XT, QL and RS contributed reagents/materials/analysis tools. JZ prepared the manuscript and all of the authors contributed to, read and approved the final manuscript.

Acknowledgements
This research was supported by the Chinese National Science and Technology Pillar Program (No: 2012BAD39B0) and the Special Fund for Innovation Team of the Chinese Academy of Agricultural Sciences (No: ASTTP-IAS07).

References
1. NRC. Air emissions from animal feeding operations: current knowledge, future needs. Washington DC: The National Academies Press; 2003.
2. van Aardenne JA, Dentener FJ, Olivier JGJ, Goldewijk C, Lelieveld J. A 1 degrees × 1 degrees resolution data set of historical. anthropogenic trace gas emissions for the period 1890–1990. Glob Biogeochem Cycles. 2001;15:909–28.
3. US EPA. National emission inventory: ammonia emissions from animal husbandry operations. Washington DC: U.S. EPA; 2004.
4. Miles DM, Branton SL, Lott BD. Atmospheric ammonia is detrimental to the performance of modern commercial broilers. Poult Sci. 2004;83:1650–4.
5. Shlomo Y. Ammonia affects performance and thermoregulation of male broiler chickens. Anim Res. 2004;53:289–93.
6. Sherlock L, McKeegan DE, Cheng Z, Wathes CM, Wathes DC. Effects of contact dermatitis on hepatic gene expression in broilers. Br Poult Sci. 2012;53:439–52.
7. Wei FX, Xu B, Hu XF, Li SY, Liu FZ, Sun QY, et al. The effect of ammonia and humidity in poultry houses on intestinal morphology and function of broilers. J Anim Vet Adv. 2012;11:3641–6.
8. Witkowska D, Sowinska J, Iwanczuk Czernik K, Mituniewicz T, Wojcik A, Szarek J. The effect of a disinfectant on the ammonia concentration on the surface of litter, air and the pathomorphological picture of kidneys and livers in broiler chickens. Archiv fur Tierzucht. 2006;49:249–56.
9. Bobermin LD, Quincozes-Santos A, Guerra MC, Leite MC, Souza DO, Gonçalves CA, et al. Resveratrol prevents ammonia toxicity in astroglial cells. PLoS One. 2012;7:e52164.
10. Cremin Jr JD, Fitch MD, Fleming SE. Glucose alleviates ammonia-induced inhibition of short-chain fatty acid metabolism in rat colonic epithelial cells. Am J Physiol Gastrointest Liver Physiol. 2003;285:105–14.
11. Andriamihaja M, Davila AM, Eklou-Lawson M, Petit N, Delpal S, Allek F, et al. Colon luminal content and epithelial cell morphology are markedly modified in rats fed with a high-protein diet. Am J Physiol Gastrointest Liver Physiol. 2010;299:1030–7.
12. Tsujii M, Kawano S, Tsuji S, Fusamoto H, Kamada T, Sato N. Mechanism of gastric mucosal damage induced by ammonia. Gastroenterology. 1992;102:1881–8.
13. Igarashi M, Kitada Y, Yoshiyama H, Takagi A, Miwa T, Koga Y. Ammonia as an accelerator of tumor necrosis factor alpha-induced apoptosis of gastric epithelial cells in Helicobacter pylori infection. Infect Immun. 2001;69:816–21.
14. Furness JB, Kunze WA, Clerc N. Nutrient tasting and signaling mechanisms in the gut. II. The intestine as a sensory organ: neural, endocrine, and immune responses. Am J Physiol. 1999;277:922–8.
15. Wang X, Yang F, Liu C, Zhou H, Wu G, Qiao S, et al. Dietary supplementation with the probiotic Lactobacillus fermentum I5007 and the antibiotic aureomycin differentially affects the small intestinal proteomes of weanling piglets. J Nutr. 2012;142:7–13.
16. Wang X, Ou D, Yin J, Wu G, Wang J. Proteomic analysis reveals altered expression of proteins related to glutathione metabolism and apoptosis in the small intestine of zinc oxide-supplemented piglets. Amino Acids. 2009;37:209–18.
17. Soler L, Niewold TA, Moreno Á, Garrido JJ. Proteomic approaches to study the pig intestinal system. Curr Protein Pept Sci. 2014;15:89–99.
18. Choi PM, Guo J, Erwin CR, Wandu WS, Leinicke JA, Xie Y, et al. Disruption of retinoblastoma protein expression in the intestinal epithelium impairs lipid absorption. Am J Physiol Gastrointest Liver Physiol. 2014;306:909–15.
19. Ahmad MK, Khan AA, Mahmood R. Alterations in brush border membrane enzymes, carbohydrate metabolism and oxidative damage to rat intestine by potassium bromate. Biochimie. 2012;94:2776–82.
20. Keszthelyi D, Troost FJ, Jonkers DM, van Eijk HM, Lindsey PJ, Dekker J, et al. Serotonergic reinforcement of intestinal barrier function is impaired in irritable bowel syndrome. Aliment Pharmacol Ther. 2014;40:392–402.
21. Li C, Li Q, Liu YY, Wang MX, Pan CS, Yan L, et al. Protective effects of Notoginsenoside R1 on intestinal ischemia-reperfusion injury in rats. Am J Physiol Gastrointest Liver Physiol. 2014;306:111–22.
22. Kleinman MT, Araujo JA, Nel A, Sioutas C, Campbell A, Cong PQ, et al. Inhaled ultrafine particulate matter affects CNS inflammatory processes and may act via MAP kinase signaling pathways. Toxicol Lett. 2008;178:127–30.
23. Huang J, Zhang Y, Zhou Y, Zhang Z, Xie Z, Zhang J, et al. Green tea polyphenols alleviate obesity in broiler chickens through the regulation of lipid-metabolism-related genes and transcription factor expression. J Agric Food Chem. 2013;61:8565–72.
24. Luo J, Zheng A, Meng K, Chang W, Bai Y, Li K, et al. Proteome changes in the intestinal mucosa of broiler (Gallus gallus) activated by probiotic Enterococcus faecium. J Proteomics. 2013;91:226–41.
25. Uni Z, Gal-Garber O, Geyra A, Sklan D, Yahav S. Changes in growth and function of chick small intestine epithelium due to early thermal conditioning. Poult Sci. 2001;80:438–45.
26. Diz AP, Truebano M, Skibinski DO. The consequences of sample pooling in proteomics: an empirical study. Electrophoresis. 2009;30:2967–75.
27. Su L, Cao L, Zhou R, Jiang Z, Xiao K, Kong W, et al. Identification of novel biomarkers for sepsis prognosis via urinary proteomic analysis using iTRAQ labeling and 2D-LC-MS/MS. PLoS One. 2013;8:e54237.
28. Olsen JV, Blagoev B, Gnad F, Macek B, Kumar C, Mortensen P, et al. Global, in vivo, and site-specific phosphorylation dynamics in signaling networks. Cell. 2006;127:635–48.
29. Hakimov HA, Walters S, Wright TC, Meidinger RG, Verschoor CP, Gadish M, et al. Application of iTRAQ to catalogue the skeletal muscle proteome in pigs and assessment of effects of gender and diet dephytinization. Proteomics. 2009;9:4000–16.
30. Ye J, Fang L, Zheng H, Zhang Y, Chen J, Zhang Z, et al. WEGO: a web tool for plotting GO annotations. Nucleic Acids Res. 2006;34:293–7.
31. Zi J, Zhang J, Wang Q, Zhou B, Zhong J, Zhang C, et al. Stress responsive proteins are actively regulated during rice (Oryza sativa) embryogenesis as indicated by quantitative proteomics analysis. PLoS One. 2013; 8:e74229.
32. Chulayo AY, Muchenje V. The effects of pre-slaughter stress and season on the activity of plasma creatine kinase and mutton quality from different sheep breeds slaughtered at a smallholder abattoir. Asian-Australas J Anim Sci. 2013;26:1762–72.
33. Dong XY, Azzam MM, Rao W, Yu DY, Zou XT. Evaluating the impact of excess dietary tryptophan on laying performance and immune function of laying hens reared under hot and humid summer conditions. Br Poult Sci. 2012;53:491–6.
34. Di Garbo A, Johnston MD, Chapman SJ, Maini PK. Variable renewal rate and growth properties of cell populations in colon crypts. Phys Rev E Stat Nonlin Soft Matter Phys. 2010;81:061909.
35. Gordon JI, Hermiston ML. Differentiation and self-renewal in the mouse gastrointestinal epithelium. Curr Opin Cell Biol. 1994;6:795–803.
36. van der Flier A, Sonnenberg A. Structural and functional aspects of filamins. Biochim Biophys Acta. 2001;1538:99–117.
37. Tu Y, Wu S, Shi X, Chen K, Wu C. Migfilin and Mig-2 link focal adhesions to filamin and the actin cytoskeleton and function in cell shape modulation. Cell. 2003;113:37–47.
38. Stossel TP, Condeelis J, Cooley L, Hartwig JH, Noegel A, Schleicher M, et al. Filamins as integrators of cell mechanics and signalling. Nat Rev Mol Cell Biol. 2001;2:138–45.

39. Iiizumi G, Sadoya Y, Hino S, Shibuya N, Kawabata H. Proteomic characterization of the site-dependent functional difference in the rat small intestine. Biochim Biophys Acta. 2007;1774:1289–98.

40. Jain RN, Samuelson LC. Differentiation of the gastric mucosa. II. Role of gastrin in gastric epithelial cell proliferation and maturation. Am J Physiol Gastrointest Liver Physiol. 2006;291:762–5.

41. Murphy EF, Hooiveld GJ, Muller M, Calogero RA, Cashman KD. Conjugated linoleic acid alters global gene expression in human intestinal-like Caco-2 cells in an isomer-specific manner. J Nutr. 2007;137:2359–65.

42. Wanzel M, Russ AC, Kleine-Kohlbrecher D, Colombo E, Pelicci PG, Eilers M. A ribosomal protein L23-nucleophosmin circuit coordinates Mizl function with cell growth. Nat Cell Biol. 2008;10:1051–61.

43. Morita M, Ler LW, Fabian MR, Siddiqui N, Mullin M, Henderson VC, et al. A novel 4EHP-GIGYF2 translational repressor complex is essential for mammalian development. Mol Cell Biol. 2012;32:3585–93.

44. Ostergaard M, Hansen GA, Vorum H, Honoré B. Proteomic profiling of fibroblasts reveals a modulating effect of extracellular calumenin on the organization of the actin cytoskeleton. Proteomics. 2006;6:3509–19.

45. Birger Y, West KL, Postnikov YV, Lim JH, Furusawa T, Wagner JP, et al. Chromosomal protein HMGN1 enhances the rate of DNA repair in chromatin. EMBO J. 2003;22:1665–75.

46. Schmidts M, Arts HH, Bongers EM, Yap Z, Oud MM, Antony D, et al. Exome sequencing identifies DYNC2H1 mutations as a common cause of asphyxiating thoracic dystrophy (Jeune syndrome) without major polydactyly, renal or retinal involvement. J Med Genet. 2013;50:309–23.

47. Skabkin MA, Skabkina OV, Dhote V, Komar AA, Hellen CU, Pestova TV. Activities of Ligatin and MCT-1/DENR in eukaryotic translation initiation and ribosomal recycling. Genes Dev. 2010;24:1787–801.

48. van Wietmarschen N, Moradian A, Morin GB, Lansdorp PM, Uringa EJ. The mammalian proteins MMS19, MIP18, and ANT2 are involved in cytoplasmic iron-sulfur cluster protein assembly. J Biol Chem. 2012;287:43351–8.

49. Fanis P, Gillemans N, Aghajanirefah A, Pourfarzad F, Demmers J, Esteghamat F, et al. Five friends of methylated chromatin target of protein-arginine-methyltransferase[prmt]-1 (chtop), a complex linking arginine methylation to desumoylation. Mol Cell Proteomics. 2012;11:1263–73.

50. Marc Rhoads J, Wu G. Glutamine, arginine, and leucine signaling in the intestine. Amino Acids. 2009;37:111–22.

51. Shen Y, Lee HY, Rawson J, Ojha S, Babbitt P, Fu YH, et al. Mutations in PNKD causing paroxysmal dyskinesia alters protein cleavage and stability. Hum Mol Genet. 2011;20:2322–32.

52. Oliver CN, Starke-Reed PE, Stadtman ER, Liu GJ, Carney JM, Floyd RA. Oxidative damage to brain proteins, loss of glutamine synthetase activity, and production of free radicals during ischemia/reperfusion-induced injury to gerbil brain. Proc Natl Acad Sci U S A. 1990;87:5144–7.

53. Aung W, Hasegawa S, Furukawa T, Saga T. Potential role of ferritin heavy chain in oxidative stress and apoptosis in human mesothelial and mesothelioma cells: implications for asbestos-induced oncogenesis. Carcinogenesis. 2007;28:2047–52.

54. Polyzos S, Kountouras J, Zavos C, Papatheodorou A, Katsiki E, Patsiaoura K, et al. Serum ferritin in patients with nonalcoholic fatty liver disease: evaluation of ferritin to adiponectin ratio and ferritin by homeostatic model of assessment insulin resistance product as non-invasive markers. Immuno-Gastroenterology. 2012;1:119–26.

55. Fong CC, Zhang Y, Zhang Q, Tzang CH, Fong WF, Wu RS, et al. Dexamethasone protects RAW264.7 macrophages from growth arrest and apoptosis induced by H_2O_2 through alteration of gene expression patterns and inhibition of nuclear factor-kappa B (NF-kappaB) activity. Toxicology. 2007;236:16–28.

56. Villani G, Greco M, Papa S, Attardi G. Low reserve of cytochrome c oxidase capacity in vivo in the respiratory chain of a variety of human cell types. J Biol Chem. 1998;273:31829–36.

57. Kadenbach B, Arnold S, Lee I, Hüttemann M. The possible role of cytochrome c oxidase in stress-induced apoptosis and degenerative diseases. Biochim Biophys Acta. 2004;1655:400–8.

58. Arnoult D, Rismanchi N, Grodet A, Roberts RG, Seeburg DP, Estaquier J, et al. Bax/Bak-dependent release of DDP/TIMM8a promotes Drp1-mediated mitochondrial fission and mitoptosis during programmed cell death. Curr Biol. 2005;15:2112–8.

59. Zaza G, Granata S, Masola V, Rugiu C, Fantin F, Gesualdo L, et al. Downregulation of nuclear-encoded genes of oxidative metabolism in dialyzed chronic kidney disease patients. PLoS One. 2013;8:e77847.

60. Sparks LM, Xie H, Koza RA, Mynatt R, Hulver MW, Bray GA, et al. A high-fat diet coordinately downregulates genes required for mitochondrial oxidative phosphorylation in skeletal muscle. Diabetes. 2005;54:1926–33.

61. Sokolović M, Wehkamp D, Sokolović A, Vermeulen J, Gilhuijs-Pederson LA, van Haaften RI, et al. Fasting induces a biphasic adaptive metabolic response in murine small intestine. BMC Genomics. 2007;8:361.

62. Yanai H, Ban T, Taniguchi T. Essential role of high-mobility group box proteins in nucleic acid-mediated innate immune responses. J Intern Med. 2011;270:301–8.

63. Zhang Q, Wang Y. High mobility group proteins and their post-translational modifications. Biochim Biophys Acta. 2008;1784:1159–66.

64. Lopez GP, Ostap EM, Shaw S. Myosin 1G is a hematopoietic-restricted protein highly enriched in lymphocyte plasma membrane/microvilli whose deficiency impairs lymphocyte activation. J Immunol. 2009;182:35.40.

65. Ding Y, Lu B, Chen D, Meng L, Shen Y, Chen S. Proteomic analysis of colonic mucosa in a rat model of irritable bowel syndrome. Proteomics. 2010;10:2620–30.

66. Collins SE, Noyce RS, Mossman KL. Innate cellular response to virus particle entry requires IRF3 but not virus replication. J Virol. 2004;78:1706–17.

67. Mazaki Y, Hashimoto S, Tsujimura T, Morishige M, Hashimoto A, Aritake K, et al. Neutrophil direction sensing and superoxide production linked by the GTPase-activating protein GIT2. Nat Immunol. 2006;7:724–31.

68. Nicholls S, Piper KP, Mohammed F, Dafforn TR, Tenzer S, Salim M, et al. Secondary anchor polymorphism in the HA-1 minor histocompatibility antigen critically affects MHC stability and TCR recognition. Proc Natl Acad Sci U S A. 2009;106:3889–94.

69. Tian Y, Pate C, Andreolotti A, Wang L, Tuomanen E, Boyd K, et al. Cytokine secretion requires phosphatidylcholine synthesis. J Cell Biol. 2008;181:945–57.

70. Comar JF, Suzuki-Kemmelmeier F, Constantin J, Bracht A. Hepatic zonation of carbon and nitrogen fluxes derived from glutamine and ammonia transformations. J Biomed Sci. 2010;17:1.

71. Essa MM, Subramanian P. Pongamia pinnata modulates the oxidant-antioxidant imbalance in ammonium chloride-induced hyperammonemic rats. Fundam Clin Pharmacol. 2006;20:299–303.

2

Analysis of protein expression in periodontal pocket tissue: a preliminary study

Emanuela Monari[1]*[ID], Aurora Cuoghi[1], Elisa Bellei[1], Stefania Bergamini[1], Andrea Lucchi[2], Aldo Tomasi[1], Pierpaolo Cortellini[3], Davide Zaffe[4] and Carlo Bertoldi[5]

Abstract

Background: The periodontal disease is caused by a set of inflammatory disorders characterized by periodontal pocket formation that lead to tooth loss if untreated. The proteomic profile and related molecular conditions of pocket tissue in periodontally-affected patients are not reported in literature. To characterize the proteomic profile of periodontally-affected patients, their interproximal periodontal pocket tissue was compared with that of periodontally-healthy patients. Pocket-associated and healthy tissue samples, harvested during surgical therapy, were treated to extract the protein content. Tissues were always collected at sites where no periodontal-pathogenic bacteria were detectable. Proteins were separated using two-dimensional gel electrophoresis and identified by liquid chromatography/mass spectrometry. After identification, four proteins were selected for subsequent Western Blot quantitation both in pathological and healty tissues.

Results: A significant unbalance in protein expression between healthy and pathological sites was recorded. Thirty-two protein spots were overall identified, and four proteins (S100A9, HSPB1, LEG7 and 14-3-3) were selected for Western blot analysis of both periodontally-affected and healthy patients. The four selected proteins resulted over-expressed in periodontal pocket tissue when compared with the corresponding tissue of periodontally-healthy patients. The results of Western blot analysis are congruent with the defensive and the regenerative reaction of injured periodontal tissues.

Conclusions: The proteomic analysis was performed for the first time directly on periodontal pocket tissue. The proteomic network highlighted in this study enhances the understanding of periodontal disease pathogenesis necessary for specific therapeutic strategies setting.

Keywords: Proteome analysis, Periodontitis, Two-dimensional gel electrophoresis, Protein identification, LC-MS/MS

Background

The periodontal ligament surrounds the tooth root, connects it to its bony socket constituting a fibrous joint named gomphosis. It contains a large variety of cells and tissues, including immune and stem cells [1, 2]. The periodontitis is a set of inflammatory disorders characterized by gingival and periodontal inflammation, periodontal attachment loss, and alveolar bone resorption, following to periodontal pocket development [3–6].

The periodontal-pathogenic microbiome ecosystems are the most proven risk factor of periodontal disease. Pathogens adhere to and grow on the tooth surfaces, and the inappropriate inflammatory response causes the loss of periodontal attachment and alveolar bone, giving rise to the periodontal pocket, the typical expression of periodontitis, that lead to tooth loss, if untreated.

Periodontitis diagnosis is based on clinical assessment only, inspecting the soft gum tissues around the teeth with a probe (i.e. a clinical examination) in order to

* Correspondence: emanuela.monari@unimore.it
[1]Department of Diagnostic, Clinical and Public Health Medicine, University of Modena and Reggio Emilia, Largo del Pozzo, 71-41124 Modena, Italy
Full list of author information is available at the end of the article

detect interproximal attach level-loss and periodontal pocket depth, in the absence of a reliable pathogenic check based on appropriate interpretation of inflammation [5, 7, 8].

Recently, the role of cytokines and other protein mediators of inflammation erupted in all their importance [9]. A modern pathogenetic model incorporating gene, protein, and metabolite data into dynamic biological processes is based on a multilevel framework that include disease-initiating and -resolving mechanisms that are regulated by innate and environmental factors [10].

Now, we can describe more effectively the basic elements of a new model of pathogenesis using emerging genomic, proteomic, and metabolomic techniques [10, 11]. Recent progress in tissue isolation, protein separation, quantification and sequence analysis utilizing novel proteomic techniques promises to bring periodontal physiology and pathology into a new era. A list of inflammation-involved proteins, cytokines, matrix expression and cellular proteins in the periodontal tissues is not currently available. Studies on periodontal diseases utilizing proteomic analysis have been performed on saliva or crevicular fluid samples [12–15], peripheral blood [16–18] or periodontal plaque samples [19], but not on the pathologic tissue of the periodontal pocket, which is the key lesion of the periodontal disease.

To overcome the lack of data, studies assessing the proteomic profile of periodontal pocket tissue and evaluating the molecular characteristics of the periodontally affected patient, are needed.

The aim of this work was to compare the proteomic profile of the pathologic interproximal gingival pocket tissue with that of interproximal gingival healthy tissue, obtained from sites where no periodontal-pathogenic bacteria were detectable.

Results

Fifteen subjects (T, test group), 3 males and 12 females, ranging in age from 20 to 64 years, average 42.82 ± 13.2 (m ± SD), and fifteen periodontally healthy subjects (C, control group), 6 males and 9 females, ranging in age from 19 to 60 years, average 44.90 ± 11.55, fulfilled study requirements (Table 1). T subjects underwent to the periodontal resective treatment, while C subjects underwent to the crown lengthening surgical treatment. All subjects followed a stringent post-operative supportive care program and achieved satisfactory clinical outcomes.

2DE (Two-dimensional gel electrophoresis) and LC-MS/MS (Liquid Chromatography-tandem mass spectrometry) analysis were performed for T and C gingival tissue samples. Figure 1 shows representative 2DE gel images of T and C tissues and the 32 protein spots identified. Primary accession number, entry name of UniProt database, MW (Molecular Weight), highest score, matches, sequences and emPAI (Exponentially Modified Protein Abundance Index) for each identified protein are reported in Table 2. Web-based bioinformatics tools (iPROClass and CateGOrize) were employed to investigate all potential localizations, molecular functions and biological processes of the identified proteins (Fig. 2). Intracellular proteins represented the most abundant population (77.3 %). Among them, 18.6 % of protein annotations were recognized as belonging to cytoplasm, 5.6 % to cytoskeleton, and nucleus 8.8 %. Other protein annotations were recognized, with smaller fractions, as belonging to other cellular component such as mitochondrion, vacuole or plasma membrane (Fig. 2a). The reported function of the identified proteins suggests that they are mainly involved in binding (46.8 %), protein binding (15.3 %), and catalytic activity (8.8 %). Other relevant functions are nucleic acid binding (6.6 %), enzyme regulator activity (4.4 %), structural molecule activity (4.4 %), antioxidant activity, lipid binding (2.9 %), nucleotide binding (2.2 %) and transferase activity (1.5 %) (Fig. 2b). In summary, the identified proteins are mainly involved in metabolism (31,9 %), transport (13,3 %) and cell organization and biogenesis (12 %) (Fig. 2c).

PDQuest spot intensity quantification identified four proteins expressed only in the pathological tissue, as shown in Fig. 3, that were selected for subsequent Western blot analysis in both T and C subjects.

Table 1 Systemic and specific adopted inclusion criteria

Systemic Inclusion Criteria	Local Inclusion Criteria
Absence of relevant medical conditions: Medical history of good health (particularly ruling out bone disease, uncontrolled or poorly controlled diabetes, unstable or life-threatening conditions, or requiring antibiotic prophylaxis were excluded).	*Defect anatomy*: Presence of at least one intrabony defect in patient with periodontal disease. Excessive gingival display or gingival margin asymmetries required a surgical correction in periodontally-healthy patients
Smoking status: Non-smokers and without a story of alcohol abuse.	*Good oral hygiene*: full-mouth plaque score (FMPS) ≤20 %
Compliance: only patients showing high levels of compliance (as assessed during the cause-related phase of therapy) were selected.	*Low level of residual infection*: full-mouth bleeding score (FMBS) ≤20 %.
Pregnancy or lactation and underage were excluded.	*Endodontic status*: Experimental teeth had to be vital or properly treated with root canal therapy.

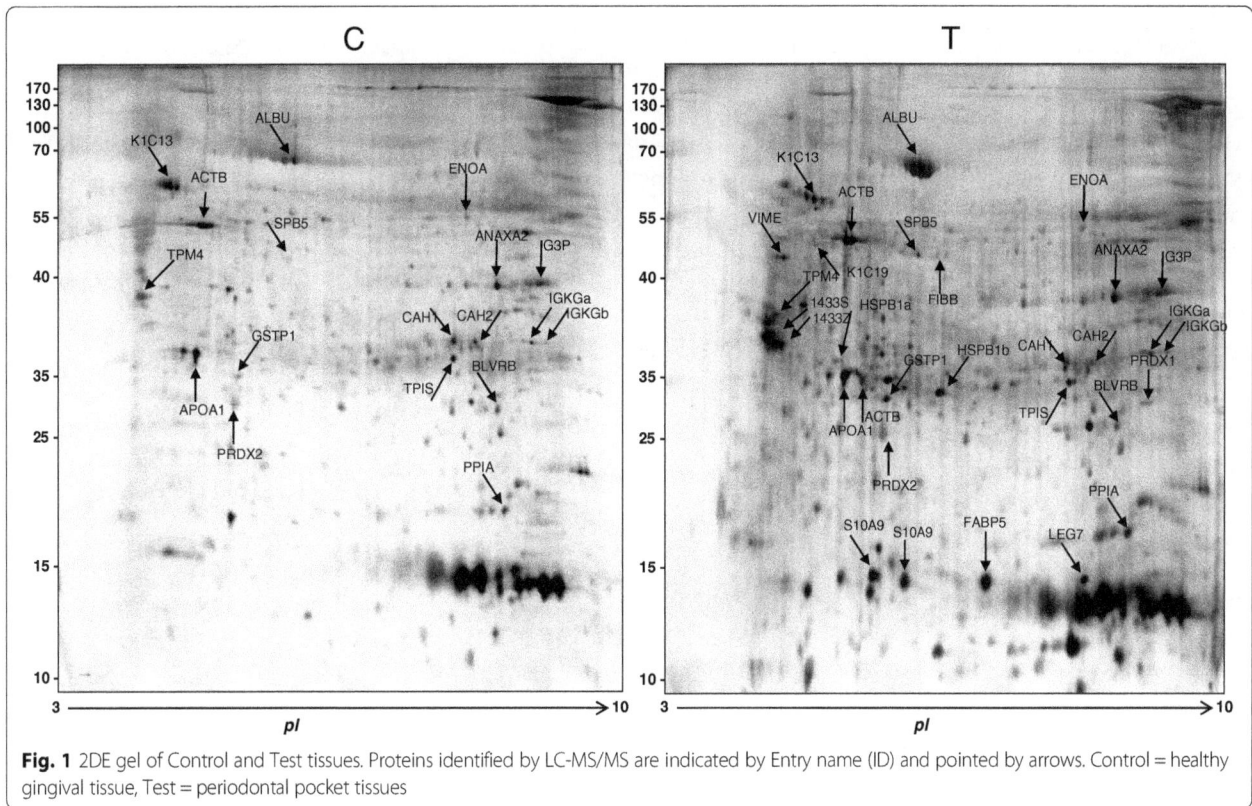

Fig. 1 2DE gel of Control and Test tissues. Proteins identified by LC-MS/MS are indicated by Entry name (ID) and pointed by arrows. Control = healthy gingival tissue, Test = periodontal pocket tissues

Representative Western blot images for S100A9, HSPB1, 14-3-3 and LEG7 and β-actin are reported in Fig. 4. Densitometric analyses of Western blot images, normalized to β-actin expression (Fig. 5) confirmed that in C specimens (control tissue), the S100A9, 14-3-3, LEG7 and HSPB1 proteins were not expressed. In pathological tissue (T specimens), the intensity signal for S100A9 and HSPB1 was much higher than the intensity signal of 14-3-3 and LEG7 proteins.

Discussion

Very little is know on the proteomic analysis of the periodontal pocket tissue [6]. Most proteomic studies have been performed on gingival crevicular fluid, saliva or blood serum, since the non-invasive nature of the collection and availability [20–23]. These studies assume the existence of a direct correlation between the periodontal disease and the biological source to be tested, however there is no direct evidence supporting such assumption. Actually, the periodontal pocket is certainly the anatomo-pathological lesion signifying for an active periodontal disease [7, 8]. Therefore, the periodontal pocket tissue is the only biologic material in which proteomic analysis enables the correct molecular assessment of the periodontal disease. Despite genomic, transcriptomic and proteomic research performed up to now, there are no available biomarkers for periodontitis diagnosis,

prognosis and treatment indication giving assistance to the clinician in the disease management. Excluding the studies performed on biologic material different from periodontal tissue, some genomic studies carried out on the gingival tissue showed an association of specific gene polymorphism to periodontitis [24–26].

In our study, proteomics was applied to comparatively analyze the protein content of interproximal tissues of healthy and periodontally affected patients, with the aim to identify differentially expressed proteins. The presence of common periodontopathogenic bacteria was excluded, focusing on disease progression related to the prevailing stimulus, avoiding in this way the possible straining effect caused by periodontopathogenic bacteria on the local protein content. In this way we could discriminate the individual component of the periodontal disease.

The differential expression level observed in 2DE, in pathological and healthy tissues, of 4 proteins, S100A9, HSPB1, LEG7 and 14-3-3, was confirmed by Western blot analysis.

S100A9, is a calcium- or zinc-binding protein involved in the regulation of inflammatory processes and immune response [27], predominantly found as calprotectin (S100A8/S100A9 etherodimer). S100A9 contribute to homeostatic processes that include cytoskeletal rearrangements during trans-endothelial migration of pro-

Table 2 List of identified proteins in periodontal pocket tissues

AC Number[a]	Entry Name[b]	Score	Mass (kDa)	pI	Matches[c]	Significant matches	Sequences[d]	Significant sequences	emPAI[e]	Description
P02768	ALBU_HUMAN	5019	71317	5.92	435	310	43	40	6.59	Serum albumin
P13646	K1C13_HUMAN	2745	49900	4.91	247	140	31	23	4.99	Keratin, type I cytoskeletal 13
P19013	K2C4_HUMAN	1509	57649	6.25	176	101	25	18	2.21	Keratin, type II cytoskeletal 4
Q01469	FABP5_HUMAN	1050	15497	6.60	95	44	21	12	16.08	Fatty acid-binding protein, epidermal
P01834	IGKC_HUMAN	829	11773	5.58	65	51	5	5	6.66	Ig kappa chain C region
P01834	IGKC_HUMAN	771	11773	5.58	62	47	5	5	5.11	Ig kappa chain C region
P06733	ENOA_HUMAN	757	47481	7.01	67	40	22	15	2.43	Alpha-enolase
P31947	1433S_HUMAN	592	27871	4.68	88	49	19	12	3.12	14-3-3 protein sigma
P08727	K1C19_HUMAN	492	44079	5.05	51	27	14	9	0.78	Keratin, type I cytoskeletal 19
P00915	CAH1_HUMAN	403	28909	6.59	46	29	8	6	1.39	Carbonic anhydrase 1
P07355	ANXA2_HUMAN	382	38808	7.57	31	17	12	7	0.74	Annexin A2
P04406	G3P_HUMAN	377	36201	5.43	20	19	5	4	0.69	Glyceraldehyde-3-phosphate dehydrogenase
Q06830	PRDX1_HUMAN	358	22324	8.27	53	26	13	6	1.10	Peroxiredoxin-1
P63104	1433Z_HUMAN	317	27899	4.7	43	17	14	8	2.31	14-3-3 protein zeta/delta
P67936	TPM4_HUMAN	282	28619	4.67	51	27	19	14	3.41	Tropomyosin alpha-4 chain
P02647	APOA1_HUMAN	260	30759	5.56	45	22	16	9	5.56	Apolipoprotein A-I
P32119	PRDX2_HUMAN	223	22049	5.66	24	15	9	8	2.10	Peroxiredoxin-2
P06702	S10A9_HUMAN	217	13291	5.71	42	20	8	4	2.94	Protein S100-A9
P47929	LEG7_HUMAN	207	15123	7.03	59	22	7	7	3.74	Galectin-7
P62937	PPIA_HUMAN	197	18229	7.68	11	7	4	3	0.97	Peptidyl-prolyl cis-trans isomerase A
P02675	FIBB_HUMAN	183	55892	8.54	5	3	5	3	0.27	Fibrinogen beta chain
P04792	HSPB1_HUMAN	177	22826	5.98	27	10	9	5	0.83	Heat shock protein beta-1
P60709	ACTB_HUMAN	144	42052	5.29	55	23	13	7	0.67	Actin, cytoplasmic 1
P04792	HSPB1_HUMAN	132	22826	5.98	38	14	7	3	0.69	Heat shock protein beta-1
P09211	GSTP1_HUMAN	124	23569	5.43	6	3	2	1	0.30	Glutathione S-transferase P
P60174	TPIS_HUMAN	118	31057	6.45	17	9	5	4	0.43	Triosephosphate isomerase
P30043	BLVRB_HUMAN	118	22219	7.13	9	5	3	1	0.28	Flavin reductase (NADPH)
P36952	SPB5_HUMAN	80	42530	5.72	4	3	2	2	0.16	Serpin B5
Q9BYX7	ACTBM_HUMAN	76	41989	5.29	2	2	2	2	0.24	Putative beta-actin-like protein 3
P08670	VIME_HUMAN	71	53676	5.06	37	5	10	3	0.20	Vimentin
P06702	S10A9_HUMAN	54	13291	5.71	17	6	3	2	0.83	Protein S100-A9
P00918	CAH2_HUMAN	53	29285	8.15	12	6	3	2	0.33	Carbonic anhydrase 2

[a] Primary accession number from UniProt database
[b] Primary entry name from UniProt database
[c] The number of peptides that matched the identified proteins
[d] The number of distinct sequences
[e] Exponentially Modified Protein Abundance Index

inflammatory phagocytes, binding to receptors as Toll-like receptor 4 and receptor for advanced glycation end products, antimicrobial, oxidant-scavenging and apoptosis-inducing activities [28, 29]. In extracellular fluid, increased levels of S100A9/S100A8 etherodimer were reported in numerous inflammation-associated conditions, such as rheumatoid arthritis, Crohn's Disease, colorectal cancer and in GCF (Gingival Crevicular Fluid) of patients suffering from gingivitis and periodontitis [30, 31]. The S100A9/S100A8 etherodimer concentration was correlated with periodontal markers of inflammation such as pocket probing depth or gingival bleeding [32, 33].

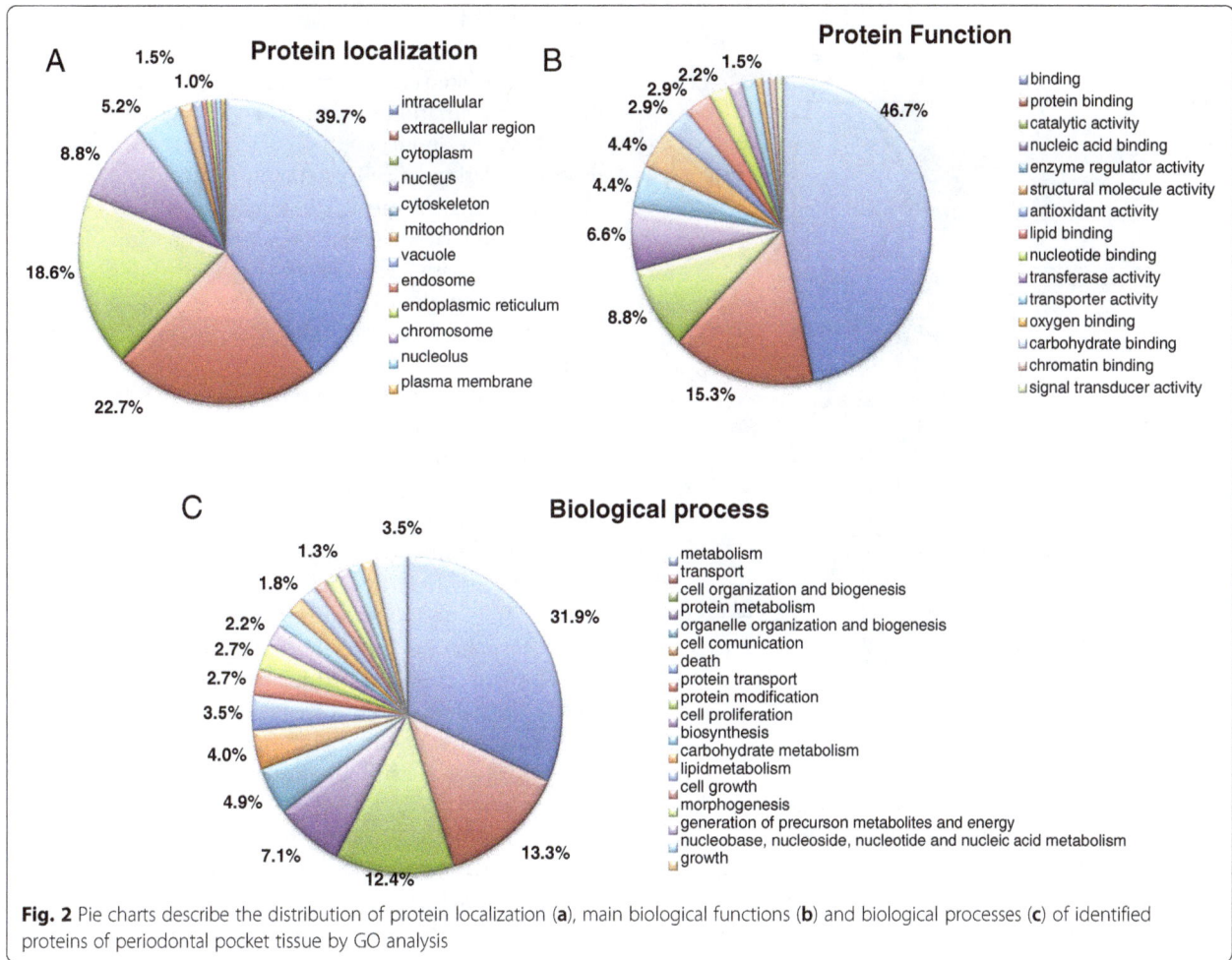

Fig. 2 Pie charts describe the distribution of protein localization (**a**), main biological functions (**b**) and biological processes (**c**) of identified proteins of periodontal pocket tissue by GO analysis

S100A9 was identified in saliva and proposed as a potential marker to monitor the progression of orthodontic treatment [33]. However, the Authors showed an apparent down-regulation of S100A9 protein, suggesting that this protein could not be involved during bone resorption in orthodontic tooth movement but it was implicated in inflammation. This protein also promotes apoptosis and modulate the inflammatory response in periodontal ligament cells so its down-regulation could suggest a suppression of inflammation [33, 34].

HSPB1 synthesis increases in response to a variety of stresses (e.g. elevated temperatures, heavy metals, toxins, oxidants, bacterial and viral infections) in order to minimize the deleterious consequences of these stimuli and provide the maximal cytoprotective effect [35–39]. In oral tissues, HSPB1 was localized in fibroblasts, odontoblasts, osteoblasts, epithelial cells, endothelial cells of the vascular wall of the dental pulp, and cells of the periodontal ligament [37, 40]. HSPB1 is also a procollagen-binding protein involved in the biosynthesis of type I collagen and major bone extracellular matrix [41]. HSPB1 could be a potential target for the periodontal

regeneration process related to cell migration, cytoskeleton maintaining and tissue preservation, also through the modulation of the immune system, and its under-expression induces differentiation abortion, in relation to cell death by apoptosis [40].

Galectin-7 (LEG7) is associated with epithelial cell migration and accelerates the re-epithelialization of wounds. In particular, LEG7 expression contributes to the tissue remodeling processes following tissue damage that involves apoptotic cell death [42]; a defective LEG7 expression could impair the healing processes. LEG7 could function in the maintenance of the normal phenotype of epithelial cell and is activated by a wide range of cellular stresses including UV and γ irradiation. LEG7 protects cell from death by functioning inside the cell and interacting with intracellular proteins. Finally, LEG7 belongs to a protein family promoting healing processes and favor immune responses [43]. HSPB1 and LEG7 would seem to be part of a protein network that play an important role in controlling cell and tissue damage, in moderating the inflammation and destructive immune response. The lack of the previously described multi-

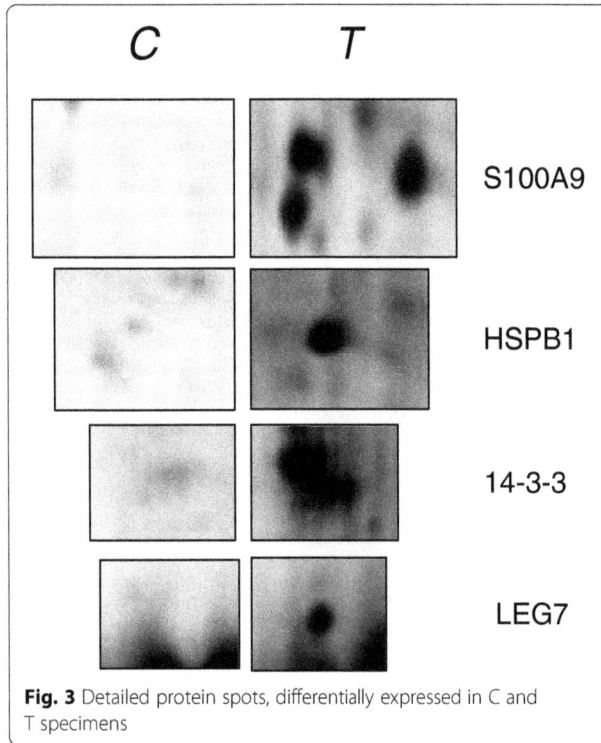

Fig. 3 Detailed protein spots, differentially expressed in C and T specimens

protective effect of these proteins, particularly in the presence of inflammation, hinders cells to protect themselves against the cytotoxicity of inflammatory mediators [44], increases their susceptibility to necrotic cell death [45] and probably does not allow an efficient immune response against the noxa [46].

The 14-3-3 protein sigma and zeta/delta are regulatory phosphorserine/threonine binding proteins involved in

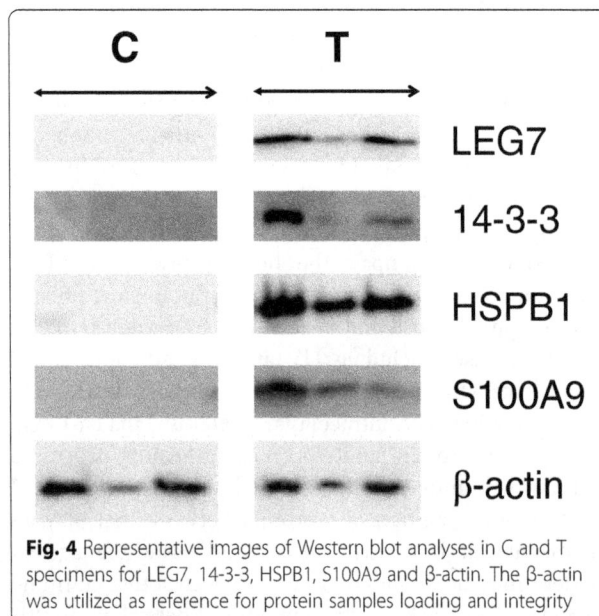

Fig. 4 Representative images of Western blot analyses in C and T specimens for LEG7, 14-3-3, HSPB1, S100A9 and β-actin. The β-actin was utilized as reference for protein samples loading and integrity

the control of several cellular events, including cell cycle checkpoint, connective tissue remodeling, apoptosis signaling, Toll-like receptor activation and TNF production during inflammation response [47, 48]. In particular the 14-3-3 sigma induces up-regulation of differentiation, down-regulation of cell proliferation and collagenase induction as matrix metalloproteinase-1 [47, 49, 50]. Wu et al. [12] found decreased 14-3-3 protein sigma in saliva of subjects with generalized aggressive periodontitis, and, analyzing the GCF, Huynh et al. [15] found 2–3 time higher level in gingivitis than in chronic periodontitis.

Our results are congruent with an inflammatory response oriented to defense and regeneration of injured tissues. S100A9, HSPB1, LEG7 and 14-3-3 proteins resulted over-expressed in periodontal pocket tissue when compared with healthy patient analogous tissue. A significant unbalancing in protein expression between healthy and pathological sites was recorded. In a previous study [6] we aimed to compare the interproximal pocket tissue with interproximal tissues at sites with normal probing depth in patients affected from periodontal disease and HSPB1, LEG7 and 14-3-3 proteins resulted significantly under-expressed. Also S100A9 resulted under-expressed, but not significantly.

In all likelihood, patients have to be regarded as affected by periodontal disease because a complex pathologic network (composed by genetic structure involving immunology and inflammatory regulation) at the root of periodontitis even if the periodontal disease is being clinically burnt out [51–54].

A limited number of biomolecular studies have been carried out on the periodontal pocket. Hence, inadequate data are available on the pathognomonic lesion of the periodontal disease. Genes related to apoptosis, antimicrobial humoral response, antigen presentation, regulation of metabolism, signal transduction and angiogenesis were found to be differently expressed in patients with periodontitis and healthy subjects, as found in trascriptomic studies [55, 56]. Moreover, with the aid of microarray technologies, cell communication pathways were shown to be down-regulated in periodontitis-affected tissues, either in cell-to-cell communications at the soft tissue level, or in cell-to-tooth signaling as a consequence of the inflammatory status of the periodontium [55, 56].

It is conjecturable that an high inflammatory background is anyway present in healthy gingival tissues of patients with periodontal disease, and the change of the inflammatory and immunologic processes could have a role in periodontal damage with the hindering of the protecting molecular network when the risk factor challenge induces the periodontal pocket.

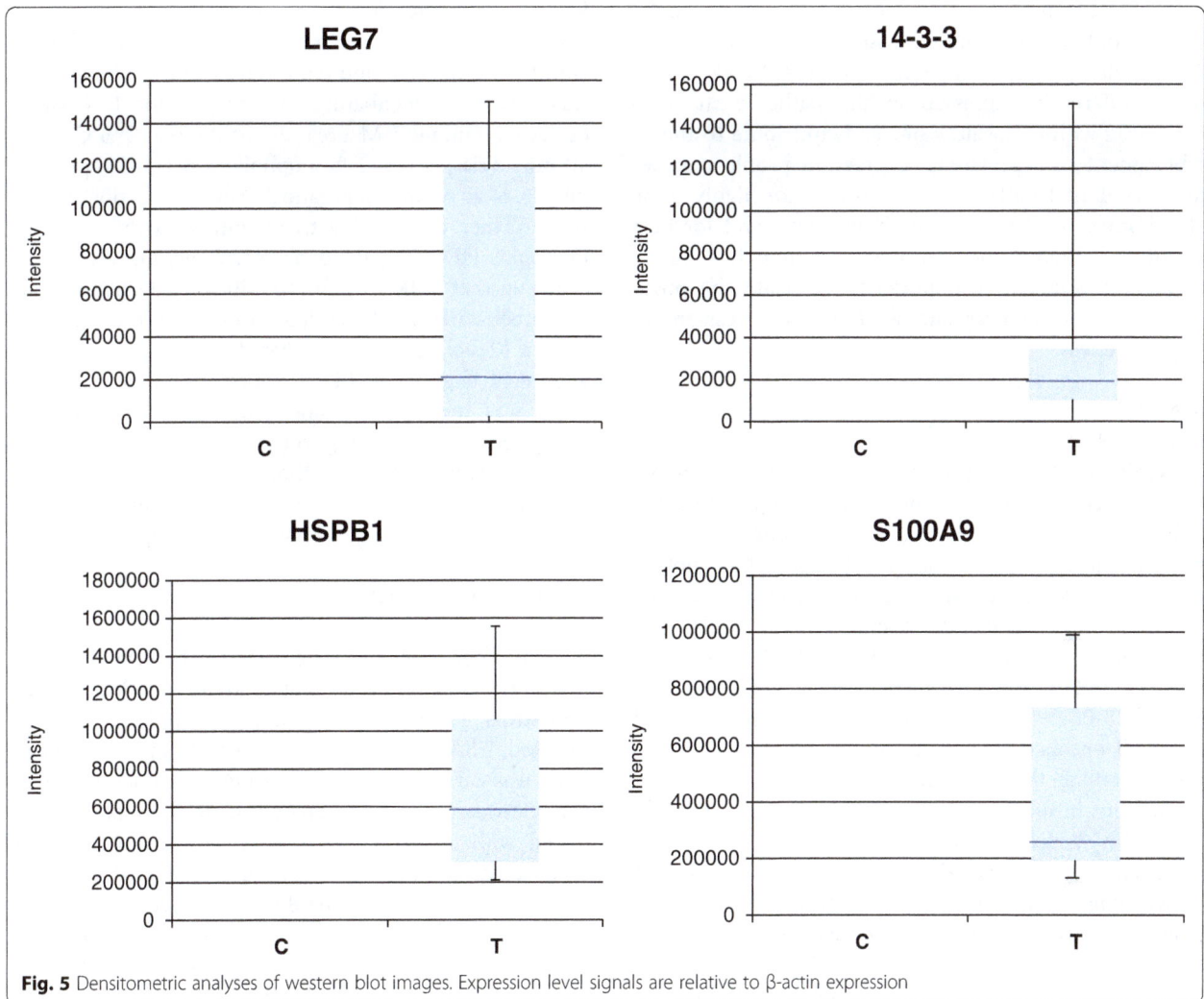

Fig. 5 Densitometric analyses of western blot images. Expression level signals are relative to β-actin expression

The identified proteins play a role in increased cell proliferation, decrease in cell tissue differentiation, impairing of metalloproteinases on connective tissue, organized action monocytes/macrophages (e.g. osteoclasts) and osteoblasts or fibroblast together with enhanced humoral and cell-mediated immunity. All these processes are consistent with a defensive response of the organism that also aims to regenerate the lost tissue after exogenous injury. Furthermore, the phagocytosis of opsonized cells, the alternative way by Toll-like receptors 2 and 4 or the T-cell activation to bacterial LPS could contribute to microbiota clearance along with the presence of chemo attractant molecules in the recruitment of immune cells. During these phases, an increasing cell migration could also play an important role in the recovery of damaged tissue as periodontal ligament or bone. It is also to consider that HSPs, whose synthesis is increased in response to a variety of stresses, and S100A9 play an important role in controlling cell and tissue damage, and in moderating inflammation and destructive immune response. HSPB1 and S100A9 resulted significantly increased, in all the tested pathological sites, while LEG7 and 14-3-3 proteins, although not expressed in normal tissue, exhibited fluctuation in expression level in periodontal lesions. These results need further studies to understand if the lack in expression of these proteins may worsen the tissue damage.

Conclusions

Most proteomic studies have been performed on different biological samples but not on periodontal pocket tissue that is the anatomo-pathological lesion signifying for the clinical diagnosis of the periodontal disease. Therefore, the periodontal pocket tissue is the only biologic material in which proteomic analysis enables the correct molecular assessment of this pathology. The establishment of a robust proteomic expression profile database for periodontal pathology would be highly desirable both to understand the pathogenesis and for periodontitis therapeutic strategies [13]. This is the first work that compares the proteomic content of periodontal pocket lesions and healthy gingival tissue of healthy subjects. The results of this study pointed out a network of proteins, differently expressed in the pathological tissue compared with the healthy one and contribute to the establishment of a proteomic expression profile database for periodontal pathology. These data are highly desirable both to understand the pathogenesis and for periodontitis therapeutic strategies, though further population studies are required to correlate the proteomic data with clinical data, as disease progression and aggressiveness.

Methods
Sample collection
Systemic healthy subjects, with and without periodontal chronic disease, examined and treated in a private dental office, were enrolled in this study, according to the protocol described in Table 1. Italian law does not require any ethical committee authorization for clinical trials performed in private dental offices, while such authorization is required for public dental health centers (DM 18/3/1998 published in the Official Gazette, GU n. 122 of 28-05-1998). Therefore, for the purposes of this study all enrolled subjects signed an informed consent form detailing the study procedures. The research was carried out in full accordance with the ethical principles of the WMA Helsinki Declaration [57].

Fifteen periodontally-healthy patients underwent crown lengthening surgery, where gingiva in excess or gingival margin asymmetries required a surgical correction [58, 59]. Periodontally-affected subjects presented at least a shallow intrabony defect suitable for treatment by osseous resective surgery.

Samples for microbial analysis were obtained from patients immediately before surgery, in order to exclude periodontopathogenic bacteria. The samples were analyzed by PCR-RT technique (GABA International AG, Lorrach, Germany), to confirm the absence of *Actinobacillus actinomycetemcomitans*, *Porphyromonas gingivalis*, *Tannerella forsythensis*, *Treponema denticola*, *Fusobacterium nucleatum ssp*, and *Prevotella intermedia*.

The surgery was performed in both periodontally-healthy and -affected patients following the completion of a preliminary cause-related treatment required for the surgical approach and having reached a full-mouth plaque score and full-mouth bleeding score lower than 20 %.

Tissue specimens of interproximal healthy tissues were harvested in fifteen periodontally-healthy patients (C, control) at sites with normal probing depth [58–60]. Tissue specimens of interproximal pocket-associated tissue were harvested in subjects affected by chronic periodontitis (T, test). The harvested tissues were immediately frozen at −80 °C, for proteomic analyses.

Sample preparation for proteomic analysis
Tissue samples were ground in small pieces in a mortar with liquid nitrogen and collected in tubes. Tissue lysate was performed incubating the samples for 1 h with a buffer containing 7 M urea, 2 M thiourea, 3 % CHAPS, 40 mM Tris pH 8.3, 1 % ampholytes pH 3–10, protease inhibitors at room temperature. After incubation, tissues were further disrupted with an ultrasonic homogenizer (Sonoplus HD 2070, Bandelin electronic, Germany), and centrifuged at 10,000 x g for 10 min at +4 °C. Supernatant was precipitated by the addition of cold acetone (dilution ratio 1:12 vol/vol) and incubated at −20 °C overnight. After centrifugation at 14,000 x g for 15 min at +4 °C, the pellet was re-suspended and the protein concentration was determined according to the Bradford method. Three pooled samples, both for pathological and healthy gingival tissues, were obtained by mixing equal protein amount of 5 different subjects and were analyzed in duplicate.

2-Dimension electrophoresis
Tissue extracted proteins were separated by 2DE following a protocol previously described [61]. Briefly, for the first dimension, 80 μg of total proteins were loaded onto IPG strips, 17 cm long, pH range 3–10 (ReadyStripTM, Bio-Rad, USA). Afterwards, the second-dimension separation was carried out at 10 °C using 12 % polyacrylamide gels. Between these two separation steps, strips were reduced with 1 % DTT, and later alkylated with 2.5 % iodoacetamide in an equilibration buffer (6 M urea, 1 % DTT, 50 mM Tris–HCl pH 8.8, 30 % glycerol, 2 % SDS). After 2-DE, the protein spots in the gels were visualized following a silver nitrate staining protocol, as previously described [62]. The silver-stained gel images were acquired using a GS-800 Calibrated Densitometer (Bio-Rad, USA) and analysed with the PDQuest 2-D Image software program, version 7.3.1 (Bio-Rad, USA).

Protein identification by LC-MS/MS analysis
Protein spots excised manually from the gels were subjected to the "in-gel" tryptic digestion as previously described [63]. Dried samples were then re-suspended in 97 % Water/3 % ACN added of 1 % formic acid (Buffer A) and analyzed by a Nano LC-CHIP-MS system, consisting of the Agilent 6520 ESI-Q-TOF, coupled with a 1200 Nano HPLC-Chip microfluidic device (Agilent Technologies Inc., USA). Four microliters of each sample were loaded into the system and transported to the Chip enrichment column (Zorbax C18, 4 mm x 5 μm i.d., Agilent Technologies) by a capillary pump, with a loading flow of 4 μL/min, using 95 % ACN/5 % water added of 0.1 % formic acid (buffer B) as mobile phase. Nitrogen was used as the nebulizing gas. A separation column (Zorbax C18, 43 mm x 75 μm i.d., Agilent Technologies), at flow rate of 0.4 μL/min, was used for peptide

separation. Protein-identification peak lists were generated using MASCOT search engine (http://mascot.cigs.unimo.it/mascot) against the UniProt Knowledgebase database (UniProt.org), specifying the following parameters: Homo sapiens taxonomy, parent ion tolerance ± 20 ppm, MS/MS error tolerance ± 0.12 Da, alkylated cysteine as fixed modification and oxidized methionine as variable modification, and two potential missed trypsin cleavages. Proteins that were identified with at least 2 or more significant peptides sequences and with the highest score hits among MASCOT search results. "High-scoring" corresponded to proteins that were above the significant threshold in Mascot searches (5 % probability of false match for each proteins above this score).

Protein functional analysis

The protein functional analysis and classification of identified proteins were performed using the iPROClass integrated database (http://pir.georgetown.edu/pirwww/dbinfo/iproclass.shtml accessed on March, 2015) for protein annotation and GO Terms Classifications Counter (CateGOrize, http://www.animalgenome.org/bioinfo/tools/countgo/ accessed on March, 2015) for clusterization according to the Gene Ontology (GO) hierarchy.

Western blot analysis

For Western blotting experiments, 3 µg of gingival tissue protein extracts were solubilized in Laemmli's buffer 1X and denatured for 5 min at 95 °C. Protein content was resolved by 12 % or 15 % SDS-PAGE and the proteins were transferred onto a nitrocellulose membrane by electroblotting. Membranes were blocked at 4 °C overnight with 5 % non-fat dry milk in PBS containing 0.1 % Tween 20. The membranes were incubated with the relevant antibodies for 2 h (1:1000 dilution), washed, and incubated with HRP coniugated Polyclonal Goat Anti-rabbit secondary antibody (1:2000 dilution, DakoCytomation, Denmark) for 1 h. The proteins were visualized using WesternSure Premium Chemiluminescent substrate (LI-COR, USA) and C-Digit Blot scanner (LI-COR, USA) according to the manufacturer's instructions. Acquired images were analyzed and compared using Image Studio Software 4.0.21 (LI-COR, USA). The sources of primary antibodies were as follows: anti-galectin 7 (ab10482), anti-HspB1 (ab1426) and anti-14-3-3 all isoforms (ab9063) were from Abcam (UK); anti- S100A9 (PA1-46489) was from Thermo scientific (USA). β-actin (ab8227, Abcam UK) was used for normalization of western blot analysis.

Abbreviations

2DE: Two-dimensional electrophoresis; LC-MS/MS: Liquid Chromatography-Mass/Mass spectrometry; MW: Molecular weight; emPAI: Exponentially modified protein abundance index; HSPB1: Heat shock protein ß-1; LEG7: Galectin-7; 1433S: 14-3-3 Protein sigma; 1433Z: 14-3-3 protein zeta/delta; S100A9: Protein S100-A9; GCF: Gingival Crevicular Fluid; CHAPS: 3-[(3-Cholamidopropyl) dimethylammonio]-1-propanesulfonate; DTT: Dithiothreitol; ACN: Acetonitrile; ESI-Q-TOF: Electrospray ionization quadrupole time-of-flight mass; SDS-PAGE: Sodium dodecyl sulfate polyacrylamide gel electrophoresis; HRP: Horseradish peroxidase.

Competing interests
The authors declare that they have no competing interests.

Authors' contributions
EM designed the study, performed sample preparation, western blotting analysis and drafted the manuscript; AC carried out ESI-QTOF-MS analysis and performed protein identification; EB and SB performed bi-dimensional gel electrophoresis and in gel digestions; AL participated to the sample collections; AT and DZ provided useful advices to improve the study and revised the manuscript; PC participated in the design of the study; CB conceived the study, participated in its design and coordination, enrolled all subject of the study and helped to draft the manuscript. All authors read and approved the final manuscript.

Acknowledgement
We thank Dr. Filippo Genovese (C.I.G.S., University of Modena and Reggio Emilia) for technical assistance during LC-MS/MS analysis. We are grateful to the "Fondazione Cassa di Risparmio di Modena, Italy" for have supported the purchase of the ESI-Q-TOF mass spectrometer.

Author details
[1]Department of Diagnostic, Clinical and Public Health Medicine, University of Modena and Reggio Emilia, Largo del Pozzo, 71-41124 Modena, Italy. [2]Private Practice, Modena, Italy. [3]European Research Group on Periodontology (ERGOPERIO), Berne, Switzerland. [4]Department of Biomedical, Metabolic and Neural Sciences, University of Modena and Reggio Emilia, Modena, Italy. [5]Department of Surgery, Medicine, Dentistry and Morphological Sciences with Transplant Surgery, Oncology and Regenerative Medicine Relevance, University of Modena and Reggio Emilia, Modena, Italy.

References
1. Bertoldi C, Bencivenni D, Lucchi A, Consolo U. Augmentation of keratinized gingiva through bilaminar connective tissue grafts: a comparison between two techniques. Minerva Stomatol. 2007;56:3–20.
2. Tripodo C, Di Bernardo A, Ternullo MP, Guarnotta C, Porcasi R, Ingrao S, et al. CD146(+) bone marrow osteoprogenitors increase in the advanced stages of primary myelofibrosis. Haematologica. 2009;94:127–30.
3. Flemmig TF. Periodontitis. Ann Periodontol. 1999;4:32–8.
4. Socransky SS, Haffajee AD. The nature of periodontal diseases. Ann Periodontol. 1997;2:3–10.
5. Bertoldi C, Pellacani C, Lalla M, Consolo U, Pinti M, Cortellini P, et al. Herpes Simplex I virus impairs regenerative outcomes of periodontal regenerative therapy in intrabony defects: a pilot study. J Clin Periodontol. 2012;39:385–92.
6. Bertoldi C, Bellei E, Pellacani C, Ferrari D, Lucchi A, Cuoghi A, et al. Non-bacterial protein expression in periodontal pockets by proteome analysis. J Clin Periodontol. 2013;40:573–82.
7. Tonetti MS, Claffey N. Advances in the progression of periodontitis and proposal of definitions of a periodontitis case and disease progression for use in risk factor research. Group C consensus report of the 5th European Workshop in Periodontology. J Clin Periodontol. 2005;32 Suppl 6:210–3.
8. Page RC, Eke PI. Case definitions for use in population-based surveillance of periodontitis. J Periodontol. 2007;78:1387–99.
9. Huang R-Y, Lu S-H, Su K-W, Chen J-K, Fang W-H, Liao W-N, et al. Diacerein: a potential therapeutic drug for periodontal disease. Med Hypotheses. 2012;79:165–7.
10. Kornman KS. Mapping the pathogenesis of periodontitis: a new look. J Periodontol. 2008;79:1560–8.
11. Gonçalves LDR, Soares MR, Nogueira FCS, Garcia C, Camisasca DR, Domont G, et al. Comparative proteomic analysis of whole saliva from chronic periodontitis patients. J Proteomics. 2010;73:1334–41.
12. Wu Y, Shu R, Luo L-J, Ge L-H, Xie Y-F. Initial comparison of proteomic profiles of whole unstimulated saliva obtained from generalized aggressive periodontitis patients and healthy control subjects. J Periodontal Res. 2009;44:636–44.

13. Gorr S-U, Abdolhosseini M. Antimicrobial peptides and periodontal disease. J Clin Periodontol. 2011;38 Suppl 1:126–41.

14. Jönsson D, Ramberg P, Demmer RT, Kebschull M, Dahlén G, Papapanou PN. Gingival tissue transcriptomes in experimental gingivitis. J Clin Periodontol. 2011;38:599–611.

15. Huynh AHS, Veith PD, McGregor NR, Adams GG, Chen D, Reynolds EC, et al. Gingival crevicular fluid proteomes in health, gingivitis and chronic periodontitis. J Periodontal Res. 2015;50:637–49.

16. Yokoyama T, Kobayashi T, Yamamoto K, Yamagata A, Oofusa K, Yoshie H. Proteomic profiling of human neutrophils in relation to immunoglobulin G Fc receptor IIIb polymorphism. J Periodontal Res. 2010;45:780–7.

17. Mizuno N, Niitani M, Shiba H, Iwata T, Hayashi I, Kawaguchi H, et al. Proteome analysis of proteins related to aggressive periodontitis combined with neutrophil chemotaxis dysfunction. J Clin Periodontol. 2011;38:310–7.

18. Yokoyama T, Kobayashi T, Ito S, Yamagata A, Ishida K, Okada M, et al. Comparative analysis of serum proteins in relation to rheumatoid arthritis and chronic periodontitis. J Periodontol. 2014;85:103–12.

19. Pham TK, Roy S, Noirel J, Douglas I, Wright PC, Stafford GP. A quantitative proteomic analysis of biofilm adaptation by the periodontal pathogen Tannerella forsythia. Proteomics. 2010;10:3130–41.

20. AlMoharib HS, AlMubarak A, AlRowis R, Geevarghese A, Preethanath RS, Anil S. Oral fluid based biomarkers in periodontal disease: part 1. Saliva. J Int oral Heal. 2014;6:95–103.

21. Trindade F, Oppenheim FG, Helmerhorst EJ, Amado F, Gomes PS, Vitorino R. Uncovering the molecular networks in periodontitis. Proteomics Clin Appl. 2014;8:748–61.

22. Guzman YA, Sakellari D, Arsenakis M, Floudas CA. Proteomics for the discovery of biomarkers and diagnosis of periodontitis: a critical review. Expert Rev Proteomics. 2014;11:31–41.

23. Gupta A, Govila V, Saini A. Proteomics - The research frontier in periodontics. J Oral Biol Craniofacial Res. 2015;5:46–52.

24. Kornman KS, Crane A, Wang HY, di Giovine FS, Newman MG, Pirk FW, et al. The interleukin-1 genotype as a severity factor in adult periodontal disease. J Clin Periodontol. 1997;24:72–7.

25. Masamatti S, Kumar A, Mehta D, Bhat K, Baron TK. Evaluation of interleukin -1B (+3954) gene polymorphism in patients with chronic and aggressive periodontitis: a genetic association study. Contemp Clin Dent. 2012;3:144.

26. Yang W, Jia Y, Wu H. Four tumor necrosis factor alpha genes polymorphisms and periodontitis risk in a Chinese population. Hum Immunol. 2013;74:1684–7.

27. Marenholz I, Heizmann CW, Fritz G. S100 proteins in mouse and man: from evolution to function and pathology (including an update of the nomenclature). Biochem Biophys Res Commun. 2004;322:1111–22.

28. Goyette J, Geczy CL. Inflammation-associated S100 proteins: new mechanisms that regulate function. Amino Acids. 2011;41:821–42.

29. Srikrishna G. S100A8 and S100A9: new insights into their roles in malignancy. J Innate Immun. 2012;4:31–40.

30. Kido J, Bando M, Hiroshima Y, Iwasaka H, Yamada K, Ohgami N, et al. Analysis of proteins in human gingival crevicular fluid by mass spectrometry. J Periodontal Res. 2012;47:488–99.

31. Kojima T, Andersen E, Sanchez JC, Wilkins MR, Hochstrasser DF, Pralong WF, et al. Human gingival crevicular fluid contains MRP8 (S100A8) and MRP14 (S100A9), two calcium-binding proteins of the S100 family. J Dent Res. 2000;79:740–7.

32. Andersen E, Dessaix IM, Perneger T, Mombelli A. Myeloid-related protein (MRP8/14) expression in gingival crevice fluid in periodontal health and disease and after treatment. J Periodontal Res. 2010;45:458–63.

33. Ellias MF, Zainal Ariffin SH, Karsani SA, Abdul Rahman M, Senafi S, Megat Abdul Wahab R. Proteomic analysis of saliva identifies potential biomarkers for orthodontic tooth movement. Sci World J. 2012;2012:1–6.

34. Zheng T, Hou J, Peng L, Zhang X, Jia L, Wang X, et al. The Pro-Apoptotic and Pro-Inflammatory Effects of Calprotectin on Human Periodontal Ligament Cells. Glogauer M, editor. PLoS One. 2014;9:e110421.

35. Leonardi R, Villari L, Caltabiano M, Travali S. Heat shock protein 27 expression in the epithelium of periapical lesions. J Endod. 2001;27:89–92.

36. Park J-H, Yoon J-H, Lim Y-S, Hwang H-K, Kim S-A, Ahn S-G, et al. TAT-Hsp27 promotes adhesion and migration of murine dental papilla-derived MDPC-23 cells through beta1 integrin-mediated signaling. Int J Mol Med. 2010;26:373–8.

37. Mitsuhashi M, Yamaguchi M, Kojima T, Nakajima R, Kasai K. Effects of HSP70 on the compression force-induced TNF-α and RANKL expression in human periodontal ligament cells. Inflamm Res. 2011;60:187–94.

38. Sreedharan R, Riordan M, Thullin G, Van Why S, Siegel NJ, Kashgarian M. The maximal cytoprotective function of the heat shock protein 27 is dependent on heat shock protein 70. Biochim Biophys Acta. 2011;1813:129–35.

39. Koromantzos PA, Makrilakis K, Dereka X, Offenbacher S, Katsilambros N, Vrotsos IA, et al. Effect of non-surgical periodontal therapy on C-Reactive Protein, Oxidative Stress, and Matrix Metalloproteinase (MMP)-9 and MMP-2 levels in patients with type 2 diabetes: a randomized controlled study. J Periodontol. 2012;83:3–10.

40. Kwon S-M, Kim S-A, Yoon J-H, Ahn S-G. Transforming growth factor beta1-induced heat shock protein 27 activation promotes migration of mouse dental papilla-derived MDPC-23 cells. J Endod. 2010;36:1332–5.

41. Chung E, Rylander MN. Response of preosteoblasts to thermal stress conditioning and osteoinductive growth factors. Cell Stress Chaperones. 2012;17:203–14.

42. Kuwabara I, Kuwabara Y, Yang R-Y, Schuler M, Green DR, Zuraw BL, et al. Galectin-7 (PIG1) exhibits pro-apoptotic function through JNK activation and mitochondrial cytochrome c release. J Biol Chem. 2002;277:3487–97.

43. Saussez S, Kiss R. Galectin-7. Cell Mol Life Sci. 2006;63:686–97.

44. Polla BS, Cossarizza A. Stress proteins in inflammation. EXS. 1996;77:375–91.

45. Leonardi R, Pannone G, Magro G, Kudo Y, Takata T, Lo ML. Differential expression of heat shock protein 27 in normal oral mucosa, oral epithelial dysplasia and squamous cell carcinoma. Oncol Rep. 2002;9:261–6.

46. Alford KA, Glennie S, Turrell BR, Rawlinson L, Saklatvala J, Dean JLE. Heat shock protein 27 functions in inflammatory gene expression and transforming growth factor-beta-activated kinase-1 (TAK1)-mediated signaling. J Biol Chem. 2007;282:6232–41.

47. Asdaghi N, Kilani RT, Hosseini-Tabatabaei A, Odemuyiwa SO, Hackett T-L, Knight DA, et al. Extracellular 14-3-3 from human lung epithelial cells enhances MMP-1 expression. Mol. Cell. Biochem. 2012;360:261–70.

48. Ben-Addi A, Mambole-Dema A, Brender C, Martin SR, Janzen J, Kjaer S, et al. IkB kinase-induced interaction of TPL-2 kinase with 14-3-3 is essential for Toll-like receptor activation of ERK-1 and −2 MAP kinases. Proc Natl Acad Sci U S A. 2014;111:E2394–403.

49. Westfall MD, Mays DJ, Sniezek JC, Pietenpol JA. The Np63 Phosphoprotein Binds the p21 and 14-3-3 Promoters In Vivo and Has Transcriptional Repressor Activity That Is Reduced by Hay-Wells Syndrome-Derived Mutations. Mol Cell Biol. 2003;23:2264–76.

50. Oh J-E, Jang DH, Kim H, Kang HK, Chung C-P, Park WH, et al. Alpha3Beta1 integrin promotes cell survival via multiple interactions between 14-3-3 isoforms and proapoptotic proteins. Exp Cell Res. 2009;315:3187–200.

51. Dowsett SA, Archila L, Foroud T, Koller D, Eckert GJ, Kowolik MJ. The effect of shared genetic and environmental factors on periodontal disease parameters in untreated adult siblings in Guatemala. J Periodontol. 2002;73:1160–8.

52. Kinane DF, Lappin DF. Immune processes in periodontal disease: a review. Ann Periodontol. 2002;7:62–71.

53. Dasanayake AP. Periodontal disease is related to local and systemic mediators of inflammation. J Evid Based Dent Pract. 2010;10:246–7.

54. Taba M, de Souza SLS, Mariguela VC. Periodontal disease: a genetic perspective. Braz Oral Res. 2012;26 Suppl 1:32–8.

55. Demmer RT, Behle JH, Wolf DL, Handfield M, Kebschull M, Celenti R, et al. Transcriptomes in healthy and diseased gingival tissues. J Periodontol. 2008;79:2112–24.

56. Abe D, Kubota T, Morozumi T, Shimizu T, Nakasone N, Itagaki M, et al. Altered gene expression in leukocyte transendothelial migration and cell communication pathways in periodontitis-affected gingival tissues. J Periodontal Res. 2011;46:345–53.

57. Puri KS, Suresh KR, Gogtay NJ, Thatte UM. Declaration of Helsinki, 2008: implications for stakeholders in research. J Postgrad Med. 2009;55:131–4.

58. Pontoriero R, Carnevale G. Surgical crown lengthening: a 12-month clinical wound healing study. J Periodontol. 2001;72:841–8.

59. Ribeiro FV, Hirata DY, Reis AF, Santos VR, Miranda TS, Faveri M, et al. Open-flap versus flapless esthetic crown lengthening: 12-month clinical outcomes of a randomized controlled clinical trial. J Periodontol. 2014;85:536–44.

60. Nethravathy R, Vinoth SK, Thomas AV. Three different surgical techniques of crown lengthening: a comparative study. J Pharm Bioallied Sci. 2013;5:S14–6.

61. Bellei E, Monari E, Bergamini S, Ozben T, Tomasi A. Optimizing protein recovery yield from serum samples treated with beads technology. Electrophoresis. 2011;32:1414–21.
62. Bellei E, Rossi E, Lucchi L, Uggeri S, Albertazzi A, Tomasi A, et al. Proteomic analysis of early urinary biomarkers of renal changes in type 2 diabetic patients. Proteomics Clin Appl. 2008;2:478–91.
63. Bellei E, Bergamini S, Monari E, Fantoni LI, Cuoghi A, Ozben T, et al. High-abundance proteins depletion for serum proteomic analysis: concomitant removal of non-targeted proteins. Amino Acids. 2011;40:145–56.

3

Quantitative succinylome analysis in the liver of non-alcoholic fatty liver disease rat model

Yang Cheng[1,2], Tianlu Hou[2], Jian Ping[2], Gaofeng Chen[2] and Jianjie Chen[1,2]*

Abstract

Background: Non-alcoholic fatty liver disease (NAFLD) is a clinical frequent disease. However, its pathogenesis still needs further study, especially the mechanism at the molecular level. The recent identified novel protein post-translational modification, lysine succinylation was reported involved in diverse metabolism and cellular processes. In this study, we performed the quantitative succinylome analysis in the liver of NAFLD model to elucidate the regulatory role of lysine succinylation in NAFLD progression.

Methods: Firstly, experimental model of NAFLD was induced by carbon tetrachloride injection and supplementary high-lipid and low-protein diet. Then series histochemical and biochemical variables were determined. For the quantitative succinylome analysis, tandem mass tags (TMT)-labeling, highly sensitive immune-affinity purification, liquid chromatography-tandem mass spectrometry techniques were applied. Bioinformatics analysis including gene ontology annotation based classification; Wolfpsort based subcellular prediction; function enrichment; protein-protein interaction network construction and conserved succinylation site motifs extraction were performed to decipher the differentially changed succinylated proteins and sites and p-value < 0.05 was selected as threshold.

Results: Totally, 815 succinylation sites on 407 proteins were identified, of which 243 succinylation acetylation sites on 178 proteins showed changed succinylation level with the threshold fold change > 1.5. Theses differentially changed succinylated proteins were involved in diverse metabolism pathways and cellular processes including carbon metabolism, amino acid metabolism, fat acid metabolism, binding and catalyzing, anti-oxidation and xenobiotics metabolism. Besides, these differentially changed succinylated proteins were prominently localized to cytoplasm and mitochondria. Moreover, 8 conserved succinylation site motifs were extracted around the succinylation sites.

Conclusions: Protein succinylation was an extensive post-translation modification in rat. The changed succinylation level in diverse proteins may disturb multiple metabolism pathways and promote non-alcoholic fatty liver disease development. This study provided a basis for further characterization of the pathophysiological role of lysine succinylation in NAFLD progression, which laid a foundation for the innovation of novel NAFLD drugs and therapies.

Keywords: Non-alcoholic fatty liver disease, Lysine succinylation, Succinylome, Bioinformatics analysis

* Correspondence: Jianjie_chen@hotmail.com
[1]Department of liver disease, Hospital for Infectious Diseases of Pudong New Area, Shanghai 201299, P. R. China
[2]Shuguang Hospital affiliated to Shanghai University of Traditional Chinese Medicine, Shanghai 201203, P. R. China

Background

Non-alcoholic fatty liver disease (NAFLD) is an insulin resistance and genetic susceptibility related metabolic stress induced liver injury, including non-alcoholic simple fatty liver (NAFL), non-alcoholic steatohepatitis (NASH) and cirrhosis [1]. With the age of onset of obesity and it induced metabolic disorder becoming younger, the injury of NAFLD on liver gets progressively worse. In 2009, investigation on 3175 adults in Shanghai indicated that the rate of NAFLD was 17.29 % [2]. However, the rate ascended to 23.3 % in Shanghai and increased rapidly in the major cities of China in 2012 [3]. NAFLD now is the second most common liver disease after viral hepatitis in China [2, 3]. A well-known two-hit hypothesis has been proposed for the demonstration of the pathogenesis of NAFLD. According to this hypothesis, excess fat accumulated in the liver firstly, then oxidative stress and lipid peroxidation led to hepatic inflammation, fibrosis and a progressive form NASH and to cirrhosis [4]. However, the underlying deep mechanisms are still unclear and approved pharmacological agents are unavailable for NAFLD [4]. Undoubtedly, NAFLD places an enormous economic burden on society and lowers the quality of one's life greatly. It's profound to study NAFLD pathogenesis and explore novel cure avenues.

Protein post-translational modifications (PTMs) are important regulatory patterns in a multiple of cellular events [5, 6]. PTMs are defined as covalent processing events that change the properties of a protein by proteolytic cleavage or by adding a modified group to one or more amino acids, thus the activity stage, localization, turnover and interactions with other proteins were changed [7]. Among all the amino acids, lysine is a frequent target to be modified, which can be subjected to a variety of PTMs including methylation, acetylation, biotinylation, ubiquitination, ubiquitin-like modifications, propionylation and butyrylation [8–11]. Increasing novel PTMs were identified recent years. Identifying various potential novel PTMs in diverse species and demonstrating the biological roles of these novel PTMs in organism have been a research hotspot, such as the study of protein propionylation, butyrylation and succinylation [9, 12, 13].

Recent studies indicated that some PTMs were closely connected with liver diseases. Kendrick et al. studied lysine acetylome in mice liver and found multiple hyper-acetylated proteins were presented in the liver of high-fat diet mice [14]. Albitar et al. reported that the ubiquitination of certain proteins was related with chronic liver disease and they developed an ubiquitin proteasome system profiling for the diagnosis of chronic liver disease [15]. However, the role of PTMs, especially some recently identified PTMs such as lysine succinylation and lysine malonylation in liver disease still need further study.

In the present study, we carried out the quantitative analysis of lysine succinylation in the liver of NAFLD rat model. A series of bioinformatics analysis were conducted to explore the molecular mechanisms of NAFLD genesis and progress, where the change of protein succinylation level may be involved. We aim to explore the effects of protein succinylation on the pathogenesis of NAFLD and probe potential diagnostic biomarkers and/or therapeutical targets of NAFLD by antibody based immunoprecipitation affinity enrichment and HPLC-MS/MS based proteome analysis.

Results

Morphological changes in NAFLD rat model

Initially, hemotoxylin and eosin (HE) staining was applied to observe the liver pathological changes in NAFLD rats. The results showed that, in control group (Fig. 1a), the structure of liver tissue was complete and the structure of sinus hepaticus was clearly observed; Cells in the tissue were with normal morphology and presented radial arrangement with the central vein as the center. In contrast, structures of liver acini hepatis in NAFLD rats were disorganized, and cells in the liver were degenerated with cytoplasm rarefaction and large number of fat vacuoles (Fig. 1b). Further oil red O staining from frozen section indicated that few lipid droplets were observed in the liver tissues of normal rats (Fig. 1c), while lots of fused lipid droplets were presented in the liver tissues of NAFLD rats (Fig. 1d).

Biochemical changes of NAFLD rat model

As shown in Table 1, the glycerin trimyristate (TG) content in the liver of NAFLD model was significantly higher than that in the liver of normal control. What's worse, the serum alanine aminotransferase (AST), alanine aminotransferase (ALT) and glutamyltranspetidase (GGT) activities in NAFLD rats also ascended significantly compared with the control group, which indicated that liver dysfunction may have occurred in the NAFLD model. Antioxidant index determination results indicated that the activity of superoxide dismutase (SOD) and the content of glutathione (GSH) declined significantly in the NAFLD group while the content of malonaldehyde (MDA) increased significantly compared with the control group, implying the liver of NAFLD rats suffered severe oxidative damage. Overall, we constructed a NAFLD rat model successfully.

General characterization of the quantitative succinylome in rat liver tissues

A total of 815 succinylation sites on 407 proteins were identified with two technical repeat experiments (686 succinylation sites in the first experiment and 692 in the second one, among which 563 sites were

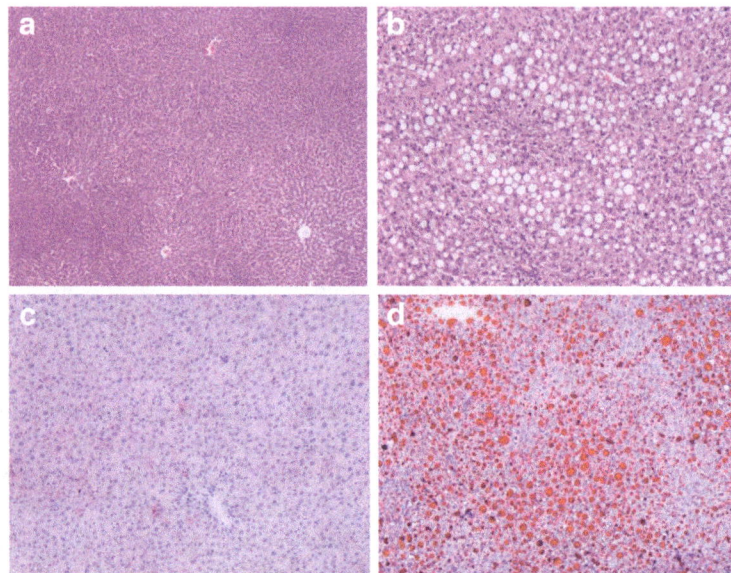

Fig. 1 Tissue morphological characteristics in the liver of NAFLD rat model and normal control. HE staining in normal control (**a**) and Non-alcoholic fatty liver disease (**b**); oil red O staining in normal control (**c**) and Non-alcoholic fatty liver disease (**d**)

identified in both experiments). With the threshold fold change > 1.5, 243 succinylation sites corresponding to 178 proteins showed different succinylation level in our two technical repeat experiments (180 up-regulated succinylated sites on 132 proteins, 94 down-regulated succinylated sites on 46 proteins, NAFLD group compared with normal control group). All the identified succinylated sites and corresponding proteins were summarized in Additional file 1.

GO annotation based classification and Wolfpsort based subcellular prediction analysis were conducted to investigate the nature of these differentially changed succinylated proteins (Fig. 2). The classification result on category of biological process showed that metabolic process and cellular process related proteins were the two largest groups suffered changed succinylation level. Molecular function analysis indicated the major differentially changed succinylated proteins were binding and catalytic activity related, whose percentage was 45.9 % and 39.6 %, respectively. Cellular component based classification analysis found cell and organelle related proteins were the main proteins with changed succinylation level while other cellular component related proteins were relatively less (Fig. 2a).

Subcellular location prediction result (Fig. 2b) showed the majority of these differentially changed succinylated proteins were localized to cytoplasm (38 %) and mitochondria (26 %), followed by nucleus (13.1 %), extracellular (6.9), endoplasmic reticulum (ER, 6.9 %), cytonucl (4.0 %), plasm (1.7 %) and peroxisome (1.1 %).

Enrichment analysis of the differentially changed succinylated proteins

As shown in (Fig. 3a), on molecular functions ontology, the top three enriched GO terms were oxidoreductase activity, cofactor binding and coenzyme binding. The markedly enriched cellular components were organelle envelope, mitochondrion and envelope (Fig. 3a), signifying mitochondrion and/or other enveloped structures may be the place where the proteins were differentially succinylatation modified with a relatively higher frequency. On the ontology of biological processes, the majority of these markedly enriched terms were concerned with the metabolism of small molecules, such as oxoacid, organic acid, carboxylic acid, amino acid and lipid (Fig. 3a).

In the KEGG pathway enrichment analysis, the top three enriched pathways were metabolic pathways, valine, leucine and isoleucine degradation and carbon

Table 1 Biochemical variable determination

	Normal control group	NAFLD group
Liver tissue TG content (mg/g)	12.4 ± 1.5	75.6 ± 13.4
Serum ALT activity (U/L)	47 ± 9	172 ± 37
Serum AST activity (U/L)	126 ± 11	295 ± 43
Serum GGT activity (U/L)	6 ± 1.5	58 ± 5.3
Liver tissue SOD activity (U/mg protein)	214 ± 35	106 ± 23
Liver tissue MDA content (mg/g protein)	19 ± 8	75 ± 23
Liver tissue GSH content (mg/g protein)	21 ± 3.7	9 ± 2.3

Notes: Each value is the mean ± SE ($n = 10$) and p value less than 0.01 was regarded as statistic significant

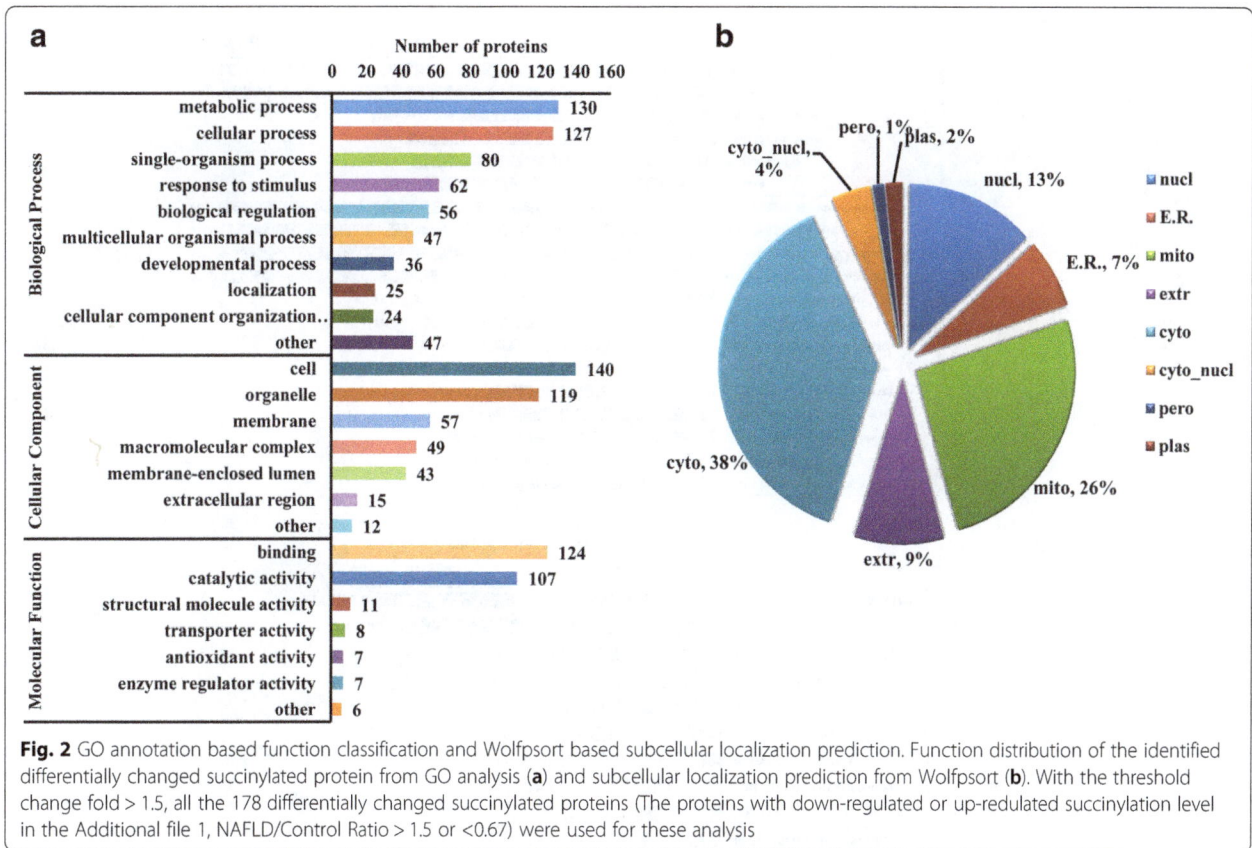

Fig. 2 GO annotation based function classification and Wolfpsort based subcellular localization prediction. Function distribution of the identified differentially changed succinylated protein from GO analysis (**a**) and subcellular localization prediction from Wolfpsort (**b**). With the threshold change fold > 1.5, all the 178 differentially changed succinylated proteins (The proteins with down-regulated or up-reduluated succinylation level in the Additional file 1, NAFLD/Control Ratio > 1.5 or <0.67) were used for these analysis

metabolism. Besides, many other amino acid metabolism and fat acid metabolism related pathways were also enriched (Fig. 3b)

Pfam domain analysis revealed that the top two significantly enriched terms were thiolase-like and thioredoxin-like fold. In addition, thiolase C-terminal and thiolase, N-terminal domain was also significantly enriched with high confidence. What's noticeable was that all the significantly enriched protein domains contained active group with sulfur or sulphydryl (Fig. 3c). We infer the activity of some sulfur /sulphydryl containing proteins/enzymes may be changed by succinylation or dysuccinylation and these changes may be involved in the progression of NAFLD.

Protein protein interaction (PPI) network analysis

To illuminate the mechanism of protein succinylation mediated NAFLD genesis and development, we constructed the PPI network for all the succinylated proteins with STRING database and Cytoscape software (Fig. 4). With MCODE plug-in toolkit, a total of 5 highly enriched interaction clusters were obtained, which were related with metabolism of xenobiotics by cytochrome P450, oxidative phosphotylation and ribosome, proteasome, fatty acid degradation and valine, leucine and isoleucine degradation.

Characterization of succinylated lysine sites in rat liver

As shown in Fig. 5a, a total of 8 definitively conserved succinylation site motifs were defined, namely K^{su} K^*D, $LK^{su}P$, $DK^{su}D$, $K^{su}P$, KK^{su}, RK^{su}, $K^{su}D$, DK^{su} (K^{su} represents the succinylated lysine and * represents a random amino acid residue). To determine whether there are specific amino acids adjacent to succinylated lysines, we examined the amino acid sequences flanking succinylation sites by heat'map (Fig. 5b). Aspartic acid (D), lysine (K) and arginine (R) were overrepresented in the −1 position of succinylation sites. D was also appeared in the +1 to +3 positions with high frequency while K and R appeared in these positions with very low frequency. Besides, a special amino acid, proline (P) appeared in the +1 position of succinylation sites frequently but seldom appeared in the −1 position. In addition, it seemed that Serine (S) was unwelcomed surrounding the succinylation sites as its frequency of occurrence was obviously lower than other amino acids in both upstream and downstream of succinylation sites.

Secondary structure analysis showed that the succinylation sites distribution was about 64.1 % in coil, 30.3 % in helix, and 5.5 % in beta-strand (Fig. 5c).

Discussion

Lysine succinylation is a newly identified PTM and impacts diverse metabolic pathways [16]. Recent protein

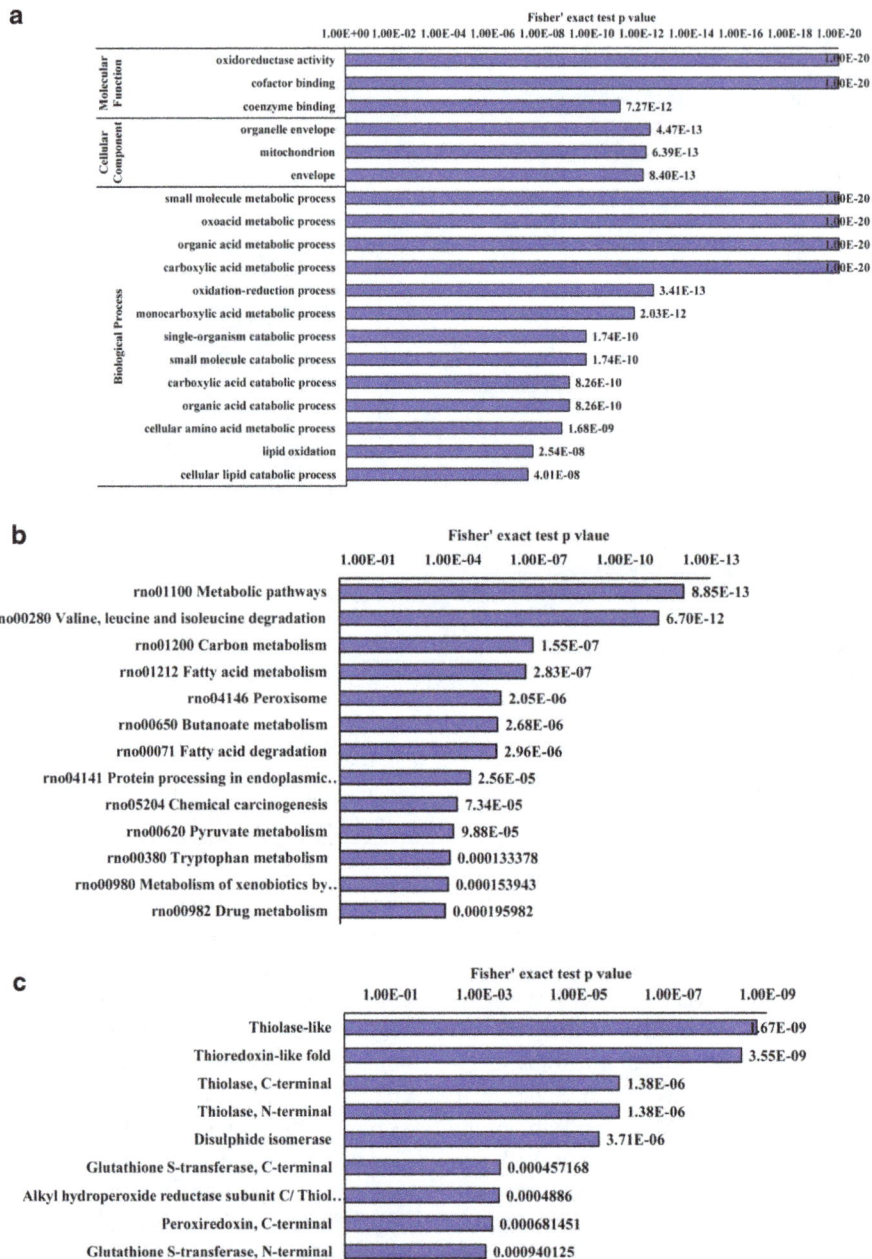

Fig. 3 Enrichment analysis of the differentially changed succinylated proteins. GO annotation based enrichment analysis (**a**), KEGG pathway enrichment analysis (**b**) and domain enrichment analysis (**c**). With the threshold change fold > 1.5, all the 178 differentially changed succinylated proteins (The proteins with down-regulated or up-redulated succinylation level in the Additional file 1, NAFLD/Control Ratio > 1.5 or <0.67) were used for these analysis. The GO enrichment analysis were performed on the ontology of molecular function, cellular component and biological process. DAVID as selected as the tool and the adjusted p-value less than 0.05 was chosen as cut-off criterion

succinylome studies in mouse liver and mouse embryonic fibroblasts (MEFs) showed that proteins succinylation level changes participated in fat acid metabolism related pathways such as fatty acid β-oxidation and long-chain fatty acid transportation [16–18]. It was well known that various fatty liver diseases including nonalcoholic simple fatty liver and alcoholic fatty liver were highly related with liver fatty acid metabolism dysfunction. Thus we investigated the quantitative protein succinylome in NAFLD rat model, with the purpose of exploring the possible roles of lysine succinylation in NAFLD progression.

Firstly, we constructed a classical NAFLD rat model according to previous report [19] and a series of histological, physiological and biochemical variables were detected. High fat and low protein diet induced over accumulation of TG, disturbance of biochemical variables and serious

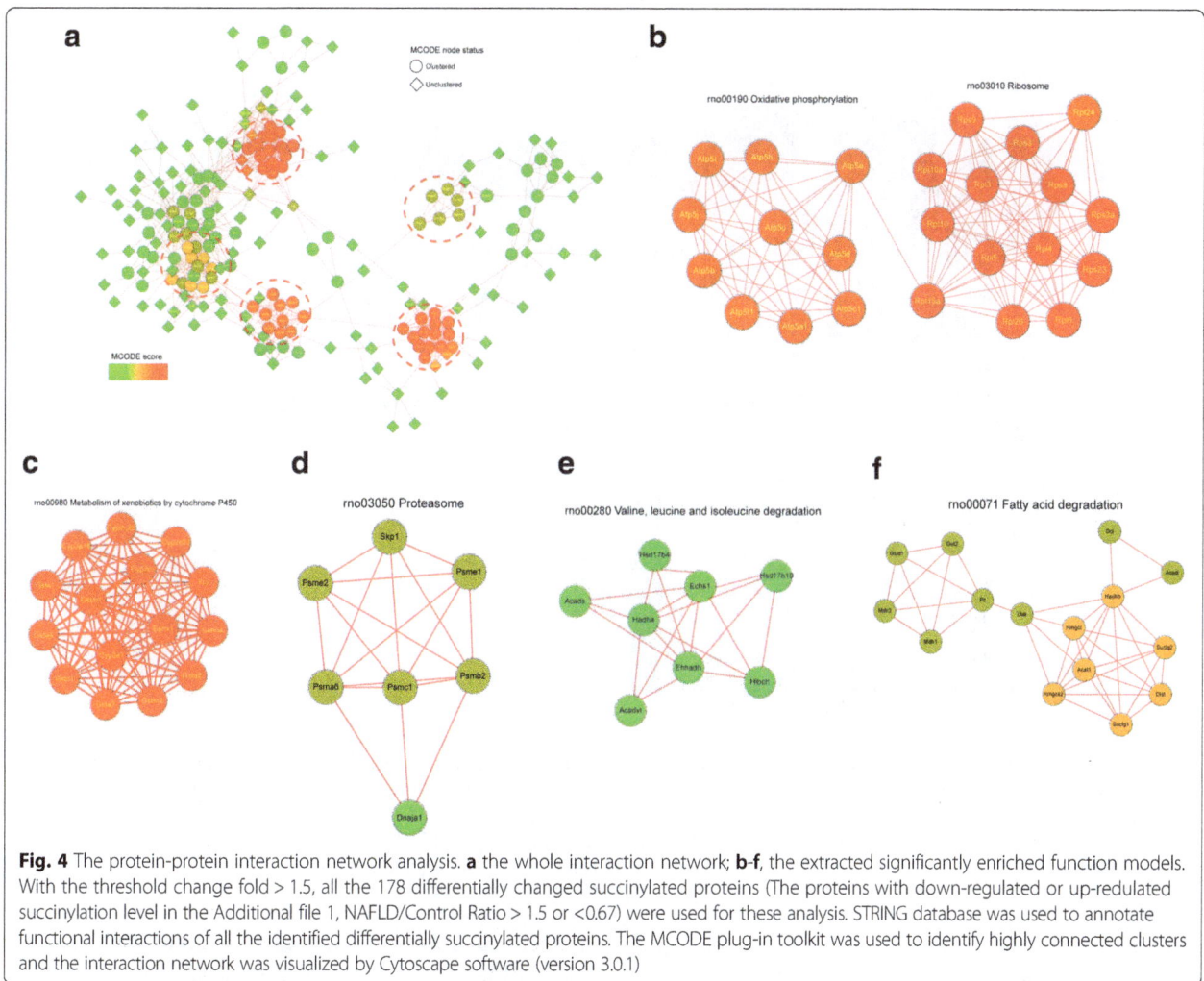

Fig. 4 The protein-protein interaction network analysis. **a** the whole interaction network; **b-f**, the extracted significantly enriched function models. With the threshold change fold > 1.5, all the 178 differentially changed succinylated proteins (The proteins with down-regulated or up-redulated succinylation level in the Additional file 1, NAFLD/Control Ratio > 1.5 or <0.67) were used for these analysis. STRING database was used to annotate functional interactions of all the identified differentially succinylated proteins. The MCODE plug-in toolkit was used to identify highly connected clusters and the interaction network was visualized by Cytoscape software (version 3.0.1)

oxidative damage (Fig. 1 and Table 1), which was consistent with previous report [19].

Lysine succinylome analysis identified 815 succinylation sites on 407 proteins in the liver tissues of NAFLD rat model and normal control. Park et al. identified 1,675 sites from 436 proteins in liver tissues of Sirt5 knockout mouse [18]. The huge difference of the volume of succinylated sites may be related with the knockout of Sirt5 as Sirt5 can remove succinyl moieties from target lysine and acts as a desuccinylase role [20]. Quantification analysis indicated that approximately two thirds proteins showed elevated succinylation level in NAFLD group compared with control group. These hyper-succinylated proteins may be involved in the genesis and development of NAFLD.

The classification result (Fig. 2a) implied that changed lysine succinylation level may have influenced many metabolic processes and cellular processes in NAFLD rat liver. GO and KEGG pathway enrichment analysis (Fig. 3a and b) further supported this hypothesis as many small molecules (oxoacid, organic acid, carboxylic

acid, amino acid, fatty acid and lipid etc.) metabolism processes and pathways were significantly enriched. Previous studies have well demonstrated the role of abnormal metabolism of aboveing metabolites in NAFAD pathogenesis [21–23]. A recent study on mouse liver lysine succinylome also revealed many lysine-succinylated proteins were predominantly involved in fatty acid metabolism, amino acid degradation and the tricarboxylic acid cycle [16]. Here we infer the succinylation level alteration (preferentially hyper-succinylation) on these molecules metabolism related proteins and/or enzymoses may be a potential promoting mechanism of NAFLD progression.

High proportion binding and catalytic related proteins (Fig. 2a) and significantly enriched cofactor binding and coenzyme binding terms (Fig. 3a) imply changed succinylation level may perturb the normal catalyzing or binding function of target proteins/enzymes. These proteins/enzymes were succinylation modified, thus their space structure, charge states and stability may be altered and they can't bind to their coenzymes or

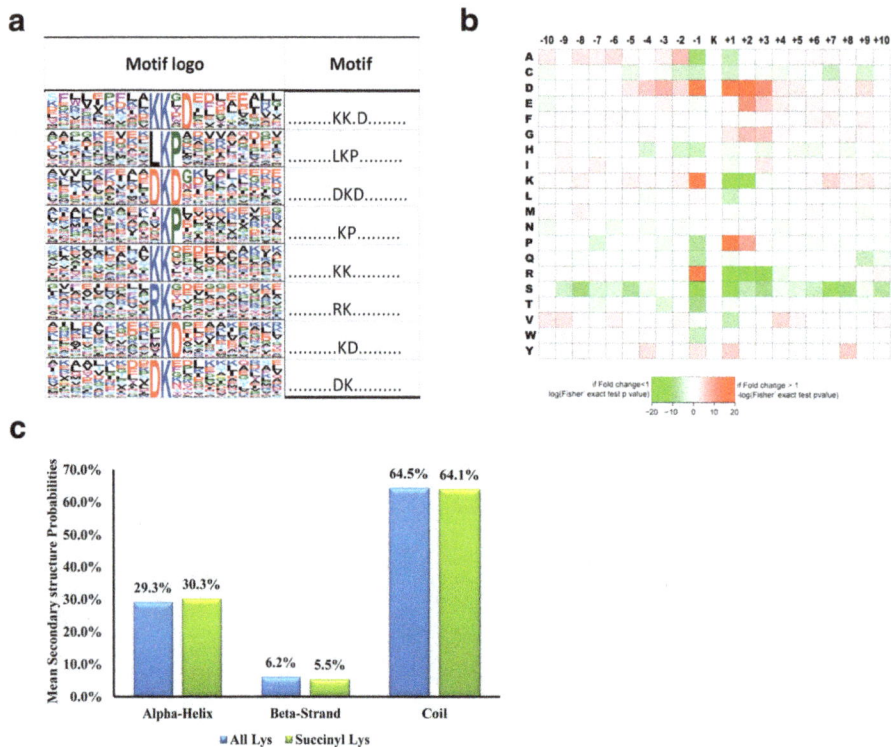

Fig. 5 Properties of the succinylated peptides. **a** Succinylation motifs and conservation of succinylation sites. **b** Heat map of the amino acid compositions of the succinylated site. **c** Distribution of succinylated lysines and all lysines in protein secondary structures. All the identified 815 succinylation sites (Additional file 1) were used for the analysis. Software motif-x was used to analysis the model of sequences constituted with amino acids in specific positions of succinyl-21-mers (10 amino acids upstream and downstream of the site). The local secondary structures of succinylated proteins were predicted by NetSurfP. P-value < 0.05 was considered significant for these analysis

cofactors successfully, which further disturbed their binding ability and catalytic activity. Succinylation disturbed substrate binding and enzyme catalytic activity had been reported in isocitrate dehydrogenase [24]. What's worse, succinylation adds a bigger structural moiety than acetylation or methylation and it is likely to lead to more significant changes in protein structure [24], which may change the protein binding or catalyzing activity seriously.

What's noticeable was that oxidoreductase activity term was the top significant enriched term in molecular function enrichment analysis; functional classification analysis also identified 7 antioxidant ability related proteins. Correspondingly, we observed aggravated reactive oxygen species (ROS) stress and deteriorated anti-oxidation enzyme system in NAFLD by biochemical variables determination. We guess the changes of succinylation level on antioxidant enzymes may be another important event in NAFLD development. The antioxidant related proteins with changed succinylation level might be potential targets for NAFLD treatment.

Subcellular location prediction result indicated cytoplasm located differentially changed succinylated proteins accounted for the largest part, and then was the

mitochondria located proteins. In a recent study, cellular compartment analysis of all the lysine-succinylated proteins identified from Sirt5 knockout (KO) liver tissue and mouse embryonic fibroblasts (MEFs) found mitochondria located succinylated proteins accounted for the largest part while cytoplasm located proteins was the second most proteins [18]. Another succinylome study in mouse liver tissue also observed the similar trends [16]. The difference of subcellular distribution may be related with the different experiment materials or experimental treatments. For example, Sirt5 is an inhibitor of lysine succinylation in mitochondria [18], Sirt5 knockout naturally induced more protein succinylation modification in mitochondria.

PPIs are crucial for most cellular processes [25]. Various PTMs can change the interactions among protein, thus metabolism processes and cellular processes were affected [7]. In the obtained 5 significantly enriched function modules, we found function module metabolism of xenobiotics by cytochrome P450, fatty acid degradation and valine, leucine and isoleucine degradation were also enriched in the preceding GO and KEGG enrichment analysis, which implied their important roles in NAFLD development.

The score of interaction clusters metabolism of xenobiotics by cytochrome P450, oxidative phosphotylation and ribosome were very high, signifying alteration of succinylation level on these proteins may contribute to the development of NAFLD. In the metabolism of xenobiotics by cytochrome P450 function module, most of these interacted proteins belonged to the Cytochrome P450 (CYP450) family or Glutathione S-transferase (GST) family. These two families participated in the metabolism of various metabolites together, especially the metabolism of secondary metabolism such as steroids, fatty acids, xenobiotics and so on [26–28]. Lysine succinylation on the members of CYP450 family and GST family may disturb various secondary metabolism and induced NAFAD. This is consistent with the aforementioned classification and enrichment analysis as various molecular metabolism related terms were significantly enriched. The major proteins in oxidative phosphotylation and ribosome cluster were subunits of mitochondrial F0F1 ATP synthase and ribosome, which imply succinylation on F0F1 ATP synthase and ribosome were important events in NAFAD development. The energy production and protein synthesis was likely to be interfered by increased lysine succinylation modification in the liver of NAFLD model.

In the motif analysis, we identified 8 definitively conserved succinylation site motifs. Previous motif and flanking amino acid sequence analysis in SIRT5 knockout mouse showed that positively charged amino acids (lysine or arginine) were strongly excluded from positions −1 and +1 of the succinylation logo [17, 18]. Our study indicated lysine and arginine were rare represented in +1 position while over represented in the −1 position of succinylation sites. Furthermore, the negatively charged amino acid aspartic acid was highly represented in the positions near succinylation sites (−1, +1, +2, +3). In the secondary structure analysis, we observed a slight bias for succinylation to occur in alpha helix regions and a slight bias against beta-strand regions and coil regions, which was similar with previous finding that succinylation sites have moderate local structural preferences for helical regions and a moderate bias against strand and coiled regions in mouse [16].

Conclusions

In conclusion, we performed the quantitative succinylome analysis in the liver of NAFLD rat model. A total of 815 succinylation sites from 407 proteins were identified, of which 243 succinylation sites corresponding to 178 proteins had shown changed succinylation level with the threshold fold change > 1.5. Bioinformatics analysis indicated that these differentially changed succinylated proteins participated in various metabolism processes and cellular processes including but not defining to carbon metabolism, amino acid metabolism, fat acid metabolism, binding and catalyzing, anti-oxidation and xenobiotics metabolism. Alteration of succinylation level on these metabolism and cellular processes related proteins may have changed many normal metabolism pathways and promoted NAFLD progression. These proteins with differential succinylation level could be the potential diagnose biomarkers/therapy targets for NAFLD treatment. Besides, the hyper-succinylated proteins were mainly localized to cytoplasm and mitochondria. Motif analysis obtained 8 conserved succinylation site motifs, namely $K^{su} K^{*}D$, $LK^{su}P$, $DK^{su}D$, $K^{su}P$, KK^{su}, RK^{su}, $K^{su}D$ and DK^{su}. This is the first quantitative succinylome analysis in the liver tissues of NAFLD rat model, which confer a novel perspective for the elucidation of the mechanism underling NAFLD genesis and development as well as the innovation of new drugs and therapy avenues to cure NAFLD.

Materials and Methods
Experiment design
The purpose of this study was to perform the quantitative lysine succinylome analysis in the NAFLD rat model. Firstly we established the rat model according to previous report. After different treatments, the liver tissue morphological observation and biochemical variables measurement were conducted in both NAFLD model group and normal control group. Then we performed the quantitative succinylation analysis in the liver of NAFLD rat model by using tandem mass tags (TMT)-labeling, antibody-based succinylated peptides affinity enrichment and nano LC-MS/MS techniques. Lastly, bioinformatics analysis were carried out for the systematic interpretation of the identified lysine succinylated sites and succinylated proteins. Two technical repeats were performed for the succinylome analysis.

Regents
All reagents unless otherwise stated were purchased from Sigma (St. Louis, America). Alanine aminotransferase (ALT), aspartate aminotransferase (AST), glutamyltranspetidase (GGT), glycerin trimyristate (TG), superoxide dismutase (SOD), reduced glutathione (GSH) and malonaldehyde (MDA) assay kit were purchased from Nanjing Jiancheng Bioengineering Institute (Nanjing, China). Anti-succinyl lysine antibody agarose beads were purchased from PTM Biolabs (Hangzhou, China).

Experimental animal model
The detailed description of the model establishment and diet for rats were given in the Additional file 2. Twenty male Wistar rats with SPF grades, weight range 170 ± 10 g were used in this study and the rat model was established according to Yao et al. [19]. The animal studies

and protocols were approved by the Experimental Animal Ethics Committee of Hospital for Infectious Diseases of Pudong New Area.

Liver tissue preparation

Four weeks after treatment, all rats were fasted but normally supplied with water for 24 h. Then rats were performed peritoneal injection with 3 % pentobarbital sodium according to injection dose to rat weight of 2 ml/Kg. Abdominal cavity of anesthetic rat was incised, and then inferior vena cava blooding sampling was performed with 10 ml syringe. After standing for 3 h at 4 °C, blood samples were centrifuged at 5000 x g for 15 min. Supernatant was saved at –80 °C for future use. Liver was incised and two segments (1.0 cm x 1.0 cm x 0.3 cm) were cut from right lobe of the liver. Two liver segments were fixed in 10 % neutral formalin and embedded in opt-imum cutting temperature compound (OCT). The remained liver samples were saved at –80 °C for future use.

Morphologic observation

To observe liver tissues morphologic changes, the HE staining [29] and red O staining [30] of liver segments was performed according to previous description.

Biochemical Variable Determination

Activities of AST, ALT and GGT in serum, and TG content, MDA content, GSH content and SOD activity in liver tissues from experimental rats were determined according to the instructions provided by the assay kit [31].

Proteomic analysis
Sample preparation

The protein extraction of the rat liver tissues was performed according to previous report [32] and digested by trypsin (Promega) with the second digestion method to ensure thorough digestion [33].

Briefly, the liver tissues were first grinded by liquid nitrogen and the powder was transferred to 5 mL centrifuge tube and precipitated with cold 10 % TCA/acetone supplemented with 50 mM DTT, 0.1 % Protease Inhibitor Cocktail Set IV and HDACinhibitor (50 mM sodium butyrate, 30 mM nicotinamide, and 3 μM Trichostatin A) for 2 h at –20 °C. After centrifugation at 20,000 g at 4 °C for 10 min, the resulting precipitate was washed with cold acetone for three times and air dried. Then the precipitate was re-suspended in lysis buffer (8 M urea, 2 mM EDTA, 10 mM DTT) and sonicated three times on ice using a high intensity ultrasonic processor (Scientz) and the remaining debris was removed by centrifugation.

The supernatant was reduced with 10 mM DTT for 1 h at 56 °C and alkylated with 55 mM IAA for 45 min at room temperature in darkness. Afterwards, the

protein was precipitated with 3 volumes of pre-chilled acetone for 30 min at –20 °C. After centrifugation, the pellet was then dissolved in 0.5 M TEAB and sonicated for 5 min. Repeat the centrifugation step and collect the supernatant. Protein content in the supernatant was determined with 2-D Quant kit according to the manufacturer's instructions. The protein was then digested with trypsin (Promega) at an enzyme-to-substrate ratio of 1:50 for 12 h at 37 °C. To insure complete digestion and improve protein identification and characterization, additional trypsin at an enzyme-to-substrate ratio of 1:100 was added, and the mixture was incubated for an additional 4 h. After trypsin digestion, peptide was desalted by Strata X C18 SPE column (Phenomenex) and vacuum-dried. Then peptides were labeled with a 2-plex TMT kit with the protocol of the manufacture.

Enrichment of succinylated lysine peptides

Enrichment of lysine succinylated peptides was implemented by immunoprecipitation according to previous report [34] and the anti-succinyl lysine antibody agarose conjugated beads (PTM Biolab) was used with a ratio of 15 μL beads/mg protein.

Briefly, 5 mg tryptic peptides was re-dissolved in NETN buffer NETN buffer (100 mM NaCl, 1 mM EDTA, 50 mM Tris–HCl, 0.5 % NP-40, pH 8.0) and incubated with anti-succinyl lysine antibody agarose conjugated beads (PTM Biolab) in a ratio of 15 μL beads/ mg protein at 4 °C overnight with gentle end-to-end rotation. After incubation, the beads were washed four times with NETN buffer and twice with purified water. The bound peptides were eluted with 1 % trifluoroacetic acid (TFA) and dried under a vacuum. The eluted peptides were cleaned with C18 ZipTips (Millipore) in accordance with the manufacturer's instructions followed by HPLC/MS/MS analysis.

LC-MS/MS

LC-MS/MS was performed according to previous report [13] and the detailed process and parameters were shown in Additional file 2.

Database Search

All the detailed parameters were shown in the Supporting Information (Additional file 2). The protein and succinylation sites were identified using MaxQuant software (Version. 1.0.13.13) and Andromeda search 172 engine (Version 1.4.1.2). Tandem mass spectra were searched against *Uniprot_Rat* database with reverse decoy database.

Succinylation Quantification

The quantification of the succinylated peptides and proteins were calculated according to the TMT reporter ion intensities with COMPASS v1.2.1.0 software [35]. All peptides with same succinylation patterns were grouped together and their reporter ion intensities were summed. The quantitative ratios were weighted and normalized by the median ratio. The manufacturer's recommended isotope correction factors were used. Based on relative quantification and statistical analysis, 1.5-fold change was set as threshold for differentially changed succinylated proteins.

Bioinformatics Analysis

The databases and softwares for bioinformatics analysis were shown in Additional file 2. When performing the bioinformatics analysis, p-value < 0.05 was considered significant.

Statistical methods

Data were processed by using SPSS 17.0. Measurement data were indicated as mean ± SEM. Comparisons between groups were tested by One -Way ANOVA analysis and statistical difference was determined when $P < 0.05$.

Additional files

Additional file 1: The summary of all the identified succinylation sites and corresponding proteins. (XLSX 71 kb)

Additional file 2: The detailed description of the experiment methods, including rat model establishment, mass spectrometric analysis procedures and parameters, bioinformatics analysis softwares, websites. (DOCX 18 kb)

Abbreviations

NAFLD: Non-alcoholic fatty liver disease; PTM: Post-translational modification; PPI: Protein-protein interaction; GO: Gene Ontology; KEGG: Kyoto Encyclopedia of Genes and Genomes; HE: Hemotoxylin and eosin; CCL4: Analytical grade carbon tetrachloride; ALT: Alanine aminotransferase; AST: Aspartate aminotransferase; GGT: Glutamyltranspetidase; TG: Glycerin trimyristate; SOD: Superoxide dismutase; GSH: Reduced glutathione; MDA: Malonaldehyde; ROS: Reactive oxidative stress; LC-MS/MS: Liquid chromatography-mass spectrometry/ mass spectrometry.

Competing interests

The authors have declared no conflicts of interest.

Authors' contribution

Yang Cheng and Jianjie Chen designed the experiments; Tianlu Hou wrote the paper; Jian Ping contributed to HPLC-MS/MS and data analysis; Gaofeng Chen contributed to histochemical, physiological and biochemical experiments. Yang Cheng and Jianjie Chen take full responsibility for the integrity of data analysis. All authors read and approved the final manuscript.

Acknowledgement

This study was supported by grants from Science and Technology Commission of Pudong New Area Shanghai (PKJ2014-Y37), three-year plan of action of traditional Chinese medicine in Shanghai (ZY3-JSFC-1-1011); Prof. Jian-jie Chen Studio (Shanghai Legendary Medical Practitioner of Traditional Chinese Medicine, ZYSNXD-CC-MZY003), The Key Discipline Project in Hepatology of State Administration of Traditional Chinese Medicine. We are grateful to Dr. Martin Simon for his critically English scientific editing of the manuscript.

References

1. Smith BW, Adams LA. Non-alcoholic fatty liver disease. Crit Rev Clin Lab Sci. 2011;48(3):97–113.
2. Fan JG, Farrell GC. Epidemiology of non-alcoholic fatty liver disease in China. J Hepatol. 2009;50(1):204–10.
3. Wong VW, Chu WC, Wong GL, Chan RS, Chim AM, Ong A, et al. Prevalence of non-alcoholic fatty liver disease and advanced fibrosis in Hong Kong Chinese: a population study using proton-magnetic resonance spectroscopy and transient elastography. Gut. 2012;61(3):409–15.
4. Ganji SH, Kashyap ML, Kamanna VS. Niacin inhibits fat accumulation and oxidative stress in human hepatocytes and regresses hepatic steatosis in experimental rat model. J Clin Lipidol. 2014;8(3):349–50.
5. Cain JA, Solis N, Cordwell SJ. Beyond gene expression: the impact of protein post-translational modifications in bacteria. J Proteomics. 2014;97:265–86.
6. Hofer A, Wenz T. Post-translational modification of mitochondria as a novel mode of regulation. Exp Gerontol. 2014;56:202–20.
7. Mann M, Jensen ON. Proteomic analysis of post-translational modifications. Nat Biotechnol. 2003;21(3):255–61.
8. Berger SL. The complex language of chromatin regulation during transcription. Nature. 2007;447(7143):407–12.
9. Chen Y, Sprung R, Tang Y, Ball H, Sangras B, Kim SC, et al. Lysine propionylation and butyrylation are novel post-translational modifications in histones. Mol Cell Proteomics. 2007;6(5):812–9.
10. Peng C, Lu Z, Xie Z, Cheng Z, Chen Y, Tan M et al. The first identification of lysine malonylation substrates and its regulatory enzyme. Mol Cell Proteomics. 2011;10(12):M111. 012658.
11. Tan M, Luo H, Lee S, Jin F, Yang JS, Montellier E, et al. Identification of 67 histone marks and histone lysine crotonylation as a new type of histone modification. Cell. 2011;146(6):1016–28.
12. Xu G, Wang J, Wu Z, Qian L, Dai L, Wan X, et al. SAHA regulates histone acetylation, butyrylation, and protein expression in neuroblastoma. J Proteome Res. 2014;13(10):4211–9.
13. Li X, Hu X, Wan Y, Xie G, Li X, Chen D, et al. Systematic identification of the lysine succinylation in the protozoan parasite Toxoplasma gondii. J Proteome Res. 2014;13(12):6087–95.
14. Kendrick A, Choudhury M, Rahman S, McCurdy C, Friederich M, Van Hove J, et al. Fatty liver is associated with reduced SIRT3 activity and mitochondrial protein hyperacetylation. Biochem J. 2011;433:505–14.
15. Qu KZ, Zhang K, Ma W, Li H, Wang X, Zhang X, et al. Ubiquitin-proteasome profiling for enhanced detection of hepatocellular carcinoma in patients with chronic liver disease. J Gastroenterol Hepatol. 2011;26(4):751–8.
16. Weinert BT, Schölz C, Wagner SA, Iesmantavicius V, Su D, Daniel JA, et al. Lysine succinylation is a frequently occurring modification in prokaryotes and eukaryotes and extensively overlaps with acetylation. Cell reports. 2013;4(4):842–51.
17. Rardin MJ, He W, Nishida Y, Newman JC, Carrico C, Danielson SR, et al. SIRT5 regulates the mitochondrial lysine succinylome and metabolic networks. Cell Metab. 2013;18(6):920–33.
18. Park J, Chen Y, Tishkoff DX, Peng C, Tan M, Dai L, et al. SIRT5-mediated lysine desuccinylation impacts diverse metabolic pathways. Mol Cell. 2013;50(6):919–30.
19. Yao J, Zhi M, Chen M. Effect of silybin on high-fat-induced fatty liver in rats. Braz J Med Biol Res. 2011;44(7):652–9.
20. Du J, Zhou Y, Su X, Yu JJ, Khan S, Jiang H, et al. Sirt5 is a NAD-dependent protein lysine demalonylase and desuccinylase. Science. 2011;334(6057):806–9.
21. Lim JS, Mietus-Snyder M, Valente A, Schwarz JM, Lustig RH. The role of fructose in the pathogenesis of NAFLD and the metabolic syndrome. Nat Rev Gastroenterol Hepatol. 2010;7(5):251–64.
22. Musso G, Cassader M, Rosina F, Gambino R. Impact of current treatments on liver disease, glucose metabolism and cardiovascular risk in non-alcoholic fatty liver disease (NAFLD): a systematic review and meta-analysis of randomised trials. Diabetologia. 2012;55(4):885–904.
23. Vernon G, Baranova A, Younossi Z. Systematic review: the epidemiology and natural history of non-alcoholic fatty liver disease and non-alcoholic steatohepatitis in adults. Aliment Pharmacol Ther. 2011;34(3):274–85.
24. Zhang Z, Tan M, Xie Z, Dai L, Chen Y, Zhao Y. Identification of lysine succinylation as a new post-translational modification. Nat Chem Biol. 2011;7(1):58–63.

25. Wu J, Vallenius T, Ovaska K, Westermarck J, Mäkelä T, Hautaniemi S. Integrated network analysis platform for protein-protein interactions. Nat Methods. 2009;6(1):75.

26. Wang LI, Giovannucci EL, Hunter D, Neuberg D, Su L, Christiani DC. Dietary intake of Cruciferous vegetables, Glutathione S-transferase (GST) polymorphisms and lung cancer risk in a Caucasian population. Cancer Causes Control. 2004;15(10):977–85.

27. Wexler D, Courtney R, Richards W, Banfield C, Lim J, Laughlin M. Effect of posaconazole on cytochrome P450 enzymes: a randomized, open-label, two-way crossover study. Eur J Pharm Sci. 2004;21(5):645–53.

28. Rendic S, Carlo FJD. Human cytochrome P450 enzymes: a status report summarizing their reactions, substrates, inducers, and inhibitors. Drug Metab Rev. 1997;29(1–2):413–580.

29. Whitington PF, Pan X, Kelly S, Melin-Aldana H, Malladi P. Gestational alloimmune liver disease in cases of fetal death. J Pediatr. 2011;159(4):612–6.

30. Takamura A, Komatsu M, Hara T, Sakamoto A, Kishi C, Waguri S, et al. Autophagy-deficient mice develop multiple liver tumors. Genes Dev. 2011;25(8):795–800.

31. Zhuo L, Liao M, Zheng L, He M, Huang Q, Wei L, et al. Combination therapy with taurine, epigallocatechin gallate and genistein for protection against hepatic fibrosis induced by alcohol in rats. Biological and Pharmaceutical Bulletin. 2012;35(10):1802–10.

32. Lundby A, Lage K, Weinert BT, Bekker-Jensen DB, Secher A, Skovgaard T, et al. Proteomic analysis of lysine acetylation sites in rat tissues reveals organ specificity and subcellular patterns. Cell reports. 2012;2(2):419–31.

33. Pan J, Ye Z, Cheng Z, Peng X, Wen L, Zhao F. Systematic analysis of the lysine acetylome in Vibrio parahemolyticus. J Proteome Res. 2014;13(7):3294–302.

34. Xie L, Liu W, Li Q, Chen S, Xu M, Huang Q, et al. First succinyl-proteome profiling of extensively drug-resistant mycobacterium tuberculosis revealed Involvement of succinylation in cellular physiology. J Proteome Res. 2015;14(1):107–19.

35. Still AJ, Floyd BJ, Hebert AS, Bingman CA, Carson JJ, Gunderson DR, et al. Quantification of mitochondrial acetylation dynamics highlights prominent sites of metabolic regulation. J Biol Chem. 2013;288(36):26209–19.

4

Analysis of the immune response of human dendritic cells to *Mycobacterium tuberculosis* by quantitative proteomics

Chiu-Ping Kuo[1†], Kuo-Song Chang[2,5†], Jue-Liang Hsu[7], I-Fang Tsai[3], Andrew Boyd Lin[4], Tsai-Yin Wei[3], Chien-Liang Wu[1,5] and Yen-Ta Lu[1,6*] (iD)

Abstract

Background: The cellular immune response for *Mycobacterium tuberculosis* (*M. tuberculosis*) infection remained incompletely understood. To uncover membrane proteins involved in this infection mechanism, an integrated approach consisting of an organic solvent-assisted membrane protein digestion, stable-isotope dimethyl labeling and liquid chromatography-tandem mass spectrometry (LC-MS/MS) analysis was used to comparatively profile the membrane protein expression of human dendritic cells upon heat-killed *M. tuberculosis* (HKTB) treatment.

Results: Organic solvent-assisted trypsin digestion coupled with stable-isotope labeling and LC-MS/MS analysis was applied to quantitatively analyze the membrane protein expression of THP-1 derived dendritic cells. We evaluated proteins that were upregulated in response to HKTB treatment, and applied STRING website database to analyze the correlations between these proteins. Of the investigated proteins, aminopeptidase N (CD13) was found to be largely expressed after HKTB treatment.

By using confocal microscopy and flow cytometry, we found that membranous CD13 expression was upregulated and was capable of binding to live mycobacteria. Treatment dendritic cell with anti-CD13 antibody during *M. tuberculosis* infection enhanced the ability of T cell activation.

Conclusions: Via proteomics data and STRING analysis, we demonstrated that the highly-expressed CD13 is also associated with proteins involved in the antigen presenting process, especially with CD1 proteins. Increasing expression of CD13 on dendritic cells while *M. tuberculosis* infection and enhancement of T cell activation after CD13 treated with anti-CD13 antibody indicates CD13 positively involved in the pathogenesis of *M. tuberculosis*.

Keywords: CD13, *M. tuberculosis*, Membrane proteomics, Antigen presentation

Background

Despite many effective treatments are available, *M. tuberculosis* remains one of the most successful pathogens on the planet, estimated to have infected nearly one-third of the human population and cause approximately 1.7 million deaths each year [1]. *M. tuberculosis* is typically transmitted via the inhalation of aerosol droplets containing the pathogen. Once inhaled, these small droplets can spread into distal lung alveoli, where they are phagocytosed by alveolar macrophages [2]. Once inside the macrophage, *M. tuberculosis* prevents its phagosome from fusing with digestive lysosomes [3], allowing the pathogen to lay dormant within its host.

While macrophages are the primary targets of the mycobacteria, *M. tuberculosis*'s ability to exist as latent infection suggests that it is also able to suppress other immune responses. Following the initial macrophage response is an acute inflammatory response, causing a rapid recruitment of dendritic cells (DCs) into the airway epithelium [4]. Normally, DCs capture the bacteria, process them, and present their antigens on their cell surfaces to various cells of the immune system. However, it is suggested that specific functions of DCs may be modulated by the mycobacteria. More specifically,

* Correspondence: ytlhl@mmh.org.tw
†Equal contributors
[1]Division of Chest Medicine, Department of Internal Medicine, Mackay Memorial Hospital, 92, Sec 2, Chungshan North Road, Taipei, Taiwan
[6]Department of Medicine, Mackay Medical College, New Taipei City, Taiwan
Full list of author information is available at the end of the article

M. tuberculosis has been show to infect DCs and disrupt their capacity to activate and induce primary immune responses in resting naïve T lymphocytes [5–7].

While *M. tuberculosis* infection of macrophages has been studied extensively, little is known about the mechanisms that the mycobacterium uses to mediate cell entry into human DCs. It is plausible that many host factors with important functions and potential therapeutic value have not yet been evaluated. Thus, a global analysis of membrane protein expression in human DCs treated with *M. tuberculosis* could potentially provide further information about the pathogenic mechanisms of tuberculosis. Unfortunately, it is challenging to run a large-scale identification and quantitation of membrane proteins, specifically due to their hydrophobic natures that retard both solubilization in aqueous buffers and downstream enzymatic digestion in a regular bottom-up protein identification pipeline [8, 9]. Recently, possible solutions including formic acid-CNBr/trypsin [10]、high pH/proteinase K [11]、detergent-assisted approach [12]、organic solvent-assisted digestion [13] and tube-gel assisted approach [14, 15] have been used in large-scale membrane proteome studies. Among these methods, the 60 % methanol-assisted trypsin digestion is relatively simple, and the use of a methanol-based buffer circumvents the need for reagents that interfere with chromatographic separation and ionization of the peptides (e.g., detergents, chaotropes, nonvolatile salts). For quantitative aspects, isotope-coded affinity tag [16], isotope coded protein labeling [17], ^{18}O labeling [18], stable isotope dimethyl labeling [19], stable isotope labeling by amino acids in cell culture [20] and isobaric tags for relative and absolute quantitation [15, 21] have been introduced for use in comparative membrane proteomics as well as in identification of membrane proteins. Due to its simplicity, effectiveness, and—most importantly—organic solvent compatibility, dimethyl labeling can be efficiently used with 60 % methanol-assisted trypsin digestion of membrane proteins [18]. Therefore, in this study, 60 % methanol-assisted trypsin digestion coupled with this stable-isotope labeling and LC-MS/MS analysis were applied to quantitatively analyze membrane protein expression in THP-1–derived DCs, professional antigen-presenting cells that link the innate and adaptive immunities.

After evaluating proteins that were upregulated in response to heat-killed *M. tuberculosis* treatment, the STRING (Search Tool for the Retrieval of Interacting Genes/Proteins) website database was utilized to analyze associations between these proteins. Of the investigated proteins, aminopeptidase N (CD13) was found to be largely expressed after HKTB treatment. CD13 is a peptidase that affects T cell response by mediating the trimming of major histocompatibility complex (MHC) class II peptides [22], and is also an adhesion molecule involved in

leukocyte transendothelial migration into inflammatory sites [23]. In addition, CD13 is involved in phagocytic processes in macrophages and DCs [24]. Recently we have reported that *M. tuberculosis* utilizes CD13 as a mediator of cell entry in human monocytes and macrophages [25]. However, little is known about CD13's role in mycobacterial interactions with dendritic cells. The results of our study suggest that besides its known functions, CD13 is used by *M. tuberculosis* as an important mediator of cell entry in human dendritic cells to impair the antigen-presenting process and prevent T cell activation.

Results

Proteomic profiling of membrane proteins from THP-1-derived DCs treated with or without HKTB

To uncover membrane proteins involved in *M. tuberculosis* infection, an integrated approach consisting of 60 % methanol-assisted membrane protein digestion, stable-isotope dimethyl labeling, LC-MS/MS analysis and database searching was used to comparatively profile the membrane protein expression of THP-1-derived DCs treated with or without HKTB. The membrane proteins of THP-1-derived DCs were extracted by the use of ReadyPrep Membrane II Protein Extraction kit according to the manufacturer's instructions. A total of 184 proteins derived from the membrane fraction of THP-1-derived DCs treated with or without HKTB were identified through this integrated approach (MascotProtein score and numbers of matched peptide were included in Additional file 1: Table S1). About 62.5 % of proteins were identified as membrane proteins from plasma, endoplasmic reticulum/Golgi, mitochondria, microsome, peroxisome, endosome and lysosome membranes according to their subcellular location shown in UniProt database. Among 49 membrane proteins (around 27 % of total proteins), which were identified as surface membrane proteins, 34 proteins were detected with expression ratios more than 2 and one protein with expression ratios less than 1 when their relative abundance was compared in treated and non-treated samples (Table 1). The ratio distributions of most identified proteins were ~2. These ratios did not converge well into a single value, which may be due to the variance of membrane protein extraction between treated and non-treated samples. Without a reliable internal standard in these identified proteins, a statistic approach was used to discriminate proteins with up or down regulations. We used the average ratio of the membrane proteins (2.48) and standard derivation of these ratios (1.41) to roughly distinguish proteins with significant expression.

To further investigate the role of those up-regulated proteins in mycobacterial infection, STRING database with known and predicted protein-protein interactions was used for functional clustering. The correlations of

Table 1 Identified major proteins increased/decreased in response to HKTB stimulation

No.	Name	Protein description	Ratio
1	GGT5	Gamma-glutamyltransferase 5 precursor	8.7
2	ANPEP	Aminopeptidase N, CD13	5.2
3	B*15	HLA class I histocompatibility antigen, B-15 alpha chain precursor	4.6
4	SEL1L	Sel-1 homolog precursor	3.7
5	SLC3A2	4F2 cell-surface antigen heavy chain	3.5
6	VAPA	Vesicle-associated membrane protein-associated protein A	3.4
7	CD74	HLA class II histocompatibility antigen gamma chain	3.4
8	C20orf3	Adipocyte plasma membrane-associated protein	3.4
9	ITGAL	Integrin alpha-L precursor	3.3
10	SLC2A5	Solute carrier family 2, facilitated glucose transporter member 5	3.3
11	GGT1	Gamma-glutamyltranspeptidase 1 precursor	3.2
12	ATP1B1	Sodium/potassium-transporting ATPase subunit beta-1	3.1
13	ITGAX	Integrin alpha-X precursor	3.0
14	TM9SF4	Transmembrane 9 superfamily protein member 4	3.0
15	CPM	Carboxypeptidase M precursor	2.9
16	CD1A	T-cell surface glycoprotein CD1a precursor (CD1a antigen)	2.9
17	PTPRC	Leukocyte common antigen precursor	2.9
18	SLC1A5	Neutral amino acid transporter B(0) (ATB(0))	2.7
19	RAB10	Ras-related protein Rab-10	2.6
20	RAP2A	Ras-related protein Rap-2b precursor	2.6
21	TCIRG1	Vacuolar proton translocating ATPase 116 kDa subunit a isoform 3	2.6
22	KIAA0090	Protein KIAA0090 precursor -	2.5
23	TBXAS1	Thromboxane-A synthase (Cytochrome P450 5A1)	2.5
24	ATP1A1	Sodium/potassium-transporting ATPase alpha-1 chain precursor	2.5
25	CD1C	T-cell surface glycoprotein CD1c precursor (CD1c antigen)	2.5
26	HLA-DR15	HLA class II histocompatibility antigen, DRB1-1 beta chain precursor	2.4
27	TECR	Synaptic glycoprotein SC2	2.4
28	CCDC47	Coiled-coil domain-containing protein 47 precursor	2.3
29	RAB5A	Ras-related protein Rab-5A	2.3
30	ITGB2	Integrin beta-2 precursor (Cell surface adhesion glycoproteins LFA-1/CR3/subunit beta)	2.2
31	ATP6V1B1	Vacuolar ATP synthase subunit B, kidney isoform	2.2
32	EFR3A	Protein KIAA0143 (Fragment)	2.1
33	ESYT1	Protein FAM62A (Membrane-bound C2 domain-containing protein)	2.0
34	SCAMP2	Secretory carrier-associated membrane protein 2	2.0
35	CAP1	Adenylyl cyclase-associated protein 1	0.82

these over-expressed membrane proteins were shown in Fig. 1. Several functional clusters including proteins involved in extracellular peptide breakdown, antigen presentation, phagosome maturation and markers of dendritic cells were derived from the STRING database. Among these proteins, GGT5 and CD13 are the top two proteins with highest expression ratios (8.7 and 5.2, respectively). To determine which protein would be much relevant in response to clinical tuberculosis, we analyze the expression of GGT5 and CD13 in circulating monocytes from patients with active tuberculosis. The trend of CD13 expression in circulating monocytes from patients with active tuberculosis is also increased as seen in the proteomic analysis. By contrast, the expression of GGT5 in circulating monocytes from patients with active tuberculosis was decreased, which was not compatible to the data seen in proteomic analysis (unpublished observations). Due to its high expression ratio in both in-vitro and clinical samples and potential immune modulatory functions, CD13 was chosen for further study.

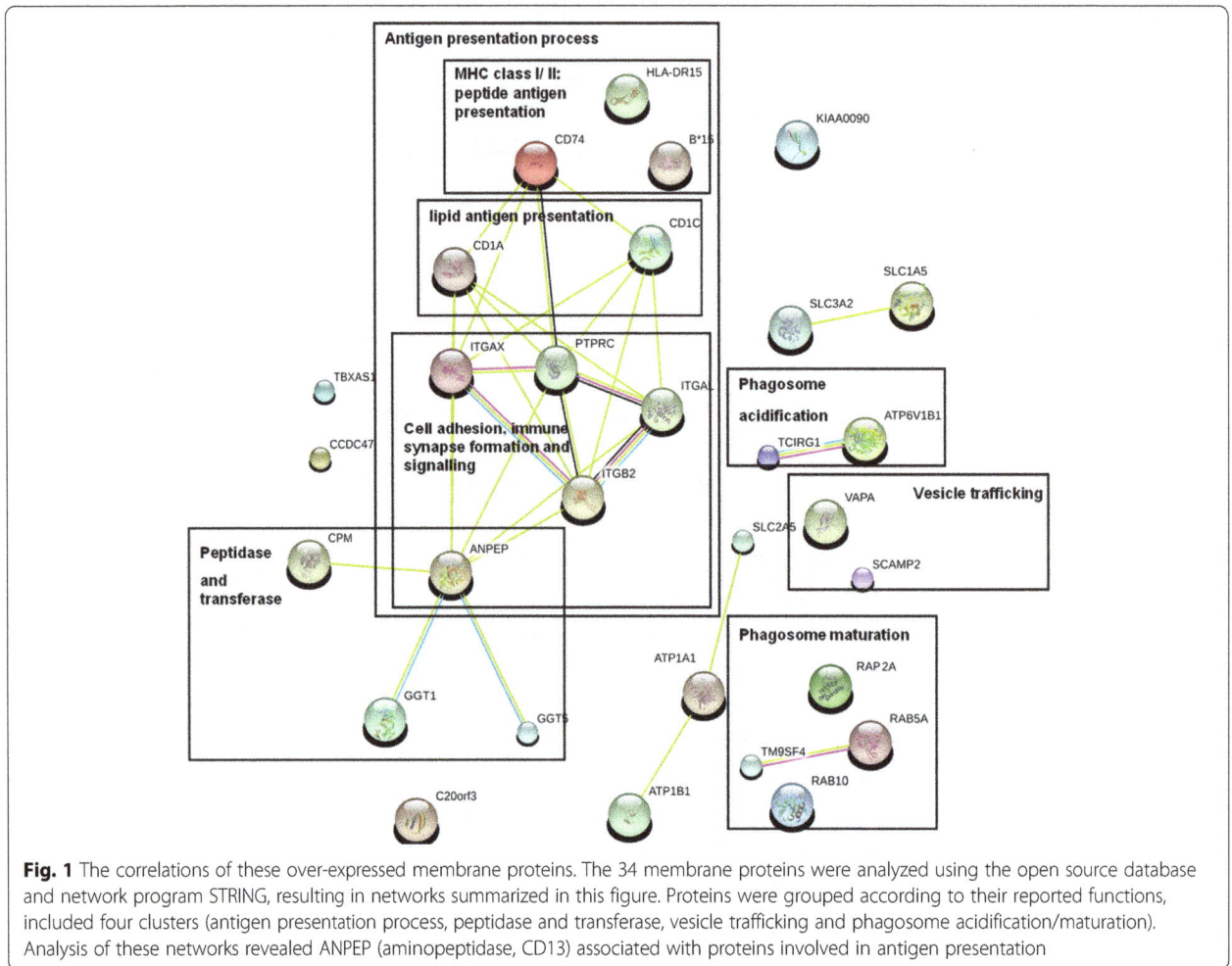

Fig. 1 The correlations of these over-expressed membrane proteins. The 34 membrane proteins were analyzed using the open source database and network program STRING, resulting in networks summarized in this figure. Proteins were grouped according to their reported functions, included four clusters (antigen presentation process, peptidase and transferase, vesicle trafficking and phagosome acidification/maturation). Analysis of these networks revealed ANPEP (aminopeptidase, CD13) associated with proteins involved in antigen presentation

The expression of CD13 was confirmed by Western blotting as shown in Fig. 2a. For THP-1-derived DCs, HKTB promoted the expression of CD 13 as compared to control (HKTB/ control ratio =3.28± 0.44), even higher than those by inflammatory stimuli, such as lipopolysaccharide (LPS) and tumor necrosis factor alpha (TNF-α). Furthermore, THP-1-derived DCs and human monocyte-derived DCs were treated with HKTB for 48 h, and surface CD13 expression was evaluated by flow cytometry (Fig. 2b). For THP-1-derived DCs, HKTB significantly increased the expression of CD13 (ratio = 1.56 ± 0.05, $P = 0.0113$). For human DCs, driven from freshly isolated PBMC, HKTB also significantly increased the expression of CD13 (ratio = 1.52 ± 0.07, $P = 0.0214$).

CD13 mediated the binding and entry of live
M. tuberculosis onto human DCs

To assess whether *M. tuberculosis* interacts with CD13, the binding and entry of live *M. tuberculosis* through surface CD13 of human DCs were examined by confocal microscopy at different time points. Figure 3 shows the results that were obtained in a representative experiment.

As *M. tuberculosis* infected human DCs (multiplicity of infection (MOI) = 10) at an indicated time, the samples were fixed by formalin addition to stop the interaction between cells and bacteria. The slides were stained and mounted for subsequent analysis of the mycobacteria-binding process through surface CD13 by confocal microscopy. In the beginning, most mycobacteria were found in the extracellular region and few of them were partially attached to cell surface (Fig. 3a, left panel). As the incubation period increased, *M. tuberculosis* were colocalized with the surface CD13 and internalized into the DC cytoplasm (Fig. 3a, middle panel). After that, most mycobacteria were found inside the DCs (Fig. 3a, right panel). The picture inside each panel shows a drawing of partial enlargement observed in confocal microscopic analysis. The lines traced on the interface of DCs/*M. tuberculosis* depicts the lineal regions of interest that were analyzed using the Lieca confocal software. As shown in each panel of Fig. 3a, analysis of the profiles of green/red channel of these cells confirmed that mycobacteria attached to the surface CD13 of human DCs and were internalized through the binding of the CD13 protein. Interestingly, we observed some

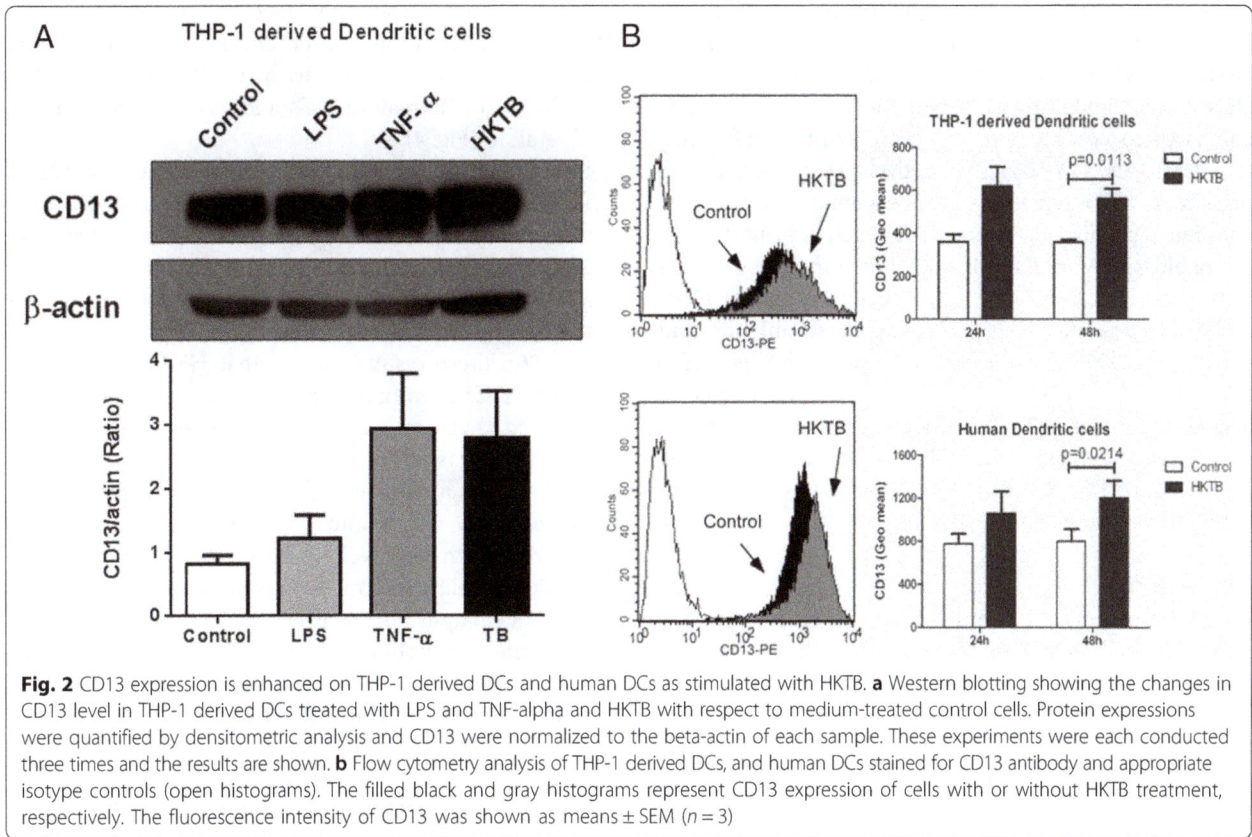

Fig. 2 CD13 expression is enhanced on THP-1 derived DCs and human DCs as stimulated with HKTB. **a** Western blotting showing the changes in CD13 level in THP-1 derived DCs treated with LPS and TNF-alpha and HKTB with respect to medium-treated control cells. Protein expressions were quantified by densitometric analysis and CD13 were normalized to the beta-actin of each sample. These experiments were each conducted three times and the results are shown. **b** Flow cytometry analysis of THP-1 derived DCs, and human DCs stained for CD13 antibody and appropriate isotype controls (open histograms). The filled black and gray histograms represent CD13 expression of cells with or without HKTB treatment, respectively. The fluorescence intensity of CD13 was shown as means ± SEM ($n = 3$)

mycobacteria still co-localized with CD13 in the cell cytoplasm after 60 min incubation whereas some mycobacteria escaped from the endocytic vesicles and resided freely (Fig. 3b). In summary, we infer that the membranous CD13 on DCs is capable of binding to live mycobacteria.

CD13 is involved in the inhibition of CD1 expression and T cell response by HKTB

It has been demonstrated that CD13 is involved in cell-surface antigen processing [22], and the CD13 on DCs

are able to selectively and efficiently degrade exogenously provided peptide antigens [26] and further influence the capability of triggering T cells. We also wanted to clarify whether CD13 was involved in the lipid antigen presenting process. For this reason, we further studied the influence of highly expressed CD13 molecules in HKTB-stimulated human DCs on CD1 expression and T cell response.

As we used HKTB to stimulate the completely differentiated DC, expressions of CD1b and CD1c decreased.

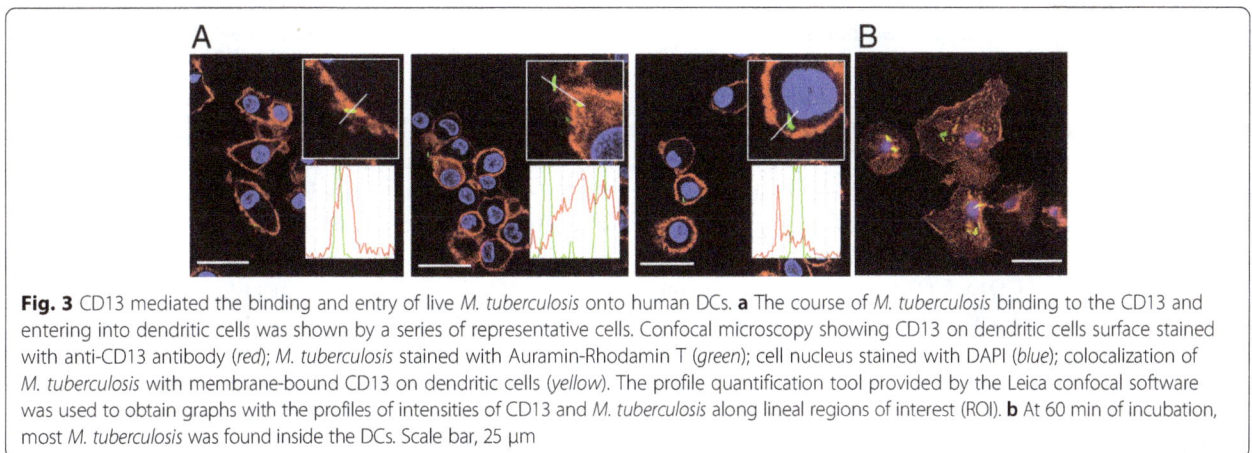

Fig. 3 CD13 mediated the binding and entry of live *M. tuberculosis* onto human DCs. **a** The course of *M. tuberculosis* binding to the CD13 and entering into dendritic cells was shown by a series of representative cells. Confocal microscopy showing CD13 on dendritic cells surface stained with anti-CD13 antibody (*red*); *M. tuberculosis* stained with Auramin-Rhodamin T (*green*); cell nucleus stained with DAPI (*blue*); colocalization of *M. tuberculosis* with membrane-bound CD13 on dendritic cells (*yellow*). The profile quantification tool provided by the Leica confocal software was used to obtain graphs with the profiles of intensities of CD13 and *M. tuberculosis* along lineal regions of interest (ROI). **b** At 60 min of incubation, most *M. tuberculosis* was found inside the DCs. Scale bar, 25 μm

Furthermore, with statistical significance, increasing stimulation with HKTB led to decreasing expressions of CD1 (Fig. 4). We found that HKTB infection of human DCs strongly inhibited CD1b, and, to a lesser extent, CD1c, compared to non-infected DCs with statistical differences ($P = 0.0013$ and $P = 0.0342$, respectively) (Fig. 4a-c). The expression of CD1 represents its non-peptide antigen presentation capability. CD1b and CD1c each present different glycolipids. LAM in the cell wall of *M. tuberculosis* is majorly presented by CD1b [27]. The expression of CD1b and CD1c are affected significantly by stimulation with LAM ($P = 0.0067$ and $P = 0.0271$, respectively) (Fig. 4f-g). We found that the affects of HKTB on CD1c could be reversed by an anti-CD13 antibody (WM15), which also subsequently enhanced the ability of DCs to trigger T cell proliferation (Fig. 4d). We also observed the same trends with LAM treatments (Fig. 4h).

Discussion

Proteins upregulated in human DCs after HKTB treatment engaged in a diverse range of cellular processes (Fig. 1 and Table 1), suggesting a complex interaction between the bacterium and its host. A proteomic analysis of membrane fraction proteins of THP-1–derived DCs revealed that proteins involved in antigen presentation (HLA class I/II histocompatibility antigens, CD1a, CD1c and CD13) were increased to confront *M. tuberculosis* infection. A cell surface binding protein, CD13 was found to be upregulated 5.2-fold upon HKTB treatment. CD13 is a large cell surface peptidase expressed on the membrane of myeloid dendritic cells [28, 29]. By using Western blot and flow cytometry, we demonstrated that CD13 expression on human DCs was upregulated under inflammatory conditions, especially those induced by HKTB (Fig. 2). However, the physiological relevance of this high expression is not fully understood yet. We have previously reported that CD13 serves as a receptor on monocyte/macrophages that binds to *M. tuberculosis* [25], as well as a possible receptor that mediates of lipid uptake (unpublished observations, [30]). Via proteomics data and STRING analysis, we demonstrated that the highly-expressed CD13 is also associated with proteins involved in the antigen presenting process, especially with CD1 proteins.

CD1 presents glycolipids such as LAM [31], the most immunogenic glycolipid antigen and a key virulence factor on mycobacterial envelopes, to a distinct group of T cells [32]. However, *M. tuberculosis* has been shown to suppress ~60-90 % of CD1a, CD1b, and CD1c expressions on differentiating monocytes [33]. The mycobacterium has also been shown to induce human monocytes to differentiate into CD1-negative DCs, which cannot present lipid antigens to specific T cells [34]. Taken together, these studies suggest that mycobacterial impairment of cellular

immune function is strongly associated with CD1 expression [35]. In our study, HKTB and LAM treatments did indeed inhibit the expressions of CD1b and CD1c—but not CD1a—on immature DCs and also decreased T cell proliferation (Fig. 4).

Interestingly, targeting CD13 helped reduce the effects of HKTB and LAM on DCs. Using neutralized antibodies to block the binding receptor CD13 partially reversed the effects of HKTB and LAM treatments on CD1c expression. Because CD1c-positive DCs are an essential group of DCs responsible for naïve T cell proliferation [36], these results suggest that HKTB impairs the ability of DCs to initiate T cell responses. Further testing confirmed that anti-CD13 antibodies also reversed the inhibitive effects of HKTB treatment on T cell proliferation. This was a simpler approach to clarify the interactions between CD13 and CD1, but our results were compatible with STRING database analysis.

Besides revealing information about antigen-presentation, proteomic analysis also revealed that proteins involved in phagosome maturation (Ras-related proteins like Rab5 and Rab 10), phagosome acidification (vacuolar proton ATPases like TCIRG1 and ATP6V1B1) and vesicle trafficking (vesicle-associated/secretory carrier-associated membrane proteins like VAPA and SCAMP2) were upregulated in HKTB-treated DCs. Rab5 and Rab 10, which are both localized to early phagosomes and are required for phagosome maturation, were upregulated by 2.3 fold and 2.6 fold, respectively. Rab10 expression is part of a host cell response during early stages of mycobacterial infection that rescues *Mycobacterium*-containing phagosomes maturation [37, 38]. Meanwhile, vacuolar proton ATPase plays a crucial role in the acidification of phagosomes and is an important initial determinant of mycobacterial Ag85B processing by macrophages [39]. The upregulation of Rab10 and vacuolar proton ATPase could reflect a host anti-bactericidal response to HKTB treatment. We also found that vesicle trafficking-associated proteins VAPA and SCAMP2 were upregulated 3.4 and 2.0-fold, respectively. Unfortunately, the roles of VAPA and SCAMP2 in intracellular transport of membranes are still not clear. Our observation that VAPA and SCAMP2 were upregulated in response to HKTB is a new finding.

Although our STRING analysis didn't reveal the direct association between CD13 and phagosome acidification-related proteins, CD13 has been reported to not only be a receptor on mycobacteria, but also a mediator that participates in phagosome acidification during *M. tuberculosis* infection [25]. As a result, further analysis is needed to clarify CD13's interactions with phagosome acidication-related proteins.

Utlimately, we infer that *M. tuberculosis* targets CD13 on dendritic cells to the antigen-presenting process and prevent T cell activation. However, anti-CD13 antibodies

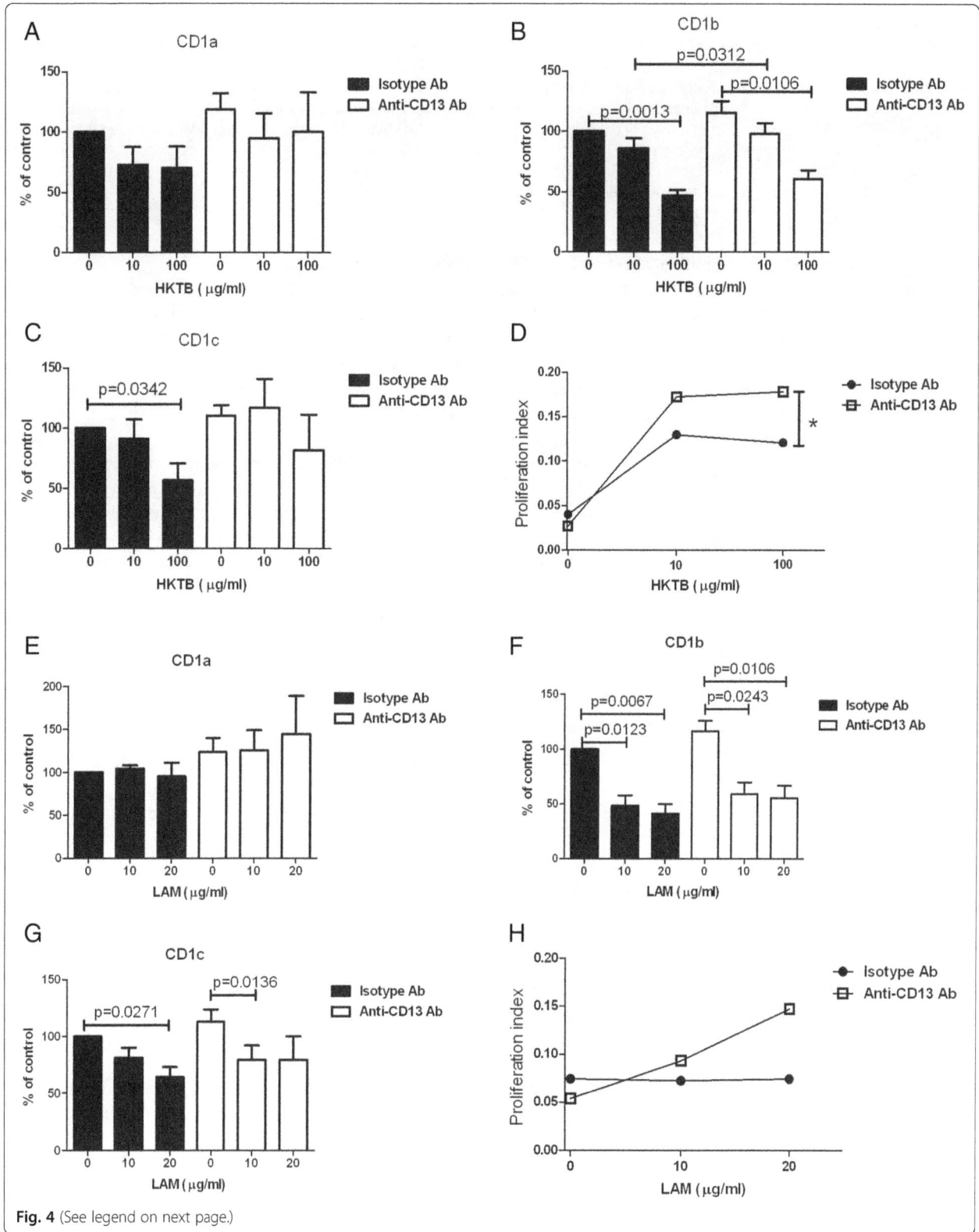

Fig. 4 (See legend on next page.)

(See figure on previous page.)

Fig. 4 CD13 is involved in the inhibition of CD1 expression and T cell response by HKTB. **a** CD1 expression on dendritic cells infected with or without HKTB for 48 h was measured by flow cytometry. The fluorescence intensity of CD1a was expressed relative to isotype antibody (100 %) and shown as means ± SEM ($n = 5$). The fluorescence intensity of CD1b expression (**b**) and CD1c expression (**c**) were shown as means ± SEM ($n = 4$). **d** CFSE-labeled T cells were cultured with dendritic cells for six days and T cell proliferation was analyzed by flow cytometry. These experiments were conducted with four times and representative result of proliferation index was shown. * $P < 0.05$. **e** CD1 expression on dendritic cells infected with or without LAM for 48 h was measured by flow cytometry. The fluorescence intensity of CD1a was expressed relative to isotype antibody (100 %) and shown as means ± SEM ($n = 5$). The fluorescence intensity of CD1b expression (**f**) and CD1c expression (**g**) were shown as means ± SEM ($n = 4$). **h** The results of T cell proliferation were representative of four different experiments

were shown to reverse the inhibitive effects of HKTB treatment on T cell proliferation. It has been reported that mycobacteria selectively target surface receptors like dendritic cell-specific intercellular adhesion molecule-3 grabbing nonintegrin (DC-SIGN) [40] to immunosuppress DC function, which prolongs the mycobacteria's survival. Our results suggest that clinical therapies targeting the CD13 receptor may help reduce *M. tuberculosis*' ability to bind to antigen presenting cells, thereby preventing the inhibition of T cell activation. However, further investigation is still needed on how exactly *M. tuberculosis* enters DCs and inhibits their cell functions through CD13.

Conclusions

These results add to a growing understanding of the interactions *M. tuberculosis* has with the human immune system. While CD13 is only one of a variety of membrane proteins involved in *M. tuberculosis* infection, evidence provided by this study may lay the groundwork for future tuberculosis target therapies.

Methods

Materials

Acetonitrile (ACN), ammonium bicarbonate, 1,4-dithiothreitol (DTT), ethylene glycol, sodium acetate and formaldehyde (37 % solution in H_2O) were purchased from *J.T. Baker* (NJ, USA). Sodium cyanoborohydride was purchased from Riedel-de Haën (Seelze, Germany). Formic acid (FA), formaldehyde-D_2 (20 % solution in D_2O), iodoacetamide (IAM), ammonium hydroxide solution (33 %), phorbol-12-myristate-13-acetate (PMA) and dimethylsulfoxide were purchased from Sigma-Aldrich (Germany, Steinheim). Sequencing grade modified trypsin was obtained from Promeg (WI, USA). RPMI 1640 media was purchased from Life Technologies Corporation (Grand Island, NY, USA). ReadyPrep Membrane II Protein Extraction kit was obtained from Bio-Rad (CA, USA). The water used in this study was obtained from Milli-Q® (Millipore) water purification system (Billerica, MA, USA). Lipoarabinomannan (LAM) from *Mycobacterium tuberculosis* Aoyama-B was purchased from Nacalai Tesque (Kyoto, Japan).

Ethics statement

This study was conducted according to the principles expressed in the International Conference on Harmonisation (ICH)/WHO Good Clinical Practice standards. Written informed consent was obtained for participation in the study, which was approved by the institutional review board of the Mackay Memorial Hospital.

Preparation of THP-1-derived dendritic cells for stimulation with HKTB

The THP-1 cell line (BCRC 60430) was obtained from the Food Industry Research and Development Institute (Hsinchu, Taiwan). THP-1 cells were grown in RPMI 1640 media containing 10 % FCS supplemented with 100 U/ml penicillin and 2 mM L-glutamine. Cells were maintained in a humidified atmosphere with 5 % $CO2$ at 37 °C. Differentiating THP-1 cells into dendritic cells was achieved by treatment of THP-1 cells with 10 ng/ml PMA, in the presence of 100 ng/ml GM-CSF and 100 ng/ml IL-4 (Peprotech Ltd) [41]. A 10 µg/ml stock solution of PMA was prepared by dissolving PMA in dimethylsulfoxide. HKTB stock solution was prepared by adding 30 ml of RPMI 1640 media into one vial of HKTB powder (100 mg, desiccated *M. tuberculosis*, purchased from BD Difco, GA, USA) and sonicated 5 min for three times. THP-1-derived DCs were then treated with HKTB (10 µg/ml) for three days.

Trypsin digestion of membrane proteins

The membrane proteins of THP-1-derived DCs were extracted by the use of ReadyPrep Membrane II Protein Extraction kit (Bio-Rad) according to the manufacturer's instructions. Briefly, 100-200 mg of wet cell pellets were lysed with lysis buffer. After the removal of insoluble material and unbroken cells, the membrane protein fraction was isolated with ice-cold Na2CO3 followed by centrifugation. Pellets from the control and experimental samples were individually dissolved and digested according to the following procedures. Proteins in the membrane fraction pellet were resuspended in 50 mM ammonium bicarbonate (200 µl, pH 8.3) by vortex and sonication. The proteins were denatured at 95 °C for 5 min then cooled at ice-bath temperature. Due to the hydrophobic nature, the membrane fraction can not be dissolved and digested well by regular digestion protocol. Organic-assisted solubilization and proteolysis were chosen in this study to dissolve and digest membrane proteins. Methanol was added into the protein solution to make a final concentration of 60 % (v/v), facilitating the solubilization of most membrane proteins. Trypsin was then immediately put into this solution in a 1/50 (w/w) trypsin-to-protein ratio followed by incubation for 14 h at 37 °C. The resulting peptide mixtures were lyophilized and stored at -20 °C for further stable-isotope labeling.

Stable-isotope dimethyl-labeling of tryptic peptides

The tryptic peptides derived from the control samples which had been dissolved in sodium acetate buffer (100 mM, pH 5-6) were mixed with formaldehyde (4 % in water, 5 µl), vortexed, and then immediately combined with freshly prepared sodium cyanoborohydride (1 M, 5 µl). The mixture was vortexed again and then allowed to react for 5 min. Ammonium hydroxide (4 % in water, 5 µl) was used to consume the excess aldehyde. Deuterium labeling of the experimental samples was performed in a similar manner except using formaldehyde-$d2$ (4 % in water, 5 µl). The tryptic digest derived from the experimental sample was diluted with sodium acetate buffer (100 mM, pH 5-6) and then labeled as described above. The labeled peptides derived from either control or experimental samples were blended and desalted by C18 Easy-Tips (MST, Taipei) for additional fractionation through strong cation exchange (SCX) chromatography.

Off-line SCX fractionation of combined mixture

The desalted peptides were redissolved in 0.1 % formic acid and fractionated using a SCX cartridge (5 µm, Vydac, CA, USA). A total of ten fractions eluted with 5, 10, 20, 30, 40, 50, 60, 100, 200, and 500 mM of sodium chloride were serially collected. The resulting fractions were desalted by EasyTipsTM (C SUN, Taipei, Taiwan) and subjected to LC-MS/MS analysis.

Mass spectrometric analysis and database searching

The dimethylated tryptic peptides were analyzed using a tandem quadrupole time-of-flight (Q-TOF) mass spectrometer (Micromass, UK) equipped with a nanoflow CapLC system (Waters). The scan range was from m/z 400 to 1600 for MS and m/z 50 to 2000 for MS/MS. For sequencing, the MS/MS spectra were obtained through a survey scan and the automated data-dependent MS analysis was performed by the dynamic exclusion feature built into the MS acquisition software. The MS/MS raw data was processed into a PKL file format using MassLynx 4.0 (Micromass, UK). The resulting PKL file was searched using the Mascot search engine v2.2 (Matrix Science, UK) with the following search parameters: (1) protein database

was set to be Swiss-Prot; (2) taxonomy was set as *Homo sapiens* (human); (3) one trypsin missed cleavage was allowed; (4) the precursor and product ion mass tolerance was set at 0.4 Da/0.2 Da; (5) carbamidomethyl (C) was chosen for fixed modification; (6) oxidation (M), deamidated (NQ), Dimethyl (K), Dimethyl (N-term), Dimethyl:2H(4) (K), Dimethyl:2H(4) (N-term) were chosen for variable modifications; (7) proteins with scores above the significance threshold ($p < 0.05$) were shown as significant hits. All MS/MS spectra of identified peptides derived from membrane proteins were further verified by manual interpretation, in particular, using a1 ion in each MS/MS spectrum to verify the N-terminus of the corresponding peptide [42]. The subcellular location and functional annotation of the identified proteins were elucidated by UniProt knowledgebase (Swiss-Prot/TrEMBL) and Gene Ontology Database. The intensity ratios of isotopic isomers were calculated based on peak area or peak intensity from selective ion chromatograms or alternatively, all of the spectra containing both mass peaks of D4- and H4-labeled peptides were combined to produce a composite MS spectrum. The ratios of the D4- and H4-labeled peptides in the composite MS spectra were calculated from the sum of the peak heights of the first three isotopic peaks. Proteins with relative ratio more than 2 were regarded as over-expressed. The accession numbers of these over-expressed surface membrane proteins from HKTB treated THP-1-derived DCs were put together and further correlated by STRING database with known and predicted protein-protein interactions (http://string-db.org/newstring_cgi/show_input_page.pl).

Western blotting

THP-1-derived DCs were treated with LPS, TNF-α or HKTB. CD13 protein levels were analyzed by Western blotting to confirm and validate significance of the proteomic findings. Each protein sample was mixed with an electrophoresis buffer containing 2 % SDS and 5 % β-mercaptoethanol and boiled for 10 min. Proteins (5 µg) were separated by electrophoresis on a 10 % SDS-polyacrylamide gel. The fractionated proteins were electroblotted onto a polyvinylidene difluoride membrane. The membranes were blocked at least 2 hrs in 5 % BSA, 0.1 % Tween 20 in TBS (TBST) and then incubated with CD13 antibodies (1:1000 dilution, Abcam, MA, USA), for 1 hr. After washing in TBST, membranes were incubated with peroxidase-conjugated secondary antibodies for 1 hr, and proteins were detected using an enhanced chemiluminescence detection system (PerkinElmer Life and Analytical Sciences, Boston, USA).

Human dendritic cells culture

Peripheral blood mononuclear cells (PBMCs) were isolated from the whole blood of healthy adult volunteers by Ficoll-Paque gradient centrifugation. Monocytes were sorted by incubating PBMCs with CD14 microbeads (Miltenyi Biotec) and then the CD14-positive cells were separated by means of a magnetic force. Dendritic cells were generated by culturing monocytes for five days in RPMI-1640 medium with 10 % FBS in the presence of 100 ng/ml GM-CSF and 100 ng/ml IL-4 (Peprotech Ltd). Monocyte-derived DCs were then treated with HKTB solution for two days.

Flow cytometry
Cells were stained with PE-conjugated anti-CD13 antibody (clone WM15, BD Pharmingen TM, San Jose, CA, USA). The mean fluorescence intensity of stained cells was measured by FACS Calibur flow cytometry and analyzed by using CellQuest software (BD Bioscience).

Live _M tuberculosis_ preparation _M. tuberculosis_ were obtained from a clinical virulent strain identified by the Mycobacteriology Laboratory, MacKay Memorial Hospital. Firstly, _M. tuberculosis_ was identified by a MPT64 rapid test, arylsulfatase test, and nitrate reductase assay. The identified strain was further confirmed by a commercial TB-PCR kit (COBAS, Taqman MTB test, Roche). Seed stocks of the collected strain were maintained in small aliquots at -80 °C. The _M. tuberculosis_ used throughout this study was prepared from the seed stocks by culturing on 7H11 agar plate for 3 weeks at 37 °C in 10 % CO2 humidified atmosphere. The culture medium used was a 7H9 broth supplemented with 0.2 % glycerol, 0.05 % Tween-80 and 10 % oleic acid-albumin-dextrose-catalase enrichment (Difco, Becton Dickinson and Company, MD, USA) to an optical density at 600 nm of 0.3 and used as the inoculum.

Confocal microscopy
Dendritic cells were cultured on 18-mm-diameter coverglass placed in 12-well culture plate and infected with _M. tuberculosis_ labeled with Auramin-Rhodamin T. At various time points, unbound bacteria were washed away with PBS and cells were fixed in 4 % formalin. CD13 were stained with Cy-Chrome 5 (Cy5)-conjugated anti-CD13 antibody, and nuclei were stained with 4'-6-Diamidino-2-phenylindole (DAPI). Samples were analyzed by Leica TCS SP5 confocal laser scanning microscopy and quantified by Leica LAS AF software. Use the Leica software profile quantification tool to manually trace a lineal region of interest in each cell that is subjected to analysis. Subsequently, obtain graphs displaying the intensity profiles of CD13/_M. tuberculosis_ (green/red channel).

T-cell proliferation assay
T cells were purified from CD14-negative cells by using the Dynal T cell negative isolation kit (Dynal Biotech Ltd; purity routinely >90 %). T cells were labeled with

5 μM CFSE dye (Invitrogen) for 10 min at 37 degree, washed once, and resuspended in RPMI-1640 medium with 10 % FBS to a final concentration of 2×10^6 cells/ml. CFSE-labebled T cell (2×10^5) were incubated with autologous antigen-loaded dendritic cells (4×10^4) (5 T cells: 1 DC) as previously described [43]. Cells were incubated for 6 days and CFSE dilution was measured by flow cytometry. The proliferation index was calculated by the formula: % of dividing cells/% of non-dividing cells.

Statistical analysis
Paired t test was used for analysis. Data are reported as the mean ± SEM. Statistical analysis was performed using Prism 3.0 software (GraphPad Software Inc.). Two-sided tests were used and a P value of < 0.05 was considered statistically significant.

Abbreviations
DCs: dendritic cells; DC-SIGN: dendritic cell-specific intercellular adhesion molecule-3 grabbing nonintegrin; HKTB: heat-killed _M. tuberculosis_; LAM: Lipoarabinomannan; LC-MS/MS: Liquid chromatography-tandem mass spectrometry; LPS: lipopolysaccharide; STRING: Search Tool for the Retrieval of Interacting Genes/Proteins; TNF-α: tumor necrosis factor alpha.

Competing interests
All authors declare that they have no competing interests.

Authors' contributions
CP, JL, CL, IF,KS and YT participated in the design of the study. JL carried out the proteomic studies and performed the LC-MS/MS analysis. IF and TY carried out the cell culture and cells' functional assays. CP, JL, TY, IF, KS and YT participated in the interpretation of data. CP, AB, and YT conceived of the study, and participated in its design and coordination and helped to draft the manuscript. KS, CL and AB provided critical revision of the manuscript for important intellectual content. All authors read and approved the final manuscript.

Acknowledgement
This project was funded with supports from Mackay Memorial Hospital (Projects # MMH-E-98008, MMH-E-99008 and MMH-102-33) and Taiwan National Science Council (Grant # NSC 99-2314-B-195-008).

Author details
[1]Division of Chest Medicine, Department of Internal Medicine, Mackay Memorial Hospital, 92, Sec 2, Chungshan North Road, Taipei, Taiwan. [2]Department of Emergency Medicine, Mackay Memorial Hospital, Taipei, Taiwan. [3]Department of Medical Research, Mackay Memorial Hospital, Taipei, Taiwan. [4]Biology Department, Case Western Reserve University, Cleveland, OH, USA. [5]Mackay Junior College of Medicine, Nursing, and Management, Taipei, Taiwan. [6]Department of Medicine, Mackay Medical College, New Taipei City, Taiwan. [7]Graduate Institute of Biotechnology, National Pingtung University of Science and Technology, Pingtung 91201, Taiwan.

References
1.   Russell DG. Who puts the tubercle in tuberculosis? Nat Rev Microbiol. 2007;5:39–47.

2. Nunes-Alves C, Booty MG, Carpenter SM, Jayaraman P, Rothchild AC, Behar SM. In search of a new paradigm for protective immunity to TB. Nat Rev Microbiol. 2014;12:289–99.

3. Vergne I, Chua J, Singh SB, Deretic V. Cell biology of mycobacterium tuberculosis phagosome. Annu Rev Cell Dev Biol. 2004;20:367–94.

4. McWilliam AS, Nelson D, Thomas JA, Holt PG. Rapid dendritic cell recruitment is a hallmark of the acute inflammatory response at mucosal surfaces. J Exp Med. 1994;179:1331–6.

5. Giacomini E, Iona E, Ferroni L, Miettinen M, Fattorini L, Orefici G, Julkunen I, Coccia EM. Infection of human macrophages and dendritic cells with Mycobacterium tuberculosis induces a differential cytokine gene expression that modulates T cell response. J Immunol. 2001;166:7033–41.

6. Wolf AJ, Linas B, Trevejo-Nuñez GJ, Kincaid E, Tamura T, Takatsu K, Ernst JD. Mycobacterium tuberculosis Infects Dendritic Cells with High Frequency and Impairs Their Function In Vivo. J Immunol. 2007;179:2509–19.

7. Hickman SP, Chan J, Salgame P. Mycobacterium tuberculosis induces differential cytokine production from dendritic cells and macrophages with divergent effects on naive T cell polarization. J Immunol. 2002;168:4636–42.

8. Speers AE, Wu CC. Proteomics of integral membrane proteins–theory and application. Chem Rev. 2007;107:3687–714.

9. Cordwell SJ, Thingholm TE. Technologies for plasma membrane proteomics. Proteomics. 2010;10:611–27.

10. Washburn MP, Wolters D, Yates 3rd JR. Large-scale analysis of the yeast proteome by multidimensional protein identification technology. Nat Biotechnol. 2001;19:242–7.

11. Wu CC, MacCoss MJ, Howell KE, Yates III JR. A method for the comprehensive proteomic analysis of membrane proteins. Nat Biotechnol. 2003;21:532–8.

12. Masuda T, Tomita M, Ishihama Y. Phase transfer surfactant-aided trypsin digestion for membrane proteome analysis. J Proteome Res. 2008;7:731–40.

13. Blonder J, Chan KC, Issaq HJ, Veenstra TD. Identification of membrane proteins from mammalian cell/tissue using methanol-facilitated solubilization and tryptic digestion coupled with 2D-LC-MS/MS. Nat Protoc. 2006;1:2784–90.

14. Lu X, Zhu H. Tube-gel digestion: a novel proteomic approach for high throughput analysis of membrane proteins. Mol Cell Proteomics. 2005;4:1948–58.

15. Han CL, Chien CW, Chen WC, Chen YR, Wu CP, Li H, Chen YJ. A multiplexed quantitative strategy for membrane proteomics: opportunities for mining therapeutic targets for autosomal dominant polycystic kidney disease. Mol Cell Proteomics. 2008;7:1983–97.

16. Ramus C, Gonzalez De Peredo A, Dahout C, Gallagher M, Garin J. An optimized strategy for ICAT quantification of membrane proteins. Mol Cell Proteomics. 2006;5:68–78.

17. Bisle B, Schmidt A, Scheibe B, Klein C, Tebbe A, Kellermann J, Siedler F, Pfeiffer F, Lottspeich F, Oesterhelt D. Quantitative profiling of the membrane proteome in a halophilic archaeon. Mol Cell Proteomics. 2006;5:1543–58.

18. Blonder J, Yu LR, Radeva G, Chan KC, Lucas DA, Waybright TJ, Issaq HJ, Sharom FJ, Veenstra TD. Combined chemical and enzymatic stable isotope labeling for quantitative profiling of detergent-insoluble membrane proteins isolated using Triton X-100 and Brij-96. J Proteome Res. 2006;5:349–60.

19. Hsu JL, Wang LY, Wang SY, Lin CH, Ho KC, Shi FK, Chang IF. Functional phosphoproteomic profiling of phosphorylation sites in membrane fractions of salt-stressed Arabidopsis thaliana. Proteome Sci. 2009;7:42.

20. Hör S, Ziv T, Admon A, Lehner PJ. Stable isotope labeling by amino acids in cell culture and differential plasma membrane proteome quantitation identify new substrates for the MARCH9 transmembrane E3 ligase. Mol Cell Proteomics. 2009;8:1959–71.

21. Mazzucchelli GD, Cellier NA, Mshviladzade V, Elias R, Shim YH, Touboul D, Quinton L, Brunelle A, Laprévote O, De Pauw EA, De Pauw-Gillet MCJ. Pores formation on cell membranes by hederacolchiside A1 leads to a rapid release of proteins for cytosolic subproteome analysis. Proteome Res. 2008;7:1683–92.

22. Larsen SL, Pedersen LO, Buus S, Stryhn A. T cell responses affected by aminopeptidase N (CD13)-mediated trimming of major histocompatibility complex class II-bound peptides. J Exp Med. 1996;184:183–9.

23. Mina-Osorio P, Winnicka B, O'Conor C, Grant CL, Vogel LK, Rodriguez-Pinto D, Holmes KV, Ortega E, Shapiro LH. CD13 is a novel mediator of monocytic/endothelial cell adhesion. J Leukoc Biol. 2008;84:448–59.

24. Villaseñor-Cardoso MI, Frausto-Del-Río DA, Ortega E. Aminopeptidase N (CD13) is involved in phagocytic processes in human dendritic cells and macrophages. Biomed Res Int. 2013;2013:562984.

25. Ho HT, Tsai IF, Wu CL, Lu YT. Aminopeptidase N facilitates entry and intracellular survival of Mycobacterium tuberculosis in monocytes. Respirology. 2014;19:109–15.

26. Dong X, An B, Salvucci Kierstead L, Storkus WJ, Amoscato AA, Salter RD. Modification of the amino terminus of a class II epitope confers resistance to degradation by CD13 on dendritic cells and enhances presentation to T cells. J Immunol. 2000;164:129–35.

27. Lawton AP, Kronenberg M. The Third Way: Progress on pathways of antigen processing and presentation by CD1. Immunol Cell Biol. 2004;82:295–306.

28. Summers KL, Hock BD, McKenzie JL, Hart DN. Phenotypic characterization of five dendritic cell subsets in human tonsils. Am J Pathol. 2001;159:285–95.

29. Collin M, McGovern N, Haniffa M. Human dendritic cell subsets. Immunology. 2013;140:22–30.

30. Kramer W, Girbig F, Corsiero D, et al. Aminopeptidase N (CD13) is a molecular target of the cholesterol absorption inhibitor ezetimibe in the enterocyte brush border membrane. J Biol Chem. 2005;280:1306–20.

31. Young DC, Moody DB. T-cell recognition of glycolipids presented by CD1 proteins. Glycobiology. 2006;16:103R–12R.

32. Van Rhijn I, Moody DB. CD1 and mycobacterial lipids activate human T cells. Immunol Rev. 2015;264:138–53.

33. Gagliardi MC, Teloni R, Mariotti S, Iona E, Pardini M, Fattorini L, Orefici G, Nisini R. Bacillus Calmette-Guérin shares with virulent Mycobacterium tuberculosis the capacity to subvert monocyte differentiation into dendritic cell: implication for its efficacy as a vaccine preventing tuberculosis. Vaccine. 2004;22:3848–57.

34. Gagliardi MC, Teloni R, Giannoni F, Mariotti S, Remoli ME, Sargentini V, Videtta M, Pardini M, De Libero G, Coccia EM, Nisini R. Mycobacteria exploit p38 signaling to affect CD1 expression and lipid antigen presentation by human dendritic cells. Infect Immun. 2009;77:4947–52.

35. Sieling PA, Jullien D, Dahlem M, Tedder TF, Rea TH, Modlin RL, Porcelli SA. CD1 expression by dendritic cells in human leprosy lesions: correlation with effective host immunity. J Immunol. 1999;162:1851–8.

36. Lozza L, Farinacci M, Bechtle M, Stäber M, Zedler U, Baiocchini A, Del Nonno F, Kaufmann SH. Communication between Human Dendritic Cell Subsets in Tuberculosis: Requirements for Naive CD4(+) T Cell Stimulation. Front Immunol. 2014;5:324.

37. Cardoso CM, Jordao L, Vieira OV. Rab10 regulates phagosome maturation and its overexpression rescues Mycobacterium-containing phagosomes maturation. Traffic. 2010;11:221–35.

38. Gutierrez MG, Mishra BB, Jordao L, Elliott E, Anes E, Griffiths G. NF-κ B activation controls phagolysosome fusion-mediated killing of mycobacteria by macrophages. J Immunol. 2008;181:2651–63.

39. Singh CR, Moulton RA, Armitige LY, Bidani A, Snuggs M, Dhandayuthapani S, Hunter RL, Jagannath C. Processing and presentation of a mycobacterial antigen 85B epitope by murine macrophages is dependent on the phagosomal acquisition of vacuolar proton ATPase and in situ activation of cathepsin D. J Immunol. 2006;177:3250–9.

40. Geijtenbeek TB, Van Vliet SJ, Koppel EA, Sanchez-Hernandez M, Vandenbroucke-Grauls CM, Appelmelk B, Van Kooyk Y. Mycobacteria target DC-SIGN to suppress dendritic cell function. J Exp Med. 2003;197:7–17.

41. Miszczyk E, Rudnicka K, Moran AP, Fol M, Kowalewicz-Kulbat M, Druszczyńska M, Matusiak A, Walencka M, Rudnicka W, Chmiela M. Interaction of Helicobacter pylori with C-type lectin dendritic cell-specific ICAM grabbingnonintegrin. J Biomed Biotechnol. 2012;2012:206463.

42. Hsu JL, Huang SY, Shiea JT, Huang WY, Chen SH. Beyond quantitative proteomics: signal enhancement of the a1 ion as a mass tag for peptide sequencing using dimethyl labeling. J Proteome Res. 2005;4:101–8.

43. Wuest SC, Edwan JH, Martin JF, Han S, Perry JS, Cartagena CM, Matsuura E, Maric D, Waldmann TA, Bielekova B. A role for interleukin-2 trans-presentation in dendritic cell-mediated T cell activation in humans, as revealed by daclizumab therapy. Nat Med. 2011;17:604–9.

5

Investigation of proteome changes in osteoclastogenesis in low serum culture system using quantitative proteomics

Qi Xiong[1†], Lihai Zhang[1†], Shaohua Zhan[2], Wei Ge[2*] and Peifu Tang[1*]

Abstract

Background: RAW 264.7 cells can differentiate into osteoclasts when cultured in medium supplemented with 1 % FBS. However, the proteomic changes in the development of RAW 264.7 cells into osteoclasts in low serum culture system have not been elucidated. Therefore, we conducted quantitative proteomics analysis to investigate proteomic changes during osteoclastogenesis in low serum culture system.

Results: Our study confirmed that mature multinucleated osteoclasts were generated in a low serum culture system, validated by upregulated expression of 15 characteristic marker proteins, including TRAP, CTSK, MMP9, V-ATPase and ITGAV. Proteomics results demonstrated that 549 proteins expressed differentially in osteoclastogenesis in low serum culture system. In-depth bioinformatics analysis suggested that the differentially expressed proteins were mainly involved in mitochondrial activities and energy metabolism, including the electron transport chain pathway, TCA cycle pathway, mitochondrial LC-fatty acid beta-oxidation pathway and fatty acid biosynthesis pathway. The data have been deposited to the ProteomeXchange with identifier PXD001935.

Conclusion: Osteoclast formation is an ATP consuming procedure, whether occurring in a low serum culture system or a conventional culture system. In contrast to osteoclasts formed in conventional culture system, the fatty acid biosynthesis pathway was upregulated in osteoclasts cultured in low serum condition.

Keywords: Osteoclast, Cell culture, Proteomics, Bioinformatics

Background

Osteoporosis is a skeletal disorder characterized by diminished bone mineral density and deteriorated bone micro-structure, which consequently leads to increased fracture risk, especially in the elder population. Epidemiological study revealed that more than 75 million people suffer from osteoporosis in United States, Europe, and Japan [1]. It is estimated that approximately 40 % of Caucasian women and 13 % of Caucasian men aged over 50 years old will suffer from osteoporosis-related fracture in the United States [2, 3]. Osteoporosis and osteoporosis-related fractures therefore have become a major public health concern, and impose enormous health care costs. Burge et al. demonstrated that annual costs for osteoporosis-related fractures were US$13.7 to US$20.3 billion, and predicted that the costs would increase to US$25.3 billion annually by 2025 in the United States [4]. Therefore, investigating the pathological mechanisms of osteoporosis will be of great help to reduce the tremendous osteoporosis related costs and promote life quality of elder populations.

Over-activation of bone resorption plays a critical role in the pathological mechanisms of osteoporosis [5]. To date, large multinucleated cells termed osteoclasts are the exclusive cells known to have the ability of bone resorption. Thus, studying the molecular mechanisms of osteoclast formation is essential for investigating pathological mechanisms of osteoporosis. Osteoclasts derive from monocyte/macrophage cells [6]. Followed activation of the receptor activator of nuclear factor B ligand (RANKL)/RANK signaling pathway by RANKL,

* Correspondence: wei.ge@chem.ox.ac.uk; pftang301@163.com
†Equal contributors
[2]National Key Laboratory of Medical Molecular Biology & Department of Immunology, Institute of Basic Medical Sciences, Chinese Academy of Medical Sciences, DongdanSantiao 5#, Dongcheng District, Beijing 100005, China
[1]Department of Orthopedics, General Hospital of Chinese PLA, Fuxing Road 28#, Haidian District, Beijing 100853, China

downstream signaling cascades are stimulated, including the IκB kinase (IKK) signaling, c-Jun N-terminal kinase (JNK) signaling, NF-κB signaling and MAPK signaling [7–10]. Subsequently, monocyte/macrophage cells differentiate and fuse into mature multinucleated osteoclasts, regulated by nuclear factor of activated T-cells, cytoplasmic, calcineurin-dependent 1 (NFATc1) and microphthalmia-associated transcription factor (MITF) [11]. However, the molecular mechanisms of osteoclast formation still need to be elucidated.

The cultivation of osteoclasts in vitro is a prerequisite for studying the molecular mechanisms of osteoclasts. Traditionally, a 10 % volume fraction of serum is used to culture osteoclasts. We and other researchers previously demonstrated that RAW264.7 cells, an osteoclast precursor cell line, could differentiate into mature bone resorbing osteoclasts when cultured both in 1 % serum and serum-deprived medium [12, 13]. Our previous study revealed that a number of proteins and molecular signaling pathways, including electron transport chain and oxidative phosphorylation pathways, were altered in RAW264.7 cells cultured in low serum compared to those cultured in the conventional 10 % culture system [13]. An et al. have performed proteomics analysis to study the changes in global protein expression during osteoclastogenesis in conventional culture system [14]. However, to the best of our knowledge, the proteomic changes that occur in RAW 264.7 cells as they differentiate into osteoclasts in a low serum culture system has not been reported. In the present study, we used TMT labeling to analyze the quantitative proteomic changes during osteoclastogenesis in medium supplemented with 1 % FBS.

Results

Conformation of osteoclastogenesis in a low serum culture system

To validate the formation of osteoclasts in medium supplemented with 1 % FBS, we performed TRAP staining and bone resorption assay. After 5 days cultivation, TRAP-positive cells with more than 3 nuclei could be observed (Fig. 1a), and the simulated bone mineral surface was remarkably resorbed (Fig. 1b). These results confirmed that formation of mature multinucleated osteoclasts could be obtained in a low serum culture system. The tartrate-resistant acid phosphatase type 5 (TRAP or ACP5; Accession Number: Q05117), cathepsin K (CTSK; Accession Number: P55097), matrix metalloproteinase-9 (MMP9; Accession Number: P41245) are characteristic osteoclast marker proteins [15]. To ensure that the large multinucleated cells generated in our study were osteoclasts, quantitative proteomic data was analyzed. Our results revealed that the three characteristic marker proteins were significantly upregulated in these differentiated cells. Compared to RAW 264.7 cells, TRAP, CTSK and MMP9 in multinucleated cells were upregulated 2.59, 5.21 and 2.82-fold respectively (Table 1). The vacuolar-type H^+-ATPase (V-ATPase) proton pump complex is typically located on the ruffled border plasma membrane of osteoclasts [16], and can also be used as biomarker for mature osteoclasts. Our results identified 11 subunits of V-ATPase; all of these subunits were significantly upregulated (Table 1). Furthermore, bone resorbing osteoclasts require integrin alpha-V (ITGAV; Accession Number: P43406) for the formation of the sealing zone between osteoclasts and bone surface [17]. Our results showed that the ITGAV was upregulated 2.53-fold (Table 1). These findings clearly suggest that the giant multinucleated cells formed in low serum culture system were osteoclasts.

Western blot analysis of the differentially expressed proteins

To validate the formation of osteoclasts and the results of the quantitative LC-MS/MS data, we performed western blot analysis of three characteristic proteins of osteoclasts (TRAP, CTSK, MMP9) and two other differentially expressed proteins (ANXA1, Histone H4). The expression

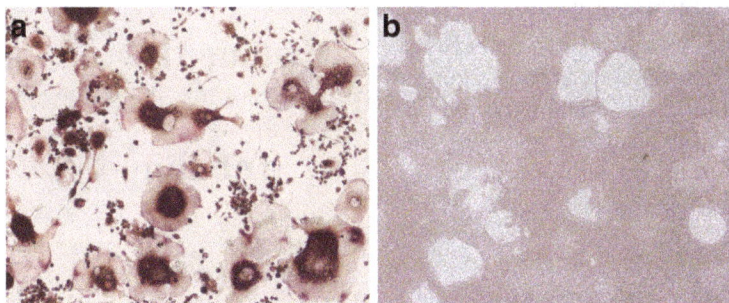

Fig. 1 Confirmation of osteoclast formation in low serum culture system using TRAP staining and bone resorption activity test. **a** TRAP-positive multinucleated osteoclasts were successfully obtained. **b** Bone mineral surface was resorbed remarkably by osteoclasts generated in low serum culture system

Table 1 Quantitative proteomics data of characteristic biomarkers for osteoclasts

Accession	Description	Score	Unique peptide	Tubulin adjust (128/127)
P55097	Cathepsin K OS = Mus musculus GN = Ctsk	168.11	7	5.21
P41245	Matrix metalloproteinase-9 OS = Mus musculus GN = Mmp9	58.12	15	2.82
Q05117	Tartrate-resistant acid phosphatase type 5 OS = Mus musculus GN = Acp5	104.03	10	2.59
Q9R1Q9	V-type proton ATPase subunit S1 OS = Mus musculus GN = Atp6ap1	25.62	5	2.50
Q8BVE3	V-type proton ATPase subunit H OS = Mus musculus GN = Atp6v1h	160.48	19	2.37
Q9CR51	V-type proton ATPase subunit G 1 OS = Mus musculus GN = Atp6v1g1	76.40	4	3.35
Q9D1K2	V-type proton ATPase subunit F OS = Mus musculus GN = Atp6v1f	67.21	9	2.45
P50518	V-type proton ATPase subunit E 1 OS = Mus musculus GN = Atp6v1e1	219.66	16	2.59
P57746	V-type proton ATPase subunit D OS = Mus musculus GN = Atp6v1d	130.06	12	2.33
Q80SY3	V-type proton ATPase subunit d 2 OS = Mus musculus GN = Atp6v0d2	134.85	13	3.46
P51863	V-type proton ATPase subunit d 1 OS = Mus musculus GN = Atp6v0d1	71.67	13	2.54
Q9Z1G3	V-type proton ATPase subunit C 1 OS = Mus musculus GN = Atp6v1c1	216.86	24	2.62
P62814	V-type proton ATPase subunit B, brain isoform OS = Mus musculus GN = Atp6v1b2	762.21	36	2.16
P50516	V-type proton ATPase catalytic subunit A OS = Mus musculus GN = Atp6v1a	658.00	38	2.12
P43406	Integrin alpha-V OS = Mus musculus GN = Itgav	17.86	4	2.53

of the three characteristic osteoclast proteins was significant upregulated on western blots, confirming the formation of osteoclasts in the low serum system. In addition, the expression of ANXA1 and histone H4 were also upregulated, as showed in our quantitative LC-MS/MS data. These findings indicated that our LC-MS/MS data and subsequent analysis were confirmed by western blotting, suggesting that the analysis results accurately reflect changes in protein expression (Fig. 2).

Quantitative analysis of proteomic changes during osteoclastogenesis in a low serum system

We performed quantitative proteomic analysis to compare the global protein expression of RAW 264.7 cells with those of the osteoclasts. Our results identified a total of 6567 proteins, of which 4656 proteins with a score of 10 or more and 2 or more unique peptides identified were selected for further analysis. Of the 4656 proteins with high confidence, 549 proteins expressed differentially when RAW 264.7 cells differentiated into osteoclasts. Among the 549 differentially expressed proteins, 98.5 % (541/549) were significantly upregulated, while only 1.5 % (8/549) of proteins were downregulated (Additional file 1: Table S1). These results indicated that a great many proteins were upregulated, and suggested that RAW 264.7 cells were highly activated in the development into osteoclasts in the low serum culture system.

Identification and classification of the differentially expressed proteins

To understand the specific proteomic changes, we performed enrichment analysis using FunRich software to

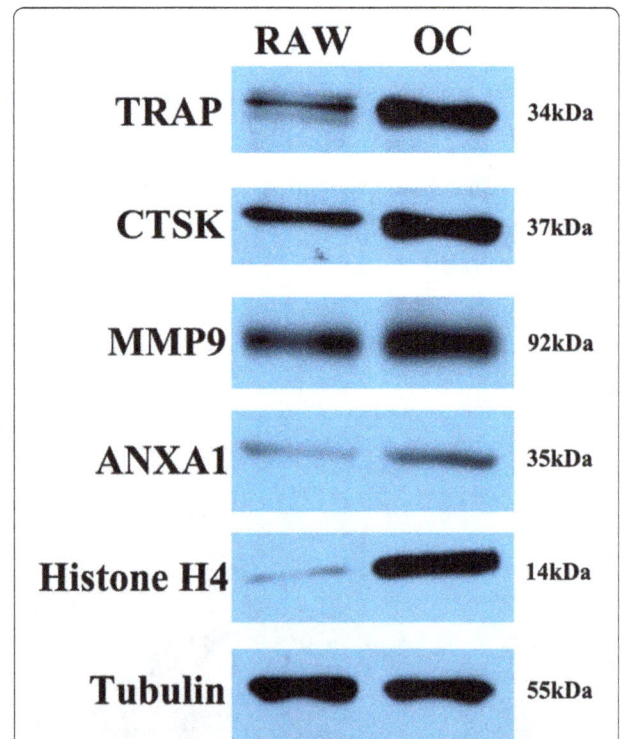

Fig. 2 Validation of the LC-MS/MS data and osteoclast formation by western blot analysis. Characteristic biomarkers of osteoclasts were seen. TRAP, CTSK, and MMP-9 were significantly upregulated in osteoclasts. Similar to the results from LC-MS/MS, ANXA1 and histone H4 were upregulated in osteoclasts compared to RAW 264.7 cells

identify and classify the differentially expressed proteins. Of the 549 upregulated proteins, 548 proteins were successfully mapped to the database. In the context of cellular component, 540 of the 541 upregulated proteins and 8 downregulated proteins matched with the database. Upregulated proteins were mostly enriched in mitochondria, mitochondrial inner membrane, mitochondrial matrix, extracellular exosome, myelin sheath and mitochondrial respiratory chain complex I (Fig. 3a). Downregulated proteins were enriched in the Rad6-Rad18 complex, GINS complex, replication fork protection complex, clathrin coat of coated pit, microtubule plus-end and XY body (Fig. 4a). These findings indicated that proteomic changes mainly occurred in mitochondria when RAW

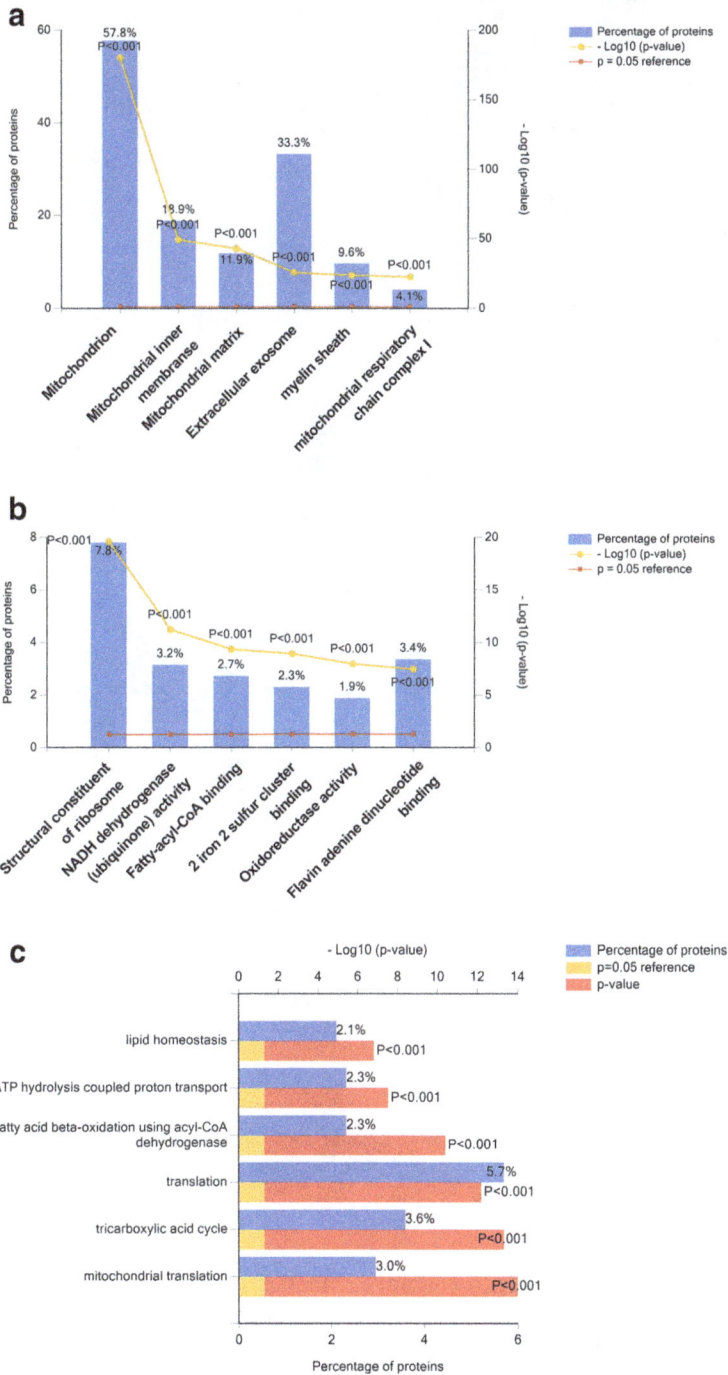

Fig. 3 Identification and classification of the upregulated proteins. **a** Column graph of cellular component enriched in upregulated proteins. **b** Column graph of molecular function enriched in upregulated proteins. **c** Bar graph of biological process enriched in upregulated proteins

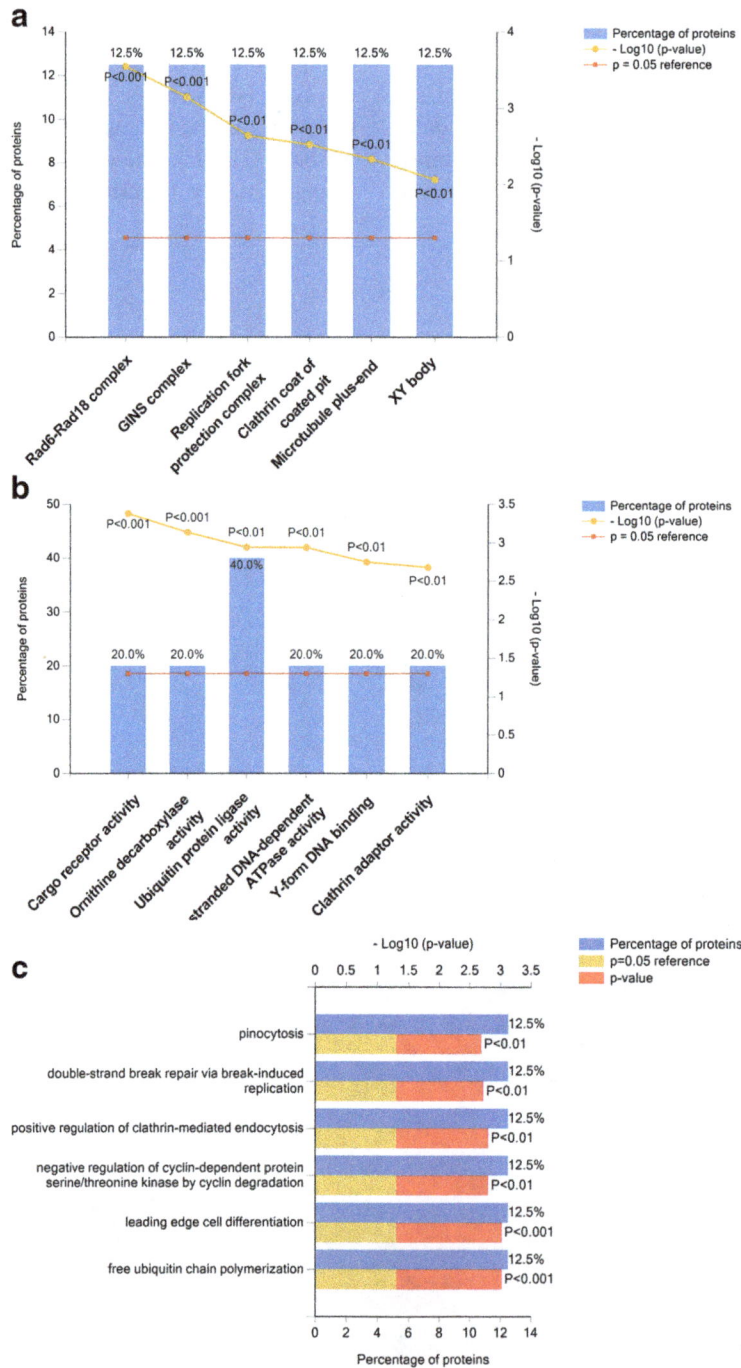

Fig. 4 Identification and classification of the downregulated proteins. **a** Column graph of cellular component enriched in downregulated proteins. **b** Column graph of molecular function enriched in downregulated proteins. **c** Bar graph of biological process enriched in downregulated proteins

264.7 cells differentiated into osteoclasts in the low serum culture system.

According to the molecular function analysis, 482 upregulated proteins and 7 downregulated proteins matched with the database. Our results revealed that upregulated proteins were mainly enriched in molecular functions of structural constituent of ribosome, NADH dehydrogenase (ubiquinone) activity, fatty-acyl-CoA binding, 2 iron 2 sulfur cluster binding, oxidoreductase activity and Flavin adenine dinucleotide binding (Fig. 3b). Downregulated proteins were enriched in cargo receptor activity, ornithine decarboxylase activity, ubiquitin protein ligase activity, single-stranded DNA-dependent ATPase activity, Y-form DNA binding, and clathrin adaptor activity and etc. (Fig. 4b).

In the context of biological process, upregulated proteins were mainly enriched in lipid homeostasis, ATP hydrolysis coupled proton transport, fatty acid beta-oxidation using acyl-CoA dehydrogenase, translation, tricarboxylic acid (TCA) cycle (also termed citric acid cycle) and mitochondrial translation (Fig. 3c). By contrast, downregulated proteins were mainly involved in biological processes of pinocytosis, double-strand break repair via break-induced replication (Fig. 4c). These findings indicated that the differentially expressed proteins were mainly involved in mitochondria and energy metabolism during osteoclastogenesis in the low serum culture system.

Functional interaction network analysis of the differentially expressed proteins

To investigate functional interaction network of the differentially expressed proteins, the ClueGO Cytoscape plugin was used. Of the 549 differentially expressed proteins, 548 proteins were successfully mapped to the REACTOME database, and one protein (Accession Number: B9EJ86) failed to map to the database. Our results revealed that the differentially expressed proteins mainly participated in

mitochondrial fatty acid beta-oxidation, mitochondrial translation, fatty acyl-CoA biosynthesis, the TCA cycle as well as respiratory election transport, among others (Fig. 5). These findings again indicated that the differentially expressed proteins were implicated in mitochondrial activities and energy metabolism activities.

Molecular pathways analysis of the differentially expressed proteins

To further investigate the interaction networks of the differentially expressed proteins, we used WebGestalt online toolkit to perform pathway enrichment analysis based on the Wikipathways database, and used the Wikipathways Cytoscape plugin to visualize the changes in the molecular pathways. Consistent with functional interaction network analysis, molecular pathway enrichment analysis demonstrated that electron transport chain pathway, TCA cycle pathway, mitochondrial LC-fatty acid beta-oxidation pathway and fatty acid biosynthesis pathway were significantly altered (Additional file 2: Table S2). In the electron transport chain, expression of 45 matched proteins were upregulated; these proteins were involved in mitochondrial complex I (NADH-Ubiquinone oxidoreductase complex), the ubiquinol-cytochrome C reductase complex and ATP synthase F1 complex (Fig. 6). In the TCA cycle pathway, 19 identified proteins were

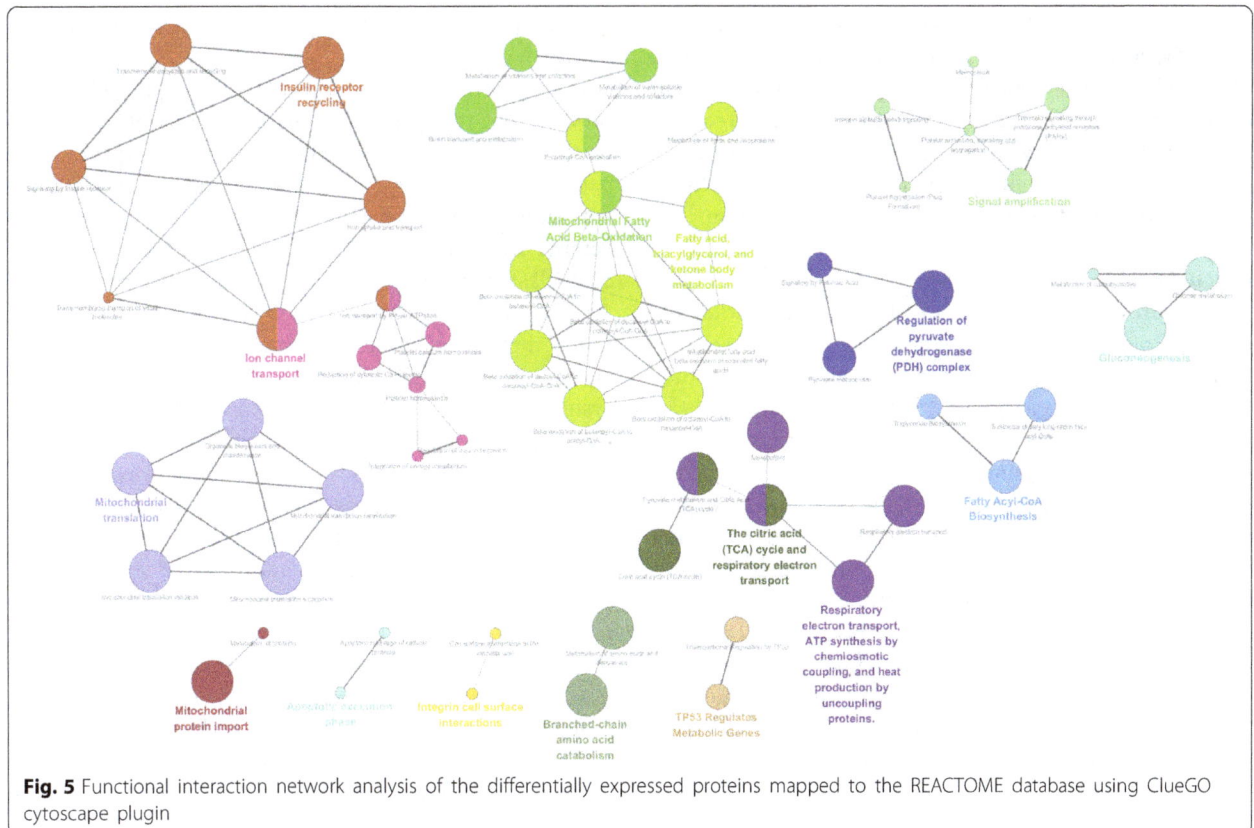

Fig. 5 Functional interaction network analysis of the differentially expressed proteins mapped to the REACTOME database using ClueGO cytoscape plugin

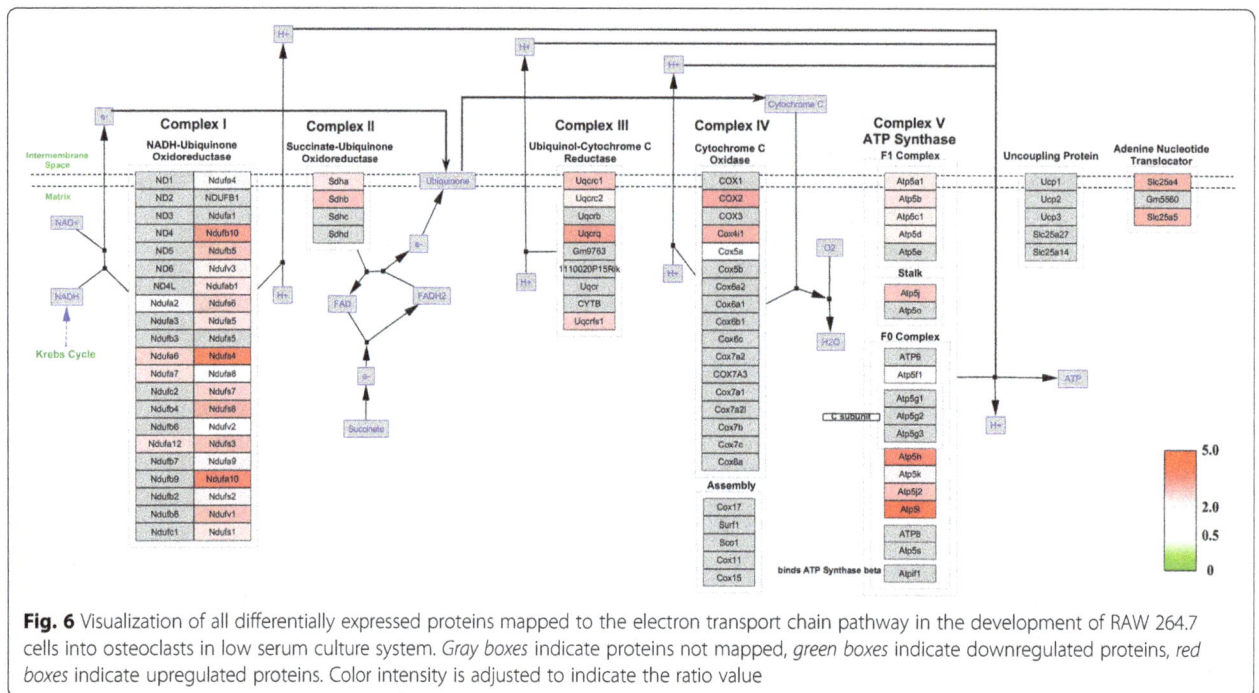

Fig. 6 Visualization of all differentially expressed proteins mapped to the electron transport chain pathway in the development of RAW 264.7 cells into osteoclasts in low serum culture system. *Gray boxes* indicate proteins not mapped, *green boxes* indicate downregulated proteins, *red boxes* indicate upregulated proteins. Color intensity is adjusted to indicate the ratio value

upregulated (Fig. 7). In the mitochondrial LC-fatty acid beta-oxidation pathway, our results showed that seven proteins, including ACADS, ACADVL, HADHA, HADH, ACADM, CPT2 and ACSL3, were upregulated, especially those involved in saturated fatty acid (Fig. 8). Similarly, in the fatty acid biosynthesis pathway, all eight identified proteins were upregulated, especially in mitochondrion (Fig. 9). Together, our results suggested that mitochondrial activity and energy metabolism were remarkably upregulated in the differentiation of RAW 264.7 cells into osteoclasts in the low serum culture system.

Discussion

In the present study, we investigated the proteomic changes during osteoclastogenesis in medium supplemented with 1 % FBS. Consistent with our previous study [13], our results confirmed that large TRAP-positive multinucleated osteoclasts with bone resorbing capacity were successfully obtained by this culturing procedure, validated by upregulation of 15 characteristic marker proteins, including TRAP, CTSK, MMP9, V-ATPase and ITGAV, three of which were also confirmed by western blot analysis. Previous study found 867 proteins (492 down-regulated proteins and 375 upregulated proteins) altered between osteoclasts and RAW264.7 cells [14], while our study found 549 proteins (541 upregulated proteins and 8 downregulated proteins) expressed differentially during osteoclastogenesis in the low serum culture system, and almost all the differentially expressed proteins were significantly upregulated. Integrated bioinformatics

analysis indicated that these differentially expressed proteins were mainly involved in mitochondrial activities and energy metabolism, including the electron transport chain pathway, TCA cycle pathway, mitochondrial LC-fatty acid beta-oxidation pathway and fatty acid biosynthesis pathway.

The electron transport chain is a complex biological process that transfers electrons from electron donors to electron acceptors [18]. The electron transport chain pathway is responsible for the synthesis of ATP, the most commonly consumed chemical energy utilized in a diversity array of cellular biological activities, including osteoclast formation [14, 19-21]. In eukaryotic cells, ATP is mainly generated in mitochondria. An et al. found that mitochondrial changes were critical in osteoclastogenesis in the conventional 10 % serum culture system [14]. Moreover, our previous study indicated that the expression of the electron transported chain in RAW264.7 cultured in low serum system was downregulated [13]. In this study, our results showed that the differentially expressed proteins were mainly located in mitochondria, suggesting changes in electron transported chain in osteoclasts formed in the low serum culture system. Similar to our results, Morten et al. found increased mitochondrial electron transport chain activity in the differentiation of human CD14 positive monocytes differentiating into osteoclasts under hypoxia conditions [22]. Moreover, Jin et al. reported that a mitochondrial complex 1 subunit Ndufs4 deletion caused systemic inflammation and osteopetrosis, and suggested that

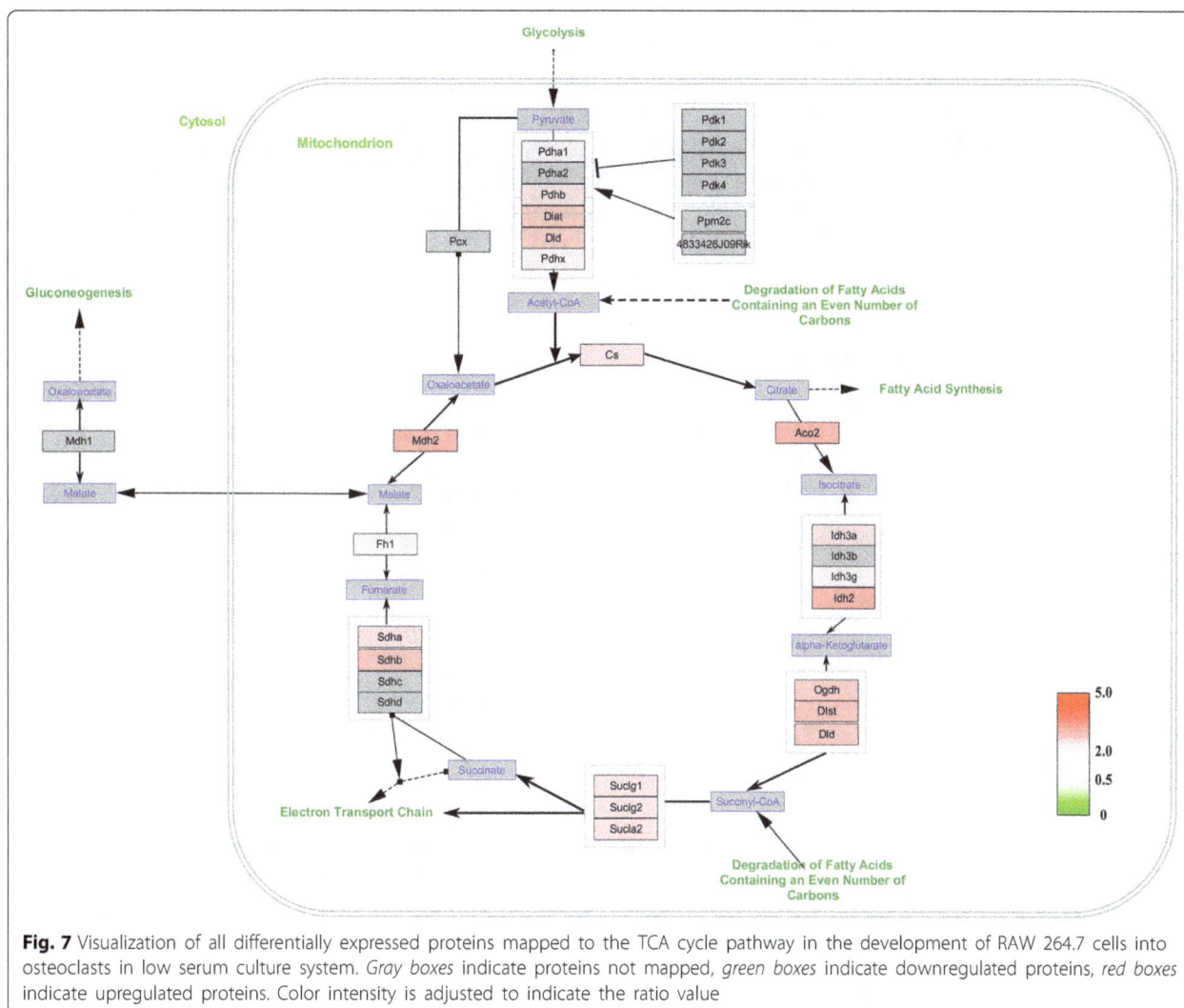

Fig. 7 Visualization of all differentially expressed proteins mapped to the TCA cycle pathway in the development of RAW 264.7 cells into osteoclasts in low serum culture system. *Gray boxes* indicate proteins not mapped, *green boxes* indicate downregulated proteins, *red boxes* indicate upregulated proteins. Color intensity is adjusted to indicate the ratio value

mitochondrial complex I promoted osteoclast differentiation, while inhibited macrophage activation [23]. In agreement with Jin et al., we found that mitochondrial complex I was activated in osteoclasts formed in low serum culture system. Taken together, these findings suggested that osteoclasts formation is an energy consuming procedure, regardless of the culture environmental condition.

ATP synthesis depends on the products of citric acid cycle (also termed TCA cycle) and fatty acid oxidation [24, 25]. Dodds et al. found that osteoclast formation was related to enhanced TCA cycle and increased fatty acid oxidation [26]. Our study supports this conclusion, as bioinformatics analysis showed differentially expressed proteins enriched in TCA cycle pathway and the mitochondrial LC-fatty acid beta-oxidation pathway. The TCA cycle is a biological process that oxidizes acetyl-CoA for ATP production [27]. Our results showed high activation of TCA cycle pathway in osteoclasts generated in medium supplemented with 1 % FBS. Fatty acids are carboxylic

acids, which have a long aliphatic tail [28]. Fatty acid oxidation mostly occurs in the mitochondria and is an energy productive biological procedure. Adamek et al. suggested that fatty acid oxidation played an important role in the energy metabolism of bone cells, including osteoblast and fibroblasts [29]. However, the role of fatty acid oxidation in osteoclasts has not been reported. In this study, our results demonstrated that differentially expressed proteins from the fatty acid oxidation pathway upregulated significantly, indicating a role of fatty acid oxidation in the development of RAW 264.7 cells into osteoclasts in the low serum culture system. Again, these findings imply that formation and bone resorption of osteoclast require high ATP production.

Since both the TCA cycle and fatty acid oxidation are associated with fatty acid biosynthesis, we also analyzed alternations of fatty acid biosynthesis pathway in osteoclastogenesis in the low serum culture system. A previous study by Cornish et al. indicated that saturated fatty acid suppressed osteoclast formation in conventional

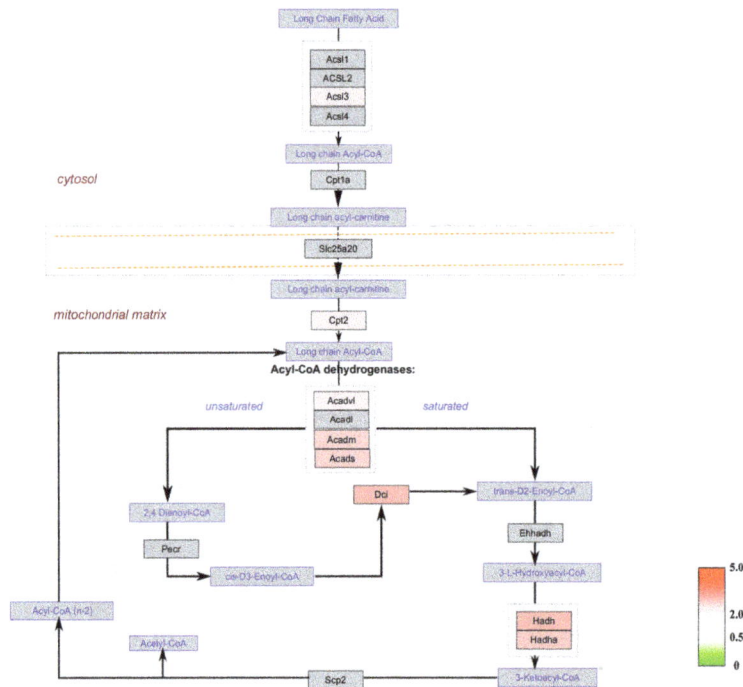

Fig. 8 Visualization of all differentially expressed proteins mapped to the mitochondrial LC-fatty acid beta-oxidation pathway in the development of RAW 264.7 cells into osteoclasts in low serum culture system. *Gray boxes* indicate proteins not mapped, *green boxes* indicate downregulated proteins, *red boxes* indicate upregulated proteins. Color intensity is adjusted to indicate the ratio value

culture system [30]. Moreover, Kim et al. suggested that a medium-chain fatty acid suppressed osteoclastogenesis by inhibiting RANKL-induced IκBα phosphorylation, p65 nuclear translocation, and NF-κB transcriptional activity. They also found that NFATc1 was inhibited by medium-chain fatty acid in osteoclastogenesis in conventional culture system [31]. In contrast, we found that fatty acid biosynthesis, mainly of saturated fatty acid, was upregulated in mitochondrion of osteoclasts formed in the low serum culture system. This difference may be because the osteoclasts were cultured in medium supplemented with low serum in our study. Therefore, our study suggested a different process of fatty acid biosynthesis in osteoclasts formed in low serum culture system from those in conventional culture system.

Additionally, cells are always pre-treated with serum free medium for 24 h or less before the start of the experiments to enhance the sensitivity of cells to drugs. We hypothesize that there might be differences between the proteomes of osteoclasts cultured in low serum system and a serum-free system for only 24 h. However, to the best of our knowledge, the effect of serum-free treatment on the osteoclast proteome has not been reported. Therefore, the comparison of effects of continuous low serum and transient serum free on osteoclasts cannot be done. Larsson et al. previously investigated the effects

of short exposure to serum-free medium on 3 T3 cells proliferation cycle, and suggested that 3T3 cells are very sensitive to serum depletion during the first part of G1; only a short exposure to serum-free medium is sufficient for the cells to leave the cell cycle [32]. However, the effects of serum-free pretreatment on osteoclast differentiation has not been reported.

Conclusion
In conclusion, our study confirmed that osteoclasts could be obtained in low serum culture system (1 % FBS), and demonstrated that osteoclast formation is an ATP consuming process, no matter if cells are cultured in a low serum culture system or a conventional culture system. Like osteoclast formation in the conventional culture system, the proteomic changes in osteoclastogenesis in the low serum culture system mainly involve mitochondrial activity and energy metabolism. Notably, our results showed that the fatty acid biosynthesis pathway was upregulated in osteoclasts cultured in low serum condition, the molecular mechanisms of which need further investigation.

Methods
RAW264.7 cell cultivation and osteoclastogenesis
RAW 264.7 cells (obtained from the Chinese Academy of Medical Sciences (Beijing, China)) were cultured at a

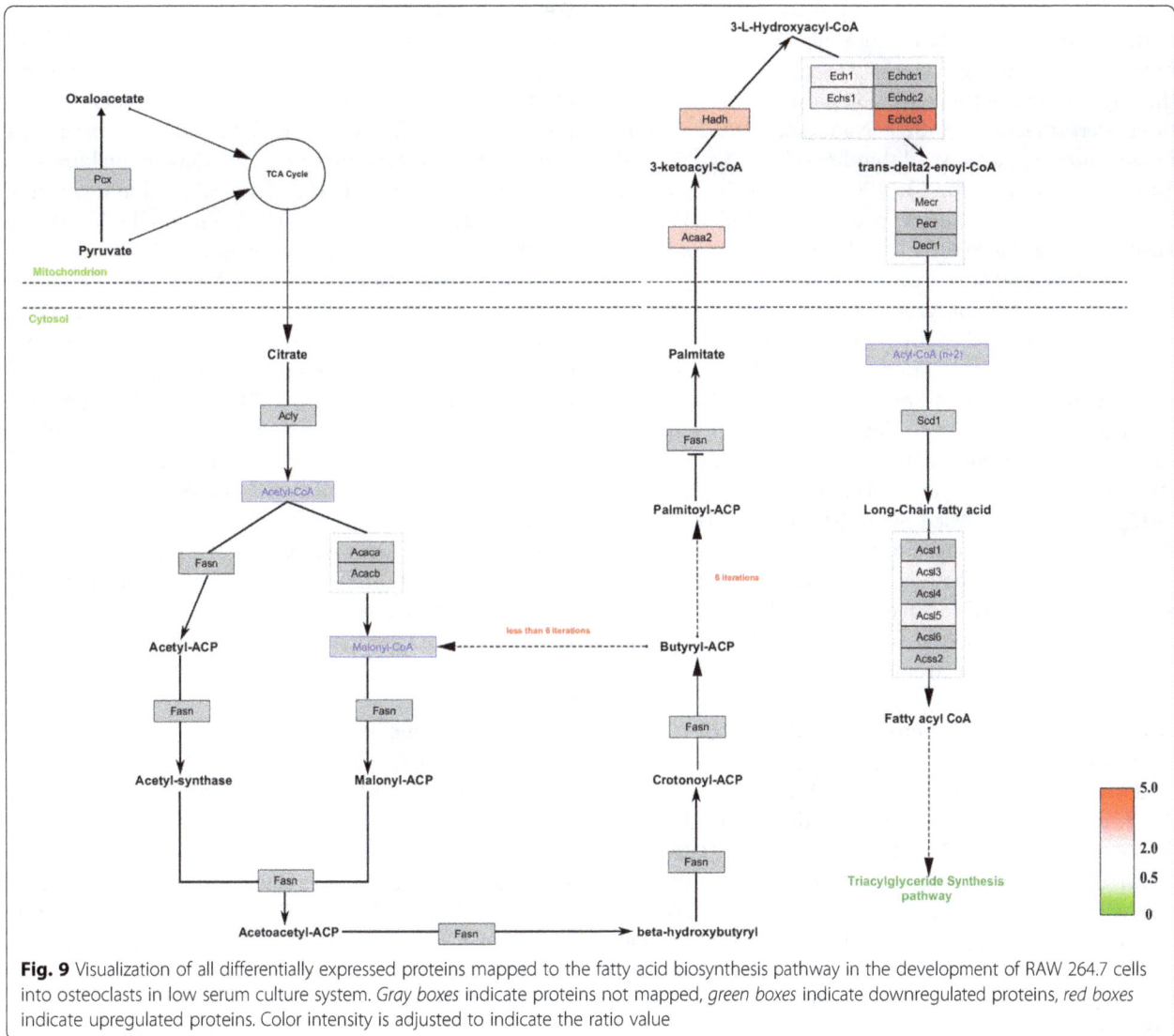

Fig. 9 Visualization of all differentially expressed proteins mapped to the fatty acid biosynthesis pathway in the development of RAW 264.7 cells into osteoclasts in low serum culture system. *Gray boxes* indicate proteins not mapped, *green boxes* indicate downregulated proteins, *red boxes* indicate upregulated proteins. Color intensity is adjusted to indicate the ratio value

density of 1.5×10^5 cells/ml in 6 well plates. The cells were divided into two groups. The cells were cultured with or without 30 ng/ml RANKL (462-TEC-010, R&D Systems) in α-MEM (11095–080, Life Technologies) supplemented with 1 % (v/v) FBS (26140079, Life Technologies) at 37 °C in a 5 % CO_2 incubator. The medium was refreshed every other day. Both sets of experiments were run in triplicated independently. On Day 5, RAW 264.7 cells and osteoclasts were harvested and pooled. Subsequently, the pooled cells were used for proteomic analysis.

TRAP staining and bone resorption assay

TRAP staining and bone resorption assays were performed to confirm the formation of mature osteoclasts as described previously [33]. Mature osteoclasts were defined as TRAP-positive cells containing three or more nuclei. TRAP staining was conducted using the Acid Phosphatase Leukocyte kit (387–1, Sigma Aldrich) according to the manufacturer's protocol. Briefly, cells were rinsed three times with cold PBS. Then, cells were fixed with Fixative Solution (Citrate Solution, acetone and 37 % formaldehyde) and rinsed thoroughly. Cells were incubated in staining solution for 1 h in a 37 °C water bath protected from light.

The Osteo Assay Surface Plate (3987, Corning) was used to evaluate the bone resorbing capacity of osteoclasts. On Day 5, following the aspiration of the medium, 100 μl of 10 % bleach solution was added to each well and incubated for 5 min. Toluidine blue staining was conducted to enhance the contrast for bone resorbing pit image analysis.

Sample preparation and TMT labeling

Sample preparation and labeling were performed as described by Xiong et al. [34]. In brief, cells were washed three times with cold PBS, and then lysed with lysis buffer (8 M urea in PBS, 1 × cocktail, 1 mM PMSF). Cell lysates were centrifuged at 16,000 × g for 10 min at 4 °C, the supernatants were collected, and protein concentrations were measured with a Nanodrop2000. Following incubation in 10 mM dithiothreitol (17131801, GE Healthcare) at 50 °C for 1 h, 100 μg of protein was incubated with 25 mM indole acetic acid (RPN6302, GE Healthcare) in the dark for 2 h. Then, trypsin/Lys-C Mix (V5072, Promega) was used to digest proteins overnight at 37 °C at a protein/protease ratio of 25:1, and the reaction was quenched by heating at 60 °C. Proteins digests were desalted, dried and solved in 200 mM triethylammonium bicarbonate buffer. TMT Isobaric Label Reagent Set (90061, Thermo Scientific) was used to label proteins according to the manufacturer's instructions. Proteins extracted from osteoclasts and RAW 264.7 cells were labeled with 0.8 mg TMT6-128 or TMT6-127, respectively. Equal amounts of the labeled protein digests from both groups were pooled, dried and solved in 0.1 % trifluoroacetic acid. Then the protein digests were finally solved in 100 μl of 0.1 % trifluoroacetic acid for mass spectrometry (MS) analysis.

High-performance liquid chromatography (HPLC)

Fractionation of pooled protein digests was performed as described by van Ulsen et al. [35]. Briefly, the pooled TMT labeled protein digests were dissolved in 100 μl 0.1 % formic acid for HPLC analysis (UltiMate 3000 UHPLC, Thermo Scientific) using an Xbridge BEH300 C18 column (4.6 × 250 mm^2, 5 μm, 300 Å, Waters). Fifty fractions were collected at 1.5 min intervals. The fractions were dried in a vacuum concentrator. Then, 20 μl 0.1 % FA was used to dissolve the fractions for LC-MS/MS analysis.

LC-MS/MS analysis

LC-MS/MS was conducted using a Q Exactive mass spectrometer. The protein digests were separated using a 120 min gradient elution at a flow rate of 0.3 μl/min using the UltiMate 3000 RSLCano System (Thermo Scientific). A directly interfaced Q Exactive Hybrid Quadrupole-Orbitrap Mass Spectrometer (Thermo Scientific) was used to analyze the protein digests. We use a home-made fused silica capillary column (75 μm × 150 mm, Upchurch, Oak Harbor, WA, USA) packed with C18 resin (300 Å, 5 μm, Varian Lexington, MA, USA) as the analytical column. Xcalibur 2.1.2 software was used with the Q Exactive mass spectrometer in data-dependent acquisition mode. Ten data-dependent MS/MS scans at 27 % normalized collision energy were performed. Thereafter, a single full-scan mass spectrum in Orbitrap (400–1,800 m/z, 60, 000 resolution) was conducted.

Western blot analysis

To validate the LC-MS/MS data and to confirm the formation of osteoclasts, western blot analysis of three characteristic osteoclast biomarkers and two other altered proteins was performed according to standard procedure with minor modifications. Equal amounts of total proteins of RAW264.7 cells and osteoclasts cultured in low serum (20 μg) were separated by SDS-PAGE on 12 % gel and transferred to nitrocellulose membranes. Membranes were blocked at room temperature for 1 h in Tris Buffer Saline with Tween 20 (TBST) with 5 % nonfat milk. Then, membranes were incubated with anti-TRAP (sc-28204), anti-CTSK (ab19027), anti-histone H4 (ab10158), anti-MMP-9 (ab137867), anti-ANXA1 (sc-12740) and anti-beta tubulin (internal control) antibody (KM9003) at 4 °C overnight. After washing in TBST for 15 min, membranes were incubated with goat anti-rabbit horseradish peroxidase (HRP)-conjugated IgG for 1 h at room temperature. Membranes were then washed three times in TBST and bands were visualized with ECL detection kits (GE Healthcare, RPN2209) according to the manufacturer's instructions.

Data analysis

Thermo Scientific Proteome Discoverer software suite 1.4 with the SEQUEST search engine and the mouse FASTA database from UniProt (released on October 16th, 2015) was used to analyze LC-MS/MS data. In the SEQUEST search engine, full trypsin specificity was selected, two missed cleavages were allowed, carbamidomethylation (C) and TMT 6-plex (K and peptide N-terminal) were set as the static modification, oxidation (M) was set as the dynamic modification, precursor ion mass tolerances were set at 20 ppm for all MS data acquired using an Orbitrap mass analyzer, and the fragment ion mass tolerance was set as 20 mmu for all MS/MS spectra acquired. Two or more unique peptides per protein had to be identified to list the protein as a hit. Proteins that scored 10 or more were selected for subsequent bioinformatics analysis. The ratio values of proteins labeled with TMT6-128 and TMT6-127 were adjusted using the beta tubulin ratio value as internal control. The thresholds for downregulation and upregulation were set at 0.5 and 2.0 respectively.

FunRich software (version 2.1.2) (www.funrich.org) was used to classify the proteins using UniProt Database (released on July 21st, 2015). The ClueGO cytoscape plugin (version 2.1.7) was used to analyze the functional interaction networks of the differentially expressed proteins. The REACTOME ontology database (released on May 5th, 2015) was used. Two-sided hypergeometric tests with Benjamini-Hochberg correction method were performed to minimize the false discovery rate. Pathways with P value of 0.01 or less were considered as significance.

The WebGestalt online toolkit (http://bioinfo.vanderbilt.edu/webgestalt/) was used to run pathway enrichment analysis and the significance level was set at 0.0001. Cytoscape software (version 3.1.1) with Wikipathways plugin was used to visualize the protein-protein interactions matching the Wikipathways database. The mass spectrometry proteomics data have been deposited to the ProteomeXchange Consortium via the PRIDE partner repository with the data identifier PXD001935 [36].

Competing interests
The authors declare that they have no competing interests.

Authors' contributions
Study design: PFT and WG. Study Conduct: QX and SHZ. Data analysis: QX, SHZ and LHZ. Data interpretation: QX and LHZ. Draft manuscript: QX and LHZ. WG and PFT take full responsibility for the integrity of the data analysis. All authors read and approved the final manuscript.

Acknowledgements
Q. Xiong, P-F Tang and L-H Zhang are supported by the National Natural Science Foundation of China (81550012). W. Ge and S-H Zhan are supported by the National Natural Science Foundation of China (81373150).

References

1. Wolf JM, Cannada LK, Lane JM, Sawyer AJ, Ladd AL. A comprehensive overview of osteoporotic fracture treatment. Instr Course Lect. 2015;64:25–36.
2. Dawson-Hughes B, Looker AC, Tosteson AN, Johansson H, Kanis JA, Melton 3rd LJ. The potential impact of the National Osteoporosis Foundation guidance on treatment eligibility in the USA: an update in NHANES 2005–2008. Osteoporos Int. 2012;23:811–20.
3. Cummings SR, Melton LJ. Epidemiology and outcomes of osteoporotic fractures. Lancet. 2002;359:1761–7.
4. Burge R, Dawson-Hughes B, Solomon DH, Wong JB, King A, Tosteson A. Incidence and economic burden of osteoporosis-related fractures in the United States, 2005–2025. J Bone Miner Res. 2007;22:465–75.
5. Feng X, McDonald JM. Disorders of bone remodeling. Annu Rev Pathol. 2011;6:121–45.
6. Udagawa N, Takahashi N, Akatsu T, Tanaka H, Sasaki T, Nishihara T, Koga T, Martin TJ, Suda T. Origin of osteoclasts: mature monocytes and macrophages are capable of differentiating into osteoclasts under a suitable microenvironment prepared by bone marrow-derived stromal cells. Proc Natl Acad Sci U S A. 1990;87:7260–4.
7. Boyce BF. Advances in osteoclast biology reveal potential new drug targets and new roles for osteoclasts. J Bone Miner Res. 2013;28:711–22.
8. Kobayashi N, Kadono Y, Naito A, Matsumoto K, Yamamoto T, Tanaka S, Inoue J. Segregation of TRAF6-mediated signaling pathways clarifies its role in osteoclastogenesis. EMBO J. 2001;20:1271–80.
9. Kim JH, Kim N. Regulation of NFATc1 in Osteoclast Differentiation. J Bone Metab. 2014;21:233–41.
10. Takayanagi H. Osteoimmunology: shared mechanisms and crosstalk between the immune and bone systems. Nat Rev Immunol. 2007;7:292–304.
11. Hu R, Sharma SM, Bronisz A, Srinivasan R, Sankar U, Ostrowski MC. Eos, MITF, and PU.1 recruit corepressors to osteoclast-specific genes in committed myeloid progenitors. Mol Cell Biol. 2007;27:4018–27.
12. Vincent C, Kogawa M, Findlay DM, Atkins GJ. The generation of osteoclasts from RAW 264.7 precursors in defined, serum-free conditions. J Bone Miner Metab. 2009;27:114–9.
13. Xiong Q, Zhang L, Xin L, Gao Y, Peng Y, Tang P, Ge W. Proteomic study of different culture medium serum volume fractions on RANKL-dependent RAW2647 cells differentiating into osteoclasts. Proteome Sci. 2015;13:16.
14. An E, Narayanan M, Manes NP, Nita-Lazar A. Characterization of functional reprogramming during osteoclast development using quantitative proteomics and mRNA profiling. Mol Cell Proteomics. 2014;13:2687–704.
15. Gallois A, Lachuer J, Yvert G, Wierinckx A, Brunet F, Rabourdin-Combe C, Delprat C, Jurdic P, Mazzorana M. Genome-wide expression analyses establish dendritic cells as a new osteoclast precursor able to generate bone-resorbing cells more efficiently than monocytes. J Bone Miner Res. 2010;25:661–72.
16. Qin A, Cheng TS, Pavlos NJ, Lin Z, Dai KR, Zheng MH. V-ATPases in osteoclasts: structure, function and potential inhibitors of bone resorption. Int J Biochem Cell Biol. 2012;44:1422–35.
17. Boyle WJ, Simonet WS, Lacey DL. Osteoclast differentiation and activation. Nature. 2003;423:337–42.
18. Cordes M, Giese B. Electron transfer in peptides and proteins. Chem Soc Rev. 2009;38:892–901.
19. Pontes MH, Sevostyanova A, Groisman EA. When Too Much ATP Is Bad for Protein Synthesis. J Mol Biol. 2015;427:2586–94.
20. Morrison MS, Turin L, King BF, Burnstock G, Arnett TR. ATP is a potent stimulator of the activation and formation of rodent osteoclasts. J Physiol. 1998;511(Pt 2):495–500.
21. Kim JM, Jeong D, Kang HK, Jung SY, Kang SS, Min BM. Osteoclast precursors display dynamic metabolic shifts toward accelerated glucose metabolism at an early stage of RANKL-stimulated osteoclast differentiation. Cell Physiol Biochem. 2007;20:935–46.
22. Morten KJ, Badder L, Knowles HJ. Differential regulation of HIF-mediated pathways increases mitochondrial metabolism and ATP production in hypoxic osteoclasts. J Pathol. 2013;229:755–64.
23. Jin Z, Wei W, Yang M, Du Y, Wan Y. Mitochondrial complex I activity suppresses inflammation and enhances bone resorption by shifting macrophage-osteoclast polarization. Cell Metab. 2014;20:483–98.
24. Fernie AR, Carrari F, Sweetlove LJ. Respiratory metabolism: glycolysis, the TCA cycle and mitochondrial electron transport. Curr Opin Plant Biol. 2004;7:254–61.
25. Senior AE. ATP synthesis by oxidative phosphorylation. Physiol Rev. 1988;68:177–231.
26. Dodds RA, Gowen M, Bradbeer JN. Microcytophotometric analysis of human osteoclast metabolism: lack of activity in certain oxidative pathways indicates inability to sustain biosynthesis during resorption. J Histochem Cytochem. 1994;42:599–606.
27. Saddik M, Gamble J, Witters LA, Lopaschuk GD. Acetyl-CoA carboxylase regulation of fatty acid oxidation in the heart. J Biol Chem. 1993;268:25836–45.
28. Layden BT, Angueira AR, Brodsky M, Durai V, Lowe WL. Short chain fatty acids and their receptors: new metabolic targets. Transl Res. 2013;161:131–40.
29 Adamek G, Felix R, Guenther HL, Fleisch H. Fatty acid oxidation in bone tissue and bone cells in culture. Characterization and hormonal influences. Biochem J. 1987;248:129–37.
30. Cornish J, MacGibbon A, Lin JM, Watson M, Callon KE, Tong PC, Dunford JE, van der Does Y, Williams GA, Grey AB, et al. Modulation of osteoclastogenesis by fatty acids. Endocrinology. 2008;149:5688–95.
31. Kim HJ, Yoon HJ, Kim SY, Yoon YR. A medium-chain fatty acid, capric acid, inhibits RANKL-induced osteoclast differentiation via the suppression of NF-kappaB signaling and blocks cytoskeletal organization and survival in mature osteoclasts. Mol Cells. 2014;37:598–604.
32. Larsson O, Zetterberg A, Engstrom W. Consequences of parental exposure to serum-free medium for progeny cell division. J Cell Sci. 1985;75:259–68.
33. Mozar A, Haren N, Chasseraud M, Louvet L, Maziere C, Wattel A, Mentaverri R, Morliere P, Kamel S, Brazier M, et al. High extracellular inorganic phosphate concentration inhibits RANK-RANKL signaling in osteoclast-like cells. J Cell Physiol. 2008;215:47–54.
34. Xiong L, Darwanto A, Sharma S, Herring J, Hu S, Filippova M, Filippov V, Wang Y, Chen CS, Duerksen-Hughes PJ, et al. Mass spectrometric studies on epigenetic interaction networks in cell differentiation. J Biol Chem. 2011;286:13657–68.
35. van Ulsen P, Kuhn K, Prinz T, Legner H, Schmid P, Baumann C, Tommassen J. Identification of proteins of Neisseria meningitidis induced under iron-limiting conditions using the isobaric tandem mass tag (TMT) labeling approach. Proteomics. 2009;9:1771–81.
36. Vizcaino JA, Deutsch EW, Wang R, Csordas A, Reisinger F, Rios D, Dianes JA, Sun Z, Farrah T, Bandeira N, et al. ProteomeXchange provides globally coordinated proteomics data submission and dissemination. Nat Biotechnol. 2014;32:223–6.

6

Removal of SDS from biological protein digests for proteomic analysis by mass spectrometry

Soundharrajan Ilavenil[1†], Naif Abdullah Al-Dhabi[2†], Srisesharam Srigopalram[1], Young Ock Kim[3*], Paul Agastian[4], Rajasekhar Baaru[5], Ki Choon Choi[1*], Mariadhas Valan Arasu[2], Chun Geon Park[3] and Kyung Hun Park[3]

Abstract

Background: Metal-organic frameworks (MOFs - MIL-101) are the most exciting, high profiled developments in nanotechnology in the last ten years, and it attracted considerable attention owing to their uniform nanoporosity, large surface area, outer-surface modification and in-pore functionality for tailoring the chemical properties of the material for anchoring specific guest moieties. MOF's have been particularly highlighted for their excellent gas storage and separation properties. Recently biomolecules-based MOF's were used as nanoencapsulators for antitumor and antiretroviral controlled drug delivery studies. However, usage of MOF material for removal of ionic detergent-SDS from biological samples has not been reported to date. Here, first time we demonstrate its novel applications in biological sample preparation for mass spectrometry analysis.

Methods: SDS removal using MIL-101 was assessed for proteomic analysis by mass spectrometry. We analysed removal of SDS from 0.5 % SDS solution alone, BSA mixture and HMEC cells lysate protein mixture. The removal of SDS by MIL-101 was confirmed by MALDI-TOF-MS and LC-MS techniques.

Results: In an initial demonstration, SDS has removed effectively from 0.5 % SDS solution by MIL-101via its binding attraction with SDS. Further, the experiment also confirmed that MIL-101 strongly removed the SDS from BSA and cell lysate mixtures.

Conclusions: These results suggest that SDS removal by the MIL-101 method is a practical, simple and broad applicable in proteomic sample processing for MALDI-TOF-MS and LC-MS analysis.

Keywords: MOFs, SDS removal, Biological sample, Proteomic analysis

Abbreviations: SDS, Sodium dodecyl sulfate; MOFs, Metal organic frameworks; HMEC, Human mammary epithelial cells; MALDI-TOF, Matrix assisted laser desorption/ionization-time of flight; BSA, Bovine serum albumin; DTT, Dithiothreitol; GO, Gene ontology; LC-MS, Liquid chromatography- mass spectrometry; FASP, Filter aided sample preparation); SPE, Solid phase extraction; CHAPS, 3-[(3-cholamidopropyl)dimethylammonio]-1-propanesulfonate

Background

Metals organic frameworks (MOF) are nano-porous compounds that contain metal ions or clusters that are connected by organic ligands to forms two or three dimensional structures. Last 10 years, MOFs are most potential compounds in hydrogen storage [1, 2], CO_2 capture [3, 4], catalysis [5, 6], sensing [7], and others [8]. It has highly porous nature that makes very attractive for catalysis applications. In addition, MOFs have large diversity in nature as compared to Zeolites due to the use of SiO4/AlO4 tetrahedral building units in the later materials. The catalytic applications of zeolites are more restricted to relatively small organic molecules (typically no larger than xylene) because due to its microporous in nature. Whereas, the size of pore, shape, dimensionality and chemical natures in MOFs have been controlled by

---

* Correspondence: kyo9128@korea.kr; choiwh@korea.kr
[†]Equal contributors
[3]Department of Medicinal Crop Research, Rural Development Administration, Eumseong, Chungbuk 369-873, Republic of Korea
[1]Grassland and Forage Division, National Institute of Animal Science, RDA, Seonghwan-Eup, Cheonan-Si, Chungnam 330801, Korea
Full list of author information is available at the end of the article

the suitable selection of their building blocks (metal and organic linker). Apart from these features, the lower acidity of the active centers in MOFs makes these materials even very attractive compared Zeolites (highly acidic centers). Also, MOFs may be changing the interactions of adsorbing reactants and the transition states or intermediaries formed inside the framework cavity between the host and guest.

The fundamental step of global proteomics experiments, particularly involving sensitive MS technique is efficient sample preparation. Extraction of total proteins from various biological sources including tissues and cultured cells using SDS-Sodium Dodecyl Sulfate, CHAPS, and Triton is well known in several biochemical studies. SDS is widely used and considered to be very beneficial due to complete cell lysis, disaggregation and efficient solubilization of the global proteome primarily hydrophobic membrane-bound components. SDS interacted with proteins by ionic and hydrophobic bonds and dissolves proteins by changing their secondary and tertiary structures [9]. Further, it plays an important role in studies of membrane proteins or aggregated proteins, because of these proteins are not soluble in other agents. In addition, SDS continuously used in protein separations from biological samples by sodium dodecyl sulfate-polyacrylamide gel electrophoresis (SDS-PAGE) method. Unfortunately, SDS in the samples can be unfavourable due to its unwanted effects in liquid chromatography and it produces large DS⁻ related signals and ion suppression effects [10, 11]. Therefore, SDS removal is an essential prerequisite for achieving higher peptide/protein coverage in the mass-spectrometric analysis. It can be accomplished by numerous techniques such as precipitation, strong cation exchange, protein and peptide level purification with pierce detergent removal cartridges and FASP II etc. [12]. Gu et al. 2011 [13] reported that the MOFs based material is very useful for biological applications. But, usage of MOFs material for removal of ionic detergent-SDS from biological samples has not been reported to date. Therefore, we tried to find out a novel method for removal of ionic detergent-SDS from biological samples by microdevices based MOFs material for proteomic analysis.

Results and discussion

Standard sample cleans up with different solid phase extraction (SPE) or desalting methods have not been much effective in SDS depletion to date; prompting the investigation of various methods for SDS removal will make several commercial products [14], Filter aided sample preparation (FASP) [15], high salt precipitation kits [16], and SDS specific binding spin cartridges are very famous methods for depletion of detergent from protein mixtures for proteomic sample preparation. However, a majority of techniques are hindered by low protein recovery, labour intensiveness, irreproducibility and incomplete SDS removal. To these address, we tried to find out a novel and efficient method for SDS removal. Here we used MOFs (MIL-101) as a binding material for SDS removal from the biological samples by micro- device based method. Our preliminary results were highly encouraged and demonstrated a novel biological application of MOF materials which could have significant value in sample processing i.e SDS depletion for MALDI-TOF and LC-MS analysis.

SDS removal from 0.5 % SDS solution using MIL-101

First, we tried to remove the SDS from 0.5 % SDS solution using MIL-101 as a binding material. Presently, this method has not been applied for removing SDS during proteomic analysis by the mass spectroscopy. Thus, we designed and used this method for SDS removable from biological protein mixture before MALDI-TOF-MS analysis. Initially, we removed the SDS from 0.5 % SDS solution with the help of MIL-101 material by centrifugation at 3000 rpm and the results exhibited that maximum concentration of SDS was removed as evidence the peak intensity of SDS at 287.89[M + H] was differed and clear in after removal of SDS (Fig. 1a & b). This result suggests that for the solution containing SDS can be removed by MIL-101 material.

SDS removal from BSA mixtures using MIL-101

Further, we analysed capability of MIL-101 for removal of SDS from bovine serum albumin. Different concentration of BSA was mixed with 0.5 % of SDS. Then BSA in SDS mixture was treated with MIL-101 for overnight and then centrifuged at 3000 rpm prior to MALDI-TOF-MS analysis. Generally, SDS acts as a potent surfactant that could be denatured the trypsin activity when digestion process. But, we observed good signalling intensity for BSA tryptic peptides. It indicated that MIL-101 has an ability to remove the SDS from single protein mixture with robust trypsin activity (Fig. 2a).

SDS removal from cell lysate protein mixture using MIL-101

Further, we planned to assess the weather MIL-101 have an ability to remove the SDS from cell lysate protein mixtures or not, because it is very complicated and little difficult to separate SDS from the biological samples as compared with single protein mixtures. So, we prepared protein extract from HMEC cells and mixed with 0.5 % boiling SDS. This mixture was incubated with 30 mg of MIL-101 slurry with end-over-end rotation overnight. Proteins were further separated from MIL-101 by centrifugation at 3000 rpm for 5 min. Supernatant containing protein mixture was processed for MALDI-TOF-MS

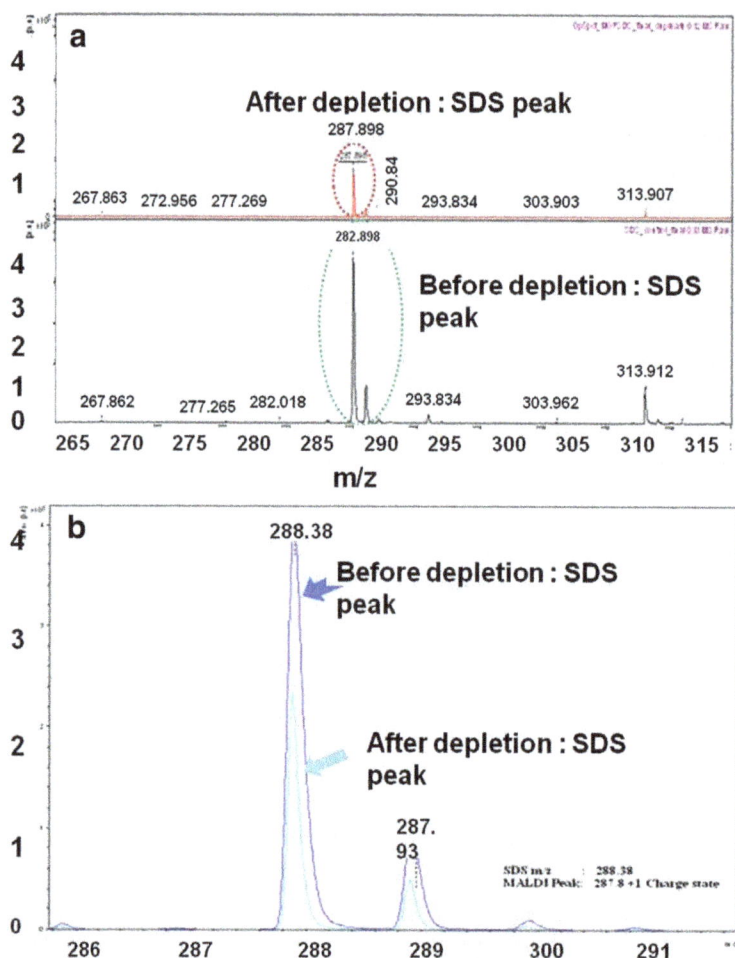

Fig. 1 a-b SDS m/z intensity, before and After SDS removal by MIL-101, data acquired on MALDI-TOF MS

analysis. This result suggests that MIL-101 has effectively removed SDS from cell lysate protein mixtures (Fig. 2b).

LC-MS analysis of tryptic peptides

Finally, the processed sample was subjected to LC-MS analysis. The result indicates that MIL-101 processed sample shows little ion suppression effect only. It may be due to trace amount of SDS was found in the samples. Overall MIL-101 effectively removed SDS from cell lysate by its binding efficiency (Fig. 3a). Figure 3b demonstrated the number of proteins with 1 or 2 unique peptide matches, Totally 750 protein were identified; among them 438 proteins having two peptide matches and 319 proteins having one peptide hits. Figure 3c demonstrated the GO (Gene Ontology) based annotations and cellular distribution of identified proteins. From our preliminary analysis, we hypothesize that MIL-101 is an attractive MOFs candidate for SDS removal from biological proteins before MS analyses.

Conclusions

The MIL-101 is an attractive MOFs candidate for removing SDS from biological samples through its electrostatic interactions. All the experimental results suggested that MIL-101 method could be a very useful method to prepare biological sample preparations for spectroscopy based proteomic analysis.

Methods

Chemical

BCA kit and HMEC cells were purchased from Thermo Fisher Scientific and American Type Culture Collection [Rockwille, MD, USA] respectively. SDS, Trypsin, BSA and DTT obtained from Sigma-Aldrich, USA. MIL-101 was from sigma product # 185361[Final concentration 100 µg/ml] and Microdevices obtained from Mobitec; product code: MobiSpin Column F (1.5 ml tubes).

SDS removal from 0.5 % SDS solution

MIL-101 (100 µg slurry) was mixed with 0.5 % SDS solution and incubated at room temperature for overnight.

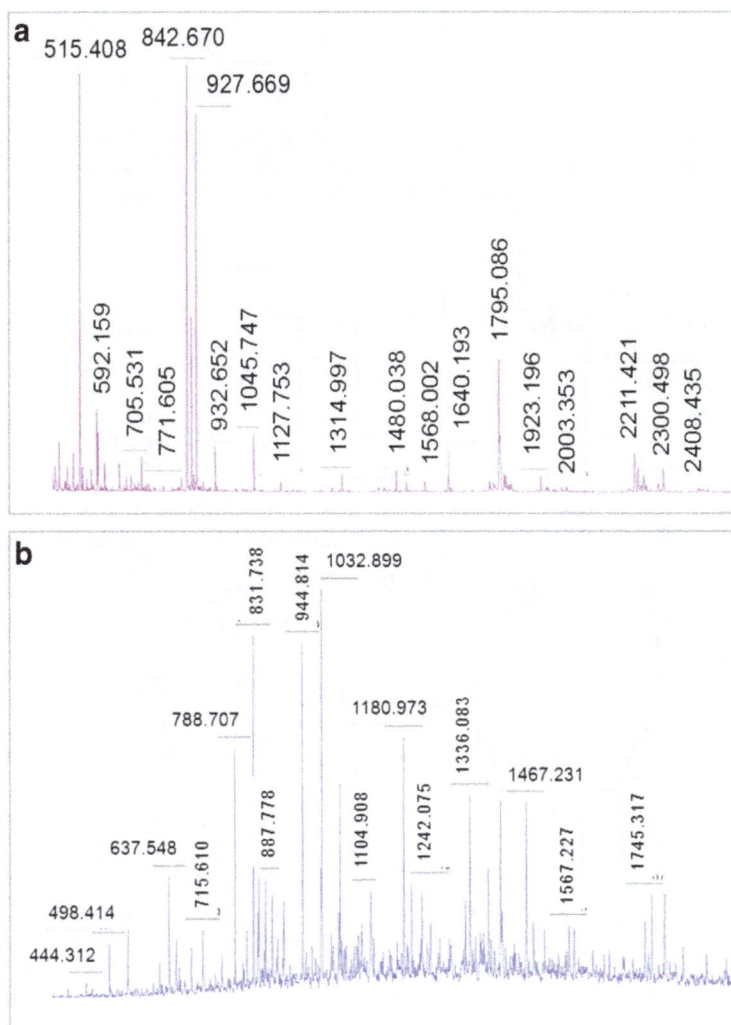

Fig. 2 a PMF profile of SDS removed BSA tryptic digest mixture using MALDI-TOF. **b** PMF profile of SDS depleted tryptic digest mixture using MALDI-TOF

Then, the sample was centrifuged at 3000 rpm for 5 min and then subjected to MALDI-TOF analysis.

SDS removal from BSA mixture

Different concentrations of BSA (5, 10, 25 μg) were mixed with the freshly prepared SDS (0.5 % SDS w/v) and was further incubated overnight with the slurry prepared using 100 μg MIL-101 with one ml of freshly prepared ice cold phosphate buffer and reduced with 10 mM DTT. Later, the sample was subjected to in-solution digestion using trypsin and peptides were desalted and then loaded into MALDI-TOF analysis.

SDS removal from cell lysate mixture

HMEC cells were lysed using 0.5 % boiling SDS and sonicated to clear the chromatin. Lysate protein was

estimated by BCA method and incubated with MIL-101 matrix (100 μg slurry) at room temperature for overnight to SDS attraction by the MOF and then separated by centrifugation. Protein mixture was in-solution digested, desalted and then subjected into MALDI-TOF and LCMS analysis. An outline of the work flow was illustrated in Fig. 4a-b.

MALDI-TOF analysis

Peptide mass fingerprinting (PMF) was conducted using MALDI-TOF-MS equipped with linear mode 20 KvA, laser shots 150 (337 nm, 50 H, N2 laser (Bruker Daltonics, Germany). For, each sample spectra required in the positive linear mode, and an average of 200 spectra that passed the accepted criteria of peak intensity was automatically selected and accumulated.

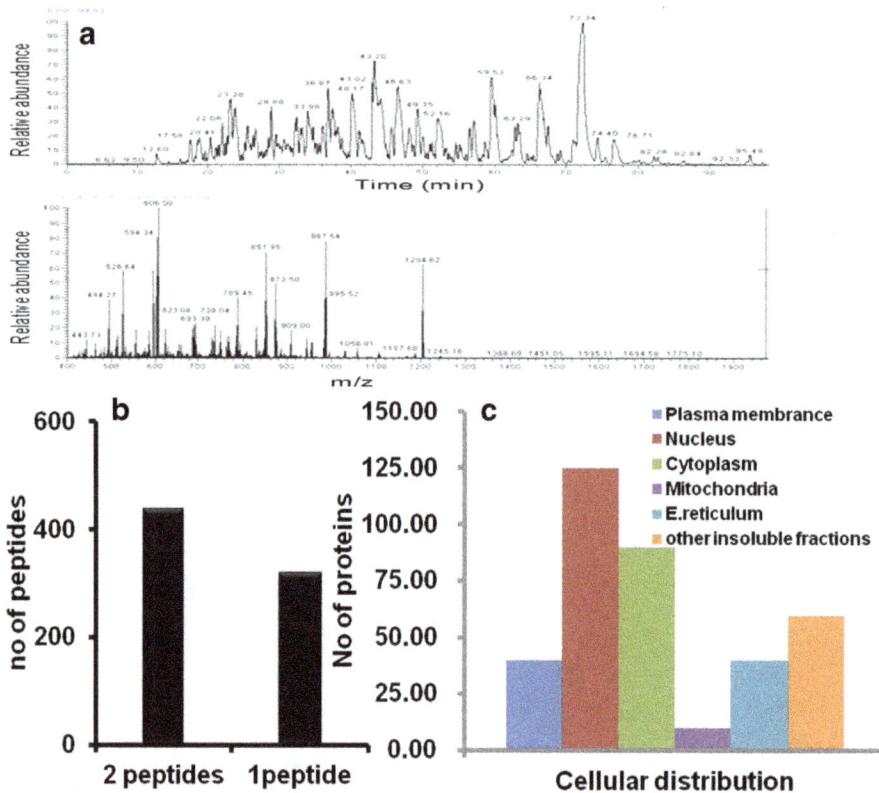

Fig. 3 a Liquid Chromatogram separations of SDS removed tryptic. **b** Number of proteins with 1 or 2 unique peptide matches. **c** GO based annotations and cellular distribution

Spectrum processing and data base searches were conducted b automatic mode with internal calibration using trypsin autolysis peaks (m/z 842.509 and m/z 2211.104). The fragmentation of selected peptide was measured using the PSD mode for MS analysis [17].

LC-MS analysis

The peptides were analysed using a reverse phase capillary column (LC system -LTQ; Thermo Scientific, San Jose, CA) prepared by slurry packing 3-µM Jupiter C18 bonded particles into a 65 cm long and 75 µM inner

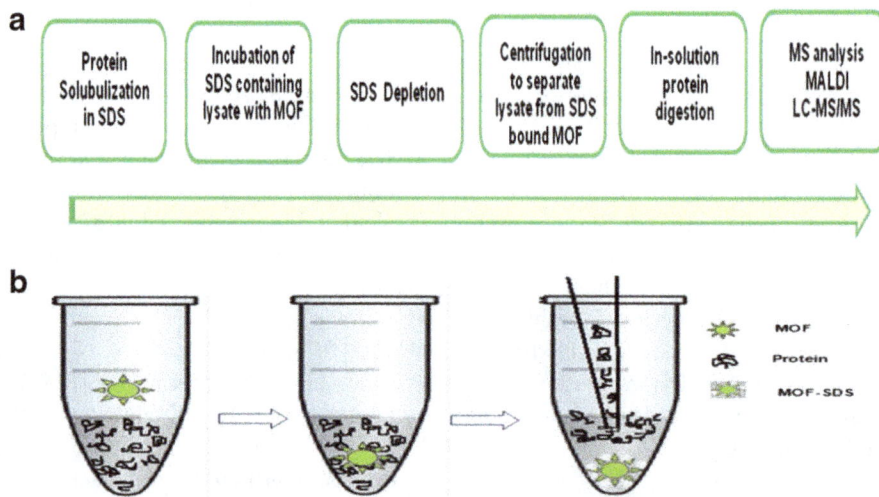

Fig. 4 a Work flow of SDS removal from a complex lysate mixture. Microdevices based **b** SDS removal using MIL-101, SDS was bound to MOF and proteins were separated by Centrifugation for downstream analysis

diameter fused silica capillary. 2.5 µg of peptides were loaded onto the column; the mobile phase was held at 100 % buffer A (0.1 % formic acid) for 20 min, followed by a linear gradient from 0 to 70 % buffer B(0.1 % formic acid in 90 % acetonitrile) for more than 85 min. Each full MS scan (m/z 400–2000) was followed by collision induced MS/MS spectra. The dynamic exclusion duration was set to 1 min; the heated capillary was maintained at 200 °C and the ESI voltage was held at 2.2 kV [18].

Acknowledgments
This study was supported by grants from the Cooperative Research Program for Agriculture Science & Technology Development (Project No. PJ009559012016), Rural Development Administration, Korea. The authors extend their sincere appreciation to the Deanship of Scientific Research at King Saud University for its funding this Prolific Research Group (PRG-1437-28).

Authors' contributions
NAA-D, YOK and MVA were planned and executed the laboratory experiments. SI, MVA, PA, RB, KCC CGP, SS, and KHP participated in data analysis, results interpretation, and chemicals arrangement. All listed authors were contributed to this research work and accepted the final manuscript for publications. All authors read and approved the final manuscript.

Competing interests
The authors declare that they have no competing interests.

Financial disclosure
The present work was supported by the Cooperative Research Program for Agriculture Science & Technology Development, Rural Development Administration, Korea (Grant No: PJ009559012016) and the authors extend their sincere appreciation to the Deanship of Scientific Research at King Saud University for its funding this Prolific Research Group (PRG-1437-28).

Author details
[1]Grassland and Forage Division, National Institute of Animal Science, RDA, Seonghwan-Eup, Cheonan-Si, Chungnam 330801, Korea. [2]Department of Botany and Microbiology, Addiriyah Chair for Environmental Studies, College of Science, King Saud University, P. O. Box 2455, Riyadh 11451, Saudi Arabia. [3]Department of Medicinal Crop Research, Rural Development Administration, Eumseong, Chungbuk 369-873, Republic of Korea. [4]Research Department of Plant Biology and Biotechnology, Loyola College, Nungambakkam, Chennai-34, Tamil Nadu, India. [5]Labmate (Asia) Pvt. Ltd, Chennai, Tamil Nadu 600015, India.

References
1. Zhao D, Yuan D, Zhou HC. The current status of hydrogen storage in metalorganic frameworks. Energy Environ Sci. 2008;1:222–35.
2. Furukawa H, Michael Miller A, Omar Yaghi M. Independent verification of the saturation hydrogen uptake in mof-177 and establishment of a benchmark for hydrogen adsorption in metal-organic frameworks. J Mater Chem. 2007;17:3197–204.
3. Titus MP. Porous inorganic membranes for CO_2 capture: present and prospects. Chem Rev. 2014;114:1413–92.
4. Liu J, Praveen K, Thallapally B, McGrail P, Daryl Brown R, Liu J. Progress in adsorption-based co2 capture by metal-organic frameworks. Chem Soc Rev. 2012;41:2308–22.
5. Farha OK, Shultz AM, Sarjeant AA, Nguyen ST, Hupp JT. Active-site-accessible, porphyrinic metal-organic framework materials. J Am Chem Soc. 2011;133:5652–5.
6. Lee JY, Farha OK, Roberts J, Scheidt KA, Nguyen ST, Hupp JT. Metal-organic framework materials as catalysts. Chem Soc Rev. 2009;38:1450–9.
7. Kreno LE, Leong K, Farha OK, Allendorf M, Van Duyne RP, Hupp JT. Metal-organic framework materials as chemical sensors. Chem Rev. 2012;112:1105–25.
8. Stroppa A, Barone P, Jain P, Perez-Mato JM, Picozzi S. Hybrid improper ferroelectricity in a multiferroic and magnetoelectric metal organic framework. Adv Mat. 2013;25:2284–90.
9. Andersen KK, Oliveira CL, Larsen KL, Poulsen FM, Callisen TH, Westh P, Pedersen JS, Otzen D. The role of decorated SDS micelles in sub-CMC protein denaturation and association. J Mol Biol. 2009;391:207–26.
10. Rundlett KL, Armstrong DW. Mechanism of signal suppression by anionic surfactants in capillary electrophoresis-electrospray ionization mass spectrometry. Anal Chem. 1996;68:3493–7.
11. Botelho D, Wall MJ, Vieira DB, Fitzsimmons S, Liu F, Doucette A. Top-down and bottom-up proteomics of SDS-containing solutions following mass-based separation. J Proteome Res. 2010;9:2863.
12. Yeung YG, Nieves E, Angeletti R, Stanley ES. Removal of detergents from protein digests for mass spectrometry analysis. Anal Biochem. 2008;382:135–7.
13. Gu ZY, Chen YJ, Jiang JQ, Yan XP. Metal-organic frameworks for efficient enrichment of peptides with simultaneous exclusion of proteins from complex biological samples. Anal Commun. 2011;47:4787–9.
14. Hengel SM, Floyd E, Baker ES, Zhao R, Wu S, Tolić LP. Evaluations of SDS depletion using an affinity spin column and IMS-MS detection. Proteomics. 2012;12:3138–42.
15. Wisniewski JR, Zougman A, Nagaraj N, Mann M. Universal sample preparation method for proteome analysis. Nat Methods. 2009;6:359–62.
16. Zhou JY, Dann GP, Shi T, Wang L, et al. Simple sodium dodecyl sulfate-assisted sample preparation method for LC-MS-based proteomics applications. Anal Chem. 2012;84:2862–8.
17. Karthik D, Ilavenil S, Kaleeswaran B, Sunil S, Ravikumar S. Proteomic analysis of plasma proteins in diabetic rats by 2D electrophoresis and MALDI-TOF-MS. Appl Biochem Biotechnol. 2012;2012(166):1507–19.
18. Zhou JY, Schepmoes AA, Zhang X, Moore RJ, Monroe ME, Lee JH, Camp DG, Smith RD, Qian WJ. Improved LC-MS/MS spectral counting statistics by recovering low-scoring spectra matched to confidently identified peptide sequences. J Proteome Res. 2010;9:5698–704.

7

Differential proteins among normal cervix cells and cervical cancer cells with HPV-16 infection, through mass spectrometry-based Proteomics (2D-DIGE) in women from Southern México

Idanya Serafín-Higuera[1], Olga Lilia Garibay-Cerdenares[2], Berenice Illades-Aguiar[2], Eugenia Flores-Alfaro[1], Marco Antonio Jiménez-López[3], Pavel Sierra-Martínez[1] and Luz del Carmen Alarcón-Romero[1,4*]

Abstract

Background: Cervical cancer (CC) is the fourth most common cancer in women worldwide with an estimated 528,000 new cases in 2012. The same year México had an incidence of 13,960 and a mortality of 4769 cases. There are several diagnosis methods of CC; among the most frequents are the conventional Pap cytology (Pap), colposcopy, and visual inspection with acetic acid (VIA), histopathological examination, tests of imaging and detection of high-risk papilloma virus (HR-HPV) with molecular tests (PCR, hybridization, sequencing). Proteomics is a tool for the detection of new biomarkers that can be associated with clinical stage, histological type, prognosis, and/or response to treatment. In this study we performed a comparative analysis of CC cells with normal cervical cells. The proteomic analysis was carried out with the fluorescent two-dimensional electrophoresis (2D-DIGE) technique to subsequently identify differential protein profiles using Decyder Software, and the selected proteins were identified by Mass Spectrometry (MALDI-TOF).

Results: The proteins that showed an increased expression in cervical cancer in comparison with normal cervix cells were: Mimecan, Actin from aortic smooth muscle and Lumican. While Keratin, type II cytoskeletal 5, Peroxiredoxin-1 and 14-3-3 protein sigma showed a decrease in their protein expression level in cervical cancer in comparison with normal cervix cells.

Conclusions: Thus, this study was successful in identifying biomarker signatures for cervical cancer, and might provide new insights into the mechanism of CC progression.

Keywords: Cervical cancer, Human Papilloma Virus 16 (HPV-16), Proteomics, 2D DIGE, Mass spectrometry

Abbreviations: ECM, Extracellular matrix; FIGO, International Federation of Gynecology and Obstetrics; HR-HPV, High-risk papilloma virus; MALDI-TOF, Matrix-Assisted Laser Desorption/Ionization-Time of Flight; OGN, Osteoglycin; Prdxs, Peroxirredoxins; ROS, Reactive oxygen species; SCC, Squamous cell carcinoma; SIL, Squamous Intraepithelial Lesion; SLRP, Small leucine rich proteoglycan

* Correspondence: luzdelcarmen14@gmail.com; luzcarmen14@gmail.com
[1]Laboratorio de Citopatología e Histoquímica, Unidad Académica de Ciencias Químico Biológicas, Universidad Autónoma de Guerrero, Chilpancingo, Guerrero, México
[4]Laboratorio de Investigación en Citopatología e Histoquímica, Unidad Académica de Ciencias Químico Biológicas Universidad Autónoma de Guerrero Avenida Lázaro Cárdenas, Ciudad Universitaria, Chilpancingo, Guerrero C.P. 39090, México
Full list of author information is available at the end of the article

Background

Cervical cancer is the fourth most common malignancy and accounts for 10–15 % of cancer-related deaths in women worldwide [1, 2]. Cervical cancer affects approximately six out of 100,000 women and accounts for approximately 275,000 deaths annually in developing countries, which corresponds to 88 % of cases worldwide [2, 3].

Human papillomavirus (HPV) is the main etiological agent of cervical cancer detected in 95 to 100 % of cases [4–6]. Although more than 150 variants of this virus exist, only certain genotypes, such as HPV 16, 18, 33, 45 and 58 are known as high-risk types (HR-HPV); low-risk HPV types (LR-HPV), mainly HPV 6 and 11, seldom cause genital tumors; however, they do cause condylomata acuminata (anogenital warts) [7]. Persistent HPV oncoprotein expression (E6/E7) in HPV infected epithelial basal cells deregulates cell division [8]. Overexpression of these viral genes causes the deregulation of cell proliferation, metabolism, apoptosis, differentiation and genomic instability, all of which may lead to consecutive stages of cervical cancer [9].

Current approaches for the prevention of cervical cancer relies mainly on the cytologic screening, known as the Pap test, often combined with the detection of high-risk human papillomaviruses (HR-HPVs) [10]. Patients with abnormal Paps undergo colposcopy with directed biopsies. If precursor lesions are identified, patients are treated by cryotherapy or loop electrosurgical excision procedure (LEEP). The treatment is effective in the prevention of cervical cancer; however, it is expensive, cumbersome, and dependent on very good infrastructure and well-trained personnel [11]. Nevertheless, the diagnosis may results in a poor outcome, which lies on the lack of valuable objective indicators for determining cervical chronic inflammation, reactive hyperplasia, and benign or malignant lesions [12].

Compared with conventional 2-DE, two-dimensional differential in-gel electrophoresis (2D-DIGE)-based quantitative proteomics has several advantages, such as higher sensitivity, accuracy, and reproducibility, which facilitate spot-to-spot comparisons, precisely because of pre-labeling of protein samples with different fluorescent dyes (Cy3, and Cy5) prior to separation by 2-DE [13]. As a result, samples labeled with different dyes are separated in the same 2D gel; moreover, the same internal standard is used in all gels to avoid inter-gel variation [12, 14]. In the present study, a differential proteomic technique was applied for the comparative analysis of cervical cancer samples infected with HPV 16 and normal cervical tissue.

Results

To carry out a comparative analysis of cervical cancer/HPV-16 and normal samples without HPV, a proteomic analysis was done. Two groups of pooled samples were used. The first one consisted of six samples from women (average age of 50.7 years) diagnosed with HPV-16, by the INNOLIPA assay, and with histopathological staging of squamous cell carcinoma (SCC) of which four cases were stage IIB, one stage IB1 and one case stage IB2. The second group consisted of four pooled samples from women (mean age of 49.8 years) with normal cytology and colposcopy and negative for HPV infection. In Fig. 1 is observed the proliferation index of the cervical tissues by immunohistochemistry. Figure 2 shows the protein profile of each pooled sample, used as a reference pattern, allowing the visualization of more than 2204 spots that were resolved for each 2D gel analyzed. The samples were pooled in two groups and analyzed by triplicated and whose profiles showed technical reproducibility.

To compare samples, Decyder software (www.gelifesciences.com) was used, which is a platform that allows the qualitative and quantitative evaluation of spots profiles from fluorescent bidimensional electrophoresis gels (DIGE) (Fig. 3).

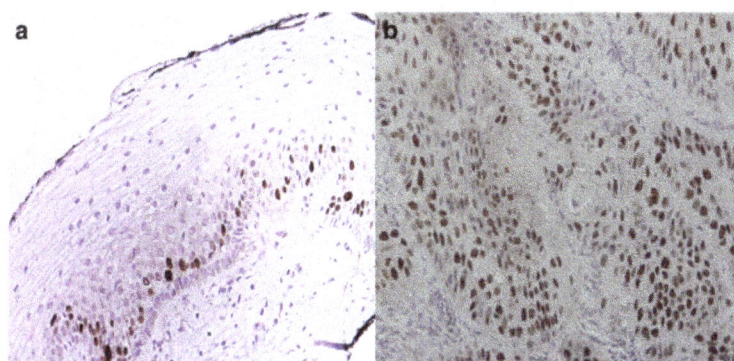

Fig. 1 Expression of Ki-67. **a** Without SIL and HPV infection tissue showing nuclear immunostaining was exclusively confined to the parabasal layers of normal epithelium (**b**) SCC and HPV-16 infection tissue showing strong immunostaining of Ki-67 in large pleomorphic nuclei of malignant squamous cells form irregular nests invading the stroma. 40 X Immunohistochemistry

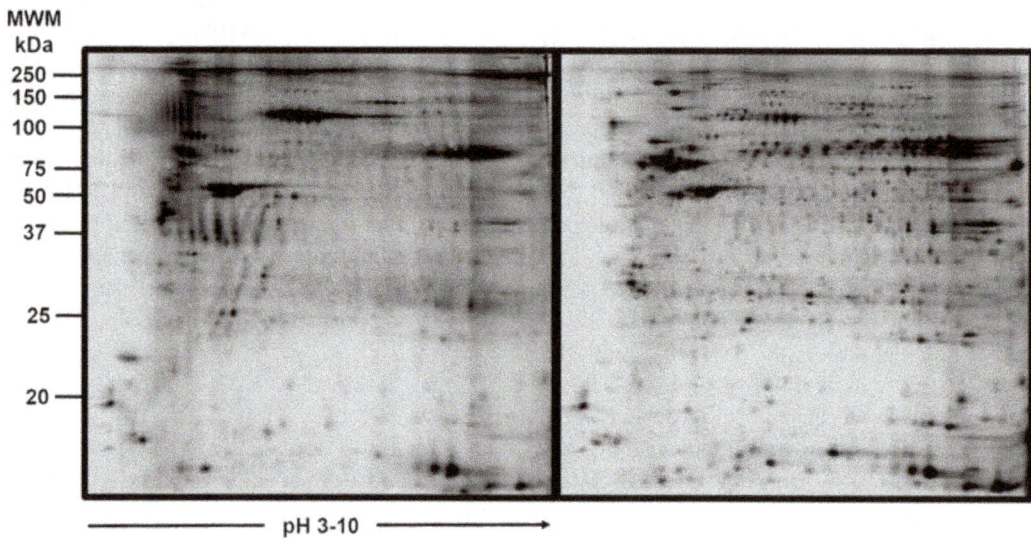

Fig. 2 Bidimensional electrophoresis of squamous cell carcinoma HPV-16 pool vs control cervical cells without HPV infection pool. Gel representative of processing samples, normal cervical cells (left), cervical cancer (rigth). More than 2000 proteins were resolved

The profile of proteins achieved under these conditions was reproducible among pools, the comparison showed different areas of differential expression (yellow spots). An area with a differential expression pattern among samples was located, from which 129 spots with a diminishing expression and 150 with increased

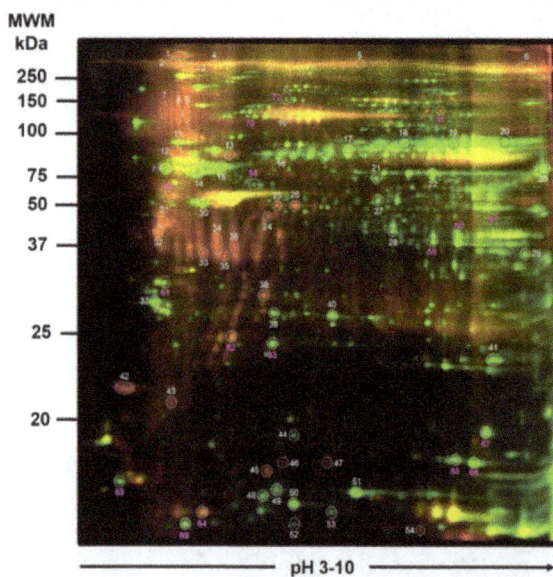

Fig. 3 Representative 2D-DIGE proteome map of control cervical cells sample vs cervical cancer sample. (Control cervical cells without HPV 16 infection, stained with Cy3 (green), and a cervical cancer group stained with Cy5 (red), and overlapping of two groups stained with Cy3 and Cy5 (yellow). The samples were processed through PAGE at 10 %

expression +/− 5 times, 53 spots showed a greater differential expression and the identification of the ten spots was achieved by Mass Spectrometry (MALDI-TOF). Figure 4 shows that Mimecan, Actin aortic smooth muscle and Lumican increased their expression in cervical cancer, in comparison with normal cervical cells, while Keratin, type II cytoskeletal 5, Peroxiredoxin-1 and 14-3-3 protein sigma showed a decrease in their protein expression pattern in cervical cancer in comparison with no SIL tissues.

Table 1 shows the identification of proteins and some biochemical characteristics as name, identification code, and chromosomal localization among others.

Analyzing the participation of the identified proteins in different cellular pathways through Reactome Database (www.reactome.org), the results suggest biosynthesis, metabolism and degradation of keratan sulfate, transcriptional regulation by TP53, metabolism of carbohydrates, intrinsic pathway for apoptosis, detoxification of reactive oxygen species, and cell cycle as the most important.

Discussion

In 2012 the CC was the second cause of cancer death in Mexican women [15], HPV-16 has been reported in studies carried out by our group, as the more common in populations of women of the southern region of Mexico since 1997 [16, 17].

Multiple genes are involved in the occurrence and development of cancers, and even tough genes carry the genetic information, proteins are the final executors of life events [1]. Due to this, the proteomic profiling offers

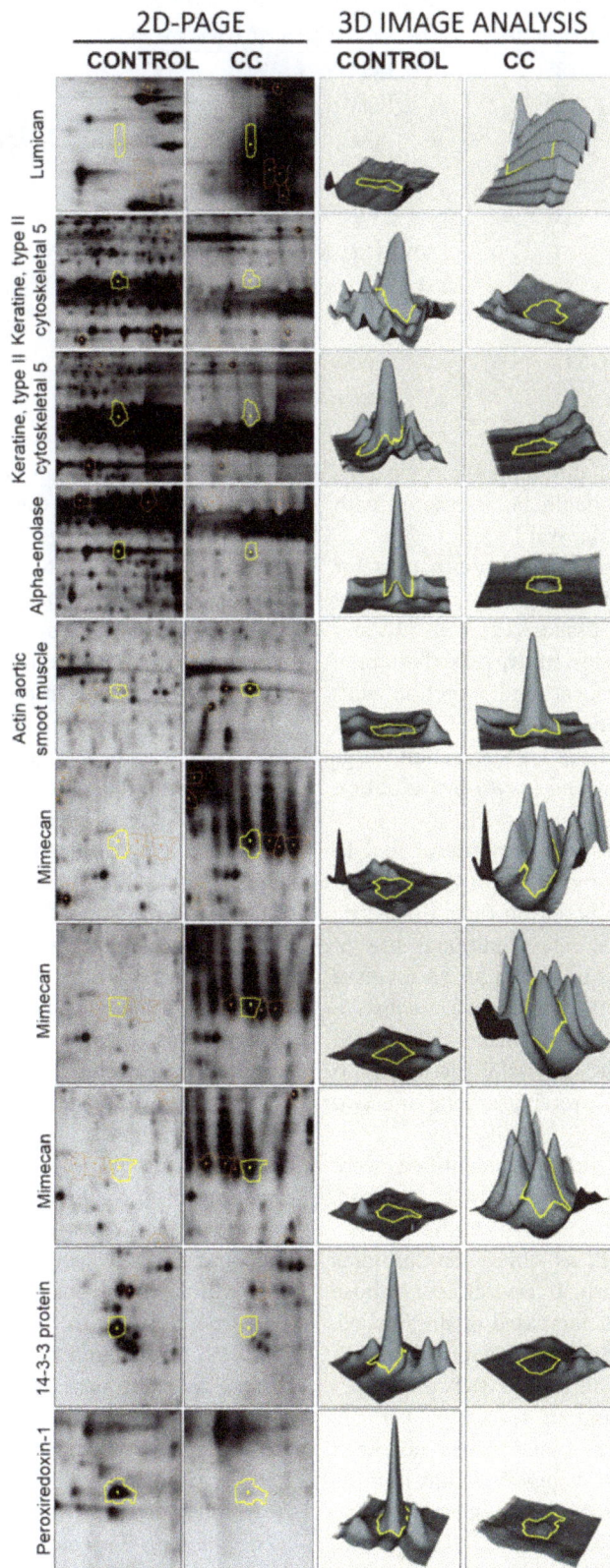

Fig. 4 Comparison of protein expression levels among normal cervical cells without HPV infection, and cervical cancer HPV16 infection groups. Decyder software allowed the tridimensional comparisons between study groups

Table 1 Differential expression of proteins in SCC/HPV-16 against control (without SIL and HPV infection)

Protein	Expression index (n+/−)	Accession number	Chromosomal localization	Gene	Peptide coverage (%)	Score
Mimecan	38.85	MIME_HUMAN	9q22	OGN	46	956
Mimecan	31.23	MIME_HUMAN	9q22	OGN	54	820
Actin, aortic smooth muscle	19.14	ACTA_HUMAN	10q23.31	ACTA2	59	311
Mimecan	17.77	MIME_HUMAN	9q22	OGN	54	447
Lumican	16.52	LUM_HUMAN	12q21.33	LUM	40	556
Peroxiredoxin-1	−8.37	PRDX1_HUMAN	1p34.1	PRDX1	79	968
14-3-3 protein sigma	−14.97	1433S_HUMAN	1p36.11	SFN	67	977
Alpha-enolase	−17.38	ENOA_HUMAN	1p36.2	ENO1	71	976
Keratin, type II cytoskeletal 5	−39.7	K2C5_HUMAN	12q13.13	KRT5	52	885
Keratin, type II cytoskeletal 5	−65.34	K2C5_HUMAN	12q13.13	KRT5	52	1 090

an option for the selection of differential protein patterns whose expression levels could be associated with progression, prognosis and/or survival.

There have been several experimental approaches for the analysis of differential protein expression in cervical cancer, such as the analysis of tissues [12], cells [1], and even plasma samples [12]. In this study, cells that come from women with cervical cancer and infection with HPV-16 were compared with normal cells of the cervix recovered from individuals without HPV infection, diagnosed through INNOLIPA assay and *in situ* hybridization with tyramide amplification.

Bidimensional electrophoresis is a very useful tool to analyze broad and complex samples, through the wide distribution of protein profiles based on their isoelectric points and molecular weights. The additional use of fluorescence during this technique allows for an increase in sensitivity in 2D DIGE, and the use of special software facilitates the differential analysis. Finally, the use of MALDI-TOF MS, allowed the successful identification of ten differentially expressed proteins among the two different groups.

In the present study, the proteins identified were mimecan (osteoglycine), aortic smooth muscle, Lumican, Peroxiredoxin-1, 14-3-3 protein sigma, Alpha-enolase, Keratin, type II cytoskeletal 5, as differential proteins among cervical cancer and normal cervical cells, whose expression patterns were either increased or diminished. Mimecan (or Osteoglycin, OGN), is a secretory protein that belongs to a family of small leucine rich proteoglycan (SLRPs). It has been found in several cancer cell lines, although its physiological function has not been completely understood [18, 19], being abundant in bone matrix, cartilage cells, and connective tissues; it is also important for collagen fibrillogenesis, cellular growth, differentiation and migration [20] and it has been involved in the pathogenesis of different cancers such as colorectal [21], laryngeal carcinoma [19] and cervical cancer.

The actin cytoskeleton is substantially altered in cancer cells as a result of the changes in the abundance of proteins, among other factors. As a result, cancer cells acquire increased motility and distinctive mechanical properties, which are important for processes such as invasion and metastasis [22]. ACTA2 is a α-smooth muscle actin whose expression is transformation sensitive to growth signals in normal cells [23].

In cases of basal cell carcinoma skin cancer, it has been found with a high expression too, both in the tumor and in the adjacent stroma and it is used as a marker of aggressiveness in pancreatic cancer, because it is presenting more aggressive histological variants [24], suggesting that the increased expression may contribute to the local invasion. It's role in the biology of the tumor is not completely known, however, it has been hypothesized that increases in the cellular motility result in an increase in the cells capacity of invasion [25].

Lumican is a keratin sulphate belonging to the SLRP family of extracellular matrix (ECM) proteins and is expressed in different forms in several tissues and organs, such as cornea, bone, cartilage, artery, skin, kidney, and lung. It has been found to have a key role both in the organization of the extracellular matrix (ECM) and as an important modulator of biological functions in breast, lung, and pancreas cancer [26, 27], and it has been correlated with lung cancer progression as well [28].

The peroxiredoxins were identified primarily by their ability to protect the protein oxidative damage induced by free radicals, such as reactive oxygen species (ROS) and reactive nitrogen species (RNS), which is currently recognizedas a promotor for cancer development [29]. The peroxirredoxins (Prdxs) are small proteins of sweep (scavening) of H_2O_2, that could prevent tumor development since the loss of Prdx1 in mice leads to premature death due to cancer [30].

Keratins are expressed in all types of epithelial cells (simple, stratified, keratinized and cornified), are important protectors of epithelial structural integrity under conditions of stress, but have also been recognized as regulators of other cellular functions, including motility,- signaling, growth and protein synthesis [31]. KRT5 Keratin, type II cytoskeletal 5 was found decreased in primary early-stage cervical squamous cell cancer tissue with pelvic lymph node metastasis (PLNM) vs without PLNM using DIGE-based proteomics [32].

Conclusion

Based on the above results, it was possible to identify differential protein patterns among cases of cervical lesions related to cervical cancer progression with HPV-16 infection in comparison with tissue without lesion and negative for HPV infection from Mexican women. These proteins could be analyzed as potential candidates for biomarkers due to their relationship with their presence in early stages of cervical cancer development, improving the understanding of viral pathogenesis and its role in the development of cervical cancer. Eventually this information could be used in the development of strategies in clinical management, diagnosis and treatment.

Methods

Study population

A cross-sectional study was conducted, which included two study groups, the first included a pool of 6 cases of cervical cancer with HPV-16 infection (cervical cancer/ HPV-16) of which four cases were stage IIB, one stage IB1 and one case stage IB2 and the second group, which served as a comparison, consisted of a pool of four cases from surgical specimens extracted by fibroids without lesions related to cervical cancer and without HPV infection, and whose previous cytological study showed normal cells with reactive changes to nonspecific inflammatory processes, with normal colposcopy. All cases had antecedents cytological that were reported according to the Bethesda System 2001 and without coninfection by

Trichomonas vaginalis, Candida sp, Actinomyces sp and *Herpes simplex virus* [33].

The diagnosis and detection of HPV in each sample was performed as follows: three different sections of each biopsy were obtained. The first embedded in paraffin and were cut to a thickness of 3 μm (one stained with hematoxylin-eosin for histopathological diagnosis, another for detecting cell proliferation antigen Ki-67 and finally another slide for detection of DNA detection HPV by *in situ* hybridization with tyramide amplification). The second section was used for detection of HPV DNA by molecular analysis (INNOLiPA, Uniparts Innogenetics) and the third was used for protein extraction. These data allowed the formation of the two comparative groups (Table 2).

The cervical cancer tissues were obtained from the Instituto Estatal de Cancerología "Arturo Beltrán Ortega", and normal cervix tissues from patients from the Hospital General "Dr. Jorge Soberón Acevedo". The histopathological diagnosis was carried out by pathologists in each institution independently, according to each clinical record. Both groups contained women with no history of prior local treatment, chronic degenerative diseases, non-smoking and non-alcoholism history, and without coinfections corroborated by colposcopic, cytological and microbiological analyses. The diagnosis was reported in accordance with the system of the International Federation of Gynecology and Obstetrics (FIGO) [34]. Ages in both groups were 37–69 years with a mean age of 50.7 years among cervical cancer patients and 49.8 among patients free-cervical. Women signed an informed consent and their data were analyzed in a confidential manner. This study was approved by the Committee of Bioethics at the Autonomous University of Guerrero, Mexico, according to the ethical guidelines of the Declaration of Helsinki 2008 [35].

Table 2 Characteristics of cervical tissues samples of comparatives groups CC/HPV-16 and control without SIL and HPV infection

Tests[a]	Pool CC/HPV-16	Pool without SIL and HPV infection (control)
Background	Cytological[b] study: Squamous cell carcinoma Colposcopic examination: cervical carcinoma Clinical diagnosis: 4 cases stage IIB, one stage IB1 and 1 case stage IB2	Cytological[b] study: normal cells with reactive changes to nonspecific inflammatory processes Normal colposcopy
Section 1 (paraffin embedding) - Histopathological diagnosis (H-E) - *in situ* hybridization with tyramide amplification - Immunohistochemistry for Ki-67	Squamous Cell Carcinoma Positive of HR- HPV Positive in large pleomorphic nuclei of malignant squamous cells form irregular nests invading the stroma	Negative Negative Positive in nuclei of parabasal cells in basal layer
Section 2 - INNO-LIPA extra (Genotyping of HVP)	HPV-16	Negative of HPV infection
Section 3 - Extraction of proteins for 2D- DIGE and MALDI-TOF	Increased expression in cervical cancer Mimecan, Actin from aortic smooth muscle and Lumican. Decreased expression: Keratin, type II cytoskeletal 5, Peroxiredoxin-1 and 14-3-3 protein sigma	

[a] Each sample of tissue fragment into three sections and all tests were performed, six samples for group CC / HPV-16 and four samples were used for the control group. For more details see section Methods
[b] Without coninfection by *Trichomonas vaginalis, Candida sp, Actinomyces sp, Herpes simplex virus* according to the Bethesda 2001 system

Immunohistochemistry for Ki-67

To evaluate the expression of Ki-67, the monoclonal antibody MIB-1 (Dako, Carpinteria, CA, USA) was used with the Cytoscan HRP/DAB immunohistochemical system of detection (Cell Marque Corporation, Hot Springs, AR, USA [now relocated to Rocklin, CA, USA]). The histological slices was deparaffinized and placed in a solution of immunoDNA Retriever (BioSB, Inc., Santa Barbara, CA, USA). Later, the primary antibody, previously diluted in accordance with the manufacturer's instructions, was added; the chromogen diaminobenzidine was added; and finally, the specimens were stained with Mayer's hematoxylin (Merck, USA). The expression in the without SIL tissues was evaluated in accordance to the distribution and localization of the positive reaction within the cells and within the depth of the epithelium. The expression of Ki-67 was considered positive when a brown ochre color was evident in the nucleus of the cells. In SCC is observed in large pleomorphic nuclei of malignant squamous cell form irregular nests invading the stroma and tissue without SIL some cells of the parabasal layers were found (Fig. 1).

Detection and genotyping of HPV DNA by in situ hybridization with tyramide amplification and INNO-LiPA Extra

In the *in situ* hybridization with a system of tyramide signal amplification (Gen Point Dako Cytomation, Carpinteria, CA, USA), a drop of test reagent (biotinylated viral DNA) with probes for 13 HR-HPV genotypes (16,18,31,33,39,45,51,52,56,58,59 and 68) was added to each slide. The slide were denatured for 10 min and subjected to hybridization for 20 h (Hybridizer Dako, Carpinteria, CA,USA). For HVP-INNO-LIPA extra, the DNA extraction from cervical tissue samples the conventional TRIZOL method (Cat. No.15596018, Invitrogen a part of life Tech. Corp) was used, according to the manufacturer's instructions. Tissues were tested for the presence of the HPV genotypes by polymerase chain reaction (PCR), using the short PCR fragment 10 (SPF10) primers, a highly sensitive method for HPV DNA detection, according to the method of Pirog et al. [36].

Protein sample labeling with CyDye

CyDye DIGE fluors (Cy2, Cy3, and Cy5) were used to label the protein extracts following the manufacturer's protocol (GE Healthcare). The internal standard pool was generated by combining equal amounts of extracts from all samples, labeled with Cy2. Protein extracts from the normal and cervical cancer cells were labeled with Cy3 and Cy5, respectively. The labeling reaction was performed on ice for 30 min in darkness, and was then quenched with 10 mM lysine for 10 min on ice under

dark conditions. The labeled samples were then mixed and prepared for the following steps.

2-D electrophoresis in polyacrylamide gel electrophoresis (SDS-PAGE)

Total extracts from normal and cancer tissues were processed according to Klose protocol [37]. Briefly, samples were diluted in rehydration buffer containing 8 M urea, 0.5 % (w/v) CHAPS, 10 mM DTT, 0.001 % bromophenol blue, and Bio-Lyte 3–10 Ampholyte (0.2 %) (Bio-Rad, Cat. No.163-1113). The protein mixture was then applied to ReadyStrip™ IPG 7 cm strips, pH 3–10 (linear). Rehydrated strips were isoelectrically focused using a PROTEAN IEF cell System. To perform the second dimension analysis, the strips were processed by 10 % SDS-PAGE. Finally the 2D gels were stained with Coomassie staining. The IPG strips were balanced in a buffer consisting of 1.5 M Tris–HCl pH 8.8, 6 M urea and 30 % v/v glycerol, 2 % p/v SDS and traces of bromophenol blue, being carried out in two steps, using in each 5 ml of balanced solution per strip; in the first one, the strips were incubated in the balanced buffer described above with 1 % w/v of DTT, remaining in this solution for 20 min in agitation at room temperature. In the second step, 2.5 % w/v of iodoacetamide was added to the balanced buffer, repeating the 20 min incubation in agitation at room temperature; once the procedure was completed, the proteins could be picked for mass spectrometry analysis.

Image analysis

The images were analyzed using the DeCyder™ 2D Differential Analysis Software (DeCyder 2D V8.0) by differential in-gel analysis (DIA) and biological variation analysis (BVA). Protein spots were marked and selected with changes in abundance ratio >1.5-fold, P values <0.05 for protein identification. The gels were subjected to Coomassie blue staining for spot visualization and picking.

Protein identification by MALDI-TOF

Commassie-stained 2D gels were scanned and digital images were compared using the Decyder Software. Each pool was run three times. The electrophoretic entities of interest were excised, alkylated, reduced, and digested up to obtain a peptide mass fingerprint. Peak lists of the tryptic peptide masses were generated using FlexAnalysis1.2vSD1 Patch 2 (Bruker Daltonics). The search engine MASCOT server 2.0 was used to compare the fingerprints against human taxonomy with the following parameters: one missed cleavage allowed, carbamidomethyl cysteine as the fixed modification and oxidation of methionine as the variable modification. Proteins with scores greater than 50 and a $p < 0.05$ were accepted.

Acknowledgments
We thank the strengthening program graduate high with file number: I010/455/2013 C- 677/2013 of UAGro. Serafín-Higuera I was recipient of a Doctoral fellowship with file number 222252 from CONACYT, México. We are grateful to Vladimir Rodríguez-Sandoval, Pathologist of Hospital General "Jorge Soberón Acevedo" Iguala, Guerrero, México for help us in diagnosis support and to Dra Patricia Talamás-Rohana for manuscript review and comments.

Funding
This project was funded by Programa de Fortalecimiento Académico del Posgrado de Alta Calidad (I010/455/2013 C-677/2013). The funders had no role in study design, data collection and analysis, decision to publish, or preparation of the manuscript.

Authors' contributions
The work presented here was carried out in collaboration between all authors. Conceived and designed the experiments: ISH, LCAR, BIA, PSM, MAJL. Performed the experiments: ISH, LCAR. Analyzed the data: ISH, LCAR, BIA, EFA, PSM, OLGC. Contributed reagents/materials/analysis tools: ISH, LCAR, BIA, EFA, PSM, OLGC, MAJL. Wrote the paper: ISH, OLGC, LCAR. All authors read and approved the final manuscript.

Authors' information
Olga Lilia Garibay-Cerdenares: Professor of CONACyT.

Competing interests
The authors declare that they have no competing interests.

Author details
[1]Laboratorio de Citopatología e Histoquímica, Unidad Académica de Ciencias Químico Biológicas, Universidad Autónoma de Guerrero, Chilpancingo, Guerrero, México. [2]Laboratorio de Biomedicina Molecular, Unidad Académica de Ciencias Químico Biológicas, Universidad Autónoma de Guerrero, Chilpancingo, Guerrero, México. [3]Instituto Estatal de Cancerología "Dr. Arturo Beltrán Ortega", Acapulco, Guerrero, México. [4]Laboratorio de Investigación en Citopatología e Histoquímica, Unidad Académica de Ciencias Químico Biológicas Universidad Autónoma de Guerrero Avenida Lázaro Cárdenas, Ciudad Universitaria, Chilpancingo, Guerrero C.P. 39090, México.

References
1. Zhao Q, He Y, Wang XL, Zhang YX, Wu YM. Differentially expressed proteins among normal cervix, cervical intraepithelial neoplasia and cervical squamous cell carcinoma. Clin Transl Oncol. 2015;17:620–31.
2. Siegel RL, Miller KD, Jemal A. Cancer statistics, 2015. CA Cancer J Clin. 2015; 65:5–29.
3. Ferlay J, Soerjomataram I, Ervik M, Dikshit R, Eser S, Mathers C, et al. GLOBOCAN 2012 v1.1, Cancer Incidence and Mortality Worldwide: IARC Cancer Base No. 11. Lyon: International Agency for Research on Cancer; 2014. http://globocan.iarc.fr accessed on 5 Sep 2015.
4. Adam ML, Pini C, Túlio S, Cantalice JCLL, Torres RA, Correia MTDS. Assessment of the Association Between Micronuclei and the Degree of Uterine Lesions and Viral Load in Women with Human Papillomavirus. Cancer Genomics Proteomics. 2015;12:67–72.
5. Cortés-Gutiérrez EI, Dávila-Rodríguez MI, Vargas-Villarreal J, Hernández-Garza F, Cerda-Flores RM. Association between Human Papilloma Virus-type Infections with Micronuclei Frequencies. Prague Med Rep. 2010;111(1):35–41.
6. Snijders PJF, Steenbergen RDM, Heideman DAM, Meijer CJLM. HPV-mediated cervical carcinogenesis: concepts and clinical implications. J Pathol. 2006;208: 152–64.
7. Paavonen J. Human papillomavirus infection and the development of cervical cancer and related genital neoplasias. Int J Infect Dis. 2007; 11(Supplement 2):S3–9.
8. Maglennon GA, McIntosh P, Doorbar J. Persistence of viral DNA in the epithelial basal layer suggests a model for papillomavirus latency following immune regression. Virology. 2011;414:153–63.
9. Van Raemdonck GAA, Tjalma WAA, Coen EP, Depuydt CE, Van Ostade XWM. Identification of Protein Biomarkers for Cervical Cancer Using Human Cervicovaginal Fluid. PLoS One. 2014;9:e106488.
10. Lynge E, Rygaard C, Baillet MV-P, Dugué P-A, Sander BB, Bonde J, Rebolj M. Cervical cancer screening at crossroads. APMIS. 2014;122:667–73.
11. Kim HS. Correction: primary, secondary, and tertiary prevention of cervical cancer. J Gynecol Oncol. 2014;25:261.
12. Guo X, Hao Y, Kamilijiang M, Hasimu A, Yuan J, Wu G, Reyimu H, Kadeer N, Abudula A. Potential predictive plasma biomarkers for cervical cancer by 2D-DIGE proteomics and Ingenuity Pathway Analysis. Tumor Biol. 2015;36: 1711–20.
13. Maurya P, Meleady P, Dowling P, Clynes M. Proteomic approaches for serum biomarker discovery in cancer. Anticancer Res. 2007;27(3A):1247–55.
14. Pressey JG, Pressey CS, Robinson G, Herring R, Wilson L, Kelly DR, Kim H. 2D-Difference Gel Electrophoretic Proteomic Analysis of a Cell Culture Model of Alveolar Rhabdomyosarcoma. J Proteome Res. 2011;10:624–36.
15. Gutiérrez JP, Rivera-Dommarco J, Shamah-Levy T, Villalpando-Hernández S, Franco A, Cuevas-Nasu L, et al. Encuesta Nacional de Salud y Nutrición 2012, Resultados Nacionales. Cuernavaca: Instituto Nacional de Salud Pública; 2012.
16. Ortiz-Ortiz J, Alarcon-Romero L, Jimenez-Lopez M, Garzon-Barrientos V, Calleja-Macias I, Barrera-Saldana H, Leyva-Vazquez M, Illades-Aguiar B. Association of human papillomavirus 16 E6 variants with cervical carcinoma and precursor lesions in women from Southern Mexico. Virol J. 2015;12:29.
17. Illades-Aguiar B, Alarcón-Romero LC, Antonio-Vejar V, Zamudio-López N, Sales-Linares N, Flores-Alfaro E, Fernández-Tilapa G, Vences-Velázquez A, Muñoz-Valle J, Leyva-Vázquez M. Prevalence and distribution of human papillomavirus types in cervical cancer, squamous intraepithelial lesions, and with no intraepithelial lesions in women from Southern Mexico. Gynecol Oncol. 2010;117:291–6.
18. Yinghong W, Yu M, Bingjian L, Enping X, Qiong H, Maode L. Differential expression of mimecan and thioredoxin domain–containing protein 5 in colorectal adenoma and cancer: A proteomic study. Exp Biomed Biol. 2007; 232:1152–9.
19. Li L, Zhang Z, Wang C, Miao L, Zhang J, Wang J, Jiao B, Zhao S. Quantitative Proteomics Approach to Screening of Potential Diagnostic and Therapeutic Targets for Laryngeal Carcinoma. PLoS One. 2014;9:e90181.
20. Zheng C, Zhao S, Wang P, Yu H, Wang C, Han B, Su B, Xiang Y, Li X, Li S, et al. Different expression of mimecan as a marker for differential diagnosis between NSCLC and SCL. Oncol Rep. 2009;22:1057–61.
21. Wang Y, Ma Y, Lü B, Xu E, Huang Q, Lai M. Differential expression of mimecan and thioredoxin domain-containing protein 5 in colorectal adenoma and cancer: a proteomic study. Exp Biol Med. 2007;232:1152–9.
22. Efremov YM, Dokrunova AA, Efremenko AV, Kirpichnikov MP, Shaitan KV, Sokolova OS. Distinct impact of targeted actin cytoskeleton reorganization on mechanical properties of normal and malignant cells. Biochim Biophys Acta. 2015;1853:3117–25.
23. Huang H-L, Yao H-S, Wang Y, Wang W-J, Hu Z-Q, Jin K-Z. Proteomic identification of tumor biomarkers associated with primary gallbladder cancer. World J Gastroenterol. 2014;20:5511–8.
24. Sinn M, Denkert C, Striefler JK, Pelzer U, Stieler JM, Bahra M, Lohneis P, Dörken B, Oettle H, Riess H, Sinn BV. α-Smooth muscle actin expression and desmoplastic stromal reaction in pancreatic cancer: results from the CONKO-001 study. Br J Cancer. 2014;111:1917–23.
25. Mercut R, Ciurea ME, Margaritescu C, Popescu SM, Craitoiu MM, Cotoi OS, Voinescu DC. Expression of p53, D2-40 and a-smooth muscle actin in different histological subtypes of facial basal cell carcinoma. Rom J Morphol Embryol. 2014;55(2):263–72.
26. Dolhnikoff M, Morin J, Roughley PJ, Ludwig MS. Expression of Lumican in Human Lungs. Am J Respir Cell Mol Biol. 1998;19:582–7.
27. Nikitovic D, Papoutsidakis A, Karamanos NK, Tzanakakis GN. Lumican affects tumor cell functions, tumor-ECM interactions, angiogenesis and inflammatory response. Matrix Biol. 2014;35:206–14.
28. Cappellesso R, Millioni R, Arrigoni G, Simonato F, Caroccia B, Iori E, Guzzardo V, Ventura L, Tessari P, Fassina A. Lumican Is Overexpressed in Lung Adenocarcinoma Pleural Effusions. PLoS One. 2015;10:e0126458.
29. Rhee SG, Woo HA, Kil IS, Bae SH. Peroxiredoxin Functions as a Peroxidase and a Regulator and Sensor of Local Peroxides. J Biol Chem. 2012;287: 4403–10.
30. Neumann CA, Fang Q. Are peroxiredoxins tumor suppressors? Curr Opin Pharmacol. 2007;7:375–80.

31. Karantza V. Keratins in health and cancer: more than mere epithelial cell markers. Oncogene. 2011;30(2):127–38.

32. Wang W, Jia H-L, Huang J-M, Liang Y-C, Tan H, Geng H-Z, et al. Identification of biomarkers for lymph node metastasis in early-stage cervical cancer by tissue-based proteomics. Br J Cancer. 2014;110:1748–58.

33. Solomon D, Davey D, Kurman R, Moriarty A, O'Connor D, Prey M, Raab S, Sherman M, Wilbur D, Wright Jr T, Young N, Forum Group Members. Bethesda 2001 Workshop. The 2001 Bethesda System: terminology for reporting results of cervical cytology. JAMA. 2002;287:2114–9.

34. Benedet JL, Bender H, Jones III H, Ngan HYS, Pecorelli S. FIGO staging classifications and clinical practice gudelines in the management of gynecologic cancers. Int J Gynecol Obstet. 2000;70:209–62.

35. Williams JR. The Declaration of Helsinki and public health. Bull World Health Organ. 2008;86:650–2.

36. Pirog EC, Kleter B, Olgac S, Bobkiewicz P, Lindeman J, Quint WGV, et al. Prevalence of Human Papillomavirus DNA in Different Histological Subtypes of Cervical Adenocarcinoma. Am J Pathol. 2000;157:1055–62.

37. Klose J, Kobalz U. Two-dimensional electrophoresis of proteins: An updated protocol and implications for a functional analysis of the genome. Electrophoresis. 1995;16:1034–59.

8

Multi-omics analysis on the pathogenicity of *Enterobacter cloacae* ENHKU01 isolated from sewage outfalls along the Ningbo coastline

Dijun Zhang, Weina He, Qianqian Tong, Jun Zhou[*] and Xiurong Su[*]

Abstract

Background: The acquisition of iron is important for the pathogenicity of bacteria and blood. Three different culture environments (Fe stimulation, blood agar plate and normal plate) were used to stimulate *Enterobacter cloacae*, and their respective pathogenicities were compared at the proteomic, mRNA and metabolomic levels.

Methods: 2D-DIGE combined with MALDI-TOF-MS/MS, RT-PCR and ^{1}H NMR were used to analyze the differential expression levels of proteins, mRNA and metabolites.

Results: A total of 109 proteins were identified by 2D-DIGE and mass spectrometry after pairwise comparison within three culture environments, clustered into 3 classes and 183 functional categories, which were involved in 23 pathways. Based on the 2D-DIGE results, multiple proteins were selected for verification by mRNA expression. These results confirmed that most of the proteins were regulated at the transcriptional level. Thirty-eight metabolites were detected by NMR, which correlated with the differentially expressed proteins under different treatment conditions.

Conclusions: The results show that culture in a blood agar plate and a suitable concentration of iron promote the pathogenicity of *E. cloacae* and that high iron concentrations may have adverse effects on growth and iron uptake and utilization by *E. cloacae*.

Keywords: *Enterobacter cloacae*, Pathogenicity, Iron, Proteomic, Metabolomic

Background

Due to antibiotics that have been overly prescribed in recent years, *Enterobacter cloacae* has emerged as an important nosocomial pathogen in neonatal units, with numerous outbreaks of infection being reported [1, 2]. *E. cloacae* occur in water, sewage, soil, food, and as commensal microflora in the intestinal tracts of humans and animals [3]. Molecular biological studies of *E. cloacae* have revealed six species, and some strains that have been phenotypically identified as *E. cloacae* are opportunistic pathogens that have been implicated as the causative agent of local and systemic infections in humans [4]. They are important nosocomial pathogens that are responsible for bacteremia, lower respiratory

tract, skin, soft tissue, urinary tract, intra-abdominal and ophthalmic infections, endocarditis, septic arthritis and osteomyelitis, especially the outbreaks of septicemia in the neonatal intensive care unit [5, 6]. This bacterium may be transmitted to neonates through intravenous fluids, total parenteral nutrition solutions and medical equipment. Common endogenous reservoirs of *E. cloacae* include the gastrointestinal tract of healthy adults and the urinary and respiratory tracts of sick patients. Sputum, secretions and pus, and urine are the most studied specimens of human *E. cloacae* infection [7].

E. cloacae is isolated from the feces of 10–70 % of neonates. Due to their relative lack of toxicity and ability to cross the blood–brain barrier, these antimicrobial agents have been increasingly used as first-line antibiotic therapy in neonates. As a result, *E. cloacae* has become super-bacteria in hospitals due to the presence of

* Correspondence: zhoujun1@nbu.edu.cn; suxiurong_public@163.com
School of Marine Science, Ningbo University, 818 Fenghua Road, Ningbo, Zhejiang Province 315211, People's Republic of China

extended-spectrum β-lactamases (ESBLs) [1]. Although *E. cloacae* complex strains are among the most common *Enterobacter* species causing nosocomial bloodstream infections in the last decade, little is known regarding their virulence-associated properties. Among the most common risk factors for developing *E. cloacae* bloodstream infections are prolonged hospitalization, the severity of the illness, and exposure to invasive procedures [4]. Additional predisposing factors are the usage of a central venous catheter, prolonged antibiotic therapy, parenteral nutrition and immunosuppressive therapy [8].

In our previous study, we obtained 98 strains of *E. cloacae* from the Ningbo sewage outfall using *rpo*B genotyping, multi-locus sequence analysis and comparative genomic hybridization. Among the 98 strains of bacteria, the following virulence genes were identified: iron regulatory protein 2 (*irp*2), ferrichrome-iron uptake receptor (*fhu*A), superoxide dismutase B (*sod*B), and Shiga-Like-Toxin A (*slt*A), with a detection rate of 35.71 % for the *fhu*A$^+$ *irp*2$^+$ *sod*B$^+$ genotype, 25.27 % for *fhu*A$^+$ *irp*2$^+$ *slt*A$^+$, 13.19 % for *irp*2$^+$, 12.09 % for *fhu*A$^+$, 9.89 % for *fhu*A$^+$ *irp*2$^+$, 8.79 % for *slt*A$^+$ *sod*B$^+$, and 8.79 % for *fhu*A$^+$ *irp*2$^+$ *sod*B$^+$ *slt*A$^+$ [9].

The ability of bacteria to acquire iron from the external environment is known to have a strong relationship with virulence [10, 11]. Iron is an essential element for most bacteria; it is utilized as the reaction center for redox enzymes and directly participates in redox reactions by switching between the Fe^{2+} and Fe^{3+} states [12]. Among the Gram-positive pathogens, iron uptake in *Staphylococcus aureus* has been investigated most extensively [13]. In a study of Gram-negative bacteria isolated from 120 neonate blood samples with clinical signs of infection, *E. cloacae* accounted for the largest population among the pathogenic bacteria [14]. The blood agar plate is one of the most important methods for cultivating *E. cloacae* and other pathogenic bacteria to study their pathogenicity [15]. Therefore, to study the pathogenicity of *E. cloacae* isolated from sewage outfall, we set out to compare the regulation of *E. cloacae* pathogenicity by blood and iron availability. We cultured *E. cloacae* in three different media, and then assessed pathogenicity by 2D-DIGE, RT-PCR and nuclear magnetic resonance (NMR) at proteomic, mRNA and metabolic levels.

Methods

Isolation, identification and culture of bacteria

E. cloacae was isolated from sewage outfalls along the Ningbo coastline (Ningbo, China) and positively identified as *Enterobacter cloacae* ENHKU01 by sequencing using universal primers (27 F: 5'-AGAGTTTGAT CCTGGCTCAG-3' and 1492R: 5'-GGTTACCTTGTT ACGACTT-3'). *E. cloacae* was cultured on blood agar

plates in the first experimental group (hereafter referred to as Y1) and in beef extract peptone medium (5 mg/mL beef extract powder, 10 mg/mL peptone, 20 mg/mL agar, all purchased from Microbial Reagent, Hangzhou, China) in the control group (hereafter referred to as Y2). In the second experimental group (hereafter referred to as Y3), 0.1 mM $FeCl_3$ (this concentration was selected from a preliminary experiment with varying concentrations of Fe^{3+}, Additional file 1) was added to the same medium for 12 h at 28 °C. All extractions and experiments were performed in a cold room at 4 °C. *E. cloacae* were washed twice with phosphate-buffered saline (PBS), and the bacteria were collected after centrifugation (6,000 rpm, 15 min, 4 °C).

Protein identification

Sample preparation and CyDye labeling

The bacteria were dissolved in 10 mL of lysis buffer (8 mol/L urea, 2 mol/L thiourea, 4 % (w/v) CHAPS, 10 mg/mL of DTT, 2.5 mg/mL of Tris), and protein was subsequently extracted by ultrasonic disruption (200 W for 10 min) on ice. Centrifugation (12,000 rpm, 30 min, 4 °C) was used to pellet the cell debris, and the supernatant was mixed with 5 times its volume of acetone (containing 10 % TCA). The proteins were precipitated for 6 h at –20 °C, and the supernatant discarded after centrifugation (12,000 rpm, 30 min, 4 °C). The pellet was resuspended in acetone and centrifuged (12,000 rpm, 30 min, 4 °C), and the precipitate was dried in a draft cupboard. The protein pellet was resuspended in rehydration buffer (8 mol/L urea, 2 mol/L thiourea, 40 mg/mL CHAPS, 10 mg/mL of DTT). Finally, the protein concentration was determined using a 2-D Quant Kit (Amersham Biosciences, USA) with BCA (2 mg/mL) as the standard. The optimal concentration of the protein sample was between 5 and 10 mg/mL.

For each sample, 30 μg of protein was mixed with 1.0 μl of diluted CyDye (1:5 diluted with dimethyl formamide from a 1 nmol/μl stock) and maintained in the dark on ice for 30 min. Samples from each pair were labeled with Cy3 and Cy5, respectively, while the same amount of the pooled standard containing equal quantities of all samples was labeled with Cy2 (Table 1). The three labelled and quenched samples were combined,

Table 1 DIGE experimental design for sample protein labeling from different treatments and internal standard

Gel No.	Cy2	Cy3	Cy5
Gel 1	Y1 + Y2 + Y3	Y1	Y2
Gel 2	Y1 + Y2 + Y3	Y2	Y1
Gel 3	Y1 + Y2 + Y3	Y3	Y2
Gel 4	Y1 + Y2 + Y3	Y1	Y3
Gel 5	Y1 + Y2 + Y3	Y3	-

and a total of 150 μg of protein was mixed and added to the rehydration buffer and 0.5 % Immobilized pH gradient (IPG) buffer (GE Healthcare, USA) to a final volume of 460 μL.

Two-dimensional gel electrophoresis

After loading the labeled samples onto 22-cm pH 4–7 linear IPG strips (GE Healthcare, USA), iso-electric focusing (IEF) was performed as follows: 12 h of rehydration at 20 °C, followed by 300 V for 45 min, 700 V for 45 min, 1,500 V for 1.5 h, 9,000 V for 27,000 VHr, and 9,000 V for 36,000 VHr. After IEF, the IPG strips were equilibrated for sodium dodecyl sulfate-polyacrylamide gel electrophoresis (SDS-PAGE) in 5 mL equilibration buffer (0.05 M Tris–HCl (pH 8.8), 6 M urea, 30 % (v/v) glycerol, 2 % (w/v) SDS and a trace amount of bromophenol blue) containing 1 % DTT for 15 min, followed by a second equilibration step of 15 min with the same buffer containing 2.5 % (w/v) iodoacetamide. The equilibrated strips were loaded on the top of 12 % SDS-polyacrylamide gels and sealed with 0.5 % (w/v) agarose. The SDS-PAGE step was performed at 15 °C in an Ettan Dalt Twelve (Amersham Biosciences, USA) electrophoresis system at 2 W/gel for 45 min, followed by 17 W/gel for approximately 4.5 h (until the bromophenol blue reached the bottom of the gel).

Image acquisition and analysis

The CyDye-labelled gels were visualized using a TyphoonTM 9400 imager (GE Healthcare, USA) with the appropriate excitation and emission wavelength filters for each dye, according to the manufacturer's recommendations. All images were processed using Imagemaster 7.0 and then analyzed with DeCyder software (GE Healthcare, USA). The intra-gel analysis was performed using the DeCyder Difference In-gel Analysis system, and inter-gel matching was performed using the DeCyder Biological Variance Analysis, Statistical analyses were conducted for each sample. The spot volume ratios that showed a statistically significant (abundance variation of at least 1.5-fold, $p < 0.05$) difference were processed for further analysis.

Protein digestion and mass spectrometric analysis

Selected protein spots were excised from the preparative gels. Each small gel plug was destained with 100 μL of ACN in 50 mM ammonium hydrogen carbonate for approximately 1 h at room temperature, and this step was repeated until the gel was colorless. After evaporation of the solvent by vacuum centrifugation, each gel plug was rehydrated with 20 μL of 0.01 mg/mL sequencing-grade modified trypsin (Promega, Madison, WI, USA), and the mixture was agitated overnight at 37 °C. The supernatants were collected, and the gel pieces were rinsed once with 5 % TFA in 50 % ACN and then twice with 2.5 %

TFA in 50 % ACN. The supernatants were then combined and lyophilized. The lyophilized peptides were dissolved in 5 mg/mL CHCA (Sigma, USA) in 50 % ACN and 0.1 % TFA. All MS/MS experiments were performed on an Autoflex speed™ MALDI-TOF-MS/MS analyzer (Bruker Daltonics, Germany). The detection conditions were as follows: UV wavelength, 355 nm; recurrence rate, 200 Hz; accelerating voltage, 20,000 V; optimal mass resolution, 1,500 Da; mass of scanning range, 700–3,200 Da. The MS data were processed by flex Analysis (Bruker Daltonics, Germany) to produce a PKL file and analyzed with the NCBI protein sequence database using BioTools (Bruker Daltonics, Germany) via the Mascot search engine.

Biological analysis

Gene ontology (GO) annotations were performed for the identified sequences by MS/MS using BLASTx in the NCBI database. Blast2GO software was then used to annotate the sequence hits by BLASTx (sequences with scores of E > 1e – 05 were discarded). The GO hierarchical terms of homologous genes from the Interpro protein databases were extracted to assign putative functions to the unique sequences. In addition, unique sequences with homology to enzymes involved in metabolic pathways were mapped in accordance with the Kyoto Encyclopedia of Genes and Genomes (KEGG) database. Enzyme commission (EC) numbers were acquired for unique sequences by WUBLASTx searching of the KEGG database. The EC numbers were then used to putatively map unique sequences to specific biochemical pathways.

Confirmation of the mRNA level by RT-PCR
RNA extraction and cDNA synthesis

Total RNA was extracted from frozen cell pellets using the RNeasy mini RNA extraction kit (Qiagen, Germany) according to the manufacturer's instructions. Contaminating genomic (gDNA) was removed using on-column DNaseI digestion performed using the DNaseI digestion kit (Qiagen, Germany). Elution of total RNA was performed using 50 μl of DNase/RNase-free H_2O, and quantified with a NanoDrop 2000 UV–vis spectrophotometer (Thermo Scientific, USA).

Total RNA (4 μg) was used as a template for reverse transcriptase reactions, which were carried out in parallel with M-MuLV Reverse Transcriptase (Sangon Biotech, Shanghai, China), following the manufacturer's instructions. Briefly, total RNA was mixed with 10 μM of random hexanucleotide primers, incubated for 5 min at 70 °C, and kept on ice for 2 min to allow hybridization. Then, RT reaction Mix (buffer 5X, 10 mM each dNTP, RNase inhibitor (20 U/μL)) and reverse transcriptase were added according to the manufacturer's instructions. After 60 min of incubation at

42 °C, the RT enzyme was heat-inactivated at 70 °C. In each case, the total reaction volume was 20 µL.

RT-PCR

Target genes associated with pathogenicity were selected based on the results of the 2D-DIGE analysis. The encoded protein sequence was matched using the NCBI database. Primers used for RT-PCR were designed using Primer3 software and are listed in Additional file 2. The amplification efficiency of the primers for the target genes and the reference gene were validated using the same program.

RT-PCR assays were performed in strip tubes (Qiagen, Germany) in a Rotor-Gene 6000 Real-Time PCR machine (Corbett, Australia) following the protocol provided with SYBR® Premix Ex TaqTM II (TaKaRa, JAPAN). Each reaction consisted of four biological replicates and was conducted in 2 µL of cDNA and 18 µL reaction mixture containing 10 µL SYBR® Premix Ex TaqTM II (2X), 0.8 µL PCR forward primer (10 µM), 0.8 µL PCR reverse primer (10 µM), 2 µL template, and 6.4 µL ddH$_2$O. Each amplification consisted of a denaturation step of 10 s at 95 °C, followed by 40 cycles of 15 s denaturation at 94 °C, 10 s annealing at 55 °C and elongation for 10 s at 72 °C, and then a single fluorescence measurement. Diethyl pyrocarbonate (DEPC)-treated water was used as the negative control.

Detection of metabolites

Metabolites were extracted from the bacterial pellets by the addition of 10 ml methanol:water = 2:1, followed by cell lysis by ultrasonic disruption at 200 W for 15 min on ice and centrifugation (12,000 rpm, 10 min, 4 °C). The supernatant was collected, and the methanol was removed with a swab in the solid phase extraction cartridge. The supernatant was then stored at –80 °C, and the metabolites were freeze-dried. The samples were then transferred onto a pre-washed ultrafiltration membrane and centrifuged (6,000 rpm, 30 min, 4 °C) twice. Filtrates were collected and mixed with ACDSS (Anachro Certified DSS Standard Solution), vortexed (10 s) and centrifuged (13,000 rpm, 2 min, 4 °C).

The ^{1}H NMR measurements were performed at 298 K on a Bruker Avance III 600 MHz spectrometer equipped with an inverse detection cryogenic probe (Bruker Biospin, Germany), which was operated at 600.13 MHz for a ^{1}H resonance frequency. A noesypr1d/noesygppr1d pulse sequence was used to determine the bacterial metabolite profiles. One hundred twenty-eight transitions were collected as 32,768 data points for each spectrum. The ^{1}H NMR signal was imported into the Chenomx NMR suite version 7.6 (Chenomx, Canada), and the data were automatically Fourier-transformed, phase-adjusted and baseline-adjusted. Metabolites from E.

cloacae were quantified using the concentration and peak area of DSS-d6 (2,2-dimethyl-2-silapentane-5-sulfonate-d6 sodium salt) as the standard.

Results

Differential expression of E. cloacae proteins

2D-DIGE was applied to analyze the changes in the proteome of E. cloacae under the different culture conditions. The gel images from the 2D-DIGE separation of E. cloacae are presented in Fig. 1. An average of 1,700 spots were detected in all five 2D-DIGE gels; 720 spots were reproducibly matched to all samples (triplicate runs), and the protein regulatory conditions and the success rate of detection by MS/MS are listed in Table 2. Among the three samples from the three culture conditions, changes greater than 1.5-fold and p-values < 0.05 were considered significant changes in protein abundance. The regulated proteins were selected for identification by MS/MS (Fig. 2), and Additional file 3 shows the MS/MS-identified proteins from E. cloacae cultured under three different conditions. A total of 109 types of protein were successfully identified by MS/MS, which identified 3 or more unique peptides with a confidence of 95 % at the protein level and 99 % at the peptide level.

GO annotation

To understand the biological functions of the differentially expressed proteins under the three treatment conditions, GO annotation was performed. GO representation of the E. cloacae clusters was categorized according to the biological process, cellular component and molecular function (Fig. 3). Each identified protein was classified according to its GO functional annotation. These differentially expressed proteins were mainly localized in the cellular outer membrane, cytoplasm, and plasma membrane and to participate in ATP binding, protein transport and transporter activity.

According to the KEGG metabolic pathway maps of E. cloacae, a total of 23 pathways were clustered into three groups together with the differentially expressed proteins (Fig. 4). Among them, the ABC transporters, citric acid cycle (TCA cycle), glycerophospholipid metabolism, purine metabolism and pyrimidine metabolism were the pathways that were most influenced by the differentially expressed proteins.

RT-PCR analysis of differentially expressed proteins

Twenty-seven genes corresponding to the protein spots that were highly differentially expressed, or related to pathogenicity, were selected for RT-PCR analysis to validate their transcript levels. The relationship between the level of protein and mRNA is displayed in Additional file 4. The RT-PCR results were consistent with those of the DIGE studies and suggested that some proteins that

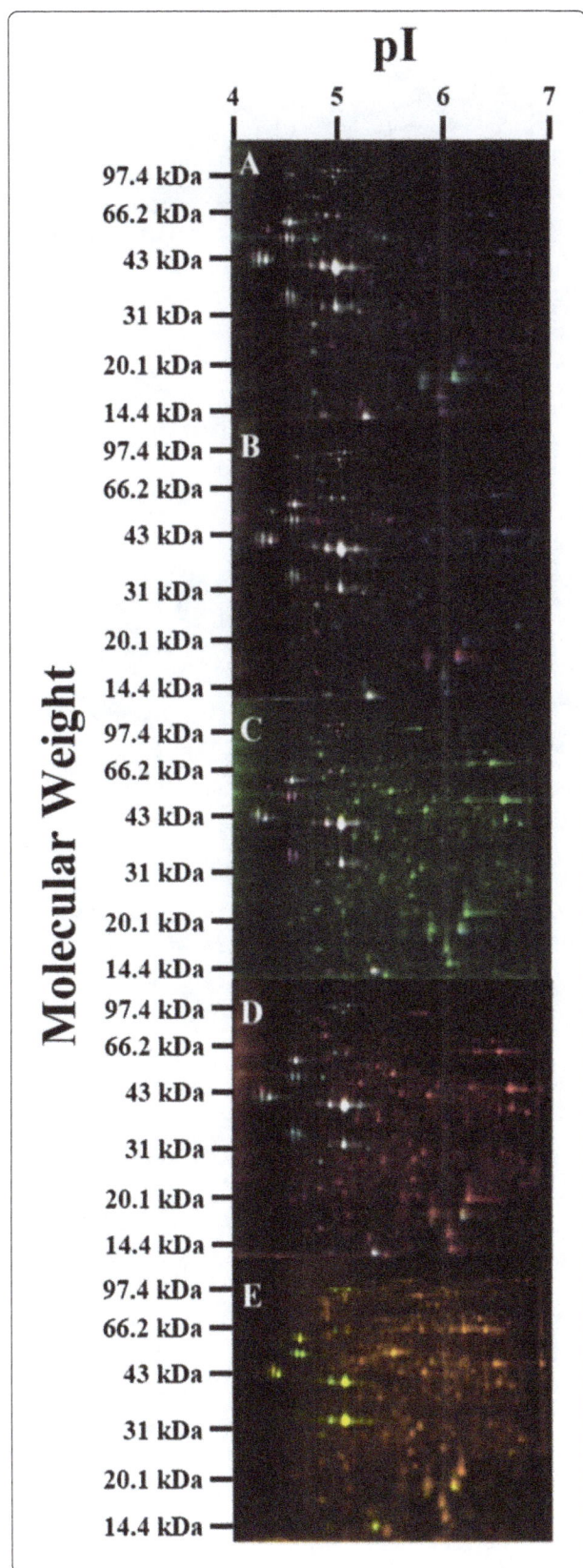

Fig. 1 Proteomic comparison of *E. cloacae* under three different culture conditions using 2D-DIGE. Protein samples (150 µg each) from total *E. cloacae* lysates were labeled with Cy-dyes and separated using 22-cm, pH 4–7 linear IPG strips. Note: A-gel1, B-gel2, C-gel3, D-gel4, E-gel5

were identified as differentially abundant were regulated at the transcriptional level (positive correlation), such as the expression of F0F1 ATP synthase subunit beta (B20), whereas others were not (negative correlation), including the type VI secretion system protein ImpC (A11). Furthermore, some proteins showed no significant correlation between the expression of protein and the gene, such as outer membrane channel protein (A15).

^{1}H NMR spectroscopic analysis of metabolites of E. cloacae

The ^{1}H NMR spectra revealed several metabolites that were modified in *E. cloacae* stimulated by the blood agar plate and Fe (Figs. 5 and 6). A total of 38 individual metabolites were detected in the three treatment groups. Among the 38 types of metabolites, 35 were detected in all treatment groups, thymine and phenylacetate were only detected in normal culture, and NAD^+ was not detected in the control group. Additionally, O-phosphocholine was not detected in the blood agar plate culture.

Three treatments effects on *E. cloacae* metabolites were emphasized during PCA and PLS-DA (Fig. 7). The PCA (principal component analysis) and PLS-DA (partial least squares discriminant analysis) showed that fumarate, acetate, ethanolamine, 2-aminoadipate, glutamate, 2-alanine, glycine, alanine and succinate made an important contribution to distinguishing among the three samples.

Comparison of pathways affected by differentially expressed proteins by the differentially expressed metabolites

The metabolome adds an additional level of information in biological systems that reflects phenotypic and functional variation. Metabolites identified by ^{1}H NMR were used to verify the pathways affected by the differentially expressed proteins. First, we classified the pathways into those that were dysregulated in only one treatment

Table 2 The condition of the different expression of proteins and the detection of MS/MS in *E. cloacae* cultured in three different media

Group	Y1 up-regulated	Y2 up-regulated	Y3 up-regulated	Success rate
Y1:Y2	35[a] (30)[b]	67 (35)	–	63.73 %
Y1:Y3	64 (45)	–	91 (76)	78.06 %
Y2:Y3	–	50 (28)	58 (49)	71.30 %

[a]The number of upregulated proteins. [b]The number of protein which detected successfully by MS/MS

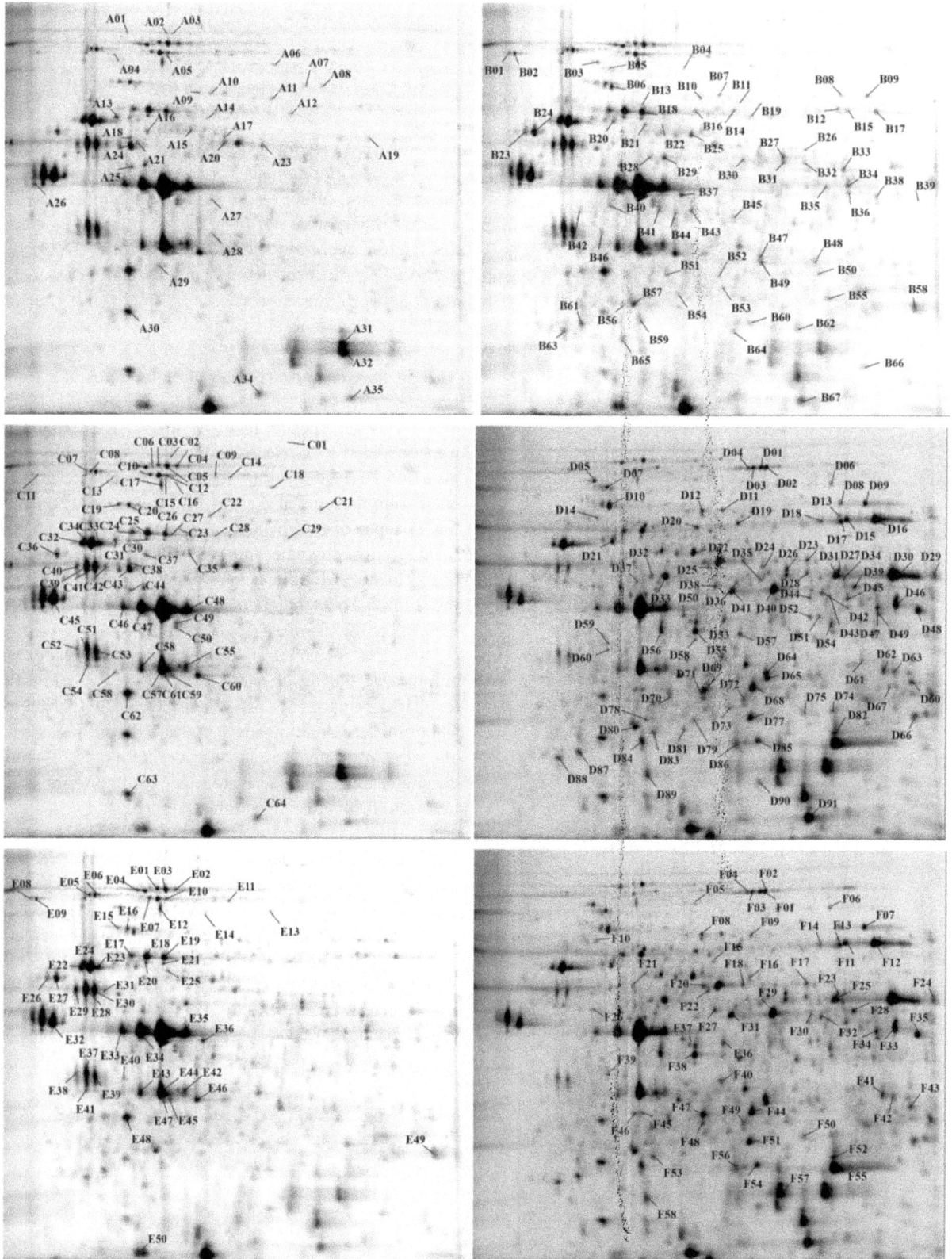

Fig. 2 Proteins exhibiting significant changes were selected for MS/MS identification. A and B denote the upregulated proteins in Y1 and Y2 and in Y1 compared with Y2, respectively. C and D denote the upregulated proteins in Y1 and Y3 and in Y1 compared with Y3, respectively. E and F denote the up-regulated proteins in Y2 and Y3 and in Y2 compared with Y3, respectively

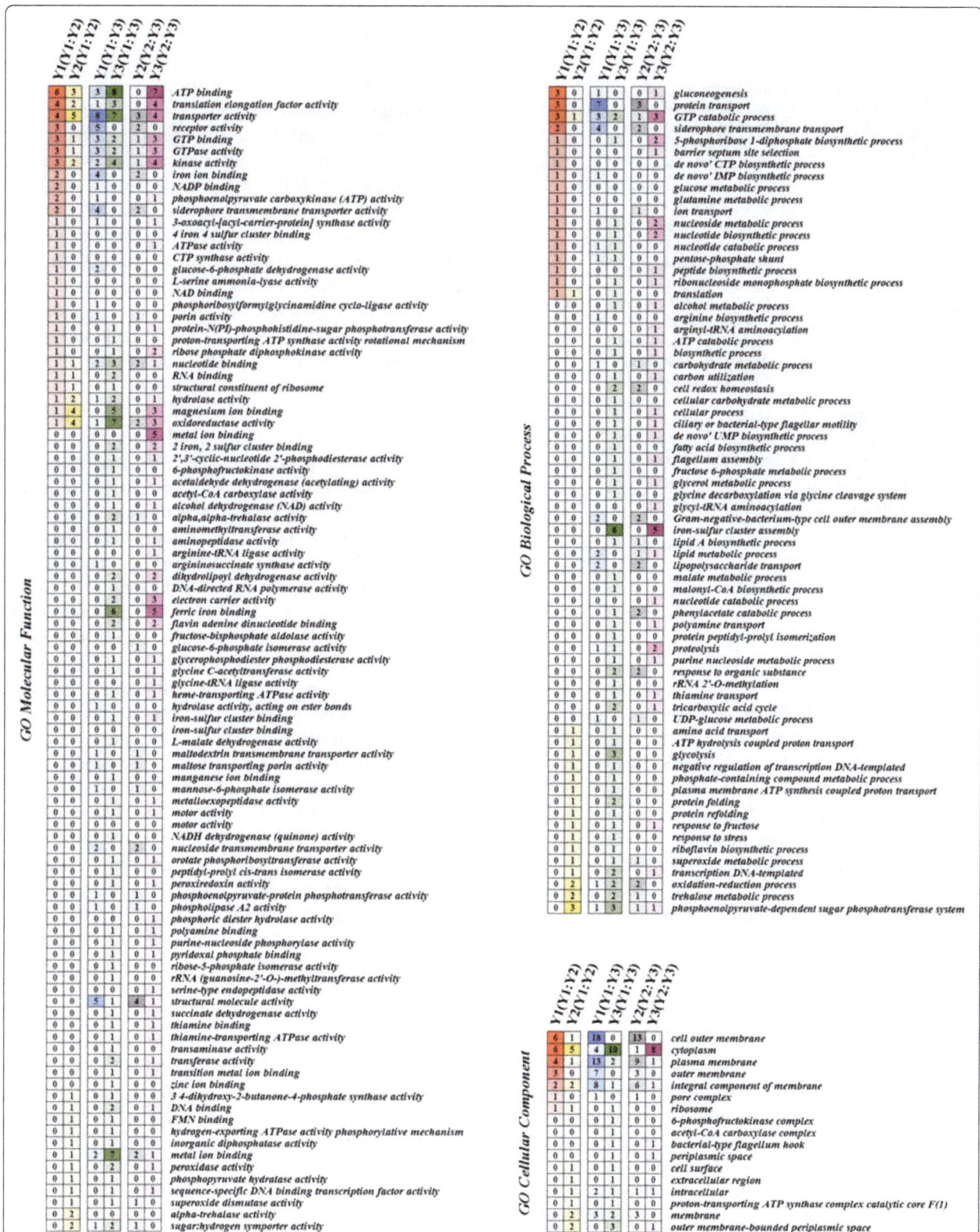

Fig. 3 GO categorization of differentially expressed proteins in *E. cloacae* cultured under three different conditions. The proteins were classified as follows: cellular component, molecular function, and biological processes, according to the GO terms. The color of the square is related to the number of times each function was clustered by proteins; a higher frequency is represented by richer shades of each respective color

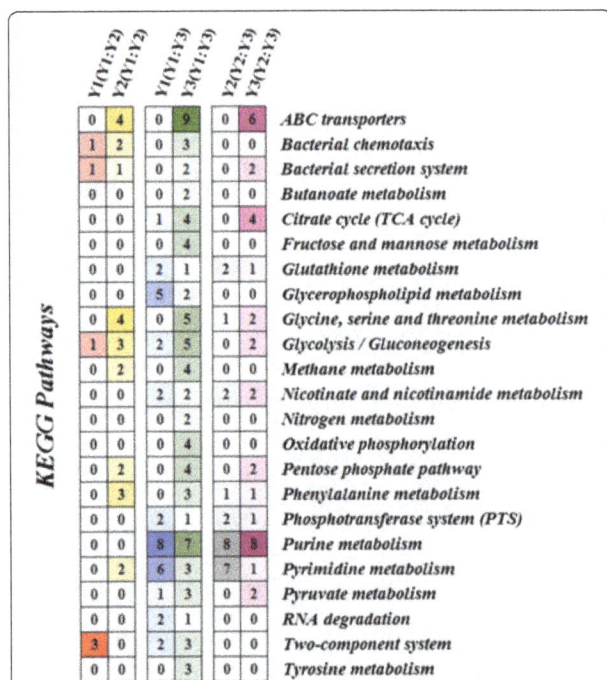

	Y1(Y1:Y2)	Y2(Y1:Y2)	Y1(Y1:Y3)	Y3(Y1:Y3)	Y2(Y2:Y3)	Y3(Y2:Y3)	KEGG Pathways
	0	4	0	9	0	6	*ABC transporters*
	1	2	0	3	0	0	*Bacterial chemotaxis*
	1	1	0	2	0	2	*Bacterial secretion system*
	0	0	0	2	0	0	*Butanoate metabolism*
	0	0	1	4	0	4	*Citrate cycle (TCA cycle)*
	0	0	0	4	0	0	*Fructose and mannose metabolism*
	0	0	2	1	2	1	*Glutathione metabolism*
	0	0	5	2	0	0	*Glycerophospholipid metabolism*
	0	4	0	5	1	2	*Glycine, serine and threonine metabolism*
	1	3	2	5	0	2	*Glycolysis / Gluconeogenesis*
	0	2	0	4	0	0	*Methane metabolism*
	0	0	2	2	2	2	*Nicotinate and nicotinamide metabolism*
	0	0	0	2	0	0	*Nitrogen metabolism*
	0	0	0	4	0	0	*Oxidative phosphorylation*
	0	2	0	4	0	2	*Pentose phosphate pathway*
	0	3	0	3	1	1	*Phenylalanine metabolism*
	0	0	2	1	2	1	*Phosphotransferase system (PTS)*
	0	0	8	7	8	8	*Purine metabolism*
	0	2	6	3	7	1	*Pyrimidine metabolism*
	0	0	1	3	0	2	*Pyruvate metabolism*
	0	0	2	1	0	0	*RNA degradation*
	3	0	2	3	0	0	*Two-component system*
	0	0	0	3	0	0	*Tyrosine metabolism*

Fig. 4 Details of the pathways that cluster with the differentially expressed proteins in the three groups. The color of the square is related to the number of regulated proteins clustered in the pathways; richer shades of each respective color indicate a higher number

group, and then we selected the metabolites that displayed statistically significant changes in abundance (at least 1.5-fold, $p < 0.05$) (Tables 3, 4, 5). Most pathways were verified by the different metabolites, such as in Y1:Y2, and ABC transporters were regulated by proteins that were upregulated in Y2. Eleven metabolites were associated with this pathway, in which 9 compounds (alanine, betaine, glycine, isoleucine, leucine, methionine, phenylalanine, valine, threonine) had a high abundance in the Y2 treatment group and 2 compounds (glutamate, 2-alanine) were highly expressed in the Y1 treatment group.

Discussion

Proteins involved in iron uptake and utilization

Transport proteins play an important role in pathogenicity. This class includes toxins, trans-envelope protein secretion systems, outer membrane protein secretion systems and outer membrane iron-siderophore receptors that function with cytoplasmic membrane ABC-type iron uptake transporters [16]. In our limited research, we were interested in investigating transport proteins related to iron absorption and transportation, and in correlating them with pathogenicity.

In the comparison of Y1 and Y2, the ferrichrome outer membrane transporter (A02, A05), L-serine ammonia-lyase (A14), and hypothetical protein EcWSU1_01016

(A08) (GO cluster analysis associated this protein with metal ion binding (GO:0046872)) were upregulated in *E. cloacae* cultured on a blood agar plate. The sheep blood used in this plate provided the iron ions required by *E. cloacae*, improving its pathogenicity.

In Y1 compared with Y3, the upregulated proteins linking iron absorption and transportation in Y2 were ferric aerobactin receptor (C08), ferrichrome outer membrane transporter (C10, C12, C15), phosphoenolpyruvate-protein phosphotransferase (C19), hypothetical protein EcWSU1_01016 (C21) and LamB type porin (C32). The expression of these 6 proteins was up-regulated more than 5-fold in Y1 compared with Y3. In contrast, the levels of 2',3'-cyclic-nucleotide 2'-phosphodiesterase/3'-nucleotidase (D06), maltose ABC transporter periplasmic protein (D30, D40, D41, D68, D69, D77), phenylacetate-CoA oxygenase, NAD(P)H oxidoreductase component (D31), 6-phosphofructokinase (D44), methionine aminopeptidase (D53), phenylacetic acid degradation protein paaC (D60), osmolarity response regulator (D62), bifunctional acetaldehyde-CoA/alcohol dehydrogenase (D63), succinate dehydrogenase iron-sulfur subunit (D66) and NADH-quinone oxidoreductase subunit E (D79) were up-regulated more than 5-fold in Y3 compared with Y2. The most up-regulated protein was maltose ABC transporter periplasmic protein (D30), with a more than 59-fold increase in expression. The DIGE results suggested that *E. cloacae* expresses more proteins to absorb and transport iron under iron-rich culture conditions. The growth curve of *E. cloacae* under different culture conditions and varying concentrations of Fe^{3+} revealed that high concentrations of Fe^{3+} had a certain inhibitory effect on growth. Although we selected a Fe^{3+} concentration that could promote the growth of *E. cloacae*, we hypothesize that it is difficult for the bacteria to take up and utilize iron from the blood agar plate culture. However, a continuous increase in the concentration of Fe^{3+} may inhibit growth as well as iron uptake and utilization (Additional file 1).

This speculation was confirmed in Y2 compared with Y3 group. We found that some of the previously mentioned proteins related to iron uptake and utilization were upregulated in the common Y2 culture, such as ferric aerobactin receptor (E05), ferrichrome outer membrane transporter (E07), and LamB type porin (E22), among others. In comparisons of the iron concentration, a greater content was detected in Y3 compared with Y1. These findings further suggest that the concentration of iron is important for the growth and pathogenicity of the bacteria.

The identification of the ferric uptake regulator (Fur) family was quite interesting. Fur plays a crucial role in bacterial metabolism, and iron deficiency is the most common nutritional stress during the process of cell

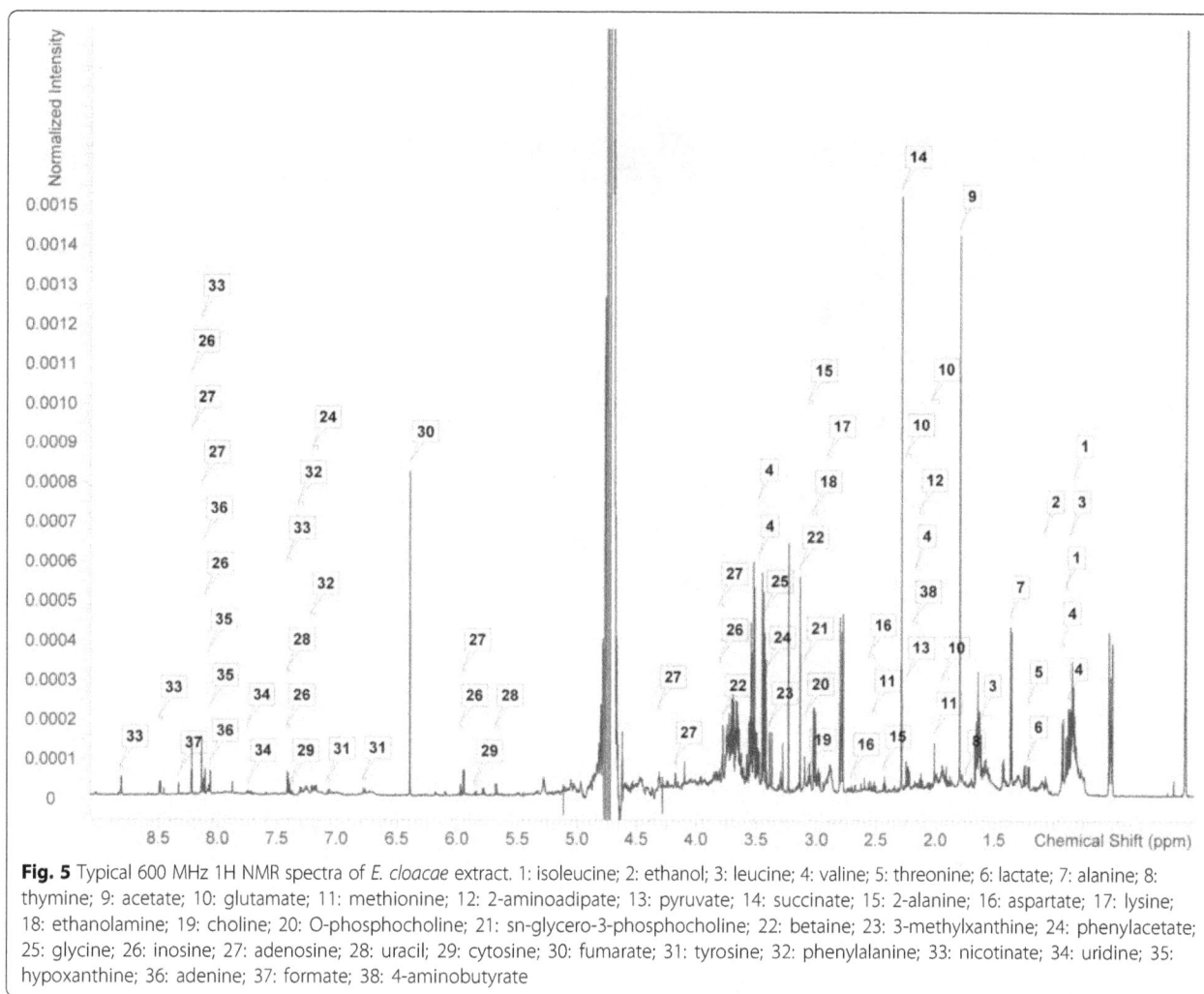

Fig. 5 Typical 600 MHz 1H NMR spectra of *E. cloacae* extract. 1: isoleucine; 2: ethanol; 3: leucine; 4: valine; 5: threonine; 6: lactate; 7: alanine; 8: thymine; 9: acetate; 10: glutamate; 11: methionine; 12: 2-aminoadipate; 13: pyruvate; 14: succinate; 15: 2-alanine; 16: aspartate; 17: lysine; 18: ethanolamine; 19: choline; 20: O-phosphocholine; 21: sn-glycero-3-phosphocholine; 22: betaine; 23: 3-methylxanthine; 24: phenylacetate; 25: glycine; 26: inosine; 27: adenosine; 28: uracil; 29: cytosine; 30: fumarate; 31: tyrosine; 32: phenylalanine; 33: nicotinate; 34: uridine; 35: hypoxanthine; 36: adenine; 37: formate; 38: 4-aminobutyrate

survival [17]. In most prokaryotic organisms, Fur controls iron metabolism and plays a role in the regulation of defenses against oxidative stress. It regulates the expression of iron-binding proteins, which depend on the concentration of iron in the cell [18].

Glycerophospholipid metabolism and ATP-binding cassette (ABC) transporters

After comparing all of the identified pathways, we observed a relationship between glycerophospholipid metabolism and ATP-binding cassette (ABC) transporters in Y1 compared with Y3. Although higher throughput protein analysis technology such as iTRAQ [19] and higher frequency NMR [20] were not used in our limited research, we still identified the relationship between the proteins and metabolites.

First, during glycerophospholipid metabolism, we located 2 dysregulated proteins using DIGE: glycerophosphodiester phosphodiesterase (EC:3.1.4.46, F32) and phospholipase A (EC:3.1.1.4, EC:3.1.1.32, E42). Part A of Fig. 8 shows that phospholipase A can catalyze two

biosynthetic processes that utilize phosphatidylcholine to synthesize 1-acyl-sn-glycero-3-phosphocholine and 2-acyl-sn-glycero-3-phosphocholine. Subsequently, lysophospholipase synthesizes sn-glycero-3-phosphocholine. In Y1, we speculate that sn-glycero-3-phosphocholine accumulated because of the increased expression level of phospholipase A and the down-regulation of glycerophosphodiester phosphodiesterase. The metabolomic results corroborated this hypothesis because the concentration of sn-glycero-3-phosphocholine in Y1 was up-regulated more than 6-fold compared with Y3.

In part B of the glycerophospholipid metabolism analysis (Fig. 8), although glycerophosphodiester phosphodiesterase was highly expressed, we did not detect a difference in sn-glycerol 3-phosphate between the Y1 and Y3 treatment groups. After searching the pathways associated with sn-glycerol 3-phosphate, we found that sn-glycerol 3-phosphate also belongs to the ABC transporter family. After sequencing the genome of *Edwardsiella tarda* EIB202, Wang et al. identified and localized a large number of ABC system components. The

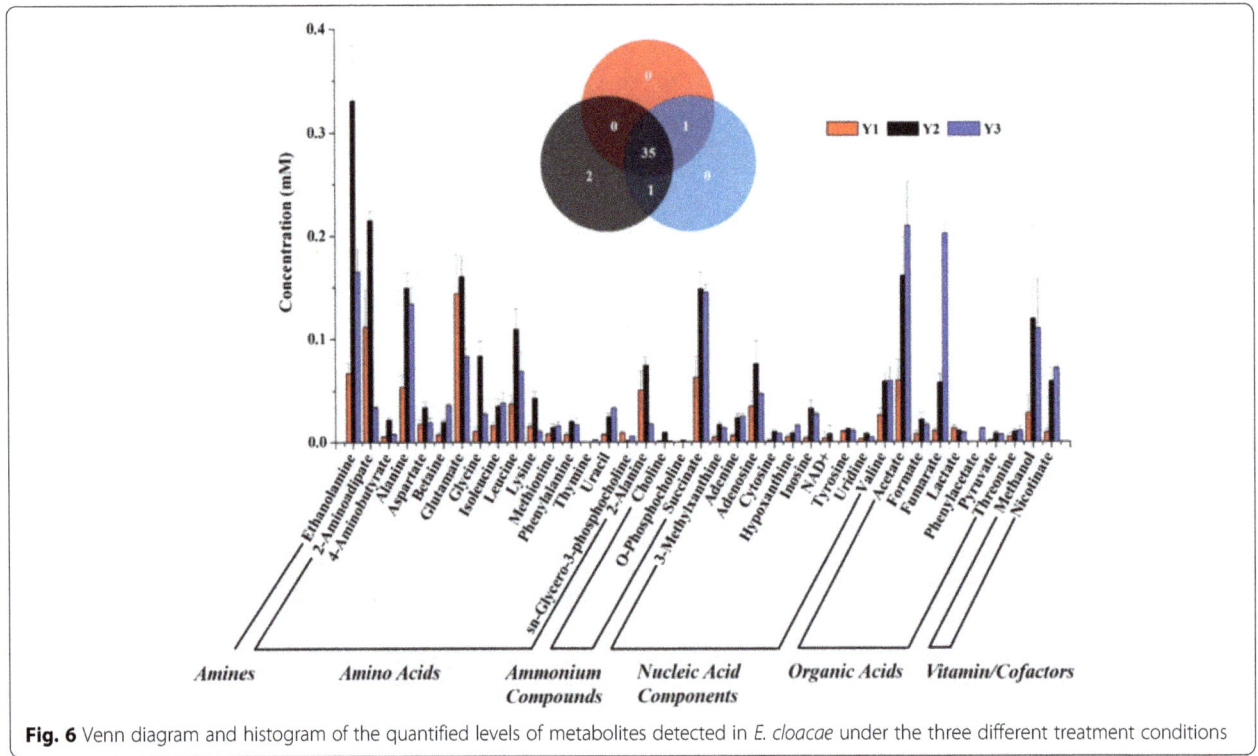

Fig. 6 Venn diagram and histogram of the quantified levels of metabolites detected in *E. cloacae* under the three different treatment conditions

ETAE_0613 and ETAE_0907 components of the ABC system are potential virulence genes, which may provide insight into the relationship between the output of virulence factors and antibiotics and the acquisition of sn-glycerol 3-phosphate [21].

The ATP-binding cassette (ABC) transporters form one of the largest known protein families and are widespread in bacteria, archaea, and eukaryotes. They couple ATP hydrolysis to the active transport of a wide variety of substrates such as ions, sugars, lipids, sterols, peptides, proteins, and drugs. ABC transporters are dedicated to the export of virulence factors under appropriate conditions such as our iron-rich culture condition. An example is provided by iron ABC uptake systems, which have long been recognized as important effectors of virulence [22]. Because iron exists primarily in the

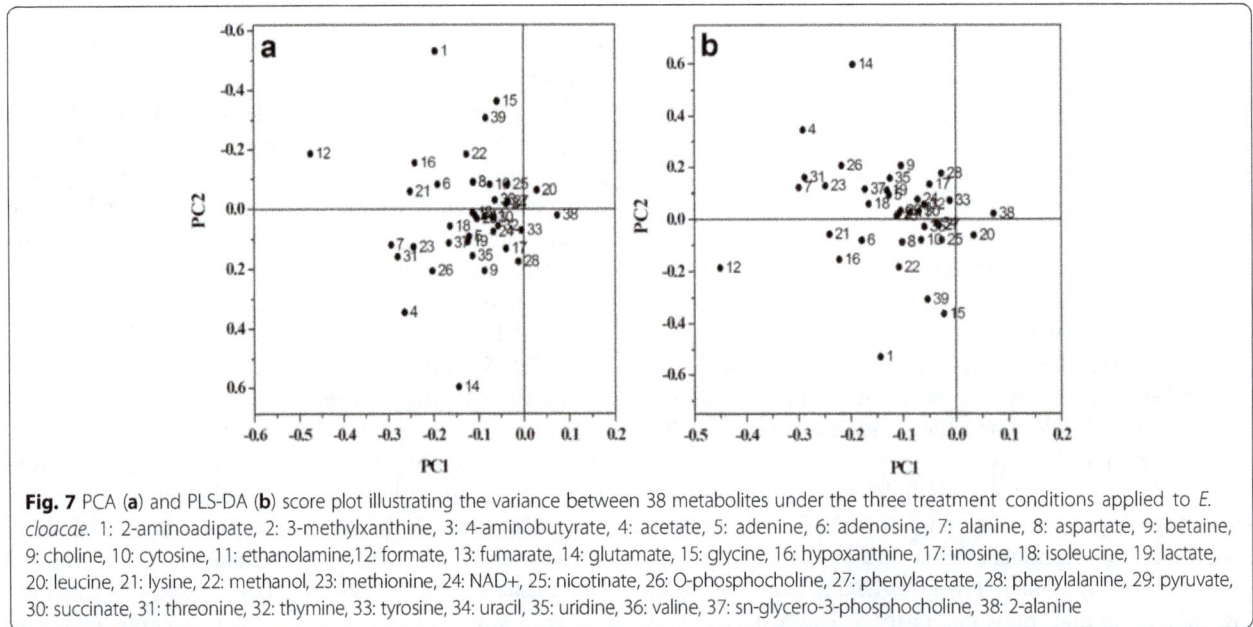

Fig. 7 PCA (**a**) and PLS-DA (**b**) score plot illustrating the variance between 38 metabolites under the three treatment conditions applied to *E. cloacae*. 1: 2-aminoadipate, 2: 3-methylxanthine, 3: 4-aminobutyrate, 4: acetate, 5: adenine, 6: adenosine, 7: alanine, 8: aspartate, 9: betaine, 9: choline, 10: cytosine, 11: ethanolamine,12: formate, 13: fumarate, 14: glutamate, 15: glycine, 16: hypoxanthine, 17: inosine, 18: isoleucine, 19: lactate, 20: leucine, 21: lysine, 22: methanol, 23: methionine, 24: NAD+, 25: nicotinate, 26: O-phosphocholine, 27: phenylacetate, 28: phenylalanine, 29: pyruvate, 30: succinate, 31: threonine, 32: thymine, 33: tyrosine, 34: uracil, 35: uridine, 36: valine, 37: sn-glycero-3-phosphocholine, 38: 2-alanine

Table 3 Validation of pathways affected by in difference expression protein by the different metabolic detected by ^{1}H NMR (Y1 compared with Y2)

The condition of regulated by proteins	Pathways	Upregulated Metabolites	
		Y1	Y2
Y1 upregulated	Two-component system	1[a]	2[b]
Y2 upregulated	ABC transporters	2[c]	9[d]
	Glycine, serine and threonine metabolism	-	4[g]
	Methane metabolism	-	5[h]
	Pentose phosphate pathway	-	1[i]
	Phenylalanine metabolism	-	4[k]
	Pyrimidine metabolism	1[e]	5[f]
Y1 and Y2 Co-regulated	Bacterial chemotaxis	-	-
	Bacterial secretion system	-	-
	Glycolysis / Gluconeogenesis	1[i]	2[j]

[a]Glutamate. [b]Succinate, Fumarate. [c]Glutamate, 2-Alanine. [d]Alanine, Betaine, Glycine, Isoleucine, Leucine, Methionine, Phenylalanine, Valine, Threonine. [e]2-Alanine. [f]Alanine, Thymine, Uracil, Cytosine, Uridine. [g]Betaine, Glycine, Pyruvate, Threonine. [h]Glycine, Acetate, Formate, Pyruvate, Methanol. [i]Lactate. [j]Acetate, Pyruvate. [k]Phenylalanine, Succinate, Fumarate, Pyruvate. [l]Pyruvate

Table 4 Verify pathways affected by in difference expression protein by the different metabolic detected by ^{1}H NMR (Y1 compared with Y3)

The condition of regulated by proteins	Pathways	Upregulated Metabolites	
		Y1	Y3
Y3 upregulated	Bacterial secretion system	-	-
	Butanoate metabolism	-	3[a]
	Nitrogen metabolism	-	1[b]
	Bacterial chemotaxis	-	1[c]
	Phenylalanine metabolism	-	4[d]
	Tyrosine metabolism	-	2[e]
	Fructose and mannose metabolism	-	-
	Methane metabolism	-	5[f]
	Oxidative phosphorylation	-	3[g]
	Pentose phosphate pathway	-	1[h]
	Glycine, serine and threonine metabolism	-	6[i]
	ABC transporters	-	13[j]
Y1 and Y3 Co-regulated	Pyruvate metabolism	-	5[k]
	Citrate cycle	-	3[l]
	Glutathione metabolism	-	1[m]
	Phosphotransferase system	-	1[n]
	RNA degradation	-	-
	Nicotinate and nicotinamide metabolism	-	5[o]
	Two-component system	-	3[p]
	Glycolysis/Gluconeogenesis	-	2[q]
	Glycerophospholipid metabolism	1[r]	2[s]
	Pyrimidine metabolism	-	4[t]
	Purine metabolism	-	5[u]

[a]Succinate, Fumarate, Pyruvate. [b]Formate. [c]Aspartate. [d]Phenylalanine, Succinate, Fumarate, Pyruvate. [e]Fumarate, Pyruvate. [f]Glycine, Acetate, Formate, Pyruvate, Methanol. [g]Succinate, NAD$^+$, Fumarate. [h]Pyruvate. [i]Aspartate, Betaine, Choline, Glycine, Pyruvate, Threonine. [j]Alanine, Aspartate, Betaine, Glycine, Isoleucine, Leucine, Lysine, Methionine, Phenylalanine, Choline, Succinate, Valine, Threonine. [k]Succinate, Acetate, Formate, Fumarate, Pyruvate. [l]Succinate, Fumarate, Pyruvate. [m]Glycine. [n]Pyruvate. [o]Aspartate, NAD$^+$, Fumarate, Pyruvate, Nicotinate. [p]Aspartate, Succinate, Fumarate. [q]Acetate, Pyruvate. [r]sn-Glycero-3-phosphocholine. [s]Ethanolamine, Choline. [t]Alanine, Uracil, Cytosine, Uridine. [u]Glycine, Adenine, Adenosine, Hypoxanthine, Inosine

insoluble Fe^{3+} form under aerobic conditions, biologically available iron in the body is found chelated by high-affinity iron-binding proteins (BPs) (e.g., transferrins, lactoferrins, and ferritins) or as a component of erythrocytes (such as heme, hemoglobin, or hemopexin) [23]. Pathogens are able to scavenge iron from these sources by secreting high affinity iron-complexing molecules called siderophores and reabsorbing them as iron-siderophore complexes [24]. For example, lactoferrin-binding protein B (LbpB) is a bi-lobed membrane-bound lipoprotein that is part of the lactoferrin receptor complex in a variety of Gram-negative pathogens [25]. Our DIGE results revealed the location of the iron complex transport system based on the differential expression of the iron-hydroxamate transporter ATP-binding subunit (EC: 3.6.3.34, F41) (using DIGE, F41 were identified as an osmolarity response regulator by MS/MS, after transformation in the *E. cloacae* subsp. cloacae ENHKU01 by BLAST, the spot was confirmed to be iron-hydroxamate transporter ATP-binding subunit, with 95 % confidence). After comparing the expression of this protein among the three treatment groups, the abundance was ranked as Y3 > Y1 > Y2. This result indicated that Fe stimulation of *E. cloacae* was greater in the medium with Fe supplementation than in the blood agar plate.

Bacterial secretion system

The host interactions of pathogenic bacteria are usually mediated via protein secretion mechanisms. Gram-negative pathogenic bacteria will transport protein to the extracellular environment or to the host cell through devices called secretion systems. To date, a least six different types of secretion systems have been discovered in Gram-negative pathogenic bacteria (I-VI secretion system). These systems can stimulate and interfere with the processes of host cells by secreting or releasing and injecting extracellular proteins or effectors [26]. Using DIGE, we located two types of secretion systems, type I and VI, based on the differential expression of the outer membrane channel protein

Table 5 Verify pathways affected by in difference expression protein by the different metabolic detected by ^{1}H NMR (Y2 compared with Y3)

The condition of regulated by proteins	Pathways	Upregulated Metabolites	
		Y2	Y3
Y3 upregulated	Bacterial secretion system	-	-
	Glycolysis/Gluconeogenesis	-	-
	Pentose phosphate pathway	-	-
	Pyruvate metabolism	-	1[a]
	Citrate cycle	-	1[b]
	ABC transporters	1[c]	6[d]
Y2 and Y3 Co-regulated	Phenylalanine metabolism	1[e]	-
	Glycine, serine and threonine metabolism	1[f]	3[g]
	Glutathione metabolism	-	2[h]
	Phosphotransferase system	-	-
	Nicotinate and nicotinamide metabolism	-	-
	Pyrimidine metabolism	-	-
	Purine metabolism	-	-

[a]Fumarate. [b]Fumarate. [c]Betaine. [d]Aspartate, Glutamate, Glycine, Leucine, Lysine, Choline. [e]Fumarate. [f]Betaine. [g]Aspartate, Glycine, Choline. [h]Glutamate, Glycine

(TolC), type VI secretion system secreted protein Hcp (hemolysin co-regulated protein) and type VI secretion system protein ImpC.

The type I secretion systems (T1SS) are responsible for the release of a variety of extracellular proteins and extracellular enzymes [27]. TolC, which we identified by DIGE, is associated with multiple drug resistance in bacteria. The expression of TolC in the three treatment groups was as follows: Y1 > Y2 > Y3. Although the majority of research investigating type I secretion systems has focused on multiple drug resistance, it can be speculated from the results of our study that the protein also has an association with pathogenicity and participates in responses to differences in iron stimulation.

In contrast, type VI secretion systems (T6SS) have a clear and strong correlation to pathogenicity, and nearly all confirmed functional T6SS are poisonous to macrophages [28]. Hcp can cross the T6SS transport channel to enter the plasma and interact with the host through the help of lipoprotein [29]. Due to protein modifications or degradation, as mentioned previously, in the group of Y1 compared with Y2, similar Hcp levels were observed. In contrast, in Y1 compared with Y3, the expression of Hcp in Y3 was almost 10-fold higher than that in Y1. Similarly, the type VI secretion system protein ImpC was approximately 2-fold higher in Y3 than in Y2. These findings

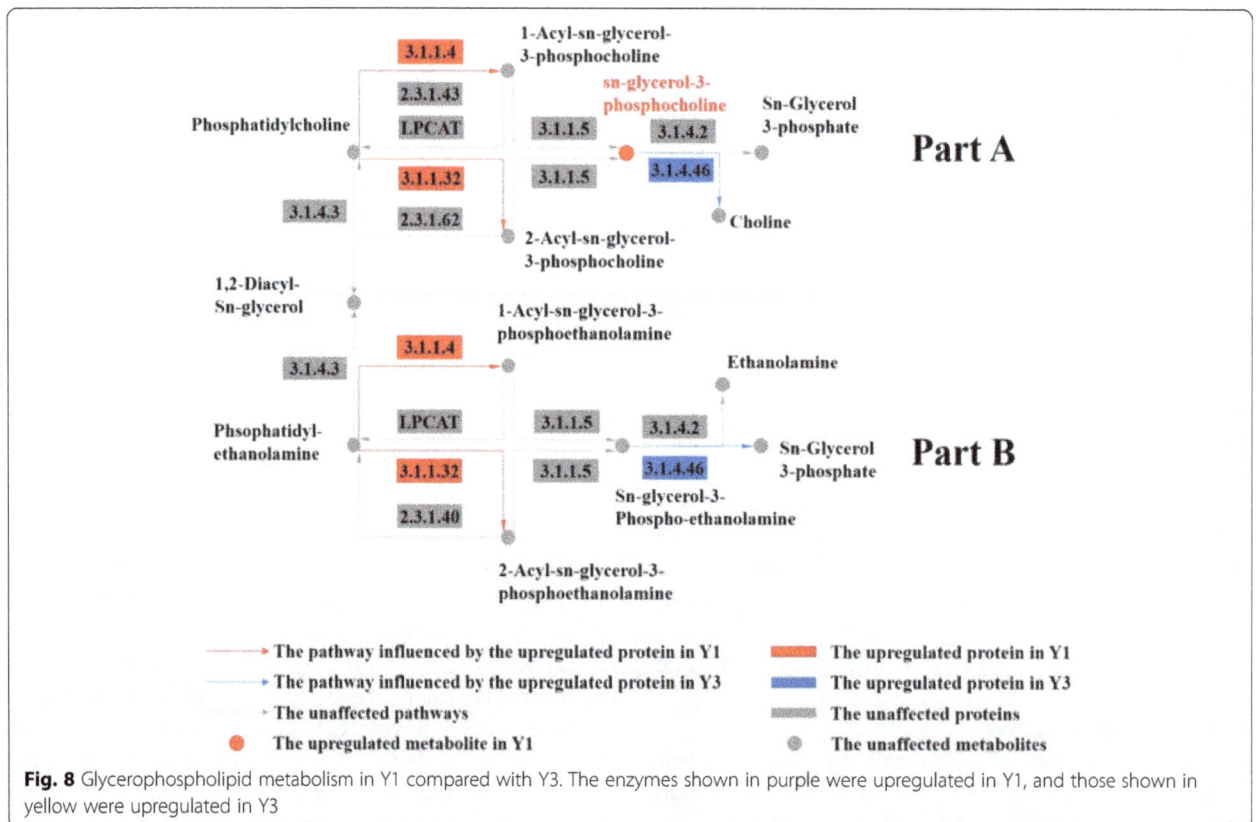

Fig. 8 Glycerophospholipid metabolism in Y1 compared with Y3. The enzymes shown in purple were upregulated in Y1, and those shown in yellow were upregulated in Y3

indicated that the secretion of Hcp was regulated by iron, which supports the research of Wang et al., who found that iron was one of the regulators of the T6SS component *evp*P in *Edwardsiella tarda* [30].

Two-component system

Two-component signal transduction systems enable bacteria to sense, respond, and adapt to changes in their environment or in their intracellular environment. In this experiment, differences in the two-component system were observed between Y1 and Y2 and between Y1 and Y3, but not between Y2 and Y3. The differentially expressed proteins were involved in resistance to the osmotic upshift (K^+). This result was consistent with our experimental design, and the major difference between the Y2 and Y3 culture conditions was the presence of iron; however, in Y1, the blood agar plate culture condition was the main difference compared with Y2 and Y3, in which the sheep blood fiber is enriched with a variety of elements, including K^+. Bacteria are sensitive to changes in the external environment, and consequently they undergo a series of mechanisms to adapt and protect themselves. Specifically, they must protect themselves against the immune response of the host during infection. In the blood agar plate condition, K^+ potentially caused a change in osmotic pressure, stressing the cells and potentially leading to cell lysis and death. As a result, the two-component system probably helped the bacteria resist the change in osmotic pressure and protected them by regulating the expression of outer membrane proteins. Although the system has mainly been reported in terms of bacterial responses to climate change, the present study also shows that the system plays a role in pathogenicity. The system can adjust the various metabolic processes of bacteria, the bacterial cell cycle, the exchange of signals between bacteria and the expression of virulence factors [31].

Conclusion

The regulation of environmental factors leads to both physiological and biochemical changes in bacteria. As a result, the pathogenicity of the bacteria also changes in response to environmental stimuli. The results of this study showed that the blood agar plate and a suitable concentration of iron ions enhanced the pathogenicity of *E. cloacae* and that very high concentrations of iron may have had an adverse effects on growth and on iron uptake and utilization by this bacteria. It is difficult to make an absolute comparison of the stimulatory effect of blood versus iron on pathogenicity. The pathogenicity of *E. cloacae* is affected by their living conditions and the condition of the bacteria.

Additional files

Additional file 1: Growth curve of *E. cloacae* cultured under six different Fe^{3+} concentrations. (PDF 116 kb)

Additional file 2: Primer sequence for RT-PCR (XLS 28 kb)

Additional file 3: Differentially expressed protiens upon three treatment as identified by MALDI-TOF-MS/MS (XLS 77 kb)

Additional file 4: The correlation of expression between protiens and genes (XLS 40 kb)

Abbreviations
2D-DIGE: Two-dimensional difference gel electrophoresis; ABC: ATP-binding cassette; ACDSS: Anachro-Certified DSS Standard Solution; DSS-d6: 2, 2-dimethyl-2-silapentane-5-sulfonate-d6 sodium salt; EC: Enzyme commission; ESBLs: Extended-spectrum β-lactamases; Fur: Ferric uptake regulator; GO: Gene ontology; Hcp: Hemolysin co-regulated protein; IEF: Iso-electric focusing; IPG: Immobilized pH gradients; Itraq: Isobaric tags for relative and absolute quantitation; NMR: Nuclear magnetic resonance; PBS: Phosphate-buffered saline; RT-PCR: Reverse transcription PCR; SDS-PAGE: Sodium dodecyl sulfate-polyacrylamide gel electrophoresis; T1SS: Type I secretion systems; T6SS: Type VI secretion systems

Acknowledgements
This work was supported by the Public Science and Technology Research Fund ocean projects (201105007, 201005016), The Ningbo Ocean Economic Innovation and Development of Regional Demonstration Projects (Research and Demonstration of Key Technology on Marine Bacteria Detection of Gene Chip Industralization), the National Natural Science Foundation of China (41306135), the K.C. Wong Magna Fund from Ningbo University and The Scientific Research Foundation of the Graduate School of Ningbo University. We are grateful for the valuable suggestions of the editors and reviewers regarding the revision of our manuscript.

Funding
This work was supported by the Public Science and Technology Research Funds ocean projects (201105007, 201005016), The Ningbo Ocean Economic Innovation and Development of Regional Demonstration Projects (Research and Demonstration of Key Technology on Marine Bacteria Detection of Gene Chip Industralization), and the National Natural Science Foundation of China (41306135). All of the above funding played roles in the study design and collection, analysis, and interpretation of data. The K.C. Wong Magna Fund from Ningbo University and The Scientific Research Foundation of the Graduate School of Ningbo University played a role in writing the manuscript.

Authors' contributions
DZ performed the 2D-DIGE and drafted the manuscript. WH performed the RT-PCR. QT performed the NMR. JZ performed the statistical analysis. XS participated in the design of the study and helped to draft the manuscript. All authors read and approved the final manuscript.

Competing interests
The authors declare that they have no competing interests.

References
1. Cascio A, Mezzatesta ML, Odierna A, Di Bernardo F, Barberi G, Iaria C, Stefani S, Giordano S. Extended-spectrum beta-lactamase-producing and carbapenemase-producing *Enterobacter cloacae* ventriculitis successfully treated with intraventricular colistin. Int J Infect Dis. 2014;20:66–7.
2. Dijk Y, Bik E, Hochstenbach-Vernooij S, Vlist G, Savelkoul P, Kaan J, Diepersloot R. Management of an outbreak of *Enterobacter cloacae* in a neonatal unit using simple preventive measures. J Hosp Infect. 2002;51:21–6.
3. Sanders W, Sanders CC. *Enterobacter* spp.: pathogens poised to flourish at the turn of the century. Clin Microbiol Rev. 1997;10:220–41.

4. Hoffmann H, Roggenkamp A. Population genetics of the nomenspecies Enterobacter cloacae. Appl Environ Microbiol. 2003;69:5306–18.

5. Yu W-L, Cheng H-S, Lin H-C, Peng C-T, Tsai C-H. Outbreak investigation of nosocomial *Enterobacter cloacae* bacteraemia in a neonatal intensive care unit. Scand J Infect Dis. 2000;32:293–8.

6. Antony B, Prasad BPMR. An outbreak of neonatal septicaemia by *Enterobacter cloacae*. Asian Pac J Trop Dis. 2011;1:227–9.

7. Zhou Q, Zhang M, Wang A, Xu J, Yuan Y. Eight-Year Surveillance of Antimicrobial Resistance among *Enterobacter Cloacae* Isolated in the First Bethune Hospital. Phys Procedia. 2012;33:1194–6.

8. Yogaraj JS, Elward AM, Fraser VJ. Rate, risk factors, and outcomes of nosocomial primary bloodstream infection in pediatric intensive care unit patients. Pediatrics. 2002;110:481–5.

9. Zhang D, Li C, Zhou J, Zhang C, Wang Z, Su X. Research of the structure diversity of bacteria form Ningbo coastal outfall and virulence genes associated with iron metabolism. Oceanologia et limnologia sinica. 2013;44:1627–35.

10. Lu F, Miao S, Tu J, Ni X, Xing L, Yu H, Pan L, Hu Q. The role of TonB-dependent receptor TbdR1 in Riemerella anatipestifer in iron acquisition and virulence. Vet Microbiol. 2013;167:713–8.

11. Olakanmi O, Kesavalu B, Abdalla MY, Britigan BE. Iron acquisition by Mycobacterium tuberculosis residing within myeloid dendritic cells. Microb Pathog. 2013;65:21–8.

12. Braun V. Iron uptake mechanisms and their regulation in pathogenic bacteria. Int J Med Microbiol. 2001;291:67–79.

13. Brown JS, Holden DW. Iron acquisition by Gram-positive bacterial pathogens. Microbes Infect. 2002;4:1149–56.

14. Mahapatra A, Ghosh S, Mishra S, Pattnaik D, Pattnaik K, Mohanty S. *Enterobacter cloacae*: a predominant pathogen in neonatal septicaemia. Indian J Med Microbiol. 2002;20:110.

15. Daniels NA, Shafaie A. Review of pathogenic Vibrio infections for clinicians. Infect Med. 2000;68:665–85.

16. Tang F, Saier MH. Transport proteins promoting *Escherichia coli* pathogenesis. Microb Pathog. 2014;71(1):41–55.

17. Fillat MF. The FUR (ferric uptake regulator) superfamily: diversity and versatility of key transcriptional regulators. Arch Biochem Biophys. 2014;546:41–52.

18. Wee S, Neilands JB, Bittner ML, Hemming BC, Haymore BL, Seetharam R. Expression, isolation and properties of Fur (ferric uptake regulation) protein of *Escherichia coli* K 12. Biol Met. 1988;1:62–8.

19. Kaltwasser B, Schulenborg T, Beck F, Klotz M, Schafer KH, Schmitt M, Sickmann A, Friauf E. Developmental changes of the protein repertoire in the rat auditory brainstem: a comparative proteomics approach in the superior olivary complex and the inferior colliculus with DIGE and iTRAQ. J Proteomics. 2013;79:43–59.

20. Masetti O, Ciampa A, Nisini L, Valentini M, Sequi P, Dell'Abate MT. Cherry tomatoes metabolic profile determined by ^{1}H-High Resolution-NMR spectroscopy as influenced by growing season. Food Chem. 2014;162:215–22.

21. Wang Q, Yang M, Xiao J, Wu H, Wang X, Lv Y, Xu L, Zheng H, Wang S, Zhao G. Genome sequence of the versatile fish pathogen *Edwardsiella tarda* provides insights into its adaptation to broad host ranges and intracellular niches. PLoS One. 2009;4:e7646.

22. Henderson DP, Payne SM. *Vibrio cholerae* iron transport systems: roles of heme and siderophore iron transport in virulence and identification of a gene associated with multiple iron transport systems. Infect Immun. 1994; 62:5120–5.

23. Köster W. ABC transporter-mediated uptake of iron, siderophores, heme and vitamin B 12. Res Microbiol. 2001;152:291–301.

24. Wandersman C, Delepelaire P. Bacterial iron sources: from siderophores to hemophores. Annu Rev Microbiol. 2004;58:611–47.

25. Morgenthau A, Beddek A, Schryvers AB. The negatively charged regions of lactoferrin binding protein B, an adaptation against anti-microbial peptides. PLoS One. 2014;9:e86243.

26. Yoshida Y, Miki T, Ono S, Haneda T, Ito M, Okada N. Functional characterization of the type III secretion ATPase SsaN encoded by salmonella pathogenicity island 2. PLoS One. 2014;9:e94347.

27. Delepelaire P. Type I secretion in gram-negative bacteria. Biochim Biophys Acta. 2004;1694:149–61.

28. Shanks J, Burtnick MN, Brett PJ, Waag DM, Spurgers KB, Ribot WJ, Schell MA, Panchal RG, Gherardini FC, Wilkinson KD. Burkholderia mallei tssM encodes a putative deubiquitinase that is secreted and expressed inside infected RAW 264.7 murine macrophages. Infect Immun. 2009;77:1636–48.

29. Shrivastava S, Mande SS. Identification and functional characterization of gene components of Type VI Secretion system in bacterial genomes. PLoS One. 2008;3:e2955.

30. Wang X, Wang Q, Xiao J, Liu Q, Wu H, Xu L, Zhang Y. *Edwardsiella tarda* T6SS component evpP is regulated by esrB and iron, and plays essential roles in the invasion of fish. Fish Shellfish Immunol. 2009;27:469–77.

31. Hoch JA. Two-component and phosphorelay signal transduction. Curr Opin Microbiol. 2000;3:165–70.

9

Follicular fluid biomarkers for human in vitro fertilization outcome: Proof of principle

Fang Chen[1], Carl Spiessens[1], Thomas D'Hooghe[1], Karen Peeraer[1] and Sebastien Carpentier[2*]

Abstract

Background: Human follicular fluid (FF) is a unique biological fluid in which the oocyte develops in vivo, and presents an optimal source for non-invasive biochemical predictors. Oocyte quality directly influences the embryo development and hence, may be used as a predictor of embryo quality. Peptide profiling of FF and its potential use as a biomarker for oocyte quality has never been reported.

Methods: This study screened FF for peptide biomarkers that predict the outcome of in vitro fertilization (IVF). Potential biomarkers were discovered by investigating 2 training datasets, consisting both of 17 samples and validating on an independent experiment containing 32 samples. Peptide profiles were acquired by nano-scale liquid chromatography coupled to tandem mass spectrometry (nano LC-MS/MS).

Results: From the training datasets 53 peptides were found as potential biomarker candidates, predicting the fertilization outcome of 24 out of the 32 validation samples blindly (81.3% sensitivity, 68.8% specificity, AUC = 0.86). Seven potential biomarker peptides were identified. They were derived from: insulin-like growth factor binding protein-5, alpha-2-antiplasmin, complement component 3, inter-alpha-trypsin inhibitor heavy chain H1, serum albumin, protein diaphanous homolog 1 and plastin-3.

Conclusions: The MS-based comprehensive peptidomic approach carried out in this study, established a novel panel of potential biomarkers that present a promising predictive accuracy rate in fertilization outcome, and indicates FF as an interesting biomarker resource to improve IVF clinic routine.

Keywords: Peptide, Biomarker, Oocyte quality, Follicular fluid

Background

Over the years, in vitro fertilization is associated with a high rate of multiple pregnancies, which presents both perinatal complications and economic complaints [1–3]. To reduce the incidence of multiple pregnancies, single embryo transfer (SET) is the only strategy [4, 5]. The selection of high quality embryos remains the major challenge in human assisted reproductive technology (ART). Worldwide, the selection of embryos has been based on morphological assessments. However, there is still a lack of evidence-based standard for ranking embryos and determining the embryo with the highest implantation potential [6]. The low rate of successful pregnancies creates the need to increase the predictive value for implantation.

Given the fact that RNA, proteins, and cellular machinery are provided by the oocyte during early zygote development, oocyte quality determines a big part of the embryo development and hence, may be used as a predictor of embryo quality [7]. Human follicular fluid (FF) has been attracting researchers' interest since it is non-invasive and easily available. FF is a product of both the transfer of blood plasma constituents that cross the blood follicular barrier and of the secretory activity of granulosa and thecal cells [8]. FF is a complex mixture of proteins, metabolites, and ionic compounds, which have been found to reflect the stage of oocyte development and the degree of follicle maturation [9–14].

* Correspondence: sebastien.carpentier@biw.kuleuven.be
[2]Facility for Systems Biology based Mass Spectrometry (SYBIOMA), KU Leuven, Leuven, Belgium
Full list of author information is available at the end of the article

It has also been previously shown that altered FF composition is associated with a diminished reproductive capacity [14, 15]. Therefore, it is reasonable to think that some biochemical characteristics of the FF reflect oocyte quality and influence fertilization [16]. In body fluids, biomarkers are often low molecular weight peptides and proteins [17, 18]. Given the complexity of the numerous independent processes involved in oocyte maturation, it is unlikely that a single biomarker can classify the oocytes [19]. Application of powerful proteomic and peptidomic technologies in reproductive medical research may significantly contribute to help diagnosis but also the comprehensive understanding of reproductive processes. Several laboratories have demonstrated the feasibility of selecting peptide/protein diagnostic biomarkers in follicular fluid [20–22].

Efforts to explore the follicular fluid proteomic signature using different proteomic approaches have been carried out by several groups [19, 23–26]. The study performed by Spitzer et al. compared protein patterns in FF from immature and mature FF using two-dimensional gel electrophoresis (2-DE) [27], reporting considerable differences in protein patterns derived from fluids of immature compared with matured follicles. Hanrieder et al. coupled isoelectric focusing to nano liquid chromatography and MALDI TOF/TOF, and identified 69 proteins in FF of women undergoing IVF [28]. Twigt et al. using SDS-PAGE and isoelectric focusing followed by LC-MS/MS identified 246 FF proteins which are involved in acute phase response and immunological function [19]. Ambekar et al. combined three different methods of protein/peptide fractionation and identified 480 proteins in FF [24]. A more recent study performed by Zamah et al. identified 742 proteins in follicular fluid from fertile woman [22].

Despite decades of efforts, comparative studies on FF with respect to IVF outcome have not yet been done yet. Specific and sensitive proteomic biomarker candidates for IVF outcome have not been found. The low-molecular-weight (LMW) subset of proteome is termed the "peptidome", including peptides and small proteins with molecular weights generally less than 10,000. At the early stage in the development of the peptidome field, the information content of peptides resides in two general categories: first, bioactive peptides and fragments shed from cells in the microenvironment, such as hormones and cytokines. Therefore, peptides may reflect the cell-to-cell communications taking place in the microenvironment [29]. The other is the cleavage products produced by enzymes and proteases as a consequence of certain physiological or pathological processes, such as apoptosis or necrosis, which may serve as reporters for biological enzymatic states of individuals [29–31]. Taking the diagnostic information carried by the peptidome into consideration, measuring panels of peptidome makers is expected to generate a higher

level of prognostic capacity. Peptides are constantly generated in vivo by active synthesis, and by proteolytic processing of larger precursor proteins, often yielding protein fragments that mediate a variety of physiological functions [32]. This study aimed to reveal if peptide profiling of individual FF could become a new non-invasive predictive biomarker for oocyte quality, and attempted to discover candidate biomarkers for fertilization. In biomarker studies, candidate biomarkers need to be validated across a large number of samples because of normal clinical or biological variability. To ensure that the discovered biomarkers are truly associated with fertilization three experiments were designed with a relatively large population. We investigated the peptide profile of human follicular fluid with successful fertilization and unsuccessful fertilization from patients undergoing in vitro fertilization using LC_MS/MS. We additionally determined the protein identities of the discovered peptide biomarkers as a first step toward understanding the pathways in which they may function.

Methods
Study design
A total number of 66 follicular fluid samples from 50 couples undergoing IVF/ICSI treatment at the Leuven University Fertility Center were analyzed. All the patients were undergoing the first or second treatment cycle with a single embryo transferred (SET) and ranged from 18 to 36 years old. Patients with repeated implantation failures (cycle rank > 2) were not considered. Patients were fully informed and consents were obtained before oocyte retrieval. The study was approved by the Commission for Medical Ethics of the University Hospital Leuven (code ML6214).

The follicular fluid samples were analyzed in 3 experiments. The first training dataset contained 17 samples (8 successfully fertilized oocytes (showing 2 pronuclei), 9 unfertilized mature oocytes), the second training dataset contained 17 samples (7 successfully fertilized oocytes, 10 unfertilized mature oocytes). The validation dataset contained 32 samples (16 successfully fertilized oocytes, 16 unfertilized mature oocytes). All the fertilized oocytes in this study resulted in implantation after SET. Samples were randomly selected regardless the treatment of insemination (IVF/ICSI). Samples of training sets and validation sets were shown in Table 1. Patients' characteristics were detailed in Table 2 and Additional file 1: Table S2.

Ovarian stimulation, oocyte retrieval and follicular fluid collection
The stimulation protocol used in this study has been published by Debrock [33]. Briefly, ovarian stimulation was carried out with gonadotropins (Menopur, Ferring, Copenhagen, Denmark; Gonal-F or Metrodin HP, Merck-

Table 1 Characteristics and treatments of the training and validation cohorts

Characteristics	Training (n = 34)	Validation (n = 32)
ICSI unfertilized mature oocytes(n)	11	7
ICSI fertilized oocytes (n)	10	7
IVF unfertilized mature oocytes(n)	8	9
IVF fertilized oocytes (n)	5	9

Serono, Geneva, Switzerland; Puregon, Organon, Oss, The Netherlands) and GnRH agonists (Buserlin acetate, Suprefact; Hoechst, Frankfurt, Germany) during a long or short protocol. The follicular response was monitored serum oestradiol levels and transvaginal ultrasound measurements. The hCG, 10,000 IU, was administered when at least three follicles reached a diameter of 17 mm. Oocyte retrieval was performed 35 h after hCG injection by ultrasound guided transvaginal aspiration. The luteal phase was supported with intravaginal application of P (600 mg/day, Utrogestan; Besins, Drogenbos, Belgium) started at the evening of the hCG injection.

Follicular fluid was aspirated and collected separately, then kept on ice immediately. Each follicle was flushed twice. To maintain a stable pH in a room atmosphere condition, commercial medium Dulbecco's phosphate-buffered saline (DPBS; Gibco, Paisley, UK) was used as flushing medium. For each FF sample, the volume and color appearance (yellow, light reddish, reddish, dark reddish and red) were recorded. Only FF of yellow or light reddish was allowed for further analysis.

The samples were centrifuged at 1500 * g for 10 min. The supernatant was transferred to a cryotube and stored in liquid nitrogen until further processing.

Table 2 Patients' characteristics

	fertilized	non-fertilized
number of patients	31	35
maternal age	31.2 ± 2.7	31.7 ± 3.7
retrieved oocytes	10.6 ± 4.7	11.0 ± 4.3
matured oocytes	9.2 ± 3.6	9.6 ± 3.3
fertilized oocytes	6.2 ± 3.1	6.4 ± 2.8
IVF	14	17
ICSI	17	18
male factor	16	15
male factor and famale factor	4	6
anovulation	4	3
endometriosis	4	4
transport	2	1
unexplained	1	6

In vitro insemination/Intracytoplasmic sperm injection

Prior to fertilization, oocytes were washed 4 times with wash medium GM501 (GM 501 Wash, Gynemed Lensahn, Germany) after retrieval in order to minimize the amount of blood/follicular fluid. Oocytes were placed separately in a 4-well dish (Nunc, Thermo Fisher Scientific, Kamstrupvej, Denmark) containing wash medium GM501 (GM 501 Wash, Gynemed Lensahn, Germany) under oil. Spermatozoa for the IVF/ICSI procedure were prepared using standard density gradient procedures (Isolate, Irvine Scientific, USA) or, in cases with very low sperm quality, diluted and centrifuged twice at 300 g for 10 min. Standard IVF/ICSI procedures were performed 2–6 h after oocyte retrieval. In the IVF procedure, oocytes were inseminated with 10,000 progressively motile spermatozoa per oocyte. In the ICSI procedure, the cumulus and corona cells were removed with hyaluronidase (conc.80 IU/m, Gynemed, Lensahn, Germany). The oocytes were injected with single sperm in a 20 µl droplet of medium. The injected oocytes were cultured individually in 20 µl culture medium (GM 501 Culture, Gynemed Lensahn, Germany) droplets under oil. On Day 1 (16–20 h after insemination/injection) fertilization was evaluated.

Peptide extraction

Follicular fluid samples (500 µl) were transferred to extraction tubes (2 ml) and mixed with an equal amount of lysis buffer containing 30 mM Tris (SIGMA, St. Louis, USA), 8 M urea (Acros Organics, New Jersey, USA), 5 mM Dithiothreitol DTT, Applichem, Darmstadt, Germany). They were vortexed thoroughly and centrifuged at 13,000 rpm for 10 min at room temperature. The supernatant was then transferred to a 3 K Da filter Microcon YM-30 filters (Millipore, Billerica, MA, USA), the filter devices were subsequently centrifuged at 14,000 g for 30 min and the flow through was collected. Iodoacetamide (IAA, SIGMA, St. Louis, USA) was added to a final concentration of 0.015 M and the samples were incubated for 30 min at room temperature in dark. Then trifluoroacetic acid (TFA) was added to a final concentration of 0.1%. The mixture was cleaned via solid phase extraction (SPE) using Pepclean C18 spin column (Thermo Scientific) according to the manufactuer's instructions and eluted in a final volume of 40 µL. Subsequently, the samples were dried in a vacuum operator. The dried sample was kept at –20°C until analysis. The peptides were dissolved in 10 µl of 0.1% formic acid (FA) and 5% acetonitrile (ACN).

MS Data Processing

Five microliters from each sample were injected and separated on an Ultimate® 3000 RSLCnano system (Dionex, Thermo Scientific, Netherlands) equipped with a

Thermo Scientific™ Acclaim™ PepMap™ RSLC Nano-Trap Column with nanoViper™ Fittings, 3 μm Particle Size. The samples were separated using a buffer A (water/0.1% FA) and buffer B (water 20%/ACN 80%/FA 0.08%) and a Thermo Scientific™ EASY-Spray™ column PepMap™ RSLC, C18, 2 μm, 100 Å, 50 μm x 150 mm using a gradient of 4 to 10% B in 12 min followed by a gradient of 10 to 35% B in 20 min, a gradient 35 to 65% B in 5 min and then a final elution and re-equilibration step at 95 and 5% buffer B respectively for 9 min. The flow-rate was set at 0.300 μL/min. The hybrid quadrupole orbitrap mass spectrometer, Q Exactive (Thermo Scientific), was operated in positive ion mode with a nanospray voltage of 1.5 kV and a source temperature of 250°C. ProteoMass LTQ/FT-Hybrid ESI Pos. Mode CalMix (MSCAL5-1EA SUPELCO, Sigma-Aldrich) was used as an external calibrant and the lock mass 445.12003 as an internal calibrant. The instrument was operated in data-dependent acquisition (DDA) mode with a survey MS scan at a resolution of 70,000 for the mass range of m/z 400–1600 for precursor ions, followed by MS/MS scans of the top 10 most intense peaks with +2, +3, +4 and +5 charged ions above a threshold ion count of 16,000 at 17,500 resolution using normalized collision energy (NCE) of 25 eV with an isolation window of 3.0 m/z, an apex trigger 5–15 s and a dynamic exclusion of 10 s. All data were acquired with Xcalibur 3.0 software (Thermo Scientific).

The LC-MS data were imported to Progenesis Nonlinear software Progenesis v4.1 (Nonlinear Dynamics, UK) for alignment and normalization to compare the different sample runs. The software selected automatically the sample run with the greatest similarity to all other runs as the reference alignment. The aligned runs containing all ion peak information from all sample files was exported as mgf and send to Mascot (version 2.2.06; database swissprot 15,720 accessions). Mass tolerance was set to 10 ppm for MS and 20 mmu for MS/MS, and no cleavage enzyme for protein digestion was chosen. Search parameters allowed for carbamidomethylation as fixed modification and oxidation of M as variable modification. Search results were evaluated by Scaffold™ (released version 4.4.5) combining mascot and X! Tandem. Only peptides with an expected value of <0.05 are reported.

Statistical analysis

The aligned and normalized peptide data were exported from Progenesis as csv format. The determination of candidate peptides in the training datasets was accomplished by partial least square discriminative analysis (PLS-DA) using the NIPALS algorithm of Statistica 8.1 (Statsoft). The analysis of variance (ANOVA) was taken over from Progenesis. The prediction capability of biomarker candidates and blind classifying was evaluated by

principle component analysis (PCA) using the NIPALS algorithm of Statistica 8.1 (Statsoft).

To evaluate the predictive capability of the biomarker candidates, receiver operating characteristic (ROC) analysis was carried out with MATLAB (2014 a).

Probability calculations were performed according to the binomial distribution. The probability of getting exactly x successes in n trials is given by the probability mass function: $b(x; n, P) = nCx * P^x * (1 - P)^{n - x}$.

Results
Training datasets

To avoid over fitting of the PLS-DA model, we firstly examined 2 training datasets of 17 samples and evaluated the differences at the peptidome level of the individual follicular fluid samples. Peptides that met both standards: (1) $p < 0.1$ with ANOVA, and (2) top 10% on variable importance ranking were selected and compared for each experiment. In the first experiment, 12,998 peptides were detected. 394 peptides met the criteria. In the second experiment, 11,216 peptides were detected, of which 7760 peptides were in common with the first experiment (Fig. 1a), and 504 peptides met the criteria. Among all the interesting candidate peptides described above, 53 were common to both training datasets (Fig. 1b). Results of partial least square discriminate analysis were shown in Fig. 2. Those peptides are listed in Table 3. Of the 53 peptides, 9 peptides were upregulated in the fertilized group and 44 were upregulated in the non-fertilized group (Table 3). The area under ROC curve

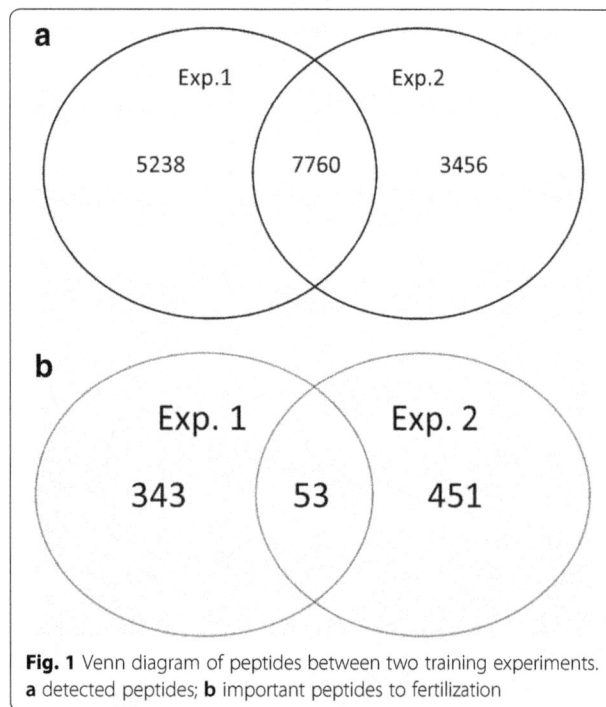

Fig. 1 Venn diagram of peptides between two training experiments. **a** detected peptides; **b** important peptides to fertilization

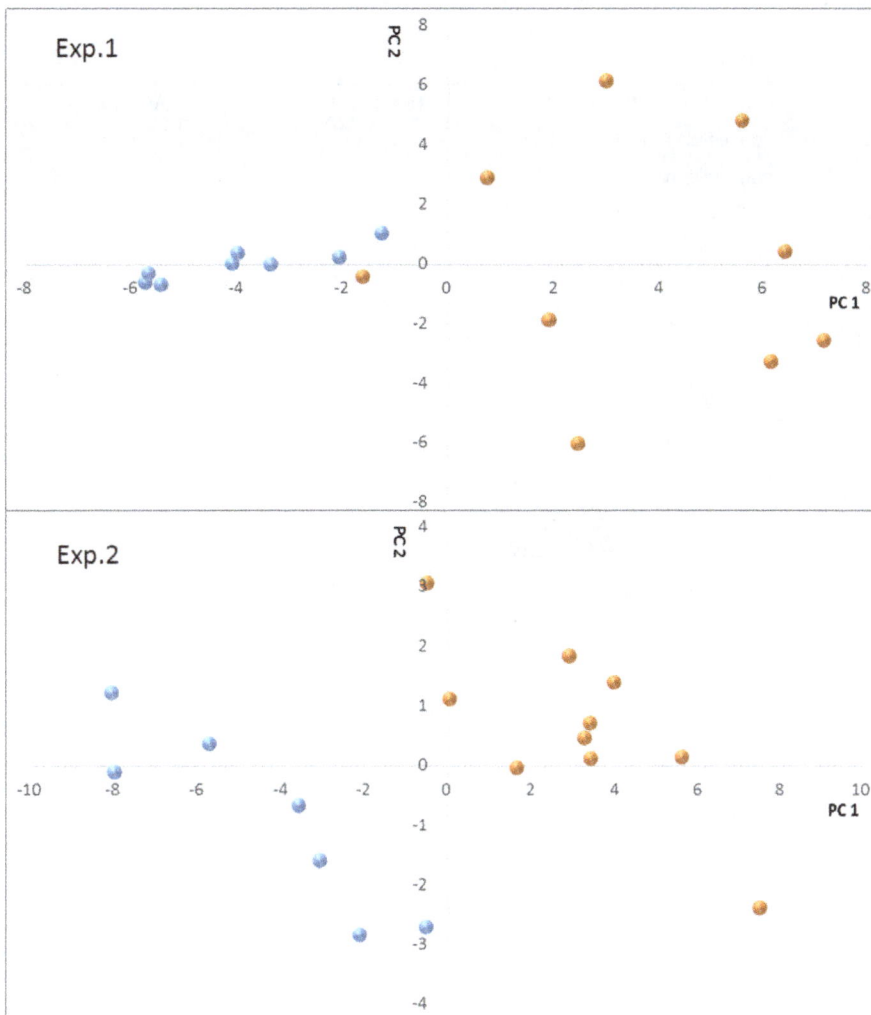

Fig. 2 Screening candidate biomarkers in FF for fertilization in training cohorts. PLS scores show evident clustering between fertilized (*blue*) and non-fertilized (*orange*) samples. Samples were separated well in the first component

(AUC) was 0.97, with the sensitivity of 0.82 and the specificity of 0.84 (Fig. 3).

External validation

The most reliable way of biomarker discovery is to test the candidates with a large cohort of samples. From the perspective of generality, 32 samples from an external study population were analyzed and we predicted the oocyte fertilization outcome blindly via principle component analysis (PCA). The 53 potential biomarkers discriminate 24 out of the 32 candidates (Fig. 4). The chance of discriminating 24/32 cases randomly is 0.25% according to the binomial distribution. As indicated above, 44/53 markers are negatively correlated to fertilization.

Receiver operation characteristic (ROC) analysis indicates that the 53 peptides panel has a high predictive ability to differentiate FF samples upon fertilization outcome, with the AUC value of 0.86, the sensitivity of 81.3%, and the specificity of 68.8%.

Peptide identification

The identification of non-tryptic peptides is challenging due to the absence of a known cleavage side and the absence of a basic amino acid at the C-terminus. In total 7102 unique peptides were identified, belonging to 159 proteins (protein false discovery rate (FDR) 0.06% and peptide FDR 0.00% (Additional file 2: Table S1 and Additional file 3: Table S3)). From the 53 peptides panel, we were able to identify 7 peptides derived from 7 different proteins (Table 3). The MS spectrum of the identified peptides is displayed in Additional file 4: Figure S1.

Subcellular and functional annotation of identified peptides

We checked the subcellular localization of all the peptides identified. We found that 5 proteins were localized

Table 3 Peptides set as biomarkers differentially detected between fertilized group and non-fertilized group

m/z	RT	Charge	Highest mean condition	max fold change		Protein	Accession number	Peptide sequence
				Exp.1	Exp.2			
401.22 ± 0.00	28.2 ± 0.11	2*	non-fertilized	8.94[b]	3.07[c]	ALBU_Human	P02768	AASQAALGL
402.23 ± 0.00	25.3 ± 0.66	1	non-fertilized	2.16[b]	2.44[c]			
409.18 ± 0.00	24.8 ± 0.22	1	non-fertilized	2.7[a]	1.99[b]			
410.17 ± 0.00	36.4 ± 0.26	1	fertilized	3.09[a]	11.07[b]			
412.34 ± 0.00	44.3 ± 0.02	1*	non-fertilized	3.36[b]	2.78[b]			
414.34 ± 0.01	43.3 ± 2.37	1*	non-fertilized	1.93[c]	2.00[c]			
416.37 ± 0.00	46.0 ± 0.15	1*	non-fertilized	2.11[c]	14.22[a]			
424.29 ± 0.01	46.5 ± 0.15	1*	fertilized	4.18[b]	2.24[a]			
426.32 ± 0.02	48.0 ± 0.77	1	fertilized	1.49[c]	2.33[b]			
438.36 ± 0.02	52.4 ± 1.14	1	non-fertilized	2.67[c]	2.76[a]			
439.28 ± 0.00	25.0 ± 0.38	1*	non-fertilized	2.88[b]	215.28[b]			
442.94 ± 0.00	27.2 ± 0.60	3	non-fertilized	21.09[b]	2.57[a]			
459.72 ± 0.00	22.1 ± 0.30	2*	non-fertilized	27.37[b]	12.84[b]			
475.01 ± 0.00	23.4 ± 0.48	4*	non-fertilized	5.63[b]	1.77[b]			
477.28 ± 0.00	47.6 ± 0.48	1*	non-fertilized	25.53[a]	25.84[c]			
478.62 ± 0.00	41.8 ± 0.00	3*	non-fertilized	3.02[b]	1.18[c]			
480.80 ± 0.00	40.5 ± 0.82	2*	fertilized	1.22[c]	1.30[c]			
490.24 ± 0.02	46.5 ± 0.25	1	fertilized	1.70[b]	8.68[c]			
494.29 ± 0.02	46.4 ± 0.97	1*	non-fertilized	549.84[a]	2.59[b]	IGBP-5_Human	P24593	FVGGAENTAHPRII
496.97 ± 0.00	43.1 ± 0.39	3	fertilized	1.36[b]	1.40[b]			
500.24 ± 0.00	26.7 ± 0.26	3	non-fertilized	341.3[c]	2.55[c]			
526.41 ± 0.02	52.1 ± 1.08	1	non-fertilized	3.59[b]	3.10[a]			
528.28 ± 0.00	34.1 ± 0.11	2*	fertilized	3.61[b]	1.94[c]	CO3_Huamn	P01024	IHWESASLL
530.29 ± 0.00	45.8 ± 1.13	1*	fertilized	1.87[c]	6.26[c]			
539.66 ± 0.00	44.7 ± 0.41	6	non-fertilized	2.08[b]	1.73[b]			
545.33 ± 0.00	44.8 ± 0.26	6	non-fertilized	Infinit[a]	2.08[b]			
546.99 ± 0.00	44.7 ± 0.42	6	non-fertilized	1.93[b]	1.60[b]			
552.30 ± 0.00	28.7 ± 0.11	5	non-fertilized	68.46[c]	2.24[b]			
552.67 ± 0.00	44.7 ± 0.40	6	non-fertilized	3.05[c]	1.60[c]			
554.33 ± 0.00	44.7 ± 0.41	6	non-fertilized	2.00[c]	1.70[b]			
558.40 ± 0.00	51.1 ± 0.31	2	non-fertilized	1795.51[a]	Infinit[a]			
570.43 ± 0.02	51.9 ± 1.06	1	non-fertilized	4.31[b]	3.77[c]			
574.68 ± 0.00	44.8 ± 0.40	6	non-fertilized	4.33[b]	2.39[c]			
576.35 ± 0.00	44.8 ± 0.41	6	non-fertilized	2.24[b]	1.94[b]			
577.28 ± 0.00	33.5 ± 0.77	1*	non-fertilized	1.42[b]	2.26[b]			
579.18 ± 0.00	44.8 ± 0.39	6	non-fertilized	2.19[b]	1.95[c]			
588.43 ± 0.01	45.5 ± 1.66	1	non-fertilized	43.29[b]	4.46[c]			
594.30 ± 0.00	24.0 ± 0.14	2*	fertilized	3.09[b]	2.90[a]	ITIH1_human	P19827	LPDRVTGVDTD
602.42 ± 0.00	51.4 ± 1.02	2	non-fertilized	164.28[c]	Infinit[a]			
609.72 ± 0.00	30.0 ± 0.00	5	non-fertilized	10.0[b]	1.91[a]			
643.33 ± 0.00	29.0 ± 1.59	2*	non-fertilized	Infinit[c]	2.59[a]	A2AP_human	P08697	MEPLGRQLTSGP
658.49 ± 0.02	51.7 ± 0.99	1	non-fertilized	7.59[b]	11.57[b]			
702.51 ± 0.02	51.6 ± 1.01	1	non-fertilized	6.36[b]	12.66[c]			

Table 3 Peptides set as biomarkers differentially detected between fertilized group and non-fertilized group *(Continued)*

703.75 ± 0.00	31.1 ± 1.97	5	non-fertilized	Infinit[a]	1.79[b]			
727.59 ± 0.00	28.0 ± 0.07	4*	non-fertilized	1266.81[a]	2.03[b]			
740.96 ± 0.00	30.6 ± 0.05	5	non-fertilized	4.35[b]	1.91[a]	PLST_HUMAN	P13797	DGETLEELMKLSPEELLLRWANFHLENSGWQ
755.34 ± 0	12.2 ± 0.36	2	non-fertilized	2.04[b]	2.16[a]			
761.97 ± 0.00	31.0 ± 0.01	5	non-fertilized	6.85[b]	2.10[a]			
805.39 ± 0.00	35.8 ± 0.01	1*	non-fertilized	Infinit[c]	7.38[b]			
859.44 ± 0.00	29.4 ± 0.04	4	non-fertilized	Infinit[a]	2.94[a]			
869.68 ± 0.00	32.7 ± 0.15	4	non-fertilized	Infinit[b]	3.92[c]			
924.74 ± 0.00	38.0 ± 0.41	3*	non-fertilized	Infinit[c]	9.77[b]			
953.51 ± 0.00	29.1 ± 0.23	3	non-fertilized	12.72[b]	3.24[c]	DIAP1_HUMAN	O60610	AEPHFLSILQHLLLVRNDYEARPQ

[a]$p < 0.01$; [b]$p < 0.05$; [c]$p < 0.1$
*:significant difference in validation experiment

at extracellular space. Insulin-like growth factor binding protein-5 (IGBP-5) is secreted by granulosa cells, four proteins are predicted to come from the circulation system. Two proteins were predicted to be localized in the intracellular region, cytoplasm and membrane.

Discussion

The morphological assessments for human embryo selection in clinic IVF routine is not fully satisfying. In the last decades, scientists have been attempting to improve the embryo selection. As the microenvironment of oocyte in vivo, the protein/peptide content of human follicular fluid has attracted researchers' interest. However, to date, research correlates protein/peptide profiling to IVF outcome has been done yet. The present study focused on the peptide profile of follicular fluid with different fertilization outcomes to screen potential

peptide biomarkers for fertilization. Our results have shown altered level of certain peptides contributed to the fertilization outcome. By external validation, we confirmed the classification capability of these peptides. We think that the findings reported identify a pool of peptides from which novel IVF-related biomarkers could be discovered.

In our study, the concentration of one peptide (m/z = 401.23) was significantly less abundant in the fertilized group and was identified as a fragment derived from serum albumin (ALBU_Human) (Table 3). In literature, albumin as a protein has been positively correlated to oocyte quality. Junko et al. [34] proposed that the biochemically reduced state of albumin in FF may play an important role in protecting oocytes from oxidative damage. Our data point towards a negative correlation at the fragment peptide level and might be related to proteolytic processing.

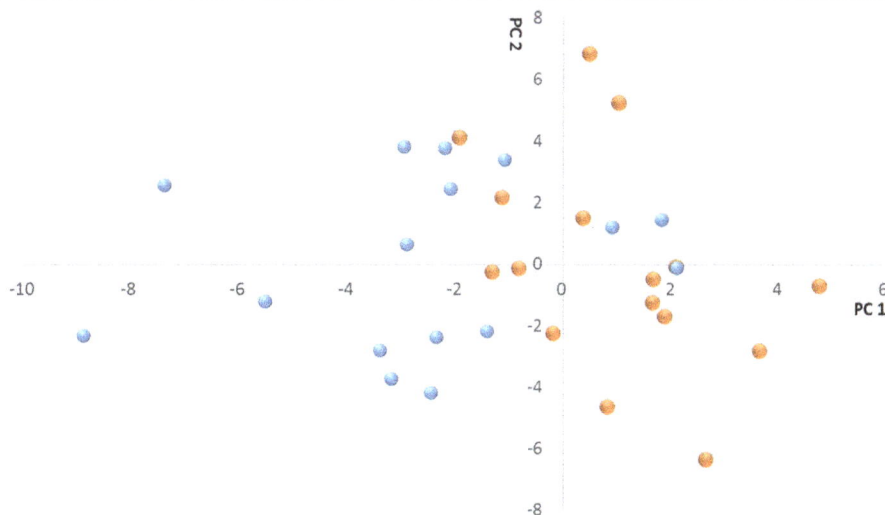

Fig. 3 ROC curve test of FF peptide biomarker candidates. *Blue*: ROC curve on training dataset; *Red*: ROC curve on validation dataset (*p*-value =0.13)

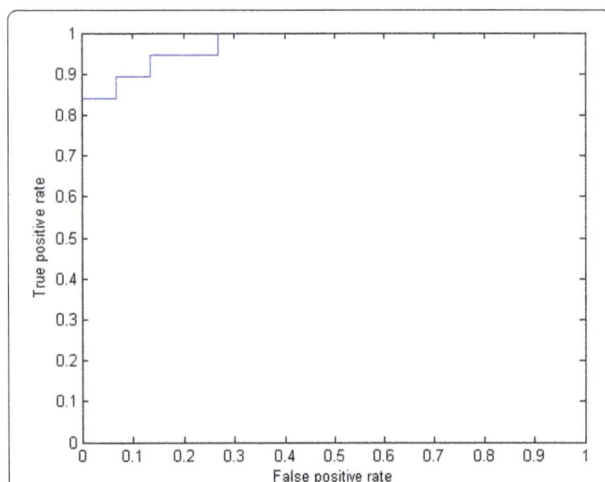

Fig. 4 Visual identification of the hFF samples using the 53 peptide biomarker candidates. PCA score plots derived from PCA of 53 important biomarker candidates of 32 hFF with fertilized oocytes or non-fertilized oocytes. PC1, explaining 36.6% of the variability is clearly associated to the outcome of fertilization

We identified m/z = 494.59 derived from insulin-like growth factor binding protein-5 (IBP-5_Human) as a second interesting negative biomarker (Table 3). IGF binding proteins (IGFBP) inhibit insulin-like growth factor (IGF) actions. The IGF system plays an important role in regulating ovarian follicular development and steroidogenesis [35] and IGFBP proteolysis is a major mechanism for regulating IGF bioavailability [36]. The gene of IGFBP-5 was reported to be highly expressed in rat primary and secondary follicles while dominant follicles were devoid of IGFBP-5 mRNA [37]. In several other studies [38–42], dominant follicles were characterized by decreased levels of low molecular weight IGF-binding proteins (IGF-2,4, and –5). The high abundance of peptide derived from IGFBP-5 may reflect the increased level of IGFBP-5 in preovulatory follicle, which reduces the activity of growth factors, or by an increased activity of its protease or both.

We identified m/z = 643.33 derived from Alpha-2-antiplasmin (A2AP_Human) as a third interesting negative biomarker (Table 3). Alpha-2-antiplasmin is a serine protease inhibitor, involved in negative regulation of plasminogen activation. Bayasula et al. confirmed the presence of A2AP in human FF [43]. The major targets of this inhibitor are plasmin. Plasmin activity is believed to be involved in physiological processes such as ovulation [44], cumulus cell layer expansion [44, 45], oocyte maturation [44, 46] and fertilization [47, 48]. Hormone-induced coordinated expression of tissue-type PA (tPA) produced mainly by granulosa cells in the prevolulatory follicles is responsible for a controlled and directed proteolysis leading to the rupture of follicles [45]. The urokinase-type PA (uPA), activates latent proteinases or

growth factors, playing an essential role in the early growing follicles during cell proliferation and migration [49]. In addition, plasmin could active pro-enzymes of the matrix metalloproteinase, which in turn, may also be involved in ovarian function and regulate follicular development [49, 50]. In sheep and rat, intra-follicular injection of A2AP suppresses ovulation of preovulatory follicles [51, 52]. Huarte et al. [48] reported that the addition of plasminogen to mouse IVF medium increased the yield of fertilized eggs, while the addition of plasmin inhibitors resulted in a significant decrease. Here, for the first time, we indicate that the differential status of A2AP may be associated with immature oocyte developmental stage and might explain the negative biomarker.

We identified m/z = 528.28 as a peptides derived from Complement component 3 (CO3_Human) in hFF (Table 3) to be present at a significantly higher level in fertilized FF group in our study. CO3 has been specifically located in ovarian follicular fluid in early 1990s [53], however, the roles of CO3 and its derivatives in FF are not yet fully known, and the association between CO3 and reproductive capability remains unclear. Gonzales et al. and Hashemitabar et al. found significantly higher concentrations of the CO3 complement in the FF of fertilized oocytes [11, 20]. In porc [54], a derivative of CO3, iCO3b was identified to positively influence oocyte maturation. Those studies are consistent with our finding as a positive biomarker. In contrast, a study assessing the FF of IVF patients showed decreased expression of complement CO3 in their fertilized group [25]. The major difference of Estes' study was the inclusion of the oocyte number as dependent variable. Instead of predicting the outcome of fertilization, it might predict the differences between good and poor responders. Additional studies are needed to establish the actual influence of the complement cascade on reproductive aging and IVF outcome.

We identified m/z = 594.29 as derived from Inter-alpha-trypsin inhibitor heavy chain H1 (ITIH1_Human) as a positive biomarker (Table 3). Proteins of the inter-alpha-trypsin inhibitor family have been identified as a serum factor responsible for the stabilization of the expanding cumulus mass [55]. These proteins do not enter the follicle until it responds to an ovulatory stimulus. The luteinizing hormone (LH) surge increases the permeability of the charge and size-selective blood follicle barrier, allowing these serum glycoproteins to enter the antral cavities of responding follicles [56, 57]. Once within the antral cavity, the inter-alpha-trypsin inhibitor (ITI) covalently crosslinked to hyaluronan (HA), driving conformational changes and thereby remodeling the morphology and physicochemical properties of HA-rich extracellular matrices. This modification is the only naturally occurring covalent modification of HA known to date. The interaction was shown to be critical to matrix stabilization in expansion of the cumulus-oocyte complex (COC) [58, 59]. Huang et al. have shown that HA interacts strongly with ITIH1 and ITIH2 in vitro, and this binding was highly resistant to ionic strength and pH [60]. In another vivo study, mice lacking intact IαI family members fail to form a stable cumulus matrix and the naked ovulated oocytes are not fertilized in vivo [58].

Conclusion

Despite progress in the treatment of IVF and advances in novel techniques, selection of embryos based on the morphologic and morphometric parameters alone is not fully satisfactory. More accurate selection of oocytes and embryos should improve success rates after IVF treatment. Follicle development and oocyte maturation, require or bring about changes to oocyte microenvironment. Given the complexity of the numerous independent processes involved in oocyte development, it is unlikely that a single biomarker can predict the result of in vitro fertilization. By characterization of the FF peptidome, we present a profile of biomarkers associated with fertilization outcome. This may offer prognostic information aiding the selection of the most viable oocytes and hence embryos. The current results confirmed our hypothesis that peptide profiling is a promising approach to screen biomarkers for fertilization potential. Comparison of our results with proteomic studies of hFF indicates that the different analytical tools each bring their own selectivity.

Abbreviations

A2AP: Alpha-2-antiplasmin; ART: Assisted reproductive technology; CO3: Complement component 3; FDR: False discovery rate (FDR); FF: Follicular fluid; IGBP-5: Insulin-like growth factor binding protein-5; ITIH1: Inter-alpha-trypsin inhibitor heavy chain H1; LC-MS/MS: Liquid chromatography-tandem mass spectrometry; LMW: Low-molecular-weight; MS: Mass spectrometry; PCA: Principle component analysis; PLS: Partial least square; SET: Single embryo transferred (SET)

Acknowledgements

We thank W. Vermaelen and K. Arat (SYBIOMA KULeuven, Belgium) for their excellent technical assistance; D. De Neubourg, C. Tomassetti and C. Meuleman for the collection of the individual follicular fluids; the embryologists and technicians at the Leuven University Fertility Center.

Funding

This study was partially funded by the Merck Serono Grant for Fertility Innovation and China Scholarship Council. The funds had no role in the study design, collection and analysis of data, data interpretation or in writing the report.

Authors' contributions

FC carried out sample processing, data analysis and writing the manuscript. CS was responsible for the design and management of the project. TD'H and KP carried out oocyte aspirations. SC designed the analysis and contributed to manuscript preparation. All authors read and approved the final manuscript.

Competing interests

The authors declares that they have no competing interests.

Author details

[1]Leuven University Fertility Centre, UZ Leuven Campus Gasthuisberg, Herestraat 49, Leuven, Belgium. [2]Facility for Systems Biology based Mass Spectrometry (SYBIOMA), KU Leuven, Leuven, Belgium.

References

1. Fauser BC, Devroey P, Macklon NS. Multiple birth resulting from ovarian stimulation for subfertility treatment. Lancet. 2005;365:1807–16.
2. Gerris JM. Single embryo transfer and IVF/ICSI outcome: a balanced appraisal. Hum Reprod Update. 2005;11:105–21.
3. Makhseed M, Al-Sharhan M, Egbase P, Al-Essa M, Grudzinskas JG. Maternal and perinatal outcomes of multiple pregnancy following IVF-ET. Int J Gynaecol Obstet. 1998;61:155–63.
4. Pandian Z, Bhattacharya S, Ozturk O, Serour G, Templeton A. Number of embryos for transfer following in-vitro fertilisation or intra-cytoplasmic sperm injection. Cochrane Database Syst Rev. 2009;7:997–1005.
5. Thurin A, Hausken J, Hillensjo T, Jablonowska B, Pinborg A, Strandell A, Bergh C. Elective single-embryo transfer versus double-embryo transfer in in vitro fertilization. N Engl J Med. 2004;351:2392–402.
6. Kovalevsky G, Patrizio P. High rates of embryo wastage with use of assisted reproductive technology: a look at the trends between 1995 and 2001 in the United States. Fertil Steril. 2005;84:325–30.
7. Sirard MA, Richard F, Blondin P, Robert C. Contribution of the oocyte to embryo quality. Theriogenology. 2006;65:126–36.
8. Fortune JE. Ovarian follicular growth and development in mammals. Biol Reprod. 1994;50:225–32.
9. Appasamy M, Jauniaux E, Serhal P, Al-Qahtani A, Groome NP, Muttukrishna S. Evaluation of the relationship between follicular fluid oxidative stress, ovarian hormones, and response to gonadotropin stimulation. Fertil Steril. 2008;89:912–21.
10. Fahiminiya S, Reynaud K, Labas V, Batard S, Chastant-Maillard S, Gerard N. Steroid hormones content and proteomic analysis of canine follicular fluid during the preovulatory period. Reprod Biol Endocrinol. 2010;8:132.
11. Hashemitabar M, Bahmanzadeh M, Mostafaie A, Orazizadeh M, Farimani M, Nikbakht R. A proteomic analysis of human follicular fluid: comparison between younger and older women with normal FSH levels. Int J Mol Sci. 2014;15:17518–40.
12. Mason HD, Willis DS, Beard RW, Winston RM, Margara R, Franks S. Estradiol production by granulosa cells of normal and polycystic ovaries: relationship to menstrual cycle history and concentrations of gonadotropins and sex steroids in follicular fluid. J Clin Endocrinol Metab. 1994;79:1355–60.
13. Monteleone P, Giovanni Artini P, Simi G, Casarosa E, Cela V, Genazzani AR. Follicular fluid VEGF levels directly correlate with perifollicular blood flow in normoresponder patients undergoing IVF. J Assist Reprod Genet. 2008;25:183–6.
14. Ocal P, Aydin S, Cepni I, Idil S, Idil M, Uzun H, Benian A. Follicular fluid concentrations of vascular endothelial growth factor, inhibin A and inhibin B in IVF cycles: are they markers for ovarian response and pregnancy outcome? Eur J Obstet Gynecol Reprod Biol. 2004;115:194–9.
15. Wu YT, Wang TT, Chen XJ, Zhu XM, Dong MY, Sheng JZ, Xu CM, Huang HF. Bone morphogenetic protein-15 in follicle fluid combined with age may differentiate between successful and unsuccessful poor ovarian responders. Reprod Biol Endocrinol. 2012;10:116.
16. Revelli A, Delle Piane L, Casano S, Molinari E, Massobrio M, Rinaudo P. Follicular fluid content and oocyte quality: from single biochemical markers to metabolomics. Reprod Biol Endocrinol. 2009;7:40.
17. Rodthongkum N, Ramireddy R, Thayumanavan S, Richard WV. Selective enrichment and sensitive detection of peptide and protein biomarkers in human serum using polymeric reverse micelles and MALDI-MS. Analyst. 2012;137:1024–30.
18. Luchini A, Fredolini C, Espina BH, Meani F, Reeder A, Rucker S, Petricoin 3rd EF, Liotta LA. Nanoparticle technology: addressing the fundamental roadblocks to protein biomarker discovery. Curr Mol Med. 2010;10:133–41.
19. Twigt J, Steegers-Theunissen RP, Bezstarosti K, Demmers JA. Proteomic analysis of the microenvironment of developing oocytes. Proteomics. 2012;12:1463–71.

20. Gonzales J, Lesourd S, Van Dreden P, Richard P, Lefebvre G, Vauthier Brouzes D. Protein composition of follicular fluid and oocyte cleavage occurrence in in vitro fertilization (IVF). J Assist Reprod Genet. 1992;9:211–6.

21. Wunder DM, Mueller MD, Birkhauser MH, Bersinger NA. Steroids and protein markers in the follicular fluid as indicators of oocyte quality in patients with and without endometriosis. J Assist Reprod Genet. 2005;22:257–64.

22. Zamah AM, Hassis ME, Albertolle ME, Williams KE. Proteomic analysis of human follicular fluid from fertile women. Clin Proteomics. 2015;12:5.

23. Angelucci S, Ciavardelli D, Di Giuseppe F, Eleuterio E, Sulpizio M, Tiboni GM, Giampietro F, Palumbo P, Di Ilio C. Proteome analysis of human follicular fluid. Biochim Biophys Acta. 1764;2006:1775–85.

24. Ambekar AS, Nirujogi RS, Srikanth SM, Chavan S, Kelkar DS, Hinduja I, Zaveri K, Prasad TS, Harsha HC, Pandey A, Mukherjee S. Proteomic analysis of human follicular fluid: a new perspective towards understanding folliculogenesis. J Proteomics. 2013;87:68–77.

25. Estes SJ, Ye B, Qiu W, Cramer D, Hornstein MD, Missmer SA. A proteomic analysis of IVF follicular fluid in women < or = 32 years old. Fertil Steril. 2009;92:1569–78.

26. Liu AX, Zhu YM, Luo Q, Wu YT, Gao HJ, Zhu XM, Xu CM, Huang HF. Specific peptide patterns of follicular fluids at different growth stages analyzed by matrix-assisted laser desorption/ionization time-of-flight mass spectrometry. Biochim Biophys Acta. 1770;2007:29–38.

27. Spitzer D, Murach KF, Lottspeich F, Staudach A, Illmensee K. Different protein patterns derived from follicular fluid of mature and immature human follicles. Hum Reprod. 1996;11:798–807.

28. Hanrieder J, Nyakas A, Naessen T, Bergquist J. Proteomic analysis of human follicular fluid using an alternative bottom-up approach. J Proteome Res. 2008;7:443–9.

29. Liotta LA, Petricoin EF. Serum peptidome for cancer detection: spinning biologic trash into diagnostic gold. J Clin Invest. 2006;116:26–30.

30. Li B, Predel R, Neupert S, Hauser F, Tanaka Y, Cazzamali G, Williamson M, Arakane Y, Verleyen P, Schoofs L, et al. Genomics, transcriptomics, and peptidomics of neuropeptides and protein hormones in the red flour beetle Tribolium castaneum. Genome Res. 2008;18:113–22.

31. Petricoin EF, Belluco C, Araujo RP, Liotta LA. The blood peptidome: a higher dimension of information content for cancer biomarker discovery. Nat Rev Cancer. 2006;6:961–7.

32. Lai ZW, Petrera A, Schilling O. The emerging role of the peptidome in biomarker discovery and degradome profiling. Biol Chem. 2015;396:185–92.

33. Debrock S, Melotte C, Spiessens C, Peeraer K, Vanneste E, Meeuwis L, Meuleman C, Frijns JP, Vermeesch JR, D'Hooghe TM. Preimplantation genetic screening for aneuploidy of embryos after in vitro fertilization in women aged at least 35 years: a prospective randomized trial. Fertil Steril. 2010;93:364–73.

34. Otsuki J, Nagai Y, Matsuyama Y, Terada T, Era S. The influence of the redox state of follicular fluid albumin on the viability of aspirated human oocytes. Syst Biol Reprod Med. 2012;58:149–53.

35. Thierry van Dessel HJ, Chandrasekher Y, Yap OW, Lee PD, Hintz RL, Faessen GH, Braat DD, Fauser BC, Giudice LC. Serum and follicular fluid levels of insulin-like growth factor I (IGF-I), IGF-II, and IGF-binding protein-1 and −3 during the normal menstrual cycle. J Clin Endocrinol Metab. 1996;81:1224–31.

36. Nyegaard M, Overgaard MT, Su YQ, Hamilton AE, Kwintkiewicz J, Hsieh M, Nayak NR, Conti M, Conover CA, Giudice LC. Lack of functional pregnancy-associated plasma protein-A (PAPPA) compromises mouse ovarian steroidogenesis and female fertility. Biol Reprod. 2010;82:1129–38.

37. Erickson GF, Nakatani A, Ling N, Shimasaki S. Localization of insulin-like growth factor-binding protein-5 messenger ribonucleic acid in rat ovaries during the estrous cycle. Endocrinology. 1992;130:1867–78.

38. Rivera GM, Fortune JE. Selection of the dominant follicle and insulin-like growth factor (IGF)-binding proteins: evidence that pregnancy-associated plasma protein A contributes to proteolysis of IGF-binding protein 5 in bovine follicular fluid. Endocrinology. 2003;144:437–46.

39. Cataldo NA, Giudice LC. Insulin-like growth factor binding protein profiles in human ovarian follicular fluid correlate with follicular functional status. J Clin Endocrinol Metab. 1992;74:821–9.

40. de la Sota RL, Simmen FA, Diaz T, Thatcher WW. Insulin-like growth factor system in bovine first-wave dominant and subordinate follicles. Biol Reprod. 1996;55:803–12.

41. Stewart RE, Spicer LJ, Hamilton TD, Keefer BE, Dawson LJ, Morgan GL, Echternkamp SE. Levels of insulin-like growth factor (IGF) binding proteins, luteinizing hormone and IGF-I receptors, and steroids in dominant follicles during the first follicular wave in cattle exhibiting regular estrous cycles. Endocrinology. 1996;137:2842–50.

42. Mihm M, Good TE, Ireland JL, Ireland JJ, Knight PG, Roche JF. Decline in serum follicle-stimulating hormone concentrations alters key intrafollicular growth factors involved in selection of the dominant follicle in heifers. Biol Reprod. 1997;57:1328–37.

43. Bayasula, Iwase A, Kobayashi H, Goto M, Nakahara T, Nakamura T, Kondo M, Nagatomo Y, Kotani T, Kikkawa F: A proteomic analysis of human follicular fluid: comparison between fertilized oocytes and non-fertilized oocytes in the same patient. J Assist Reprod Genet. 2013, 30:1231–1238.

44. Liu YX, Ny T, Sarkar D, Loskutoff D, Hsueh AJ. Identification and regulation of tissue plasminogen activator activity in rat cumulus-oocyte complexes. Endocrinology. 1986;119:1578–87.

45. Liu YX. Plasminogen activator/plasminogen activator inhibitors in ovarian physiology. Front Biosci. 2004;9:3356–73.

46. Dow MP, Bakke LJ, Cassar CA, Peters MW, Pursley JR, Smith GW. Gonadotropin surge-induced up-regulation of the plasminogen activators (tissue plasminogen activator and urokinase plasminogen activator) and the urokinase plasminogen activator receptor within bovine periovulatory follicular and luteal tissue. Biol Reprod. 2002;66:1413–21.

47. Smokovitis A, Kokolis N, Taitzoglou I, Rekkas C. Plasminogen activator: the identification of an additional proteinase at the outer acrosomal membrane of human and boar spermatozoa. Int J Fertil. 1992;37:308–14.

48. Huarte J, Vassalli JD, Belin D, Sakkas D. Involvement of the plasminogen activator/plasmin proteolytic cascade in fertilization. Dev Biol. 1993;157:539–46.

49. Ny T, Wahlberg P, Brandstrom IJ. Matrix remodeling in the ovary: regulation and functional role of the plasminogen activator and matrix metalloproteinase systems. Mol Cell Endocrinol. 2002;187:29–38.

50. Robker RL, Russell DL, Espey LL, Lydon JP, O'Malley BW, Richards JS. Progesterone-regulated genes in the ovulation process: ADAMTS-1 and cathepsin L proteases. Proc Natl Acad Sci U S A. 2000;97:4689–94.

51. Tsafriri A, Bicsak TA, Cajander SB, Ny T, Hsueh AJ. Suppression of ovulation rate by antibodies to tissue-type plasminogen activator and alpha 2-antiplasmin. Endocrinology. 1989;124:415–21.

52. Murdoch WJ. Regulation of collagenolysis and cell death by plasmin within the formative stigma of preovulatory ovine follicles. J Reprod Fertil. 1998;113:331–6.

53. Perricone R, Pasetto N, De Carolis C, Vaquero E, Piccione E, Baschieri L, Fontana L. Functionally active complement is present in human ovarian follicular fluid and can be activated by seminal plasma. Clin Exp Immunol. 1992;89:154–7.

54. Georgiou AS, Gil MA, Alminana C, Cuello C, Vazquez JM, Roca J, Martinez EA, Fazeli A. Effects of complement component 3 derivatives on pig oocyte maturation, fertilization and early embryo development in vitro. Reprod Domest Anim. 2011;46:1017–21.

55. Chen L, Mao SJ, McLean LR, Powers RW, Larsen WJ. Proteins of the inter-alpha-trypsin inhibitor family stabilize the cumulus extracellular matrix through their direct binding with hyaluronic acid. J Biol Chem. 1994; 269:28282–7.

56. Powers RW, Chen L, Russell PT, Larsen WJ. Gonadotropin-stimulated regulation of blood-follicle barrier is mediated by nitric oxide. Am J Physiol. 1995;269:E290–298.

57. Hess KA, Chen L, Larsen WJ. The ovarian blood follicle barrier is both charge- and size-selective in mice. Biol Reprod. 1998;58:705–11.

58. Zhuo L, Yoneda M, Zhao M, Yingsung W, Yoshida N, Kitagawa Y, Kawamura K, Suzuki T, Kimata K. Defect in SHAP-hyaluronan complex causes severe female infertility. A study by inactivation of the bikunin gene in mice. J Biol Chem. 2001;276:7693–6.

59. Fulop C, Szanto S, Mukhopadhyay D, Bardos T, Kamath RV, Rugg MS, Day AJ, Salustri A, Hascall VC, Glant TT, Mikecz K. Impaired cumulus mucification and female sterility in tumor necrosis factor-induced protein-6 deficient mice. Development. 2003;130:2253–61.

60. Huang L, Yoneda M, Kimata K. A serum-derived hyaluronan-associated protein (SHAP) is the heavy chain of the inter alpha-trypsin inhibitor. J Biol Chem. 1993;268:26725–30.

10

Merging clinical chemistry biomarker data with a COPD database - building a clinical infrastructure for proteomic studies

Jonatan Eriksson[1,4*], Simone Andersson[2], Roger Appelqvist[1,4], Elisabet Wieslander[1], Mikael Truedsson[5], May Bugge[5], Johan Malm[1,3], Magnus Dahlbäck[1,4], Bo Andersson[4], Thomas E. Fehniger[1,4^] and György Marko-Varga[1,4,6]

Abstract

Background: Data from biological samples and medical evaluations plays an essential part in clinical decision making. This data is equally important in clinical studies and it is critical to have an infrastructure that ensures that its quality is preserved throughout its entire lifetime. We are running a 5-year longitudinal clinical study, KOL-Örestad, with the objective to identify new COPD (Chronic Obstructive Pulmonary Disease) biomarkers in blood. In the study, clinical data and blood samples are collected from both private and public health-care institutions and stored at our research center in databases and biobanks, respectively. The blood is analyzed by Mass Spectrometry and the results from this analysis then linked to the clinical data.

Method: We built an infrastructure that allows us to efficiently collect and analyze the data. We chose to use REDCap as the EDC (Electronic Data Capture) tool for the study due to its short setup-time, ease of use, and flexibility. REDCap allows users to easily design data collection modules based on existing templates. In addition, it provides two functions that allow users to import batches of data; through a web API (Application Programming Interface) as well as by uploading CSV-files (Comma Separated Values).

Results: We created a software, DART (Data Rapid Translation), that translates our biomarker data into a format that fits REDCap's CSV-templates. In addition, DART is configurable to work with many other data formats as well. We use DART to import our clinical chemistry data to the REDCap database.

Conclusion: We have shown that a powerful and internationally adopted EDC tool such as REDCap can be extended so that it can be used efficiently in proteomic studies. In our study, we accomplish this by using DART to translate our clinical chemistry data to a format that fits the templates of REDCap.

Keywords: Proteomics, COPD, Clinical study, Biomarkers, Proteomics, Biobanking, Bioinformatics, EDC

Background

Today, the integration and compilation of the large volume of data generated from clinical studies is challenging. This data has to be available to be used in companion diagnostic and prognostic tests which in turn are crucial at every level of clinical decision making. A multi-fold of patient samples is stored in biobanks for use in future medical research projects that measure the quantitative and qualitative read-outs of gene and protein expression associated with disease processes (1-2). Furthermore, biobanking laws that regulate data processing, integration, traceability, and confidentiality are implemented on a national level.

In Sweden, all institutions storing biological samples and clinical data are required to declare their inventories to the National Board of Health and Welfare. Sweden had 651 biobanks registered with the Board in 2007, the majority deployed at University hospitals and Regional Medical Authorities. The samples come from many

* Correspondence: jonatan.eriksson@bme.lth.se
^Deceased
[1]Centre of Excellence in Biological and Medical Mass Spectrometry, Biomedical Centre D13, Lund University, 221 84 Lund, Sweden
[4]Clinical Protein Science & Imaging, Biomedical Centre, Department of Biomedical Engineering, Lund University, BMC D13, 221 84 Lund, Sweden
Full list of author information is available at the end of the article

forms of specimen including tissues, cells, cell lines, genomic material (DNA), blood, blood-plasma, and urine.

Project organization is essential to face the challenges in clinical studies, i.e., patient recruitment, appointment scheduling, blood samples collection etc. Additionally, other parts of the study are also important, such as sample storage, analysis, evaluation of the results, and compilation of data into a useful database. The latter part of the study may continue for a long time after clinical data collection is completed which is why a high quality database is crucial.

REDCap is a web-based EDC (Electronic Data Capture) tool developed by Vanderbilt University, TN, USA first released in 2004 [1]. REDCap is a secure application designed to support data capture in research studies with a user-friendly interface for data entry, audit trial to track data manipulation and export procedures, import data from external sources, and export procedures for data download to common statistical packages. We selected REDCap as the EDC tool for the current clinical study due to a version controlled update of the software, a massive user support, easy export/import of data using CSV-files (Comma-Separated Values), a large community of users, and the possibility to easily design different data collection instruments for various types of data input, e.g., text and numerical values.

KOL-Örestad is a 5-year longitudinal clinical study created to identify new biomarkers of COPD (Chronic Obstructive Pulmonary Disease) in blood. The biomarkers can be used to *diagnose* the disease, *predict* exacerbations, and to *classify* disease stages of COPD [2]. An early diagnosis would be beneficial to the patient, health-care system, and society due to lower medical costs and reduced sick leave. In the present study we describe how to transfer clinical chemistry data from the hospital information system to the KOL-Örestad database with a simple and secure EDC solution. Throughout the course of the study, blood samples are drawn from the participants at a private health-care clinic (Örestadskliniken), encoded, and sent to a public clinical chemistry laboratory. The analysis of these samples results in large number of numerical data values that need to be imported into the REDCap database. To enter that amount of data manually is time-consuming and error-prone. Instead, taking an approach that is automatic and indifferent to variable identifiers and order of variables is preferred.

In proteomic research it is critical that sample quality is preserved during storage, that sample inventories are maintained, and that the data produced by mass spectrometry or similar techniques is readily available for statistical analysis. This puts requirements on both the biobanks the samples are stored in as well as on inventory software and EDC tools.

Methods
Clinical study outline
The study group consists of approximately 300 study participants between the ages of 35 and 80 years. Two-hundred are diagnosed with COPD of one of the four stages of GOLD and 100 are healthy with normal lung function, both smokers/ex-smokers ($n = 50$) and never-smokers ($n = 50$). Participants undergo health examinations and blood sample collection every 6 months for a total of 5 years. Health examination includes a spirometer test, physical examination, and questionnaire [2]. Because the old classification system was in use at the initiation of the study, we will continue to use GOLD 1 to 4 based solely on post bronchodilator spirometry values.

Three 5 mL tubes of blood are sampled and analyzed for a set of predetermined disease biomarkers at a standard clinical chemistry laboratory and four tubes are aliquoted to 70 µL samples (plasma, serum, & whole blood) and stored in a biobank at -80° C for protein biomarker analysis by mass spectrometry [3–8]. Each aliquot has a unique 2D barcode which may be traced back to the participant. The study has been approved by the regional ethical review board in Lund (DRN 2013/480) and is registered at ClinicalTrials.gov under the official title Biomarkers of Early Chronic Obstructive Pulmonary Disease (COPD) in Smokers - Longitudinal Study (U.S. National Institutes of Health, 2014).

Data sources
The study consists of different data input sources where each source has multiple variables of data that are to be stored in the same database. One data source is the primary health-care clinic where all visits are made. At each visit, the study participants fill out a questionnaire by hand and the responses are entered manually into REDCap by members of the research team. Additionally, the results from the physical examination and spirometry test are also entered manually into REDCap. The tubes from blood collection are labeled with barcodes identifying the study participant with a specific code and are entered into REDCap by a barcode scanner. The data is coded by a unique identifier (Study ID) for each participant and the full identification is solely available to the health-care personnel. However, full traceability from sample to Study ID will be accessible within the system.

Database structure
The data is entered into REDCap at different locations and time points. Some data is entered manually and

some is entered using the import function of REDCap. Preferably, all data should be entered automatically or electronically, e.g., through an automated direct input, digital scanning of the paper format, or import file from a data generating source. Currently, this is not possible in the present study; however, this is our goal.

Data import to REDCap

The web interface of REDCap provides an import module that screens the data for errors before committing it to the database; e.g., errors such as incorrectly formatted data values or numerical values that lie beyond pre-defined bounds can be caught at this stage. Both batches of data as well as single data values can be imported through this module. Data batches are imported from CSV files whose format is required to match that which REDCap expects. Consequently, when an instrument generates a CSV files containing biomarker data it has to be translated, either manually or automatically, to this expected format. In addition to the import functions of the web user-interface, REDCap provides a web API that allows users to programmatically import and retrieve data to and from REDCap through HTTP methods. Lastly, the back-end of REDCap, a MySQL database, can

Table 1 Current data collection instruments, data format and input method to the KOL-Örestad database.

Source	Format	Input method
Blood samples	number, text, date, check box	manual, barcode scanner
Physical examination	numerical value	manual input from patient journal
Spirometry	number	manual input from paper
Questionnaire	number, text, check box	manual input from paper
CC analysis	number	import from data file
MS analysis	number, text	import from data file

The qualitative and quantitative output of the MS analysis (as well as its corresponding meta-data, e.g., MS platform, instrument setttings, Sequence Database, etc.) is stored in REDCap whereas instrument raw files are stored in a separate database

be directly modified; however, this bypasses the above mentioned data screening [1].

Results

The study participants have their blood drawn, fill out questionnaires, and undergo physical examinations at the private health-care clinic. Members of the research team enter data from questionnaires and physical examinations manually into REDCap. The blood samples from each participant are sent both to BMC to be stored in a biobank and to a clinical chemistry facility in Malmö for disease biomarker analysis. The data generated by the biomarker analysis is stored in REDCap. Mass spectrometry (MS) analysis is performed on the biobank samples and the produced data is stored in a separate, dedicated, database. Results from analysis of the raw MS data, e.g., PSMs (Peptide-to-Spectrum Matches) or FDRs (False Discovery Rates), as well as data related to the experimental setup can also be stored in REDCap. The flow of samples and data can be viewed in Fig. 1. The data collected within the study, the data formats, and the input methods can be viewed in Table 1.

At the start of the study all data was imported manually into REDCap through its web interface. We

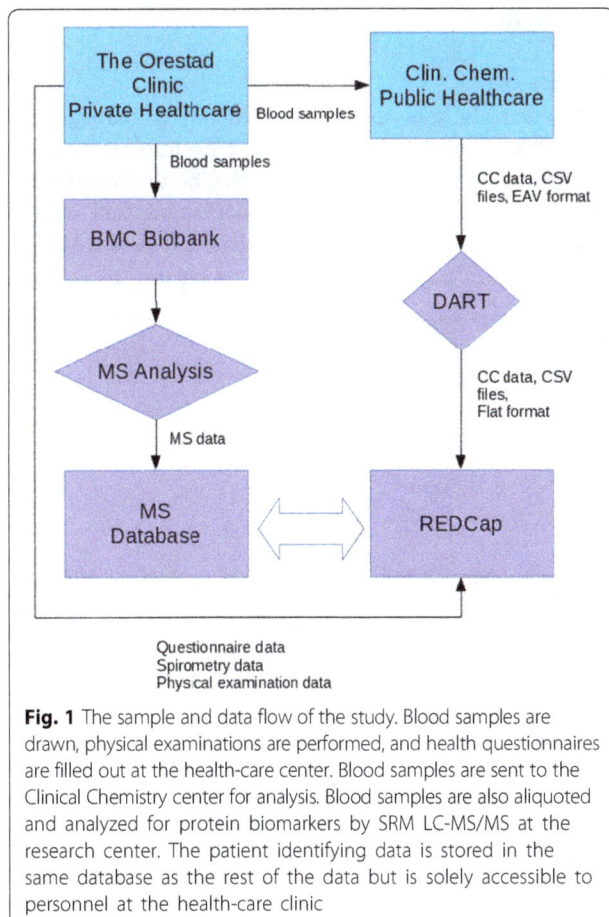

Fig. 1 The sample and data flow of the study. Blood samples are drawn, physical examinations are performed, and health questionnaires are filled out at the health-care center. Blood samples are sent to the Clinical Chemistry center for analysis. Blood samples are also aliquoted and analyzed for protein biomarkers by SRM LC-MS/MS at the research center. The patient identifying data is stored in the same database as the rest of the data but is solely accessible to personnel at the health-care clinic

Table 2 Example of laboratory instrument output, the data is modeled as EAV

pat_id	visit_id	exacerbation_id	Date	Attribute	Value
100001	visit_1		150321	calcium	6.2
100001	visit_1		150321	leukocytes	6.5
100002	visit_1		150321	calcium	2.45
100002	visit_1		150321	leukocytes	6.5
100001	visit_2		150917	calcium	2.51
100001	visit_2		150917	leukocytes	6.2
100002		exacerbation_1	150613	calcium	2.44
100002		exacerbation_1	150613	leukocytes	8.1

Table 3 Example of a flat table exported from the REDCap database with biomarker data yet to be entered

patient_id	redcap_event_name	N	ca	leuco	hb	mono
100001	visit_1					
100002	visit_1					
100001	visit_2					
100002	exacerbation_1					

chose to focus on automating the clinical chemistry data (CC data) importation since it involves the largest volume of data. The clinical chemistry laboratory instruments used to perform the biomarker analysis in the study automatically generate CSV files containing CC data. These files are formatted differently than those exported from and imported to REDCap. The instrument-generated files represent the data according to the EAV(Entity-Attribute-Value) model whereas the REDCap files mirror the flat table representation of the *DATA* table of the REDCap MySQL back-end. Furthermore, if the attribute and entity identifiers differ between the instrument and REDCap files the instrument files are not accepted by the batch import function of REDCap [1]. We created a program, DART (Data Rapid Translation), that automates the translation between the two formats. It extracts the biomarker data from the instrument-generated CSV file, re-formats it, and inserts it into the CSV file exported from REDCap. It does this by first translating the instrument-generated file to an intermediate hierarchical format which subsequently is translated to the format REDCap expects. DART is not exclusive to our particular pair of CSV files but can be configured to work with others as well.

DART was created using the C++ programming language. A minimal Graphical User Interface (GUI) was created using the wxWidgets library that allows the users to configure DART for different export/import formats, i.e., what the hierarchical structures look like and the mapping of the variable identifiers. DART is currently used in the study.

An example of biomarker data and how it can be represented in a CSV file is shown in Table 2. The corresponding REDCap CSV file is shown in Table 3. The list of disease biomarkers used in Tables 2 and 3 is truncated. At the five levels of the hierarchy of the example data are, from top to bottom; the study id, the study participants, the visits as well as exacerbations of the participants, the biomarkers, and the numerical data value belonging to each biomarker. The hierarchy of the example biomarker data is illustrated in Fig. 2. The variables are not required to have the same identifiers in the REDCap and biomarker CSV files. The hierarchy is inherently the same in the file exported from REDCap, only formatted differently. Granted that the hierarchy of the data, the relationship between the format of the biomarker file and that of the file exported from REDCap, and the mapping of variable identifiers are specified, DART can transfer the data.

Discussion

In Sweden, there is no standardized way to manage and transfer clinical data. This makes it desirable to use an already adopted and well tested tool. The decision to use REDCap as the primary data management tool in this study was motivated by its accessibility, due to a simple web interface, to a broad user spectrum. It is used

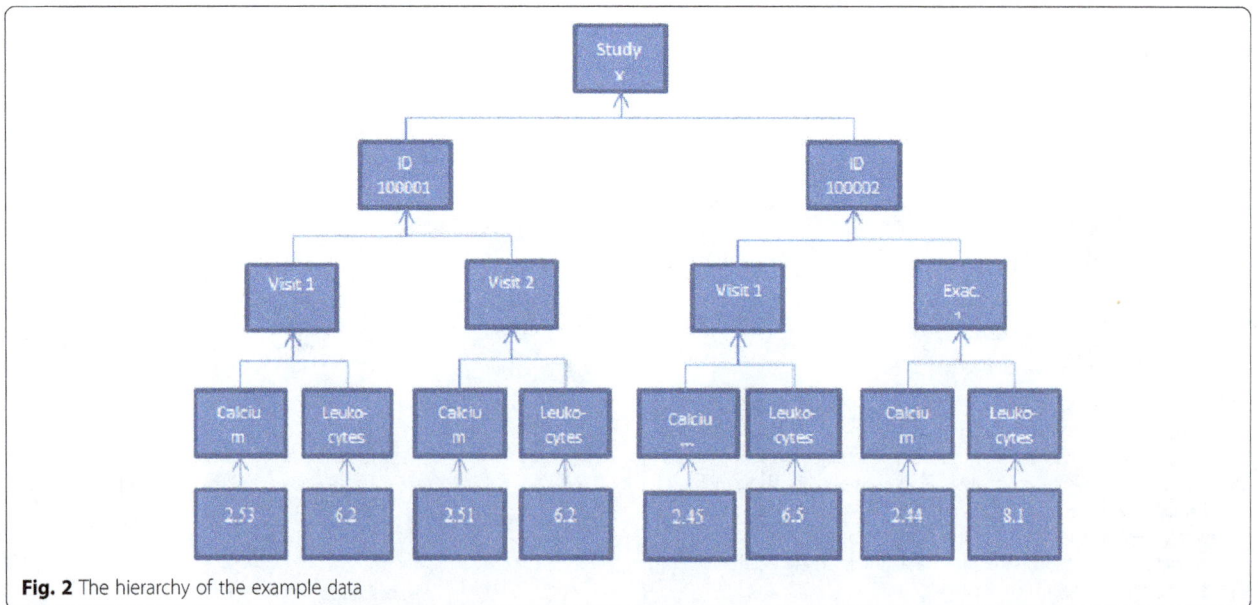

Fig. 2 The hierarchy of the example data

globally, particularly by non-profit research organization, in clinical studies [9]. The chosen approach where the database is exported as a CSV file, modified, and then imported back again, does not interfere with the built-in quality-control of data that REDCap provides. The alternative approach where the underlying database of RED-Cap is directly modified, while more straight-forward, circumvents this quality-control, and is thus more unsafe with respect to data integrity.

The combination of REDCap and DART is a flexible tool for the management of clinical study data. Some initial work has to be done when configuring the program for a particular import/export pair. Currently DART is solely used to import data into REDCap, it could, however, in a future iteration, also perform other tasks such as data analysis and/or presentation of data.

Conclusion

It is crucial to store both data from sample analyses as well as the biological samples themselves in an organized and secure way during clinical studies. Furthermore, the data and samples have to be readily accessible to researchers and health-care personnel. In the KOL-Örestad study we achieve this by using REDCap to manage and store the data and our biobank to store the samples and their aliquots.

Abbreviations
API: Application programming interface; COPD: Chronic obstructive pulmonary disease; CSV: Comma separated values; DART: Data rapid translation; EDC: Electronic data capture; GOLD: Global initiative for chronic obstructive lung disease; GUI: Graphical user interface

Acknowledgements
Not applicable.

Funding
Region Skåne, health-care project funding. This work was partly supported by grants from the National Research Foundation of Korea, funded by the Korean government (NRF-2015K1A1A2028365).

Authors' contributions
MT, MB: patient recruitment and physical examination, MB, MT, JM: clinical input, JM, TEF, EW, GMV, MD: clinical study outline, JE, SA, GMV, MD, RA, EW, JM, BA: manuscript preparation, JE, SA, MD, RA, BA: study logistics and programming. All authors read and approved the final manuscript.

Competing interests
The authors declare that they have no competing interests.

Author details
[1]Centre of Excellence in Biological and Medical Mass Spectrometry, Biomedical Centre D13, Lund University, 221 84 Lund, Sweden. [2]Encap Security, Øvre Slottsgate 7, 0157 Oslo, Norway. [3]Section for Clinical Chemistry, Department of Translational Medicine, Lund University, Skåne University Hospital Malmö, 205 02 Malmö, Sweden. [4]Clinical Protein Science & Imaging, Biomedical Centre, Department of Biomedical Engineering, Lund University, BMC D13, 221 84 Lund, Sweden. [5]Örestadskliniken, 217 67, Eddagatan 4, 217 67 Malmö, Sweden. [6]First Department of Surgery, Tokyo Medical University, 6-7-1 Nishishinjiku Shinjiku-ku, Tokyo 160-0023, Japan.

References
1. Marko-Varga G, Boja ES, Rodriguez H, Baker M, Fehniger TE. J. Proteome Res, accepted "Biorepository Regulatory Frameworks: Building Parallel Resources that both Promote Scientific Investigation and Protect Human Subjects ". J Proteome Res. 2014;12:5319–24.
2. Khleif SN, Doroshow JH, Hait WN. AACR-FDA-NCI Cancer Biomarkers Collaborative Consensus Report: advancing the use of biomarkers in cancer drug development. Clin Cancer Res. 2010;16:3299–318.
3. Harris PA, Taylor R, Thielke R, Payne J, Gonzalez N, Conde JG. Research electronic data capture (REDCap) - A metadata-driven methodology and workflow process for providing translational research informatics support. J Biomed Inform. 2009;42(2):377–81.
4. Mikael T, Johan M, Barbara Sahlin K, May B, Elisabet W, Magnus D, Roger A, Fehniger TE, György M-V. Biomarkers of early chronic obstructive pulmonary disease (COPD) in smokers and former smokers. Protocol of a longitudinal study. Clin Trans Med. 2016;5:9.
5. Malm J, Fehniger TE, Danmyr P, Végvári A, Welinder C, Lindberg H, et al. Biobanking work flow standardization-developments providing sample integrity. J Proteomics. 2013;95:38–45.
6. Malm J, Danmyr P, Nilsson R, Appelqvist R, Végvári A, Marko-Varga G. Blood sample standardization developments for large scale biobanking. J Proteome Res. 2013;12:3087–92.
7. Fehniger TE, Boja ES, Rodriguez H, Baker M, Marko-Varga G. Four areas of engagement requiring strengthening in modern proteomics today. J Proteome Res. 2014;13(12):5310–8.
8. Malm J, Linberg H, Erlinge D, Appelqvist R, Yokaleva M, Welinder C, et al. Semi-automated biobank sample processing with 384 high density sample tube robot used in cancer and cardiovascular studies. Clin Transl Med. 2015; 4(1):67.
9. Franklin J, Guidry A, Brinkley J. A Partnership approach for Electronic Data Capture in small-scale clinical trials. J Biomedical Informatics. 2011; 44:S103–8.

11

Water deficit mechanisms in perennial shrubs *Cerasus humilis* leaves revealed by physiological and proteomic analyses

Zepeng Yin[1,2,3†], Jing Ren[4†], Lijuan Zhou[1], Lina Sun[1], Jiewan Wang[1], Yulong Liu[5] and Xingshun Song[1,2*]

Abstract

Background: Drought (Water deficit, WD) poses a serious threat to extensively economic losses of trees throughout the world. Chinese dwarf cherry (*Cerasus humilis*) is a good perennial plant for studying the physiological and sophisticated molecular network under WD. The aim of this study is to identify the effect of WD on *C. humilis* through physiological and global proteomics analysis and improve understanding of the WD resistance of plants.

Methods: Currently, physiological parameters were applied to investigate *C. humilis* response to WD. Moreover, we used two-dimensional gel electrophoresis (2DE) to identify differentially expressed proteins in *C. humilis* leaves subjected to WD (24 d). Furthermore, we also examined the correlation between protein and transcript levels.

Results: Several physiological parameters, including relative water content and Pn were reduced by WD. In addition, the malondialdehyde (MDA), relative electrolyte leakage (REL), total soluble sugar, and proline were increased in WD-treated *C. humilis*. Comparative proteomic analysis revealed 46 protein spots (representing 43 unique proteins) differentially expressed in *C. humilis* leaves under WD. These proteins were mainly involved in photosynthesis, ROS scavenging, carbohydrate metabolism, transcription, protein synthesis, protein processing, and nitrogen and amino acid metabolisms, respectively.

Conclusions: WD promoted the CO_2 assimilation by increase light reaction and Calvin cycle, leading to the reprogramming of carbon metabolism. Moreover, the accumulation of osmolytes (i.e., proline and total soluble sugar) and enhancement of ascorbate-glutathione cycle and glutathione peroxidase/glutathione s-transferase pathway in leaves could minimize oxidative damage of membrane and other molecules under WD. Importantly, the regulation role of carbohydrate metabolisms (e. g. glycolysis, pentose phosphate pathways, and TCA) was enhanced. These findings provide key candidate proteins for genetic improvement of perennial plants metabolism under WD.

Keywords: *Cerasus humilis*, Proteomics, ROS, Water deficit, Perennial shrubs, qRT-PCR

Background

Among potential abiotic stresses, drought (water deficit, WD) is considered to have the largest effect on agricultural productivity, which limits plant growth, distribution and crop yield worldwide [1, 2]. It is estimated that the droughty terrestrial areas will redouble by the end of the 21st century [3]. Thus, it is extremely urgent to determine the mechanisms of plant respond to drought and improve the drought tolerance ability.

During drought stress, a series of metabolic alterations occur, including overproduction of reactive oxygen species (ROS), photoinhibition, denaturation of some proteins such as chloroplast proteins, damage to biofilm structure and functions, and inhibition in protein synthesis [1, 4, 5]. Plants employed series of strategies in response to WD, such as morphology, and physiology metabolisms. The imbalanced light energy conversion and carbon fixation in photosynthetic system may cause accumulation of ROS in plant cells, leading to photo-oxidation damages [2]. An excess of ROS production can lead to oxidative stress in

---

* Correspondence: sfandi@163.com

†Equal contributors

[1]Department of Genetics, College of Life Science, Northeast Forestry University, Harbin 150040, People's Republic of China

[2]State Key Laboratory of Tree Genetic sand Breeding, Northeast Forestry University, Harbin 150040, People's Republic of China

Full list of author information is available at the end of the article

plants and negatively impact the normal function of cells [4], which can damage DNA, lipids and proteins [6, 7]. ROS scavenging ability and subsequent injury-reducing effects may correlate with the tolerance to WD [8]. Both enzymatic and non-enzymatic defense systems have evolved in plants for scavenging and detoxifying ROS. The main non-enzymatic antioxidants in plants are soluble ascorbate and glutathione [9]. ROS scavenging enzymes such as ascorbate peroxidase (APX), superoxide dismutase (SOD), catalase (CAT) and peroxidase (POD) also play a very important role. In addition, the accumulation of various osmolytes, such as proline and soluble sugar, play an important protective role during WD, which result in the decrease of osmotic potential [10].

Although there are some researches in woody plants responses to drought in morphological and physiological level [11, 12], few studies were reported on molecular metabolisms. It has been shown that the genes/proteins were induced by WD either directly connected to stress response or implicated in the regulation of gene expression and signal transduction [13–15]. These studies provided important information for understanding WD-responsive gene functions. However, changes at the mRNA and metabolite levels do not always reflect changes at the protein level [16, 17]. Thus, studying the protein level changes in response to WD is important in woods. Proteomics technologies allow a high-throughput and systemic overview of the cellular physiology in a holistic manner to underscore the underlying metabolic and regulatory mechanisms [18]. It has been reported that WD altered the abundance of proteins involved in carbohydrate and energy metabolism, cellular detoxification, protein processing and degradation, signal transduction, and cell wall strengthening. Most of the previous work on WD related proteomics was performed on annual crops [19]. However, very limited proteomic information on perennial shrubs responses to WD is available. Perennial plants may express stress-responsive proteins associated with long-term adaptation or stress survival, as they must endure/persist through the stress period, unlike annual crops which produce seeds and may die in the case of severe WD [20, 21]. Thus, developing an adaptation mechanism is critical for the survival of perennial plants in WD environments.

Chinese dwarf cherry (*Cerasus humilis* (Bge.) Sok.), a species of perennial shrubs, originates in the north of China [22–24]. As most perennial dwarf shrubs in the world, *C. humilis* have the characteristics of being drought-, saline-, alkali-, cold- and sterile resistant, all of which endow the species considerable adaptabilities [23, 24]. Our previous studies showed that exogenous small aliphatic amines (spermidine and spermine) and microorganisms (e.g. photosynthetic bacteria) can alleviate the WD-induced oxidative stress in *C. humilis* [24]. Besides, the increased expression of *vde*, a gene encoding violaxanthin de-epoxidase in xanthophyll-cycle, confers a great capacity of photoprotection and thus contributes to the survival *C. humilis* against WD [23]. Although some physiological data and certain genes already give us some information, comparative proteomic analyses can deepen our understanding of plant stress acclimation/tolerance acquisition by providing a detailed picture of functional proteins in *C. humilis* under WD conditions.

Our objectives were to discover the WD-responsive characteristics of *C. humilis* by using combined physiological and comparative proteomic approaches. We also aimed to determine the role of ROS during this process. To address these questions, we (1) evaluated changes in gas exchange, the enzymes of antioxidants related to ROS, and the osmolytes protections, (2) carried out 2-DE based proteomics analyses the different proteins at different time point after WD, (3) investigated the expression changes of some raleted genes by qRT-PCR. These study lead to a better understanding of molecular mechanisms in perennial shrubs under WD.

Methods

Plant material and treatments

Cuttings of *C. humilis* obtained from HuaiRou district, Hebei province, China, were used in the present study. Each cutting was transplanted into the container (35 × 35 × 25 cm) filled with organic soil, irrigated regularly by Hoagland solution under a 12 h photoperiod at temperatures ranging from about 17–25 °C, photosynthetic photon flux density (PPFD) of 600 $\mu mol\ m^{-2}\ s^{-1}$ and the relative humidity of 70–75% in the greenhouse. Seedlings at the 25–35 leaf stage were divided into two groups: well-watered plants were irrigated by water every two days (control), and water-deficit (WD) plants did not receive water. All measurements of physiological parameters were carried out on the youngest fully expanded leaves, with at least ten plants per-treatment.

Determination of Relative water content (RWC), MDA, and relative electrolyte leakage (REL)

RWC was calculated as follows: RWC = [(FW - DW)/(SW - DW)] × 100. Fresh weight (FW) was measured immediately after harvesting, and saturate weight (SW) was measured right after saturation state immersed in distilled water. Then, the samples were oven-dried at 80 °C for 15 min, then vacuum-dried at 60 °C to constant weight and the DWs were recorded. The REL was measured by an electrical conductivity method [25].

Measurement of photosynthesis and chlorophyll fluorescence

Net photosynthetic rates (Pn), stomata conductance (Gs), and intercellular CO_2 (Ci) of leaves were determined during 8:30–11:30 h using a gas-exchange system (LI-6400; LICOR

Biosciences, Lincoln, USA) on 24 days WD treatment. The photosynthetically active radiation was 1000 $\mu molL\ m^{-2}\ s^{-1}$ (saturation light). The ambient CO_2 concentration was $360 \pm 10\ \mu mol\ mol^{-1}$, and the air temperature and humidity were about 24 °C and 50%. Measurements were repeated at least five times for each treatment and the averages were recorded.

The maximum photochemical efficiency of PSII (Fv/Fm) was measured using a pulse modulation chlorophyll fluorometer (FMS-2, Hansatech, UK) after 30 min dark adaptation [26].

Determination of total soluble sugar, proline, H_2O_2 content and O_2^- generation rate

Total soluble sugar and proline contents were determined using an anthrone reagent and ninhydrin reaction, respectively, as previously described [27].

To evaluate the levels of ROS in leaves, H_2O_2 content and O_2^- generation rates were measured. Briefly, leaf tissue was ground in 0.1% trichloroacetic acid. The homogenate was centrifuged at 15,000 g for 15 min at 4 °C and the supernatant was collected for H_2O_2 measurement. H_2O_2 content was determined spectrophotometrically after reaction with potassium iodide [28], and O_2^- generation rates were measured using a hydroxylamine oxidization method [29].

Determination of antioxidant enzyme activities

The activity of antioxidant enzyme assays, including superoxide (SOD), catalase (CAT), peroxidase (POD), ascorbate peroxidase (APX), monodehydroascorbate reductase (MDHAR), dehydroascorbate reductase (DHAR), glutathione reductase (GR), glutathione S-transferase (GST), glycolate oxidase (GO) and glutathione peroxidase (GPX), was assayed following the method of [30] and [24]. In all enzyme preparations, protein was quantified following by [31] using bovine serum albumin as a standard.

Protein sample preparation, 2DE, and image analysis

Protein extraction and two-dimensional electrophoresis (2DE) separation were performed as previously described, with minor modifications [30]. Briefly, treated leaves (1g) was ground in liquid nitrogen, and total soluble proteins were extracted at 4°C in 8 mL of extraction buffer containing 100 mM Tris-HCl buffer (pH 8.8),10 mM EDTA, 0.9 M sucrose, and 0.4% mercaptoethanol. Homogenates were centrifuged at 15 000g for 15 min at 4°C, and the supernatants were added to 5 vol of 100 mM ammonium sulfate/methanol. Samples were maintained at –20 °C for 4 h and then were centrifuged at 20 000 g for 15 min at 4 °C. The resulting pellets were washed with 80% acetone containing at –20 °C for 1 h, and the 100% acetone wash once after centrifugation. The final pellets were vacuum-dried and dissolved in 7 M urea, 40 mM DTT, 4% (w/v) CHAPS, and

2% (w/v) ampholyte (pH 3 – 10). Samples in ampholyte were vortexed thoroughly for 1 h at room temperature and then were centrifuged at 35 000 g for 20 min at 20 °C. Supernatants then were collected for 2DE experiments [32]. Protein concentration was determined using a Quant-kit according to manufacturer's instructions (GE Healthcare, USA). Extracted proteins were first separated by isoelectric focusing (IEF) using gel strips (pH 4-7 linear, 13 cm) (GE Healthcare,USA). Following IEF, proteins were separated by sodium dodecyl sulfate-polyacrylamide gel electrophoresis (SDS-PAGE) using 12.5% (w/v) polyacrylamide. Gel strips then were rehydrated in 450 μL of dehydration buffer containing 1600 μg of total proteins and a trace of bromophenol blue for 26 h. Gel strips were focused at 80 kV/h and 20 °C using the PROTEAN IEF system (Bio-Rad,USA) and then were equilibrated for 15 min in equilibration buffer (6 M urea, 0.5 M Tris [pH 8.8], 2% [w/v] SDS, 30% [v/v] glycerol). Gel strips then were placed over 12.5% (w/v) SDS-PAGE gels for 2DE. Gel electrophoresis was performed at 25 mA for 5 h. Gels were stained using Coomassie Brilliant Blue (CBB). After staining, gels were scanned using an ImageScanner III (GE Healthcare, USA) at a resolution of 300 dpi and 16-bit grayscale pixel depth. The images were analyzed with ImageMaster 2D software (version 6.0) (GE Healthcare, USA). The average vol% values were calculated from three technical replicates to represent the final vol% values of each biological replicate. The volume of each spot changed more than 1.5-fold among the treatments and a $p < 0.05$ were considered to be differentially expressed spots.

Protein identification and database searching

Protein spots displaying significant changes in abundance were excised manually from colloidal CBB stained 2DE gels using sterile pipette tips. Briefly, spots cut out of the gels were destained twice with 100 mM NH_4HCO_3, 50% ACN at 37 °C for 20 min in each treatment. After dehydration with 100% ACN and drying, the gel pieces were pre-incubated in 10-20 μL of trypsin solution (10 ng /μL) for 1h. Then adequate digestion buffer (40 mM NH_4HCO_3, 10% ACN) was added to cover the gel pieces, which were extracted using Milli-Q water followed by double 1h extraction with 50% ACN and 5% TFA. The combined extracts were dried in SpeedVac concentrator (Thermo Scientific) at 4 °C. The samples were then subjected to mass spectrometry [30].

The MS spectra were acquired using a MALDI-TOF-MS/MS (AB SCIEX TOF/TOF: trademark: 5800 system). A Mass standard kit (Applied Biosystems, USA) and a standard BSA digest (Sigma-Aldrich, USA) were used for MS and MS/MS calibrations and fine-tuning the resolution and sensitivity of the system as previously described [33]. The mass error was below 30 ppm at both MS and MS/MS mode and the resolution was more than 25 000. The MS/MS spectra were searched against the NCBInr protein databases (http://www.ncbi.nlm.nih.gov/)(5,222,402 sequence entries in NCBI) using Mascot software (Matrix Sciences, London, UK). The taxonomic category was green plants. The searching criteria were according to [34]. The searching criteria include mass tolerance for precursor ions of 0.3 Da, mass tolerance for fragment ions of 100 ppm, one missed cleavage allowed, carbamidomethylation of cysteine as a fixed modification, and oxidation of methionine as a variable modification. To obtain high confident identification, proteins had to meet the following criteria: the top hits on the database searching report, a probability-based MOWSE score greater than 43 (p-value < 0.01), and more than two peptides matched with nearly complete y-ion series and complementary b-ion series present.

Protein classification and hierarchical cluster analysis

To determine the functions of identified proteins, we searched against the NCBI database (http://www.ncbi.nlm.nih.gov/) (accessed on 8 August 2012) and UniProt database (http://www.ebi.uniprot.org/). By integrative analysis of all the information collected from aforementioned processed, each protein was classified into certain functional category defined by us. Log (base 2) transformed ratios were used forhierarchical clustering analysis using Cluster 3.0 available onthe Internet (http://bonsai.hgc.jp/~mdehoon/software/cluster/software.htm), and the results were visualized using Java TreeView (http://jtreeview.sourceforge.net/).

Protein subcellular location prediction

The subcellular location of the identified proteins was predicted using five internet tools: (1) YLoc (http://abi.inf.uni-tuebingen. de/Services/YLoc/webloc.cgi), confidence score ≥0.7; (2) LocTree3 (https://rostlab.org/services/loctree3/), expected accuracy ≥80%; (3) ngLOC (http://genome.unmc.edu/ngLOC/ index.html), probability ≥80%; (4) TargetP (http://www.cbs.dtu.dk/services/TargetP/), reliability class ≤3; (5) Plant-mPLoc (http://www.csbio.sjtu.edu.cn/bioinf/plant-multi/) no threshold value in Plant-mPLoc. Only the consistent predictions from at least two tools were accepted as a confident result. For the inconsistent prediction results among five tools, subcellular localizations for corresponding proteins were predicted based on literatures.

Total RNA extraction, reverse transcription and qReal-Time (qRT)-PCR analysis

Based on the findings of proteomic analysis, we chose 12-responsive proteins which might be the underlying regulator of WD tolerance for qRT-PCR analysis for the verification of proteomic data. Total RNA from abdominal adipose tissue was isolated using Trizol reagent. One-microgram total RNA was performed in reverse transcription with Revert Aid Reverse Transcriptase (Fermentas) and Oligo d (T) primers (TaKaRa). Reverse transcription conditions for each cDNA amplification were 65 °C for 5 min, 37 °C for 52 min, and 70 °C for 15 min. Real-time RT-PCR was carried out using the Real-time PCR System (Roche Light Cycler 480 II, Switzerland) and SYBR Premix Ex Taq (TaKaRa). The primers used for the PCR are listed in Additional file 1: Table S1. Values represent the mean of three biological replicates and two technical replicates. The relative gene expression levels were calculated by the $2^{-\Delta\Delta t}$ method [35].

Data analysis

All results were presented as means ± standard error (SE) of at least three replicates. Results were analyzed by one-way ANOVA using the statistical software SPSS 17.0 (SPSS Inc. Chicago, IL, USA). Posthoc comparisons were tested using the LSD test at a significance level of $p < 0.05$.

Results
Effects of WD on photosynthesis

The photosynthesis indexes of *C. humilis* under WD were analyzed. After 24 days of treatment, Pn, Gs and Ci in *C. humilis* leaves under WD were significantly lower than those under control conditions (Table 1). In addition, chlorophyll fluorescence parameters were monitored to determine the performance of photosystem II (PSII) photochemistry. Fv/Fm were not significantly altered after 6 days WD, but were significantly reduced by 7 and 14% at 12 and 24 days WD, respectively, as compare with controls (Table 1).

Table 1 Effects of Pn, Gs, Ci and Fv/Fm of *C. humilis* under WD for 0, 6, 12, and 24 d

Treatment time (d)	Pn (μmol $CO_2 \cdot m^{-2} \cdot s^{-1}$)		Gs (μmol $H_2O_2 \cdot m^{-2} \cdot s^{-1}$))		Ci (μmol $CO_2 \cdot m^{-2} \cdot s^{-1}$)		Fv/Fm	
	Control	WD	Control	WD	Control	WD	Control	WD
0	22.3 ± 0.4a	21.6 ± 0.6a	187 ± 22a	192 ± 23a	352 ± 7a	338 ± 26a	0.838 ± 0.03a	0.829 ± 0.013a
6	21.9 ± 0.4a	18.6 ± 0.8b	199 ± 11a	177 ± 16a	345 ± 9a	321 ± 12ab	0.837 ± 0.05a	0.796 ± 0.018ab
12	22.6 ± 0.7a	15.3 ± 0.9c	189 ± 12a	116 ± 9c	342 ± 12a	311 ± 21bc	0.830 ± 0.06a	0.774 ± 0.017b
24	20.8 ± 1.1a	13.6 ± 0.1d	187 ± 33a	86 ± 11d	354 ± 14a	288 ± 17c	0.822 ± 0.06a	0.716 ± 0.022c

Each value is the mean ± SE of five independent experiments. The different small letters indicate significant difference ($p<0.05$) under WD

Effects of WD on leaf relative water content (LRWC) and membrane lipid peroxidation

WD resulted in a gradual decrease in LRWC in *C. humilis*, although there were no significantly changes of LRWC in WD treated plants in the early treatment period (6 d), as compared with controls. At 12 and 24 d of WD, LRWC was decreased to 85 and 71% of controls, respectively (Fig. 1a).

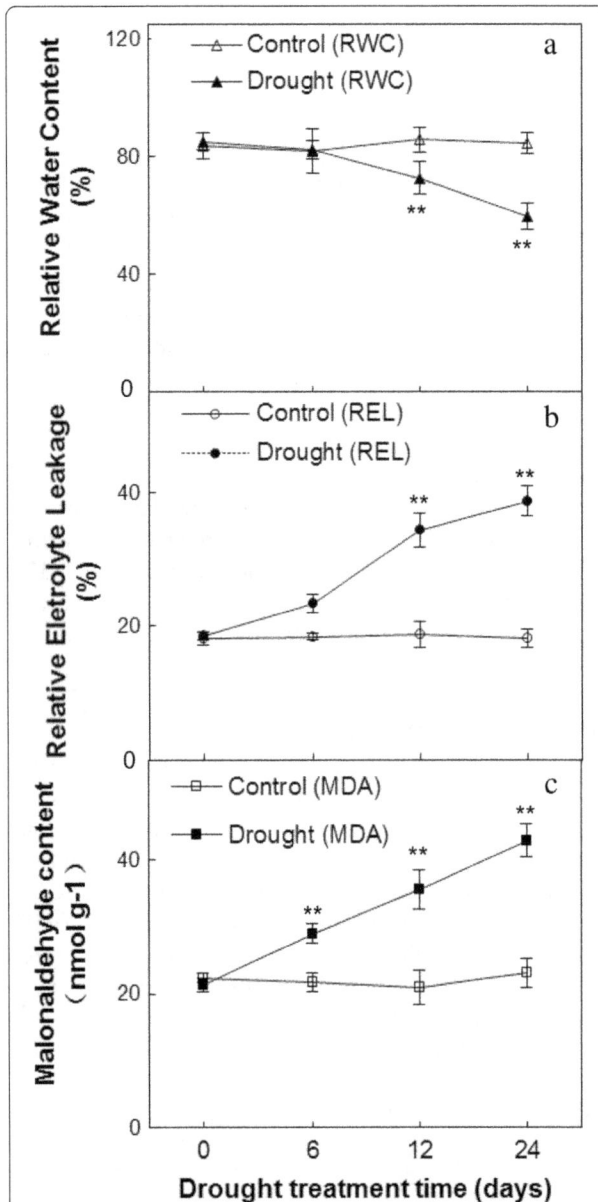

Fig. 1 Effects of water deficit on leaf relative water content (LRWC) (a), relative electrolyte leakage (REL) (b), and malondialdehyde (MDA) (c) in *C. humili* leaves. Values are presented as means ± SE (*n* = 3). The values were determined under control and water deficit treatment at 0, 6, 12, and 24 days. * and ** indicate values that differ significantly from controls at $p \leq 0.05$ and $p \leq 0.01$, respectively, according to LSD test

To determine the levels of membrane integrity and permeability of the cell membrane in *C. humilis* seedlings during WD, the status of REL and MDA were monitored. Leaf REL and MDA content increased slightly in the early WD period (6 d) (Fig. 1b, c), and *C. humilis* plants had a significantly high REL level by the end of WD treatment (24 d). These imply that the membrane integrity of WD plants was destroyed and the electrolyte inside of cells was come out to some extents.

Effects of WD on soluble sugar and proline

Under WD, plants could synthesize compatible low-molecular weight organic solutes, such as proline and soluble sugar to alleviate WD-induced osmotic stress. In this study, the soluble sugar content increased significantly during the whole WD period compared to controls (Fig. 2). The proline content did not display significantly changes in the early treatment period, while they showed dramatic increases upon 12 and 24 d of WD (Fig. 2). The accumulation of soluble sugar and proline contents may contribute to maintain osmotic balance under WD.

Effects of WD on H_2O_2 content, the generation rate of O_2^- and antioxidant enzyme activities

In the present study, the rate of O_2^- generation and H_2O_2 contents were increased gradually on time for *C. humilis* leaves subjected to WD (Fig. 3a), suggesting that oxidative stress was occurred. The activity of SOD, a key enzyme of scavenging ROS, was enhanced in *C. humilis* leaves exposed to WD (Fig. 3b), which might enable the dismutation of superoxide into oxygen and H_2O_2 timely. Plants possess a complex mechanism of enzymatic antioxidants (e.g. POD, CAT, GPX, APX, MDHAR, DHAR, GR, and GST) that protect cells from oxidative damage by scavenging ROS. Thus, their activities were examined in this study. The activities of POD, CAT, APX, DHAR and GST were significantly increased in the whole WD period (Fig. 3b, c, d, e, f). The activities of GPX, MDHAR, and GR were increased after 12 d of WD conditions (Fig. 3c, d, e). In contrast, the GOX activities were decreased under WD (Fig. 3f), although the decrease extents were less significant as compared to other enzymes. In the whole level, the induction of enzymes activities under WD implies that they play a crucial role in defending against the oxidative stress and cell damage induced by WD.

Identification, functional categorization and subcellular location of different changed proteins in response to WD

Based on the results of biochemical assays, we chose 0, 12, and 24 d WD samples for proteomic analysis. Total soluble proteins from control and WD-treated *C. humilis* leaves were separated and imaged, respectively (Fig. 4). Approximately, 700 Coomassie Brilliant Blue-stained

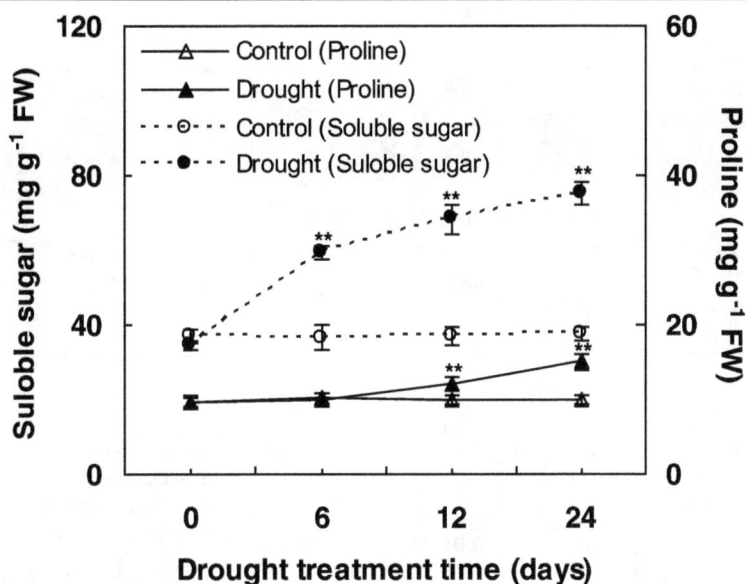

Fig. 2 Effects of water deficit on leaf soluble sugar and proline contents in *C. humili*leaves. Values are presented as means ± SE. The values were determined under control and water deficit treatment at 0, 6, 12, and 24 days. * and ** indicate values that differ significantly from controls at $p \leq 0.05$ and $p \leq 0.01$, respectively, according to LSD test

protein spots were detected on the pI 4-7 gels. Only the protein spots that exhibited reproducible changes under WD (average fold changes >1.5 or <0.6, p-value < 0.05) were retained for further analysis. A total of 46 protein spots showed significant changes in WD samples as compared to control leaves (Table 2). The differentially expressed protein spots were successfully identified by using tandem MS, of which 46 spots represented 43 unique proteins (Table 2, Additional file 2: Table S2). Further examination of their electrophoretic patterns indicated that their inferred mass or isoelectric point values differed, perhaps owing to post-translational modification or degradation. To further examine the differentially expressed proteins, the identified proteins were assigned to Gene Ontology terms (Table 2, Additional file 2: Table S2). The 46 IDs were classified into 8 categories (Fig. 5a), covering a wide range of biological processes, which include photosynthesis (26%), stress and defense (13%), carbohydrate and energy metabolism (15%), transcription related proteins (7%), protein synthesis and turnover (11%), amino acid metabolism (24%), cell wall related (2%), and cell division (2%), respectively. In total, 22 significantly changed proteins were predicted to be localized in chloroplast, 10 in cytoplasm, two in mitochondria, two in nucleus, eight in secreted, and two in vacuole (Fig. 5b, Additional file 3: Table S3).

Hierarchical clustering of different changed proteins in response to WD

To study protein expression characteristics in each functional category, hierarchical clustering analysis was

performed that yielded two main clusters (Fig. 6). Cluster I included 40 IDs (Fig. 6), the levels of which were increased under WD. These proteins were divided into two subclusters: Subcluster I-1 contained the proteins mainly increased significantly under 24 d WD, while subcluster I-2 contained the proteins mainly increased under 12 d WD. The proteins in cluster I covered seven function categories. The remaining six IDs were grouped into cluster II, representing decreased proteins under WD. Cluster II contained two subclusters: Subcluster II-1 contained 3 IDs induced either at 12 and 24 d of WD, while subcluster II-2 included 3 proteins significantly increased only at 12 d of WD (Fig. 6). When examined proteins in each cluster, most of increased IDs were involved in photosynthesis and ROS scavenging and were mainly included in cluster I, whereas carbohydrate and cell division were decreased and present in cluster II.

Homologous gene expression of different changed proteins

To determine whether changes of gene transcription levels correlated with changes of protein levels, a qRT-PCR analysis of 12 genes was performed (Fig. 7). The results demonstrated that nine genes, namely carbonic anhydrase (CA), Ribose 5-phosphate isomerase (Rpi), dehydroascorbate reductase (DHAR), glutathione peroxidase (GPX), malic enzyme (ME), transketolase 1 (TK), heat shock protein 70 (HSP 70), 20S proteasome, and acidic endochitinase (Echi) showed consistent expressional trends with their homologous proteins (Fig. 7, Table 2). They were involved in photosynthesis, stress and defense, carbohydrate and

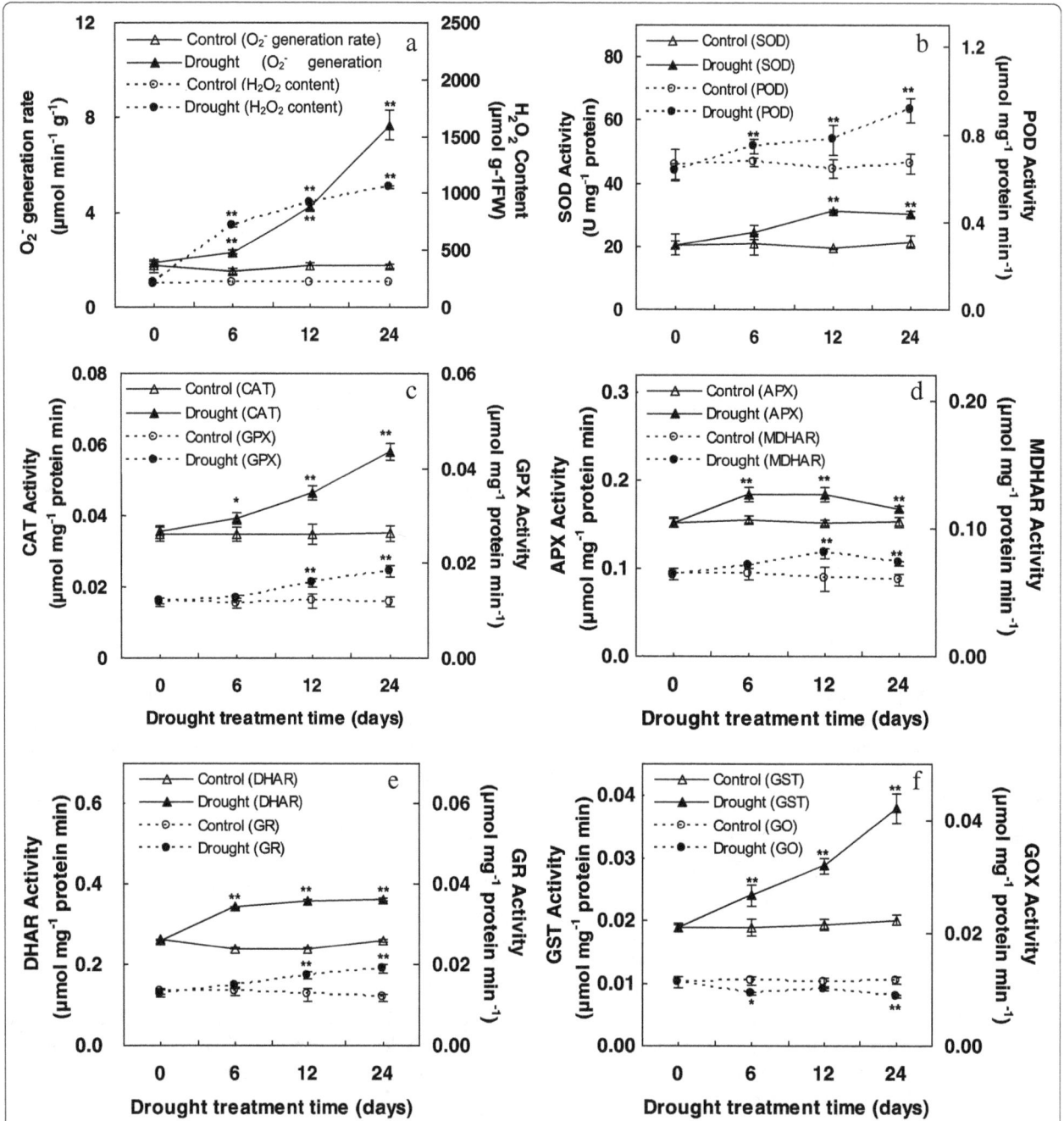

Fig. 3 Effect of water deficit on the activities of antioxidant-related enzymes in *C. humili* leaves. **a** O_2^- generation rate and H_2O_2 content; **b** superoxide dismutase (SOD) and peroxidase (POD); **c** catalase (CAT) and glutathione peroxidase (GPX); **d** ascorbate peroxidase (APX) and monodehydroascorbate reductase (MDHAR); **e** dehydroascorbate reductase (DHAR) and glutathione reductase (GR); **f** glutathione S-transferase (GST) and glycolate oxidase (GO). Values are means ± SE. based on four independent determinations after plants were treated for 0, 6, 12, and 24 days of water deficit, and *bars* indicate standard deviations. * and ** indicate values that differ significantly from controls at $p \leq 0.05$ and $p \leq 0.01$, respectively, according to LSD test

energy metabolism, protein synthesis and turnover, and cell wall related, respectively. In addition, three genes, namely cytosolic ascorbate peroxidase (APX19), triose-phosphate isomerase (TIM), and RNA recognition motif (RRM) appeared opposite expressional trends with

homologous proteins (Fig. 7, Table 2). They were involved in stress and defense, carbohydrate and energy metabolism, and transcription, respectively. The results of correlation analysis indicate that the aforementioned metabolic processes were modulated by post-transcriptional and/or

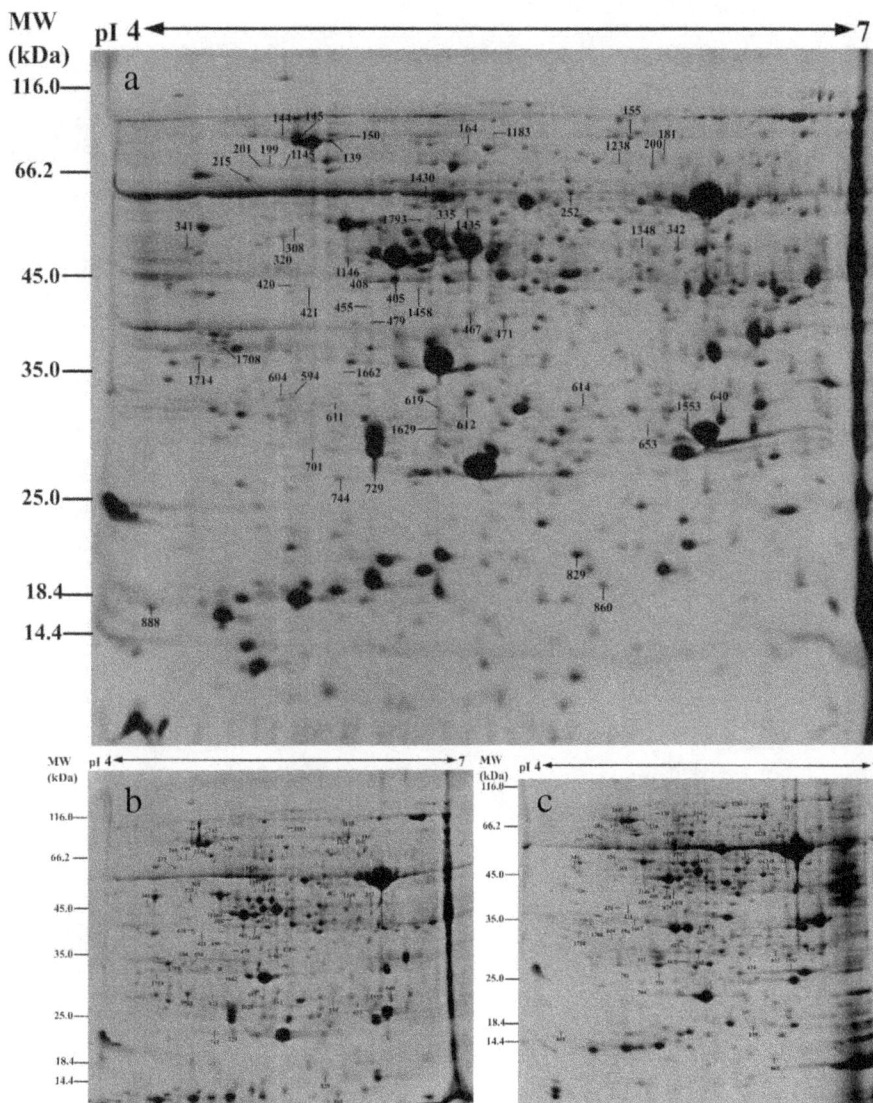

Fig. 4 Representative 2-DE gel image of proteins in response to water deficit from *C. humili*leaves. **a** control; **b** 12 days after water deficit; **c** 24 days after water deficit. Proteins were separated on 13 cm IPG strips (pI 4-7 linear gradient) using IEF in the first dimension, followed by 12.5% SDS-PAGE gels in the second dimension. The 2-DE gel was stained with Coomassie Brilliant Blue. Molecular weight (MW) in kDa and pI of proteins are indicated on the left and top of the gel, respectively. A total of 46 differentially expressed proteins identified by MALDI-TOF-MS/MS were marked with numbers on the gel, and detailed information can be found in Additional file 2: Table S2 and Table 1

post-translational regulation during WD. The inconsistent abundances of transcripts and proteins in WD also support the notion that pre-synthesized mRNA and proteins would function in the process of WD.

Discussion

WD is the main limiting factor for plants growing in arid areas. Biochemical, physiological and molecular influences on plants are wide-spread during WD and can be divided into three aspects: growth control, stress damage control and osmotic homeostasis [2, 36]. An integrated proteomics, biochemical, physiological and

morphological approach was used for our research to investigate these aspects of WD responses in *C. humilis*.

Photosynthetic acclimation to WD

In *C. humilis* seedlings, Gs and Ci were declined gradually under WD, indicating that the decrease in Pn (Table 1) may be the result of a stomatal limitation [37]. Based on our determination of chlorophyll fluorescence, Fv/Fm significantly declined in response to WD at 12 and 24 days (Table 1), suggesting that WD causes photoinhibition of PSII. To our surprise, we found that 10 out of 12 light reaction and Calvin cycle-related proteins, including a carbonic anhydrase (CA), seven ribulose-1,5-bisphosphate

Table 2 Relative protein content changes in *C. humilis* leaves under water deficit

Spot No. [a]	Protein name [b]	Subcellular location (c)	Species [e]	gi Number [e]	Thr. MW (Da) /pI [g]	Exp. MW (Da) /pI [g]	Sco [h]	QM [i]	V%±SE [j] 0, 12, 24 (d)
	Photosynthesis (12)								
653	Carbonic anhydrase (CA)	Chl	Arabidopsis lyrata subsp.	297336434	29,171/ 6.1	40,359/ 6.21	77	3	
1146	Ribulose-1,5-bisphosphate carboxylase/oxygenase large subunit (RuBisCO LSU)	Chl	Heterolepis aliena	125857653	51,796/ 6	80,159/ 5.72	285	12	
308	Ribulose-1,5-bisphosphate carboxylase/oxygenase large subunit (RuBisCO LSU)	Chl	Malus x domestica	415852	48,217/ 8.2	60,254/ 4.74	76	7	
1145	Ribulose-1,5-bisphosphate carboxylase/oxygenase large subunit (RuBisCO LSU)	Chl	Bongardia chrysogonum	7240504	52,269/ 6.13	79,338/ 4.71	237	13	
201	Ribulose-1,5-bisphosphate carboxylase/oxygenase large subunit (RuBisCO LSU)	Chl	Vanilla palmarum	39655299	49,077/ 6.33	78,475/ 4.67	289	10	
199	Ribulose-1,5-bisphosphate carboxylase/oxygenase large subunit (RuBisCO LSU)	Chl	Loesenerielia sp. Chase	9909955	52,296/ 6.04	80,473/ 4.69	212	11	
215	Ribulose-1,5-bisphosphate carboxylase/oxygenase large subunit (RuBisCO LSU)	Chl	Cremastosperma leiophyllum	54660772	51,530/ 6.78	75,683/ 4.61	512	15	
252	Ribulose-1,5-bisphosphate carboxylase/oxygenase large subunit (RuBisCO LSU)	Chl	Astripomoea grantii	21633967	52,502/ 6.57	67,432/ 5.86	761	20	
611	Chlorophyll a-b binding protein 3,chloroplastic (CAB)	Chl	Glycine max	255647962	27,959/ 5.29	36,452/ 4.93	83	4	
455	Ribose 5-phosphate isomerase (Rpi)	Chl	Vitis vinifera	225451267	33239/ 5.22	37,324/ 5.08	105	4	
408	Sedoheptulose-1,7-bisphosphatase, chloroplastic (SBPase)	Chl	V. vinifera	225466690	42,898/ 5.95	45,424/ 5.14	147	4	
405	Sedoheptulose-1,7-bisphosphatase, chloroplastic (SBPase)	Chl	V. vinifera	225466690	42,898/ 5.95	45,234/ 5.21	238	5	
	Stress and defense (6)								
1662	Copper/zinc superoxide dismutase copper chaperone precursor (SOD)	Sec	Populus trichocarpa	222841882	34,529/ 5.47	42,342/ 4.93	104	3	
1435	Peroxidase 1 (POD)	Sec	Phaseolus lunatus	73913500	32,325/ 8.07	59,243/ 5.46	289	6	
640	Cytosolic ascorbate peroxidase APX19 (APX)	Cyt	Fragaria x ananassa	5442416	27,364/ 5.69	29,253/ 6.39	207	3	
1629	Dehydroascorbate reductase 1 (DHAR)	Sec	Malus x domestica	225380890	29,487/ 8.78	34,435/ 5.38	97	2	
744	Glutathione peroxidase 1 (GPX)	Chl	V. vinifera	225426405	19,447/ 5.01	28,453/ 4.94	149	2	
829	Pathogenesis-related protein Bet v I family (Bet V I)	Cyt	Malus x domestica	4590388	17,531/ 5.47	17,352/ 5.88	139	4	
	Carbohydrate and energy metabolism (7)								
612	Triosephosphate isomerase (TIM)	Chl	Fragaria x ananassa	300659132	33,733/ 7.64	37,333/ 5.48	393	6	
614	Triosephosphate isomerase (TIM)	Chl	Vitis vinifera	300659132	33,733/ 7.64	37,833/ 5.91	263	4	
619	Triosephosphate isomerase (TIM)	Chl	Stellaria longipes	300659296	27,710/ 5.54	34,362/ 5.34	120	2	
1183	Beta-hexosaminidase-like protein (Hexose)	Chl	V. vinifera	225450263	64,148/ 5.25	80,345/ 5.58	131	3	
467	Mercapto-pyruvate sulfurtransferase (MST)	Cyt	Brassica napus	253720703	41,711/ 5.82	43,282/ 5.49	292	6	
181	Malic enzyme (ME)	Cyt	Ricinus communis	223546686	65,487/ 5.98	68,463/ 5.33	469	9	
155	Transketolase 1 (TK)	Chl	Capsicum annuum	3559814	80,398/ 6.16	84,483/ 6.11	262	9	
	Transcription related (3)								
1793	RNA recognition motif (RRM)	Sec	V. vinifera	297737424	21,608/ 5.56	67,394/ 6.28	106	3	
860	RNA recognition motif (RRM)	Nuc	Prunus avium	34851124	17,374/ 7.82	17,363/ 5.99	1,03 0	13	
1708	Nucleic acid binding protein (NABP)	Nuc	Zea mays	162463757	33,154/ 4.6	36,713/ 4.53	269	3	
	Protein synthesis and turnover (5)								
888	60S acidic ribosomal protein (RP)	Cyt	Prunus dulcis	111013714	11,408/ 4.3	18,323/ 4.23	296	5	
150	Heat shock protein 70 (HSP70)	Mit	P. trichocarpa	222872861	75,412/ 5.24	79,492/ 6.14	634	13	
164	Heat shock protein 70 (HSP70)	Mit	Phaseolus vulgaris	257310566	72,721/ 5.95	74,523/ 5.01	648	12	
420	Heat shock protein 70 (HSP70)	Cyt	Malus x domestica	257307291	71,570/ 5.17	68,532/ 4.78	114	7	
479	20S proteasome	Cyt	Medicago truncatula	217072126	24,915/ 8.74	45,384/ 5.09	704	9	
	Amino acid metabolism (11)								
594	Cysteine protease (CPs)	Vac	Prunus armeniaca	2677828	39,855/ 6.41	37,442/ 4.78	78	4	
200	Acetohydroxyacid synthase, partial (AHAS)	Chl	G. max	255689393	70,384/ 6.96	73,483/ 6.36	239	3	
341	Ankyrin repeat domain protein (ANK)	Sec	G. max	17645766	29,128/ 4.5	53,234/ 4.31	268	3	
729	Ankyrin repeats (ANK)	Cyt	G. max	255646471	38,027/ 4.53	34,832/ 5.42	271	5	
1430	Aspartate aminotransferase (AST)	Cyt	P. trichocarpa	224074105	53,334/ 5.27	32,392/ 5.31	577	8	
1533	Asparagine synthetase (AS)	Sec	Vigna radiata	2970051	25,750/ 5.62	37,564/ 6.31	295	5	
342	Glutamate-1-semialdehyde 2,1-aminomutase (GSA)	Chl	Cucumis melo subsp. Melo	307136451	50,339/ 6.16	57,322/ 6.27	604	6	
1348	Glutamate-1-semialdehyde 2,1-aminomutase (GSA)	Chl	C. Melo	307136451	50,339/ 6.16	57,478/ 6.63	549	6	
335	Glutamine synthetase (GS)	Cyt	Spiraea nipponica	227478341	47,619/ 6.77	56,425/ 6.15	695	12	
145	Choline dehydrogenase (ChDH)	Vac	Prunus dulcis	32482411	61,168/ 4.89	66,832/ 4.81	881	12	
471	Thiosulfate sulfurtransferase (TST)	Sec	Datisca glomerata	4406372	41,528/ 6.51	47,892/ 5.64	218	5	
	Cell wall related (1)								
604	Acidic endochitinase (Echi)	Sec	Arabidopsis thaliana	166664	33,533/ 5.08	37,456/ 4.73	94	4	
	Cell division (1)								
1458	FtsZ-like protein	Chl	V. vinifera	297737/5.0 8	37,442/ 5.19	40,456/ 5.53	82	4	

[a]Assigned spot number as indicated in Fig. 4. [b]The name and functional categories of the proteins using MALDI TOF-TOF MS. [c]Protein subcellular localization predicted by softwares (YLoc, LocTree3, Plant-mPLoc, ngLOC, and TargetP). Only the consistent predictions from at least two tools were accepted as a confident result listed in Additional file 3: Table S3. Chl, chloroplast; Cyt, cytoplasm; Mit, mitochondria; Nuc, nucleus; sec, secreted; vac, vacuole. [d]The plant species that the peptides matched. [e]Database accession numbers from NCBInr. [f, g]Theoretical (e) and experimental (f) mass (kDa) and pl of identified proteins. Experimental values were calculated using Image Master 2D Platinum Software. Theoretical values were retrieved from the protein database. [h]The Mascot score obtained after searching against the NCBInr database. [i]The number of unique peptides identified for each protein. [j] The mean values of protein spot volumes relative to total volume of all the spots. Four water deficit (0 d, 12 d, and 24 d) were performed. Error bars indicate mean ± standard error (SE)

carboxylase/oxygenase large subunits (RuBisCO-LSU), a Ribose 5-phosphate isomerase (Rpi) and a chlorophyll a-b binding protein (CAB) were accumulated under WD (Table 2, Fig. 8). Among them, Chlorophyll a-b binding protein (CAB), PSI light harvesting chlorophyll binding protein, is the intrinsic transmembrane antenna proteins (Lhca's) occurring in the reaction center of PSI. PSI is known to be the most efficient light converter in nature

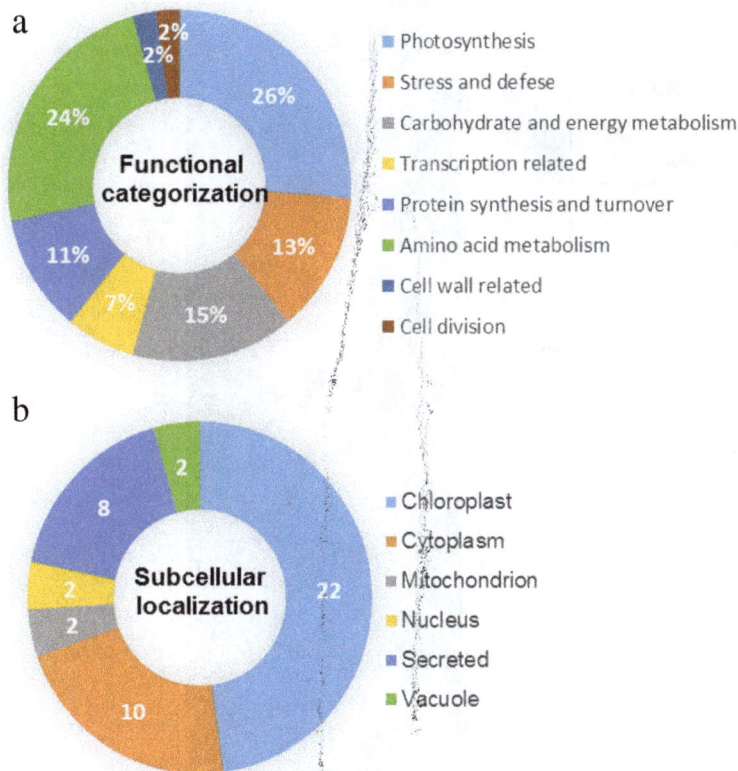

Fig. 5 Functional categorization and subcellular localization of the differentially expressed proteins at different water deficit time points of *C. humili* leaves. **a** A total of 46 DEPs were classified into eight functional categories. The percentage of proteins in different functional categories is shown in the pie. **b** Subcellular localization categories of the identified proteins. The numbers of proteins with different locations are shown

since pigments in the PSI are not being quenched and energy transfer to the electron donor is very rapid [38]. Plastocyanin functions as an electron transfer agent between cytochrome f and P700$^+$ from PSI [39]. We speculated that the increased CAB could transfer more excitation energy to the reaction center, and the accumulation of plastocyanin can donate more electrons to PSI. Finally, CAB with a higher abundance may help to minimize the energy loss caused by a reduction of PSII efficiency in the *C. humilis* response to moderate WD. In addition, CAs and RuBisCO were increased under 12 and 24 d WD in our study (Table 2). This indicated that WD promoted the CO_2 assimilation, leading to the reprogramming of carbon metabolism. Carbonic anhydrase (CA), at the donor side of PSII, can aid to increase the concentration of CO_2 within the chloroplast, which induces the carboxylation rate of RuBisCO [40]. RuBisCO is an important enzyme involved in the first major step of carbon fixation [41, 42]. It has been reported that overexpression of the CA in *Arabidopsis* resulted in an increase in plant biomass [41]. Previous proteomics studies have also revealed that the increase of Calvin cycle-related and light reaction proteins in *Cynodon dactylon* [18] and *Malus* [28].

Regulation of osmostasis and redox homeostasis to cope with WD

In our results, the accumulation of osmolytes in *C. humilis* enhanced the positive response to WD (Fig. 1b, c; Fig. 2). The regulation of osmostasis and redox homeostasis are critical for WD tolerance [43]. And WD affects cell membrane integrity and membrane lipid composition, resulting in the changes of REL and MDA content [36, 44]. This often happened in other species, such as *Amygdalus mira (Koehne) Yü et Lu* [44] and rice [45]. To cope with osmotic imbalance and protect membrane, diverse compatible osmolytes were accumulated in cells. On the other hand, proline is involved in radical scavenging, which has an important defensive role on resisting the WD-induced oxidative stress [46]. Total soluble sugar also acts as a crucial array of prevention and signals that is helpful to sense and control photosynthetic activity and ROS balance [47].

To minimize the damaging effects of ROS, plants have evolved various antioxidant enzymes defense pathways [48–50]. In this study, we found that the activities of several enzymes such as SOD, POD, CAT, GPX/GST, and four key enzymes of ascorbate–glutathione (AsA-

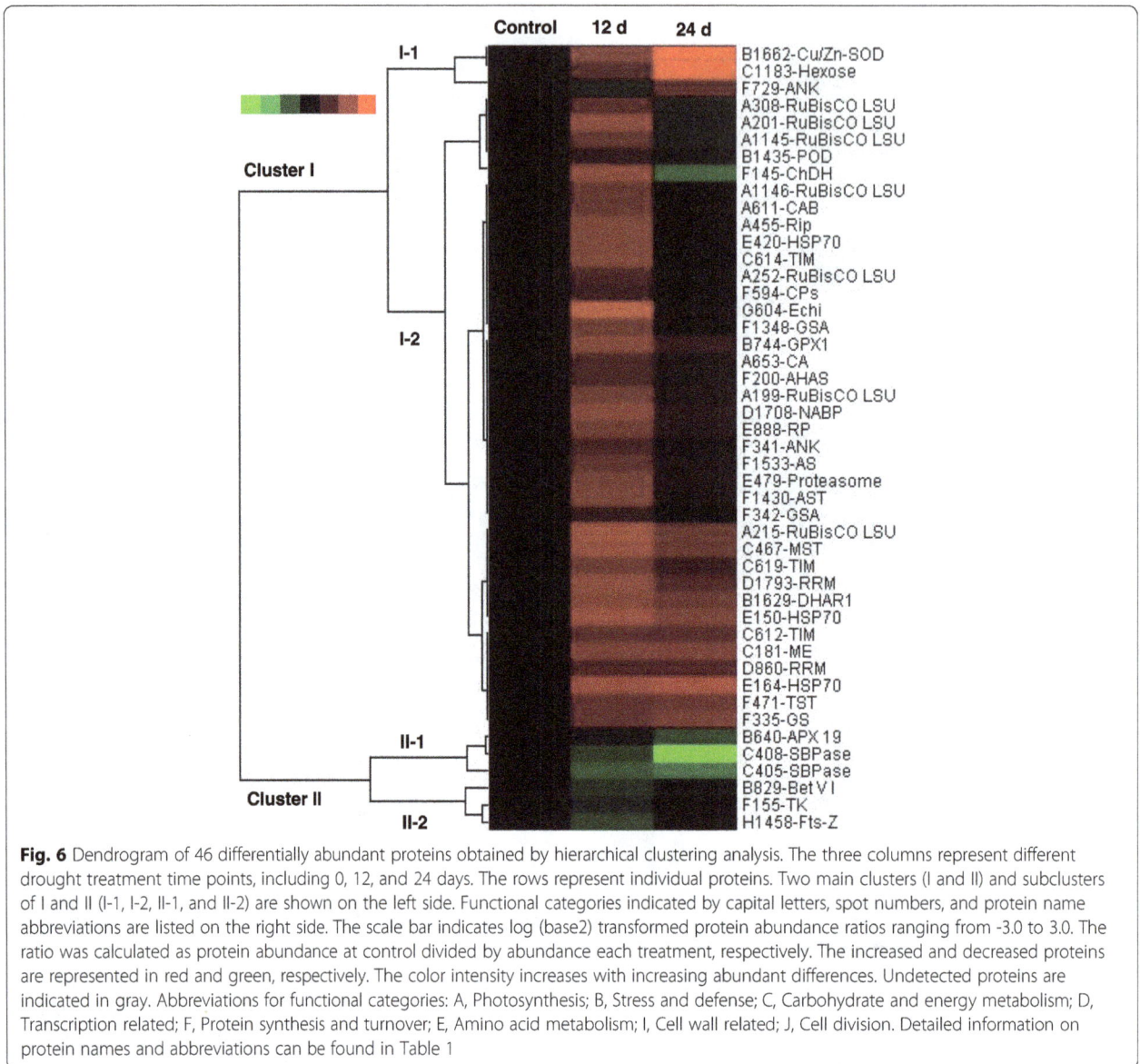

Fig. 6 Dendrogram of 46 differentially abundant proteins obtained by hierarchical clustering analysis. The three columns represent different drought treatment time points, including 0, 12, and 24 days. The rows represent individual proteins. Two main clusters (I and II) and subclusters of I and II (I-1, I-2, II-1, and II-2) are shown on the left side. Functional categories indicated by capital letters, spot numbers, and protein name abbreviations are listed on the right side. The scale bar indicates log (base2) transformed protein abundance ratios ranging from -3.0 to 3.0. The ratio was calculated as protein abundance at control divided by abundance each treatment, respectively. The increased and decreased proteins are represented in red and green, respectively. The color intensity increases with increasing abundant differences. Undetected proteins are indicated in gray. Abbreviations for functional categories: A, Photosynthesis; B, Stress and defense; C, Carbohydrate and energy metabolism; D, Transcription related; F, Protein synthesis and turnover; E, Amino acid metabolism; I, Cell wall related; J, Cell division. Detailed information on protein names and abbreviations can be found in Table 1

GSH) cycle, were affected by WD in *C. humilis* (Fig. 3). All these implied that AsA-GSH cycle and GPX/GST pathway were enhanced in *C. humilis* for WD tolerance [51]. Interestingly, the activity, expression, and abundance of some enzymes were inconsistent. Our proteomics results showed that the abundances of SOD, POD, DHAR, and GPX were increased, except that the abundances of APX were decreased. However, qRT-PCR results found that the expression of APX, DHAR, and GPX were increased (Fig. 7). Enzyme abundances may be inconsistent with their activities and expressions, as the activity is also modulated by the protein conformation and post-translational modifications. This indicated that ROS scavenging enzymes in *C. humilis* seedlings were modulated at both translational and post-translational levels for WD tolerance.

WD-responsive transcription, protein synthesis, and protein processing

In response to WD, plants exhibit quick switches from metabolic quiescent state to active state. In this study, our proteomics results revealed that two RNA recognition motif (RRM) and a nucleic acid binding protein (NABP) were increased at 12 and 24 days treatment. Besides, a ribosomal protein was increased at 12 d WD (Table 2). These indicated that RNA processing and protein synthesis-regulated metabolic increase would be a positive response to WD. During gene expression, RNA processing by RNA chaperone is critical for keeping the proper RNA structure and function in response to WD [52].

We found three HSP 70s were WD-responsive in *C. humilis* at 12 and 24 d WD (Table 2). Protein folding and processing were active for preventing WD-induced

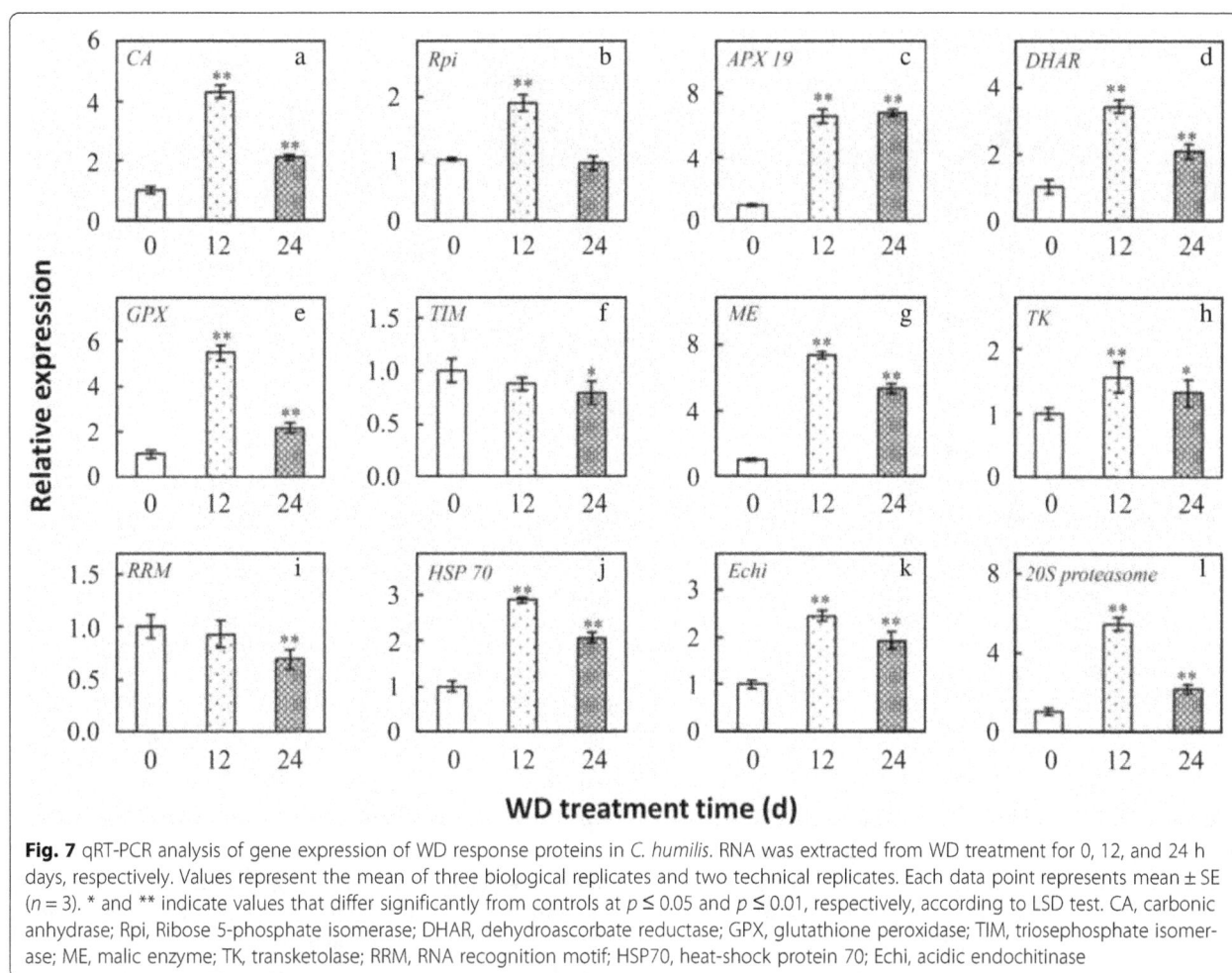

Fig. 7 qRT-PCR analysis of gene expression of WD response proteins in *C. humilis*. RNA was extracted from WD treatment for 0, 12, and 24 h days, respectively. Values represent the mean of three biological replicates and two technical replicates. Each data point represents mean ± SE ($n = 3$). * and ** indicate values that differ significantly from controls at $p \leq 0.05$ and $p \leq 0.01$, respectively, according to LSD test. CA, carbonic anhydrase; Rpi, Ribose 5-phosphate isomerase; DHAR, dehydroascorbate reductase; GPX, glutathione peroxidase; TIM, triosephosphate isomerase; ME, malic enzyme; TK, transketolase; RRM, RNA recognition motif; HSP70, heat-shock protein 70; Echi, acidic endochitinase

denature and incomplete aggregation [45], and HSP 70 belongs to the conservative family of molecular chaperones, and exists in all cells and organs, assisting protein folding, aggregation, translocation, and degradation [53]. Thus, the abundance of HSP 70 was significantly elevated in order to eliminate misfolded proteins. It has been also also revealed that 26S proteasome increased in response to drought stress (Table 2), which is important to remove abnormal or damaged proteins and to control the levels of certain regulatory proteins during drought stress. Similar study also found in *Hordeum vulgare* [54] and *Medicago sativa* [55], respectively. These findings indicate that the enhancement of the ubiquitin/26S proteasome system is important for plants to cope with drought.

Enhancement of carbohydrate supply and other specialized metabolism under WD

In this study, five IDs were affected by WD, including three triosephosphate isomerases (TIM) involved in glycolysis, one transketolase (TK) involved in pentose phosphate pathways (PPP), and malic enzymes (ME)

involved in TCA cycle (Table 2 and Fig. 8). The regulation of carbohydrate metabolism is an important strategy for plants to respond to WD [56]. TIM catalyzes the interconversion of glyceraldehyde-3-phosphate to dihydroxyacetone phosphate [57]. TK, key enzymes of the reductive and oxidative pentose phosphate pathways, are responsible for the synthesis of sugar phosphate intermediates [58]. Malic enzymes (MEs) involved in malate dehydrogenase (MDH) system catalyze the L-malate decarboxylation reaction through their oxidation, and MDH catalyzethe interconversion of malate and oxaloacetate in areversible reaction of the TCA cycle have an important role in biochemical adaptation of plants to stress [59]. In addition, TK is considered to participate in pentosephosphate pathway. The pentose phosphate pathway is important to maintain carbon homoeostasis, to provide precursors for nucleotide and amino acid biosynthesis, and to provide reducing molecules for defeating oxidative stress Stincone et al. [60]. TK is the key enzyme of the non-oxidative branch of the pentose phosphate pathway of carbohydrate transformation [61]. Accumulation of TK would promote the enhancement of the non-

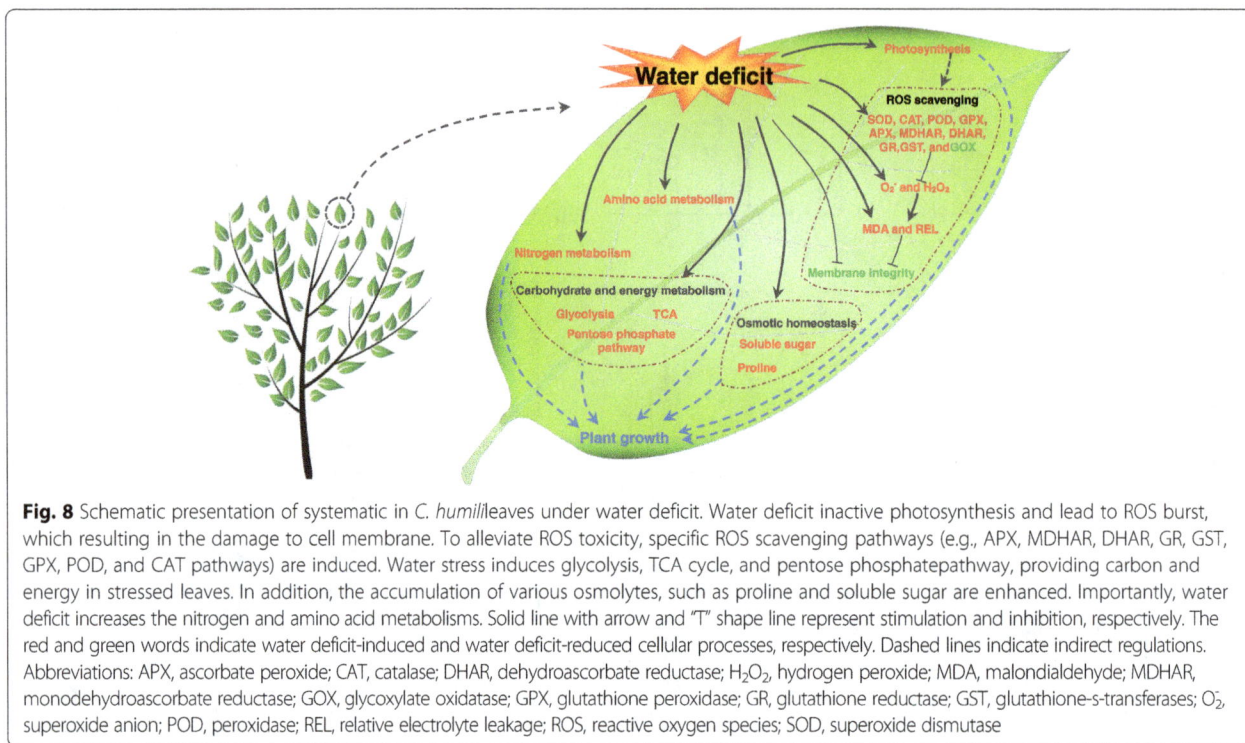

Fig. 8 Schematic presentation of systematic in *C. humili*leaves under water deficit. Water deficit inactive photosynthesis and lead to ROS burst, which resulting in the damage to cell membrane. To alleviate ROS toxicity, specific ROS scavenging pathways (e.g., APX, MDHAR, DHAR, GR, GST, GPX, POD, and CAT pathways) are induced. Water stress induces glycolysis, TCA cycle, and pentose phosphatepathway, providing carbon and energy in stressed leaves. In addition, the accumulation of various osmolytes, such as proline and soluble sugar are enhanced. Importantly, water deficit increases the nitrogen and amino acid metabolisms. Solid line with arrow and "T" shape line represent stimulation and inhibition, respectively. The red and green words indicate water deficit-induced and water deficit-reduced cellular processes, respectively. Dashed lines indicate indirect regulations. Abbreviations: APX, ascorbate peroxide; CAT, catalase; DHAR, dehydroascorbate reductase; H_2O_2, hydrogen peroxide; MDA, malondialdehyde; MDHAR, monodehydroascorbate reductase; GOX, glycoxylate oxidatase; GPX, glutathione peroxidase; GR, glutathione reductase; GST, glutathione-s-transferases; O_2, superoxide anion; POD, peroxidase; REL, relative electrolyte leakage; ROS, reactive oxygen species; SOD, superoxide dismutase

oxidative branch, yielding ribose 5-phosphate for the synthesis of nucleic acids and amino acids accompanied by the production of NADPH, which is critical to maintain redox balance under stress situations [62]. Besides, our results are consistent with previous proteomic studies in *Phaseolus vulgaris* [63], rice [64], and wheat [65] under WD. Previously, enhancement of the glycolysis, PPP, and TCA has been found in several species in response to various stress conditions [66], which would provide more glyceraldehydes-3-phosphate (G3P), glucose-6-phosphate (G6P), NADPH, and erythrose-4-phosphate (E4P) that could be used to produce more ATP for maintaining the basic metabolism under stress [66, 67]. This might be a strategy for plant to enhance the ability of seedlings to survive under WD, which makes it possible for the cell to adapt to its metabolic needs.

Rapid nitrogen and amino acid metabolisms are essential for WD

Nitrogen assimilation is affected by abiotic stress in plants. In this process, exogenous absorbed nitrate is transformed to ammonium by nitrate reductase (NR) and nitrite reductase (NIR), and then assimilated by glutamine synthetase (GS) and glutamate synthase (GOGAT) into amino acids [18, 68]. Our proteomics results revealed that 11 IDs were increased under WD (Table 2 and Fig. 8). Among them, the up-regulated 3-mercaptopyruvate sulfurtransferase (MST) implied that the efficiency of cellular redox state was increased. Related research showed that it catalyzes

pyruvate transsulfuration from 3-mercaptopyruvate, which transamined from cysteine and also contributes to maintain the cellular redox state [69]. Similarly, the increased glutamine synthetase could be involved in the osmotic stress response, as glutamine synthetase is related to proline biosynthesis [70]. In addition, the accumulation of asparagine synthetase (AS) and aspartate aminotransferase (AST) in the present study might aid to nitrogen and amino acid metabolisms, leading to resist to WD response. AS and AST are key enzymes involved in carbon and nitrogen distribution [71]. AS is necessary for the production of nitrogen-rich amino acid asparagine, which is the primary of nitrogen metabolism [72–74]. Similarly, the previous study found that influence of over-expression of cytosolic AST on amino acid metabolism and defence responses against *Botrytis cinerea* infection in *Arabidopsis thaliana* [75].

Conclusion

This study is the first proteomic analysis in *C. humilis* response to WD to our knowledge. In general, we found some responsive pathways are pivotal under WD as shown in Fig. 8 by integrative analysis of all the results from physiological and proteomic analysis in *C. humilis*. Firstly, photosynthesis is strongly affected by WD. The increase of Calvin cycle-related and light reaction proteins may help to convert more light energy and minimize the energy loss caused by a reduction of PSII efficiency. In addition, the reestablishment of osmostasis and redox homeostasis, reprogramming of nuclear and chloroplast gene expression and protein processing are positive for *C. humilis* to WD. Furthermore, nitrogen and amino acid metabolisms lead to resist to WD response. These findings provide important information for understanding WD-responsive mechanisms in *C. humilis* seedlings.

Additional files

Additional file 1: Table S1. Primers used for the quantitative real-time RT-PCR analysis. (XLSX 9 kb)

Additional file 2: Table S2. Differentially expressed proteins and the sequences of peptides identified in *C. humilis* under WD using MALDI TOF-TOF MS. (XLSX 48 kb)

Additional file 3: Table S3. The subcellular localization prediction of the differentially expressed proteins identified in *C. humilis* under WD. (XLSX 30 kb)

Abbreviations
2-DE: Two dimensional electrophoresis; ACN: Acetonitrile; DTT: Dithiothreitol; GAPDH: Glyceraldehyde 3-phosphatedehydrogenase; MALDI-TOF/TOF: Matrix-assisted laser desorption/ionization time-of-flight/time-offlight; RuBisCO: Ribulose bisphosphate carboxylase/oxygenase; TCA: Tricarboxylic acid; TFA: Trifluoroacetic acid

Acknowledgments
We acknowledge Shanghai AB Sciex Asia Pacific Application Support Center for its assistance in original data processing. This work was supported by Fundamental Research Funds for the Central Universities (2572015DA02), the National Natural Science Foundation of China (31170569, J1210053) and the Innovation Project of State Key Laboratory of Tree Genetics and Breeding (Northeast Forestry University).

Funding
This work was supported by Fundamental Research Funds for the Central Universities (2572015DA02), the National Natural Science Foundation of China (31170569, J1210053), the Innovation Project of State Key Laboratory of Tree Genetics and Breeding (Northeast Forestry University), and the Open Research Fund for Key Laboratory of Dairy Science (No.2015KLDSOF-06, Northeast Agricultural University).

Authors' contributions
ZY designed the study and draft the manuscript; JR participated in analyzed data and drafted the manuscript; JR and LZ performed 2-DE and carried out MS analysis and performed protein identification. JR and YL carried out the photosynthesis and qRT-PCR analysis. LS and JW conceived the study analyzed experiments. XS conceived of, designed, and coordinated the study, and assisted with the writing of the manuscript. All authors read and approved the final manuscript.

Competing interests
The authors declare that they have no competing interests.

Author details
[1]Department of Genetics, College of Life Science, Northeast Forestry University, Harbin 150040, People's Republic of China. [2]State Key Laboratory of Tree Genetic sand Breeding, Northeast Forestry University, Harbin 150040, People's Republic of China. [3]Horticulture Department, College of Horticulture, Shenyang Agricultural University, No. 120 Dongling Road, Shenhe District, Shenyang 110866, People's Republic of China. [4]College of Food Science; Key Laboratory of Dairy Science, Ministry of Education, Synergetic Innovation Center of Food Safety and Nutrition, Northeast Agricultural University, Harbin, Heilongjiang 150030, People's Republic of China. [5]Forest Engineering and Environment Research Institute of Heilongjiang Province, No. 134 Haping Road, Nangang District, Harbin, Heilongjiang 150081, People's Republic of China.

References
1. Bray EA. Plant responses to water deficit. Trends Plant Sci. 1997;2:48–54.
2. Zhu JK. Salt and drought stress signal transduction in plants. Annu Rev Plant Biol. 2002;53:247–73.
3. Farah D, Ashurosh KP, Sanjay R, Ashwarya M, Ruchi S, Sharma YK, Pramod AS, Vivek P. Physiological and proteomic responses of cotton (*Gossypium herbaceum* L.) to drought stress. Plant Physiol Biochem. 2012;53:6–18.4.
4. Sircelj H, Tausz E, Grill M. Detecting different levels of drought stress in apple trees (*Malus domestica* Borkh.) with selected biochemical and physiological parameters. Sci Hortic. 2007;133:362–9.
5. Shuvasish C, Piyalee P, Lingaraj S, Sanjib KP. Reactive oxygen species signaling in plants under abiotic stress. Plant Signal Behav. 2013;8(4):23681–66.
6. Reddy R, Chaitanya V, Vivekanandan M. Drought-induced responses of photosynthesis and antioxidant metabolism in higher plants. J Plant Physiol. 2004;161:1189–202.
7. Upadhyaya H, Khan MP. Hydrogen peroxide induces oxidative stress in detached leaves of *Oryza sativa* L. Gen Appl Plant Physiol. 2007;33:83–95.
8. Miller G, Suzuki N, Ciftci-Yilmaz S, Mittler R. Reactive oxygen species homeostasis and signalling during drought and salinity stresses. Plant Cell Environ. 2010;33:453–67.
9. Faize M, Burgos L, Faize L, Piqueras A, Nicolass E. Involvement of cytosolic ascorbate peroxidase and Cu/Zn-superoxide dismutase for improved tolerance against drought stress. J Exp Bot. 2011;62:2599–613.
10. Good AG, Zaplachinski ST. The effects of drought stress on free amino acid accumulation and protein synthesis in *Brassica napus*. Physiol Plant. 1994;90:9–14.
11. Villar-Salvador P, Planelles R, Oliet J, Peñuelas-Rubira JL, Jacobs DF, Gonzalez M. Drought tolerance and transplanting performance of holm oak (*Quercus ilex*) seedlings after drought hardening in the nursery. Tree Physiol. 2004;24: 1147–55.
12. Corcuera L, Morales F, Abadia A, Gil-Pelegrin E. Seasonal changes in photosynthesis and photoprotection in a Quercus ilex subsp. ballota woodland located in its upper altitudinal extreme in the *Iberian Peninsula*. Tree Physiol. 2005;25:599–608.
13. Martínez F, Arif A, Nebauer SG, Bueso E, Ali R, Montesinos C, et al. A fungal transcription factor gene is expressed in plants from its own promoter and improves drought tolerance. Planta. 2015;242:39–52.
14. Hu W, Huang C, Deng X, Zhou S, Chen L, Li Y, et al. *TaASR1*, a transcription factor gene in wheat, confers drought stress tolerance in transgenic tobacco. Plant Cell Environ. 2013;36:1449–64.
15. Jensen MK, Lindemose S, De Masi F, Reimer JJ, Nielsen M, Perera V, et al. *ATAF1* transcription factor directly regulates abscisic acid biosynthetic gene NCED3 in *Arabidopsis thaliana*. FEBS Open Bio. 2013;3:321–7.
16. Washburn MP, Koller A, Oshiro G, Ulaszek RR, Plouffe D, Deciu C, Winzeler E, Yates JR. Protein pathway and complex clustering of correlated mRNA and protein expression analyses in Saccharomy cescerevisiae. Proc Natl Acad Sci. 2003;100:3107–12.
17. Deyholos MK. Making the most of drought and salinity transcriptomics. Plant Cell Environ. 2010;33:648–54.
18. Wang XL, Cai XF, Xu CX, Wang QH. Dai SJ. Drought-Responsive Mechanisms in plant leaves revealed by proteomics. 2016;17:1706.
19. Gazanchian A, Hajheidari M, Sima NK, Salekdeh GH. Proteome response of *Elymus* elongatum to severe water stress and recovery. J Exp Bot. 2007;58:291–300.
20. Huang B, DaCosta M, Jiang Y. Research advances in mechanisms of turfgrass tolerance to abiotic stresses: From physiology to molecular biology. Plant Sci. 2014;33:141–89.
21. Shi H, Ye T, Chan Z. Comparative proteomic responses of two bermudagrass (*Cynodon dactylon* (L.). Pers.) varieties contrasting in drought stress resistance. Plant Physiol Biochem. 2014;82:218–28.
22. Song XS, Shang ZW, Yin ZP, Ren J, Sun MC, Ma XL. Mechanism of xanthophyll-cycle-mediated photoprotection in *Cerasus humilis* seedlings under water stress and subsequent recovery. Photosynthetica. 2011;49:523–30.
23. Yin ZP, Shang ZW, Wei C, Ren J, Song XS. Foliar sprays of photosynthetic bacteria improve the growth and anti-oxidative capability on Chinese Dwarf Cherry Seedlings. J Plant Nutr. 2012;35:840–53.
24. Yin ZP, Li S, Ren J, Song XS. Role of spermidine and spermine in alleviation of drought-induced oxidative stress and photosynthetic inhibition in Chinese dwarf cherry (*Cerasus humilis*) seedlings. Plant Growth Regul. 2014;209–18.
25. Gong M, Li Y-J, Chen S-Z. Abscisic acid-induced thermotolerance in maize seedlings is mediated by calcium and associated with antioxidant systems. J Plant Physiol. 1998;153:488–96.

26. Lu CM, Jiang GM, Wang BS, Kuang TY. Photosystem II photochemistry and photosynthetic pigment composition in saltadapted halophyte Artimisia anethifolia grown under outdoor conditions. J Plant Physiol. 1998;160:403–8.

27. Porcel R, Ruiz-Lozano JM. Arbuscular mycorrhizal influence on leaf water potential, solute accumulation, and oxidative stress in soybean plants subjected to drought stress. J Exp Bot. 2004;55:1743–50.

28. Ibrahim MH, Jaafar HZE. Primary, secondary metabolites, H₂O₂, malondialdehyde and photosynthetic responses of *Orthosiphon stimaneus benth.* to different irradiance levels. Molecules. 2012;17:1159–76.

29. Yan B, Dai Q, Liu X, Huang S, Wang Z. Flooding-induced membrane damage, lipid oxidation and activated oxygen generation in corn leaves. Plant Soil. 1996;179:261–8.

30. Suo J, Zhao Q, Zhang Z, Chen S, Cao J, Liu G, et al. Cytological and proteomic analyses of *Osmunda cinnamomea* germinating spores reveal characteristics of fern spore germination and rhizoid tip growth. Mol Cell Proteomics. 2015;14:2510–34.

31. Bradford MM. A rapid and sensitive method for the quantitation of microgram quantities of protein utilizing the principle of protein-dye binding. Anal Biochem. 1976;72:248–54.

32. Wang X, Chen S, Zhang H, Shi L, Cao F, Guo L, et al. Desiccation tolerance mechanism in resurrection fern-ally *Selaginella tamariscina* revealed by physiological and proteomic analysis. J Proteome Res. 2010;9:6561–77.

33. Zhu M, Simons B, Zhu N, Oppenheimer DG, Chen S. Analysis of abscisic acid responsive proteins in *Brassica napus* guard cells by multiplexed isobaric tagging. J Proteomics. 2010;73:790–805.

34. Yu J, Chen S, Zhao Q, Wang T, Yang C, Diaz C, et al. Physiological and proteomic analysis of salinity tolerance in *Puccinellia tenuiflora*. J Proteome Res. 2011;10:3852–70.

35. Livak KJ, Schmittgen TD. Analysis of relative gene expression data using real-time quantitative PCR and the 2(-Delta Delta C(T)) method. Methods. 2001;25:402–8.

36. Merewitz EB, Gianfagna T, Huang B. Protein accumulation in leaves and roots associated with improved drought tolerance in creeping bentgrass expressing an ipt gene for cytokinin synthesis. J Exp Bot. 2011;62:5311–33.

37. Sharkey TD, Bernacchi CJ, Farquhar GD, Singsaas EL. Fitting photosynthetic carbon dioxide response curves for C3 leaves. Plant Cell Environ. 2007;30: 1035–40.

38. Van Amerongen H, Croce R. Light harvesting in photosystem II. Photosynth Res. 2013;116:251–63.

39. Farkas D, Hansson Ö. Thioredoxin-mediated reduction of the photosystem i subunit PsaF and activation through oxidation by the interaction partner plastocyanin. Federation of European Biochemical Societies. 2011;585:1753–8.

40. Badger MR, Price GD. The Role of Carbonic Anhydrase in Photosynthesis. Plant Physiol Plant Mol Biol. 1994;45:369–92.

41. Feller U, Anders I, Demirevska K. Degradation of rubisco and other chloroplast proteins under abiotic stress. Plant Physiol. 2008;34:5–18.

42. Spreitzer RJ, Salvucci ME. Rubisco: structure, regulatory interactions, and possibilities for a better enzyme. Plant Biol. 2002;53:449–75.

43. Brossa R, Pinto-Marijuan M, Francisco R, Lopez-Carbonell M, Chaves MM, Alegre L. Redox proteomics and physiological responses in *Cistus albidus* shrubs subjected to long-term summer drought followed by recovery. Planta. 2015;241:803–22.

44. Cao Y, Luo Q, Tian Y, Meng F. Physiological and proteomic analyses of the drought stress response in *Amygdalus Mira (Koehne) Yü et Lu* roots. BMC Plant Biology. 2017;17–53.

45. Jin K, Gang C, Mi-Jeong Y, Ning Z, Daniel D, John EE, Hongbo S, Sixue C. Comparative proteomic analysis of Brassica napus in response to drought stress. J Proteome Res. 2015;14:3068–81.

46. Kaul S, Sharma S, Mehta I. Free radical scavenging potential of L-proline: evidence from in vitro assays. Amino Acids. 2008;34:315–20.

47. Couée I, Sulmon C, Gouesbet G, El Amrani A. Involvement of soluble sugars in reactive oxygen species balance and responses to oxidative stress in plants. J Exp Bot. 2006;57:449–59.

48. Noctor G, Mhamdi A, Foyer CH. Update on the physiology of reactive oxygen metabolism during drought the roles of reactive oxygen metabolism in drought: not so cut and dried. Plant Phisiol. 2014;164:1636–48.

49. Zhang H, Han B, Wang T, Chen S, Li H, Zhang Y, et al. Mechanisms of plant salt response: Insights from proteomics. J Proteome Res. 2012;11:49–67.

50. Jithesh MN, Prashanth SR, Sivaprakash KR, Parida AK. Antioxidative response mechanisms in halophytes: their role in stress defence. J Genet. 2006;85:237–54.

51. Gechev TS, Benina M, Obata T, Tohge T, Sujeeth N, Minkov I, et al. Molecular mechanisms of desiccation tolerance in the resurrection glacial relic *Haberlea rhodopensis*. Cell Mol Life Sci. 2013;70:689–709.

52. Jabeen B, Naqvi S, Mahmood T, Sultana T, Arif M, Khan F. Ectopic expression of plant RNA chaperone offering multiple stress tolerance in E. coli. Mol Biotechnol. 2017;59:66–72.

53. Wang W, Vinocur B, Shoseyov O, Altman A. Role of plant heat-shock proteins and molecular chaperones in the abiotic stress response. Trends Plant Sci. 2004;9:244–52.

54. Ghabooli M, Khatabi B, Ahmadi FS, Sepehri M, Mirzaei M, Salekdeh GH. Proteomics study reveals the molecular mechanisms underlying water stress tolerance induced by Piriformospora indica in barley. J Proteom. 2013;94: 289–301.

55. Aranjuelo I, Molero G, Erice G, Avice JC, Nogues S. Plant physiology and proteomics reveals the leaf response to drought in alfalfa (Medicago sativa L.). J Exp Bot. 2011;62:111–23.

56. Keller F, Ludlow MM. Carbohydrate metabolism in drought-stressed leaves of pigeonpea (*Cajanus cajan*). J Exp Bot. 1993;44:1351–9.

57. Zaffagnini M, Fermani S, Costa A, Lemaire SD, Trost P. Plant cytoplasmic GAPDH: redox post-translational modifications and moonlighting properties. Front Plant Sci. 2013;4:450.

58. Bernacchia G, Schwall G, Lottspeich F, Salamini F, Bartels D. The transketolase gene family of the resurrection plant *Craterostigma plantagineum*: differential expression during the rehydration phase. EMBO J. 1995;14:610–8.

59. Babayev H, Mehvaliyeva U, Aliyeva M, Feyziyev Y, Guliyev N. The study of NAD-malic enzyme in *Amaranthus cruentus* L. under drought. Plant Physiol Biochem. 2014;81:84–9.

60. Stincone A, Prigione A, Cramer T, Wamelink MMC, Campbell K, Cheung E, et al. The return of metabolism: Biochemistry and physiology of the pentose phosphate pathway. Biol Rev. 2015;90:927–63.

61. Kochetov GA, Solovjeva ON. Structure and functioning mechanism of transketolase. Biochim Biophys Acta - Proteins Proteomics. 1844;2014:1608–18.

62. Pang Q, Zhang A, Zang W, Wei L, Yan X. Integrated proteomics and metabolomics for dissecting the mechanism of global responses to salt and alkali stress in *Suaeda corniculata*. Plant Soil. 2016;402:379–94.

63. Zadražnik T, Hollung K, Egge-Jacobsen W, Meglič V, Šuštar-Vozlič J. Differential proteomic analysis of drought stress response in leaves of common bean (*Phaseolus vulgaris* L.). J Proteomics. 2013;78:254–72.

64. Salekdeh GH, Siopongco J, Wade LJ, Ghareyazie B, Bennett J. A proteomic approach to analyzing drought- and salt-responsiveness in rice. F Crop Res. 2002;76:199–219.

65. Bazargani MM, Sarhadi E, Bushehri AAS, Matros A, Mock HP, Naghavi MR, et al. A proteomics view on the role of drought-induced senescence and oxidative stress defense in enhanced stem reserves remobilization in wheat. J Proteomics. 2011;74:1959–73.

66. Henkes S, Sonnewald U, Badur R, Flachmann R, Stitt M. A small decrease of plastid transketolase activity in antisense tobacco transformants has dramatic effects on photosynthesis and phenylpropanoid metabolism. Plant Cell. 2001;13:535–51.

67. Fahrendorf T, Ni W, Shorrosh BS, Dixon RA. Stress responses in alfalfa (*Medicago sativa* L.). XIX. Transcriptional activation of oxidative pentose phosphate pathway genes at the onset of the isoflavonoid phytoalexin response. Plant Mol Biol. 1995;28:885–900.

68. Kosová K, Vítámvás P, Prášil IT, Renaut J. Plant proteome changes under abiotic stress - contribution of proteomics studies to understanding plant stress response. J Proteomics. 2011;74:1301–22.

69. Shibuya N, Mikami Y, Kimura Y, Nagahara N, Kimura H. Vascular endothelium expresses 3-mercaptopyruvate sulfurtransferase and produces hydrogen sulfide. J Biochem. 2009;146:623–6.

70. Silveira JAG, Viégas RDA, da Rocha IMA, Moreira ACDOM, Moreira RDA, Oliveira JTA. Proline accumulation and glutamine synthetase activity are increased by salt-induced proteolysis in cashew leaves. J Plant Physiol. 2003; 160:115–23.

71. Xu G, Fan X, Miller AJ. Plant nitrogen assimilation and use efficiency. Annu Rev Plant Biol. 2012;63:153–82.

72. Hwang IS, An SH, Hwang BK. Pepper asparagine synthetase 1 (*CaAS1*) is required for plant nitrogen assimilation and defense responses to microbial pathogens. Plant J. 2011;67:749–62.

73. Boaretto LF, Carvalho G, Borgo L, Creste S, Landell MGA, Mazzafera P, et al. Water stress reveals differential antioxidant responses of tolerant and non-tolerant sugarcane genotypes. Plant Physiol Biochem Elsevier Masson SAS. 2014;74:165–75.

12

Characterisation of the circulating acellular proteome of healthy sheep using LC-MS/MS-based proteomics analysis of serum

Saul Chemonges[1][*] (ID), Rajesh Gupta[2], Paul C. Mills[1], Steven R. Kopp[1] and Pawel Sadowski[2]

Abstract

Background: Unlike humans, there is currently no publicly available reference mass spectrometry-based circulating acellular proteome data for sheep, limiting the analysis and interpretation of a range of physiological changes and disease states. The objective of this study was to develop a robust and comprehensive method to characterise the circulating acellular proteome in ovine serum.

Methods: Serum samples from healthy sheep were subjected to shotgun proteomic analysis using nano liquid chromatography nano electrospray ionisation tandem mass spectrometry (nanoLC-nanoESI-MS/MS) on a quadrupole time-of-flight instrument (TripleTOF® 5600+, SCIEX). Proteins were identified using ProteinPilot™ (SCIEX) and Mascot (Matrix Science) software based on a minimum of two unmodified highly scoring unique peptides per protein at a false discovery rate (FDR) of 1% software by searching a subset of the Universal Protein Resource Knowledgebase (UniProtKB) database (http://www.uniprot.org). PeptideShaker (CompOmics, VIB-UGent) searches were used to validate protein identifications from ProteinPilot™ and Mascot.

Results: ProteinPilot™ and Mascot identified 245 and 379 protein groups (IDs), respectively, and PeptideShaker validated 133 protein IDs from the entire dataset. Since Mascot software is considered the industry standard and identified the most proteins, these were analysed using the Protein ANalysis THrough Evolutionary Relationships (PANTHER) classification tool revealing the association of 349 genes with 127 protein pathway hits. These data are available via ProteomeXchange with identifier PXD004989.

Conclusions: These results demonstrated for the first time the feasibility of characterising the ovine circulating acellular proteome using nanoLC-nanoESI-MS/MS. This peptide spectral data contributes to a protein library that can be used to identify a wide range of proteins in ovine serum.

Keywords: Sheep serum, Ovine circulating acellular proteome, nanoLC-nanoESI-MS/MS, Gene ontology, Protein pathway analysis, Sheep serum proteomics, Proteogenomics data

Background

There is currently no publicly available reference mass spectrometry-based circulating acellular proteome data for sheep. However, the well-defined serum proteome of humans permits analysis and interpretation of a range of physiological changes and disease states [1, 2]. To date, the serum proteome of sheep is largely extrapolated from cattle, which can be inaccurate despite a 97% similarity in protein coding sequences [3] and different promoters driving the expression of specific proteins [4]. Characterisation of the serum proteome of sheep would therefore be useful to quantify disease in this species.

Sheep are a major production species, providing meat and wool, plus are used in a range of biotechnological and translational studies [5–9]. Despite this, relatively little is known about the responses of sheep to a range of physiological and pathological events, including the effects of breed differences in these responses. There is therefore a need to comprehensively characterise the proteins in ovine serum for better quantitative assessment of disease

---
* Correspondence: s.chemonges@uq.edu.au
[1]School of Veterinary Science, The University of Queensland, Gatton, Australia
Full list of author information is available at the end of the article

and any alternations in physiology and pathology. Blood is relatively easily collected from sheep [10–14], but comparatively only a small number of proteins have been identified, limiting the capacity to assess disease [10, 15]. One problem to date is that protein sample preparation in published studies on sheep have been inadequate and have generally ignored the full conventions for reporting identified proteins from samples [16, 17]. Consequently, data are lacking on optimised sample preparation approaches for shotgun proteomics workflows using more than one protein sequence search engine to explore the circulating acellular proteome of sheep. For example, the number of proteins identified by single laboratories using gel fractionation followed by MS from human plasma has been in the region of nearly 300 protein identifications (IDs) [18]. In 2005, liquid chromatography tandem mass chromatography (LC-MS/MS) data from multiple sample preparation techniques and protein sequence search engines for the Human Plasma Proteome Project (HPPP) from 18 laboratories worldwide collectively identified 3,020 plasma proteins based on a minimum of 2-high-scoring peptides [19, 20]. This number of protein IDs from HPPP studies was subsequently revised to 889 [19, 21]. A study that used high performance liquid chromatography (RP-HPLC) and LC-ESI-MS/MS to analyse and define the human baseline plasma proteome identified 200 proteins [22]. More recently, protein expression profiles of human plasma proteins using one-dimensional sodium dodecyl sulfate polyacrylamide gel electrophoresis (1D SDS–PAGE) coupled with nanoLC–ESI–MS/MS in a single laboratory identified 253 proteins after desalting of the peptides [23]. A similar approach to that used in the preceding study was considered attractive to be used in exploring the circulating acellular proteome of sheep.

The present study used nano liquid chromatography nano electrospray ionisation tandem mass spectrometry (nanoLC-nanoESI-MS/MS) to analyse peptides derived from healthy sheep serum samples following 1D SDS–PAGE and in-solution digestion.

Methods
Overview of methods
This study used universal protein extraction techniques detailed hereinafter to comprehensively define the serum proteome of healthy sheep. Because of the genome of sheep being incompletely sequenced or annotated, proteins were identified by matching tryptic peptides against a composite protein sequence database of sheep, goat and ox using ProteinPilot™ Software (SCIEX) in the first instance in order to capture homologous sequences. The inclusion of protein sequences from related species is a helpful strategy when exploring and establishing foundation proteogenomics data to identify known or novel genes of the non-model study subject — in this case

sheep [24–30]. Mascot [31] (Matrix Science) search was subsequently conducted using a sheep-only protein sequence database to identify high-scoring proteins and PeptideShaker [32] (CompOmics, VIB-UGent) to verify protein identifications from the primary search data.

Animal care, sample collection, storage and preparation
Serum samples of healthy adult female Merino sheep (n = 6) with ear tag identification numbers 473, 413, 463, 471, 476 and 478 belonging to an experimental colony at Queensland University of Technology (QUT) and the Australian Red Cross Blood Service (ARCBS) were obtained for the development and optimisation of a comprehensive proteomic approach for interrogating the circulating acellular proteome. The sheep were reared according to established standard operating procedures, described elsewhere [33]. Sample aliquots of 500 µL were stored in 1.5 mL Eppendorf tubes at -80 °C at the ARCBS, Brisbane. The samples were transferred to the wet laboratory at the Molecular Genetics Research Facility (MGRF) within Central Analytical Research Facility (CARF), QUT for processing. The processed samples were analysed by nanoLC-nanoESI-MS/MS at the Proteomics and Small Molecule Mass Spectrometry laboratory at CARF, QUT.

Sample preparation for protein analysis
Frozen sheep serum samples were thawed on ice and then centrifuged at 13,000 g at 4 °C for 20 min. The sediment and top layer comprising mainly of lipids and suds were discarded, retaining the supernatant. The protein concentration in the supernatant was determined with bicinchoninic acid (BCA) protein assay kit (BCA Protein Assay Kit, Pierce™) according to the manufacturer's instructions using a spectrophotometer (NanoDrop 2000, Thermo Scientific). The supernatant was then either directly analysed or concentrated by acetone precipitation of proteins. In some experiments, a protease inhibitor cocktail tablet (Roche) was added into the sample after thawing, according to the manufacturer's instructions.

Acetone precipitation of proteins
Proteins in serum were precipitated by adding 4 × (v:v %) of cold (-20 °C) acetone and then incubated at -20 °C for 16 h, prior to centrifugation at 4,000 g for 2 min. The supernatant was discarded. The pellet was washed with cold acetone and the suspension was centrifuged at 4,000 g for 5 min at 4 °C. The supernatant was discarded and this procedure was repeated one more time. The pellet was then dissolved in freshly prepared 8 M urea in 25 mM ammonium bicarbonate (NH_4HCO_3) (Sigma-Aldrich) buffer. The mixture was centrifuged at 4,000 g for 5 min at 4 °C, the supernatant was kept and the insoluble sediment was discarded. The protein concentration of the supernatant was determined using the BCA method [34].

1D SDS-PAGE

The universal 1D SDS-PAGE procedure used to fractionate proteins was based on its established description [35] and subsequent refinements [36–39]. The detailed description is provided in Additional file 1.

The gels were stained with Coomassie brilliant blue (EZ-Run™, Protein Gel Staining Solution, Fisher Scientific) according to the manufacturer's instructions and then photographed using a handheld camera (5.7-inch Quad HD Super AMOLED®, Samsung; or New 8-megapixel iSight camera with 1.5 μ pixels with Optical image stabilisation, iPhone 6, Apple Inc.).

Gel bands from entire single lanes were excised into 12 approximately equal portions into a clean 1.5 mL Eppendorf tube and de-stained using 50% acetonitrile (ACN) (Optima®, Fisher Scientific) in 25 mM NH_4HCO_3 accompanied by agitation at 750 rpm for 20 min at RT. This procedure was repeated and alternated with washing the gel bands with 25 mM NH_4HCO_3 buffer. Once de-stained, final washing of the gel bands was performed using LC-MS grade water followed by incubation for 20 min at RT. The water was discarded and the gel bands were cut into approximately 1 mm^3 pieces using a 10 uL pipette tip. Gel bands were dehydrated by adding 100% ACN and agitating at 750 rpm for 10 min at RT prior to drying in a vacuum centrifuge (SpeedVac Concentrator Christ® cat. No. RVC 2-33 IR), for 10 min.

In-gel proteins were reduced in order to break disulphide bonds and alkylated to prevent the bonds reforming as originally described elsewhere [40]. Briefly, freshly prepared 10 mM DTT (Sigma-Aldrich) in 25 mM NH_4HCO_3 buffer was added sufficiently to cover the vacuum dried gel pieces and agitated at 750 rpm for 45 min at 56 °C. Twice the amount of DTT as of freshly prepared 55 mM iodoacetamide (IAM) (Sigma-Aldrich) in 25 mM NH_4HCO_3 buffer was added to the sample and agitated for 30 min at RT in the dark. The reagents were washed off with 25 mM NH_4HCO_3 buffer with agitation for 5 min at RT, before centrifuging briefly and discarding the supernatant. Gel bands were then dehydrated using 100% ACN and agitated at 1400 rpm for 10 min at RT. The entire supernatant was discarded prior to drying the gel pieces in a vacuum centrifuge as above for 20 min.

Vacuum-dried gel pieces were incubated on ice for 5 min before adding 0.005 μg/μL solution of freshly prepared ice-cold working solution of trypsin (Trypsin Gold, Mass Spectrometry Grade, Promega) in 50 mM NH_4HCO_3 buffer enough to cover the dry gel pieces [41] and left incubating for a further 30 min until the entire enzyme solution had entered the gel pieces. Gel pieces were then covered in 50 mM NH_4HCO_3 buffer and left to incubate for 16 h at 37 °C on an agitator at 300 rpm. Digestion was stopped by adding 100 μL of 5% formic acid (FA) (Sigma-Aldrich). Peptide extraction was performed by agitating the gel pieces at 1,000 rpm for 15 min at RT. The peptide-containing supernatant was collected into a clean 0.5 ml low binding Eppendorf tube. Gel pieces were further washed by adding 5% FA in 50% ACN and agitating at 1,000 rpm for 15 min, before collecting the supernatant. Gel bands were further extracted by adding 100% ACN and agitation at 1,000 rpm for 15 min at RT. The entire supernatant was collected and then completely vacuum-dried prior to reconstitution in 10 μL of 0.1% trifluoroacetic acid (TFA) (Sigma-Aldrich) in 2% ACN followed by desalting of peptides.

In-solution digestion of proteins

The method adapted here was based on the one established by Villén and Gygi [42]. Briefly, a known quantity of serum or plasma protein sample was thawed on ice at 4 °C after which freshly prepared 20 mM DTT (equal v:v% of sample) was added, vortexed and briefly centrifuged. The mixture was diluted fivefold with 25 mM NH_4HCO_3 buffer (v:v% of sample) to dilute down urea concentration below 1 M, followed by adding an equivalent (v:v% of sample) of aqueous 70 mM $CaCl_2$. Trypsin was then added at enzyme to substrate (protein concentration of sample) ratio of 1:50. The contents were incubated for 16 h at 37 °C and then cooled to RT. Digestion was stopped by adding 50 μL of 10% TFA before vacuum concentrating the contents to dryness. The dried peptides were reconstituted in aqueous 0.1% TFA in 2%ACN, and followed by desalting of peptides.

Desalting of tryptic peptide digests

It is often necessary to remove salts and particulate matter including excess trypsin from peptide digests prior to analysis to prevent blockage of nanoLC columns and also to reduce noise artefacts of MS spectra [43–45]. Desalting of tryptic peptide digests was optimised and performed using either octadecyl carbon chain (C_{18}) pipette tips (ZipTip® Pipette Tips, Millipore, or Pierce C_{18} Tips, Thermo Fisher Scientific) depending on the filter capacity according to manufacturer's instructions. Briefly for the C_{18} tips, the desalting pipette tip was conditioned using a solution of 50% ACN/0.05% trifluoroacetic acid (TFA) in LC-MS grade water (Optima®, Fisher Scientific) and then equilibrated with 2% ACN/ 0.1% TFA in LC-MS water. After carefully and gently pipetting the entire sample up and down for at least 10 times, the membrane was washed with 2% ACN/0.1% TFA in LC-MS water. The peptides were eluted using 70% ACN/0.1% TFA in LC-MS water, vacuum dried and reconstituted in 10 uL of 2% ACN/0.1% FA in LC-MS water and transferred into a polypropylene autosampler vial for nanoLC-nanoESI-MS/MS analysis.

nanoLC-nanoESI-MS/MS

Chromatography

Peptide spectral data from approximately 400 ng – 1 µg of injected tryptic peptides per sample were generated using nanoLC-nanoESI-MS/MS on a TripleTOF® 5600+ System (SCIEX) instrument. Peptides were separated by performing reversed-phase chromatography using an Eksigent ekspert™ nanoLC 400 System directly coupled to the MS/MS instrument. The LC platform was setup in a trap and elute configuration with a 10 mm × 0.3 mm trap cartridge packed with ChromXP C18CL 5 µm 120 Å material and a 150 mm × 75 µm analytical column packed with ChromXP C18 3 µm 120 Å (Eksigent Technologies, Dublin, CA). The mobile phase solvents were composed of mobile phase A: water/0.1% FA; mobile phase B: ACN/0.1% FA; and mobile phase C: water/2% ACN/0.1% FA. Trapping was performed in mobile phase C for 5 min at 5 uL/min followed by an elute configuration across a 90 min gradient using two mobile phases A and B. To minimise retention time drift, the analytical column was maintained at 40 °C.

Data dependent acquisition (DDA)

The DDA mode of the instrument was set to obtain high resolution (30,000) TOF-MS scans over a mass range of 350–1350 m/z, followed by up to 40 (top 40) high sensitivity MS/MS scans of the most abundant peptide ions per cycle. The selection criteria for the peptide ions included intensity greater than 150 cps and charge state of 2–5. The dynamic exclusion duration was set at 12 s to account for the difference in chromatographic peak width matching to the peaks in the chromatogram. Each survey (TOF-MS) scan lasted 0.25 s and the product ion (MS/MS) scan lasted 0.05 s resulting in a total cycle time of 2.3 s. The ions were fragmented in the collision cell using rolling collision energy, and CES was set to 5. The collected peptide ion fragmentation spectra were stored in .wiff format (SCIEX).

Data processing

Primary protein sequence database search for protein identification

The acquired MS/MS data from the instrument were extracted and annotated with amino acid sequences from a custom built database using the Paragon™ Algorithm: 5.0.0.0, 4767 [46] (ProteinPilot™ Software 5.0, Revision Number: 4769, SCIEX, USA.). The custom composite database (62,025 sequences; 29,099,284 residues) used in Paragon™, with added common contaminants was assembled in FASTA format downloaded on 29th July, 2015 from a repository of non-redundant and predicted protein sequences of *Ovis aries*, *Bos taurus* and *Capra hircus* sourced from UniProtKB (Universal Protein Resource Knowledgebase - http://www.uniprot.org/). Another sheep

(*Ovis aries*) only custom database (27,393 sequences, 13,114,569 residues) with added contaminants from The common Repository of Adventitious Proteins, cRAP (http://www.thegpm.org/crap/) was assembled in FASTA format (26 Jul, 2016) from UniProtKB was used for sheep protein validation. For ProteinPilot™ searches, the following settings were selected: Sample type: Identification; Cys Alkylation: Iodoacetamide; Digestion: Trypsin; Instrument: TripleTOF 5600+; Special Factors: Urea denaturation; Species: None; Search effort: Thorough ID; ID Focus: Amino acid substitution; Results Quality: Detected protein threshold [Unused ProtScore (Conf)] ≥ 0.05 with false discovery rate (FDR) selected. Annotations were only retrieved from UniProt during composite searches. The automatically generated Excel spreadsheet (Microsoft® Excel 2010, Microsoft Corporation) report in ProteinPilot™ output was manually inspected for FDR cut-off protein yields and then meticulously curated to filter out contaminants, protein identifications with 0 (zero) unused confidence scores, proteins with reversed (nonsense) sequences and redundant protein IDs. Only proteins identified at FDR ≤1% with ≥ 2 peptides were considered for protein lists and for visual comparative analysis in the first instance and further downstream analysis.

The .group file data in ProteinPilot™ were exported as calibrated Mascot generic format (.mgf) and mzIdentML (.mzid) format files. The .mgfs were further reformatted by an mgf repair tool (SCIEX) to recalibrate .mgf files so that they can be parsed to recognise the boundaries between original files and avoid collisions in spectrum identifiers, prior to loading via a Daemon application to Mascot search engine (Matrix Science, London, UK; version 2.5.1) [31]. Mascot was set up to search the same custom database that was used in ProteinPilot™ with the following search parameters: type of search: MS/MS ion search; enzyme: trypsin; fixed modifications: Carbamidomethyl (C); variable modifications: deamidated (NQ), oxidation (M); mass values: monoisotopic; protein mass: unrestricted; peptide mass tolerance: ± 10 ppm; fragment mass tolerance: ± 0.01 Da; max missed cleavages: 1; instrument type: ESI-QUAD-TOF, and the auto-decoy search option was selected. Protein identifications were made at a significance threshold of $p < 0.05$ or target decoy of 1% FDR. Peak list and identification data from the search were exported in a .dat format for further processing. Protein lists were exported in csv format for immediate data evaluation and curation to remove contaminants in Excel spreadsheet. Only proteins identified with 2 or more peptides were included for further evaluation.

Secondary protein sequence database search for protein identification and validation

The .mgf, .dat and .mzIdentML (from ProteinPilot™) files were also loaded for protein identification and validation using PeptideShaker [32]. Peak lists obtained from MS/MS spectra were identified using Mascot [31]. Protein identification was conducted against a concatenated target/decoy [47] version of the *Ovis aries* (27,284; 99.5%) complement of the UniProtKB, 27,411 (target) sequences. The decoy sequences were created by reversing the target sequences in SearchGUI. The identification settings were as follows: Trypsin with a maximum of 1 missed cleavages; 10.0 ppm as MS1 and 0.5 Da as MS2 tolerances; fixed modifications: carbamidomethylation of C (+57.021464 Da), variable modifications: deamidation of N (+0.984016 Da), deamidation of Q (+0.984016 Da), oxidation of M (+15.994915 Da), pyrolidone from E (−18.010565 Da) and pyrolidone from Q (−17.026549 Da), fixed modifications during refinement procedure: carbamidomethylation of C (+57.021464 Da). All algorithm-pecific settings are listed in the Certificate of Analysis available in the data files.

Peptides and proteins were inferred from the spectrum identification results using PeptideShaker version 1.13.0 [32]. Peptide Spectrum Matches (PSMs), peptides and proteins were validated at a 1.0% False Discovery Rate (FDR) estimated using the decoy hit distribution. All validation thresholds are listed in the Certificate of Analysis available in the data files. Post-translational modification localisations were scored using the D-score [48] and the phosphoRS score [49] with a threshold of 95.0 as implemented in the compomics-utilities package [50]. Protein identification reports were exported in .xlsx format for evaluation and curation in Excel spreadsheet. Only proteins identified with 2 or more validated peptides were included for further evaluation.

Protein lists were presented in spreadsheet and charts were made (Microsoft® Excel™ 2010, Microsoft Corporation). Data were visualised using BioVenn Software [51], where appropriate.

The mass spectrometry data along with the identification results were deposited to ProteomeXchange Consortium [52] via the proteomics identifications (PRIDE) partner repository [53] with the dataset identifiers PXD004989 and 10.6019/PXD004989 with the following data access details: Reviewer account details: Username: reviewer99399@ebi.ac.uk; Password: QBFFTGzl

Analytical samples, experimental layout and data collection

In order to characterise the serum proteome of sheep, two universal sample preparation strategies for shotgun proteome analysis [54] were employed in three paired sets of experiments (first, second and third), using in-gel and in-solution protein digestion of serum samples. This was followed by peptide analysis by nanoLC-nanoESI-MS/MS using the method described above.

1D SDS-PAGE of normal sheep serum workflow

As a pilot study, an acetone precipitated serum sample obtained from one sheep (Sheep ID 473) was processed and subjected to 1D SDS-PAGE to ascertain the feasibility of obtaining protein identification data as a basis for constructing a peptide spectral library in future (First in-gel digestion). In order to determine the optimum amount of serum protein to load, 2, 10 and 22 µg of protein were run in separate wells of the same gel. To determine the amount of protein that needed to be loaded on a gel for protein bands to be visualised after using EZ-Run protein stain, 250, 500 and 2500 fmol of bovine serum albumin (BSA) protein were loaded in separate wells of another gel and run.

In order to increase the protein coverage, a fraction of acetone precipitated serum sample from Sheep ID 473 was subjected to 1D SDS-PAGE in two gels run concurrently (second in-gel digestion). One gel was loaded with 50 µg and 100 µg of protein in adjacent lanes and the second gel was also loaded with 50 µg, 100 µg and 50 µg in adjacent lanes.

In order to determine the effect of the quantity of protein loaded, acetone precipitation and a protease inhibitor on protein coverage, pooled serum samples from six healthy sheep (Sheep IDs 413, 463, 471,473, 476 and 478) were processed and subjected to 1D SDS-PAGE in three gels (third in-gel digestion). The samples utilised consisted of crude protein (200 µg and 100 µg) on one gel and then 100 µg of acetone precipitated serum protein with or without a protease inhibitor (Roche) and 100 µg of crude serum in a second gel. A third gel was loaded and run identically as the second gel.

In-solution digestion of sheep serum workflow

As a pilot study, 10 µg of acetone precipitated serum sample obtained from one sheep was subjected to in-solution digestion to ascertain the feasibility of obtaining protein identification data as a basis for protein quantitation in future (first in-solution digestion). In order to determine the effect of using unfractionated sample on protein coverage, a fraction of 20 µg of crude serum sample from the sheep used in the first in-gel digestion was subjected to in-solution digestion and analysed (second in-solution digestion). A third experiment utilised 100 µg of pooled crude serum samples from all six sheep (Sheep IDs 473, 413, 463, 471, 476 and 478) for in-solution digestion in order to determine the effect of using a higher quantity of protein substrate on protein coverage (third in-solution digestion).

Results

The results of the first, second and third in-gel digestions are presented in Figs. 1, 2 and 3, respectively. The details of the individual gels are provided in the figure captions.

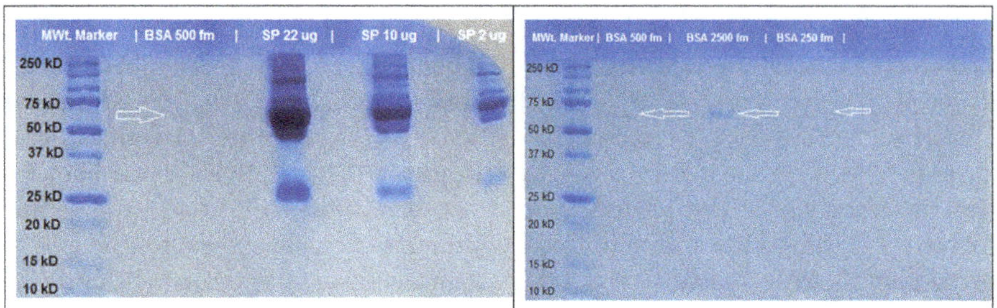

Key: kD = kiloDalton; BSA= bovine serum albumin; sheep SP= serum protein; ug=μg; MWt= Molecular weight marker fm=fmol.

Fig. 1 Coomassie-stained 1D SDS-PAGE gels used in first in-gel digestion. Fractions of acetone precipitated serum protein from a healthy sheep were loaded alongside bovine serum albumin (BSA) in Gel A. Gel A suffered a handling artefact to the *top right corner* of the gel. The *leftmost* well of both gels were loaded with 4 μL of a protein molecular weight standard (Precision Plus Protein™ Dual Xtra, Bio-Rad Laboratories). One well in Gel A was loaded with 500 fm of BSA standard; other three wells were loaded with 22 μg, 10 μg and 2 μg each of sheep serum protein sample. After the molecular weight standard, the other three wells in Gel B were loaded with 500 fm, 2500 fm and 250 fm each of BSA standard. *Arrows* show BSA standard

Except for Gel B of in Fig. 1, the protein sample lanes of all the other gels were subjected to in-gel digestion followed by nanoLC-nanoESI-MS/MS to identify proteins. The protein ID results of the first, second and third in-gel and in-solution digestions are summarised in Table 1. The detailed results are presented in the accompanying spreadsheet Microsoft® Excel™ file [see Additional file 2]. Protein IDs were obtained using ProteinPilot™ [55] to search a Uni-ProtKB composite database of *Ovis aries, Bos taurus and Capra hircus* with a results quality of FDR ≤1%; ≥ 2 peptides for a protein to be considered confidently identified as the highest scoring member of the protein group. The Pro Group™ Algorithm in ProteinPilot™ assigned one protein the best confidence possible (unused score) among protein isoforms, which enabled protein subset differentiation, as well the suppression of false positives for protein-grouping analysis [55]. The results were therefore based on protein group identifications presented as protein identifications (IDs).

In the present set of experiments, proteins were identified by using peptide signatures to search custom-built protein sequence databases. Protein ID confidence was determined by the number of proteins that were assuredly accepted as correct, having been identified as described elsewhere [56, 57]. Overall, a total of 267 confident and unique protein groups were identified using ProteinPilot™ by searching a composite *Ovis aries,*

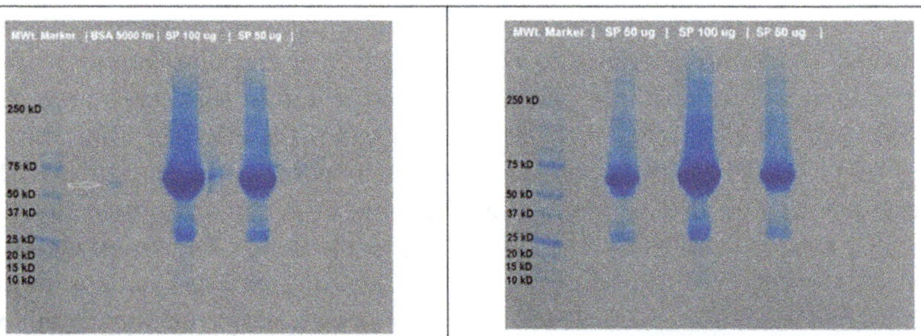

Key: kD = kiloDalton; BSA= bovine serum albumin; sheep SP= serum protein; ug=μg; MWt= Molecular weight; fm=fmol. The white arrow in gel A points at the BSA standard band.

Fig. 2 Coomassie-stained 1D SDS-PAGE gels used in the second in-gel digestion. Fractions of acetone precipitated serum protein samples from a healthy sheep were used. One well in Gel A was loaded with BSA standard (*arrow*), and two other wells were loaded with 100 μg and 50 μg of protein each. Three wells in Gel B were loaded with 50 μg, 100 μg and 50 μg each of protein

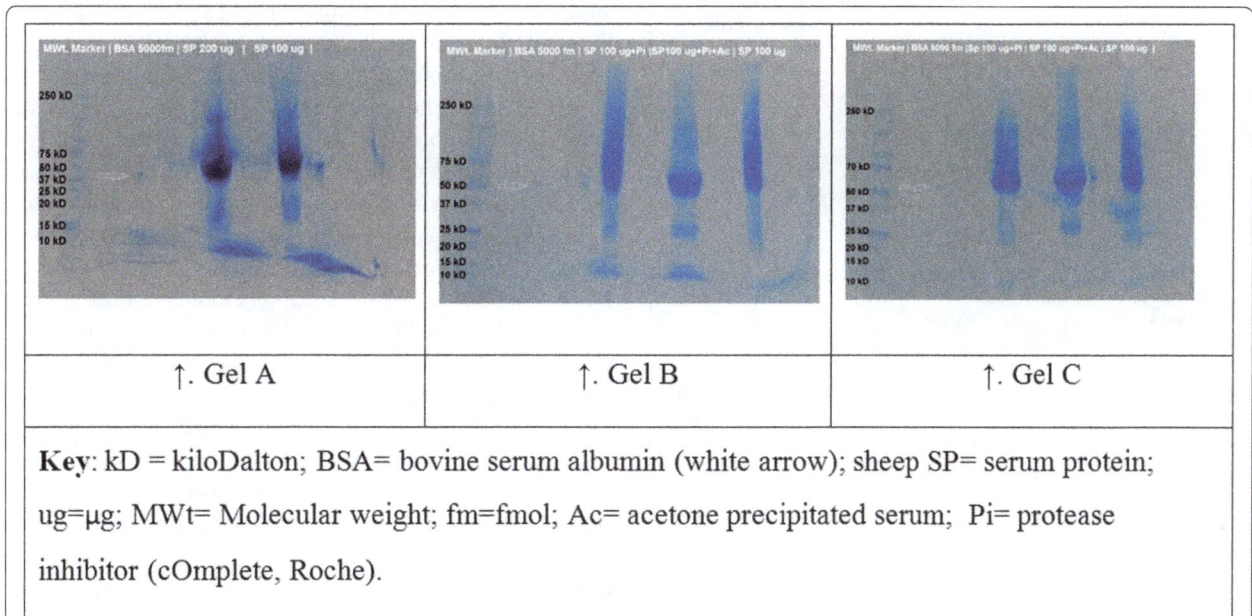

| ↑. Gel A | ↑. Gel B | ↑. Gel C |

Key: kD = kiloDalton; BSA= bovine serum albumin (white arrow); sheep SP= serum protein; ug=μg; MWt= Molecular weight; fm=fmol; Ac= acetone precipitated serum; Pi= protease inhibitor (cOmplete, Roche).

Fig. 3 Coomassie-stained 1D SDS-PAGE gels used for the third in-gel digestion. Fractions of pooled serum protein samples from six healthy sheep were used. In Gel A, one well was loaded with 200 μg of crude serum protein and the other well was loaded with 100 μg. Three 100 μg of serum protein for Gel B and C were loaded identically and treated as follows: one well was loaded with crude serum with a protease inhibitor (SP 100 μg + Pi), the second well had acetone precipitated serum and a protease inhibitor (SP 100 μg + Pi + Ac); the third well was loaded with crude serum only (SP 100 μg)

Bos taurus and Capra hircus UniProtKB database after combining all the three in-gel digestion workflows (first, second and third in-gel digestions) from a total quantity of 1,284 μg of serum protein obtained from six healthy sheep. The UniProtKB entries for the identified proteins are presented in Additional file 2.

In-solution digestion

A composite ProteinPilot™ search of all the three in-solution digestion workflow samples comprising of 130 μg of serum protein yielded a total of 102 protein IDs. The UniProtKB entries for these proteins are presented in Additional file 2.

A comparison between the protein identification list derived from combined first, second and third in-gel digestion (in-gel digestion workflow) and that of combined first, second and third in-solution digestion (in-solution digestion workflow) in BioVenn Software [51] is presented in Fig. 4. The UniProtKB entries of

the 17 proteins that were exclusive to the in-solution digestion workflow (i.e. proteins were not detected by in-gel workflow) are A0A0F6QNP7, W5PSQ7, W5QH45, W5NQW9, G5E604, W5PZF0, W5NWX6, Q1KZF3, W5PJZ2, W5QDP8, W5PDR7, W5PN97, W5PXI6, F1N3Q7, C6ZP49, G3N346 and Q3SYR8.

A combined ProteinPilot™ search of the pilot data from one sheep and the additional data from five sheep for both in-gel and in-solution digestion workflows using a composite *Ovis aries, Bos taurus and Capra hircus* protein sequence database yielded an overall outcome of 274 protein IDs. The details of these protein IDs and their peptide sequences are presented in table format in Additional file 2. Based on comparison with previous studies and protein database resources [10, 14, 15, 24, 25, 58–67], a table was drawn from the preceding references listing the details of 67 known, 207 novel and 83 disease-associated serum proteins identified is presented as Additional file 3. The known proteins are those that

Table 1 The number of proteins identified by ProteinPilot™ Software from in-gel and in-solution digestion of healthy sheep serum samples by searching a composite *Ovis aries, Bos taurus and Capra hircus* protein sequence database

Experiment →	First digestion		Second digestion		Third digestion	
Digestion type	In-gel	In-sol	In-gel	In-sol	In-gel	In-sol
Serum protein source	Ac	Ac	Ac	Crude	Ac + Crude	Crude
Total quantity of protein analysed	34 μg	10 μg	350 μg	20 μg	900 μg	100 μg
Number of protein IDs	120	25	241	100	182	32

Key: *In-sol* In-solution; *Ac* Acetone precipitated; *IDs* Identifications

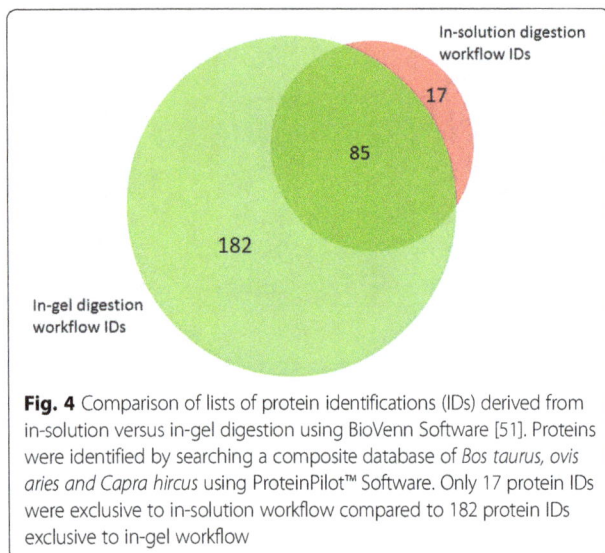

Fig. 4 Comparison of lists of protein identifications (IDs) derived from in-solution versus in-gel digestion using BioVenn Software [51]. Proteins were identified by searching a composite database of *Bos taurus, ovis aries* and *Capra hircus* using ProteinPilot™ Software. Only 17 protein IDs were exclusive to in-solution workflow compared to 182 protein IDs exclusive to in-gel workflow

have been cited in the literature and also have a confirmed status in UniProtKB. Novel proteins constitute those that previously appeared as predicted and proteins that had hitherto been inferred by homology. Disease-associated proteins refer to proteins that are expressed or alter during pathology in sheep and other species.

Combined protein identifications from 1D SDS-PAGE and in-solution digestion of serum using ProteinPilot™ and Mascot database search engines and PeptideShaker search

Protein yields of a composite search of all the sample data from the three workflows (first, second and third in-gel and in-solution digestion) using a sheep-only UniProtKB database optimised for PeptideShaker Software were as follows: ProteinPilot™: 245 IDs and Mascot: 379 IDs. A secondary analysis using PeptideShaker of this same entire dataset yielded 133 IDs (1% FDR and ≥ 2 unique validated peptides). The details of these protein IDs are provided in Additional file 4. Again, based on comparison with previous studies and protein database resources [10, 14, 15, 24, 25, 58–67], and using the 379 Mascot protein IDs, a table was drawn from the preceding references to list the details of 77 known, 302 novel and 83 disease-associated serum proteins identified using this sheep only database in Additional file 5.

The 379 protein IDs from Mascot search were used as a benchmark for further downstream analysis. Every sheep protein ID made in Mascot was mapped to a distinct gene. Of all the 379 protein IDs made by searching the sheep-only UniProtKB database, only 74 proteins had been annotated based on sequence similarity to other species, whilst 305 proteins were uncharacterised. Of the 74 annotated proteins, only annexin A2 (P14639), serum albumin (P12303), transthyretin (B3SV56), nuclear receptor

subfamily 1 group D member 1 (A2SW69) and insulin-like growth factor-binding protein 2a (Q29400) had been reviewed and therefore included in the Swiss-Prot subset of UniProtKB. The unreviewed, but named proteins included apolipoprotein E, fibulin-1, angiotensinogen, monocyte differentiation antigen CD14, plasminogen, pentaxin (pentraxin), alpha-1-antitrypsin transcript variant 1, histone H2B, alpha-1-acid glycoprotein, amine oxidase, beta-A globin chain, thyroxine-binding globulin, alpha-2-HS-glycoprotein, C-X-C motif chemokine, histone H3, coagulation factor IX, histone H4, factor H, prothrombin, clusterin, L-lactate dehydrogenase, cGMP-dependent protein kinase, antithrombin-III, gelsolin isoform b, ceruloplasmin, VH region chain, conglutinin 1, DNA polymerase, proteasome subunit alpha type, tubulin beta chain, proteasome subunit alpha type, proteasome subunit alpha type, fibrinogen alpha chain, aspartate aminotransferase, phosphodiesterase, chitinase-3-like protein 1, superoxide dismutase [Cu-Zn], uricase, glyceraldehyde-3-phosphate dehydrogenase, carbonic anhydrase 2, adiponectin, olfactory receptor, histone H2A, alpha-mannosidase, centromere protein C, importin subunit alpha, 14-3-3 protein sigma, AP complex subunit beta, carboxypeptidase, oxysterol-binding protein, growth hormone receptor variant H, condensin complex subunit 2, large tumour suppressor-like 1 protein,protein-tyrosine-phosphatase, peptidyl-prolyl cis-trans isomerase (PPIase), proteasome subunit alpha type, dipeptidase, proteasome subunit beta type, tubulin alpha chain, proteasome subunit alpha type, fructose-1,6-bisphosphatase 1, polypeptide N-acetylgalactosaminyltransferase, arginase, adenylyl cyclase-associated protein, protein-serine/threonine kinase, transaldolase, MHC class II antigen, glutathione peroxidase and corneodesmosome protein.

Gene ontology (GO) – term analysis of proteins identified in serum of healthy sheep

The 379 proteins identified by a composite Mascot search of the first, second and third in-gel and in-solution digestion of serum proteins from healthy sheep were subjected to gene ontology (GO) analysis using Protein ANalysis THrough Evolutionary Relationships (PANTHER) classification tool [68]. In the PANTHER tool, the gene entries were analysed by aligning them to *Bos taurus* as the closest organism analogous to sheep because *Ovis aries* entries were not available. The PANTHER analysis resulted into 349 bovine aligned gene entries listed in Additional file 6.

The results of GO-term analysis of molecular function, biological process, cellular component, protein class and pathway analysis of the detected proteins are provided in Fig. 5. Looking at the molecular function domain of the proteins alone based on the GO term results (Fig. 5a), catalytic activity was dominant of the 264 function hits.

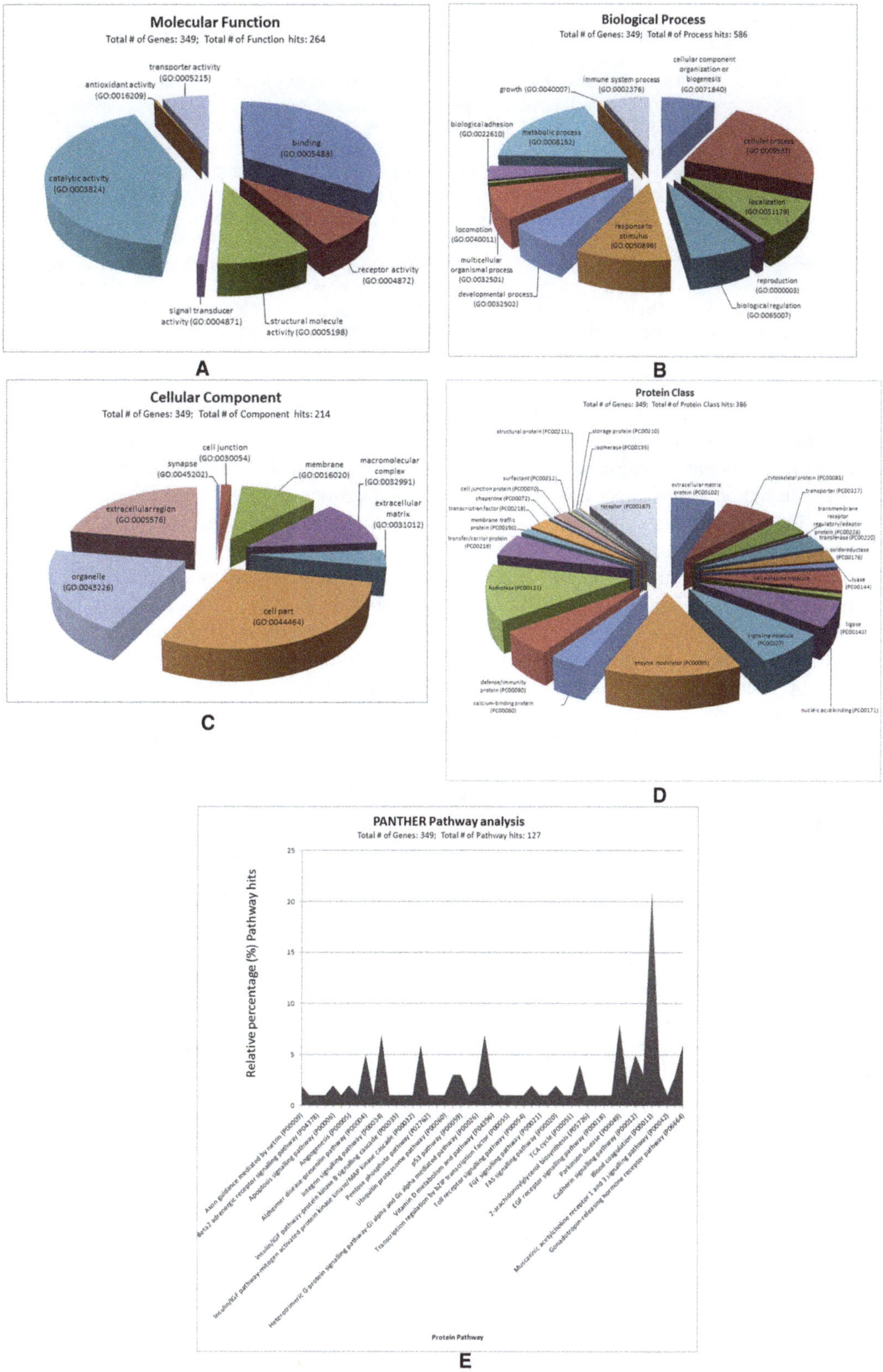

Fig. 5 (See legend on next page.)

(See figure on previous page.)

Fig. 5 Gene Ontology (GO) terms and pathway analysis of sheep proteins identified by Mascot database search engine. Protein data were derived from combined in-gel and in-solution workflows of serum protein samples from healthy sheep aligned to *Bos taurus* gene entries. Up to 349 genes were resolved from a total number of 379 sheep protein identifications (IDs) by the Protein Analysis THrough Evolutionary Relationships (PANTHER) classification tool [68]. The GO domain of molecular function had 264 function hits (**a**); biological process had 586 process hits (**b**); cellular component had 214 component hits (**c**); protein class had 386 class hits (**d**). There were 127 protein pathway hits with 49 prominent pathways, but only 25 are displayed (**e**). The other 24 protein pathway names not displayed with their codes and percentage contribution in parenthesises were: Axon guidance mediated by Slit/Robo (P00008)(0.80%), Beta1 adrenergic receptor signalling pathway (P04377)(0.80%), JAK/STAT signalling pathway (P00038) (0.80%), Ionotropic glutamate receptor pathway (P00037)(0.80%), Alzheimer disease-amyloid secretase pathway (P00003)(0.80%), Phenylethylamine degradation (P02766)(0.80%), Phenylalanine biosynthesis (P02765)(0.80%), Inflammation mediated by chemokine and cytokine signalling pathway (P00031)(4.70%), Asparagine and aspartate biosynthesis (P02730)(0.80%), Huntington disease (P00029)(2.40%), Heterotrimeric G-protein signalling pathway-Gq alpha and Go alpha mediated pathway (P00027)(0.80%), Wnt signalling pathway (P00057)(5.50%), Glycolysis (P00024)(0.80%), General transcription regulation (P00023)(0.80%), T cell activation (P00053)(1.60%), TGF-beta signalling pathway (P00052)(0.80%), Tyrosine biosynthesis (P02784) (0.80%), Plasminogen activating cascade (P00050) (3.10%), Endothelin signalling pathway (P00019)(0.80%), DNA replication (P00017)(0.80%), Cytoskeletal regulation by Rho GTPase (P00016)(1.60%), Nicotinic acetylcholine receptor signalling pathway (P00044)(2.40%), B cell activation (P00010)(2.40%) and CCKR signalling map (P06959)(2.40%)

From the protein IDs that had names, at least 27 of them were specifically classified as enzymes from protein database searches. It is evident from these results that there is a hierarchy in the biological processes of the 586 process hits (Fig. 5b). The cellular component GO domain (Fig. 5c) for serum from healthy sheep had 214 hits in total. The protein class GO domain (Fig. 5d) had 386 class hits, with enzyme modulation topping the list. Among the 49 prominent protein pathways that were displayed in PANTHER from the analysed genes, 14 were represented by over 3.0% contribution to the revealed pathway pool (Fig. 5e).

Discussion

This study reports the development of a proteomics baseline profile of healthy sheep serum by analysing peptides derived from in-solution digestion and 1D SDS-PAGE using nanoLC-nanoESI-MS/MS. The major outcome was that 379 proteins were identified, compared for example to 42 proteins from serum of sheep with mild respiratory disease during peripartum period [10] and a single protein (serum amyloid A) in sheep with scrapie [15]. Both of these cited earlier sheep studies used two dimensional (2-DE) surface enhanced laser desorption/ionisation time of flight mass spectrometry (SELDI-TOF MS) and LC-MS/MS. In species other than sheep, 490 proteins were identified in human sera using multidimensional separation coupled with MS [2], while 340 low molecular weight proteins were identified in human sera using SELDI-TOF MS analysis and LC-MS/MS [69]. There is also a report that assessed three different lots of foetal bovine serum by NanoLC-MS/MS analysis in which 79, 90, and 91 proteins were identified [70]. The preceding study recognised that there is variability in the protein content of different lots of foetal bovine serum – a commonly used growth medium for cell cultures, which affects the consistency of cell growth. The lot with a higher number of protein IDs was associated with higher cell growth rate [70]. Identification of these

proteins is important clinically to determining health or altered physiology, such as stress [10].

The use of 1D SDS-PAGE in this study facilitated serum protein samples to be fractionated to reduce protein complexity prior to nanoLC-nanoESI-MS/MS analysis [71]. The first in-gel digestion experiment enabled the determination of the quantity of protein from samples and the amount of the BSA standard that needed to be loaded onto the gel to ensure that protein bands were visible and clearly defined (Fig. 1). Loading a larger quantity of protein onto the gel was necessary to discover as many proteins as possible using DDA [72]. However, the 2 μg lane yielded 41 protein IDs in the first in-gel digestion (Fig. 1), while the 10 μg-lane yielded 20 protein IDs and the 22 μg lane yielded 121 protein IDs. The 10 μg lane was analysed initially and the 2 μg and 22 μg lanes were analysed 6 weeks later once the extractions had been optimised and the instrument tuned.

The second in-gel digestion (Fig. 2) increased the protein coverage by loading more protein into the gel wells using a fraction of the acetone precipitated serum sample used in the 1st in-gel digestion. The 100 μg (2 replicates) and 50 μg (3 replicates) protein loads in the 2nd in-gel digestion workflow yielded comparable numbers of protein IDs for each of the loaded quantity of protein. This suggests that reproducibility of the amount of protein loaded into the gel lanes had been achieved [71]. The second in-gel digestion was an improvement of the 1st in-gel digestion by having replicate and having increased quantities of loaded protein per lane, using the same serum sample of 1st in-gel digestion from Sheep ID 473.

The 1D SDS-PAGE preparation of one gel in the third in-gel digestion had a number of visual artefacts (Fig. 3). The distortion in the 10-15kD region of Gel A could have been attributed to a defect in the gel possibly due to inconsistency in gel polymerisation creating artefact bands [35], overloading and/or the presence of a pocket between the gel and the cassette housing that allowed the protein samples to leak out the gel [73]. This could have also contributed to the low number of protein yields made from this gel (200 µg: 40 protein IDs; 100 µg: 38 protein IDs), compared to the 100 µg × 2 lanes in Gels B and C that yielded 114 protein IDs [see Additional file 2]. A couple of variables were also introduced in this experiment, in addition to the quantity of proteins loaded on to the gel wells as planned. The analysis of fractionated crude serum that had a protease inhibitor (cOmplete, Roche) yielded a higher number of protein IDs (162 IDs), compared to the acetone precipitated sample that also had the protease inhibitor (143 IDs). This suggests that a considerable number of proteins were present in the acetone precipitation supernatant that was discarded. The discardment of the supernatant from acetone precipitation is a routine practice during generic or universal sample preparation for proteomic analysis [74].

As for the in-solution digestion workflow, the number of protein identifications from analysing 100 µg of crude serum protein was low when compared with 20 µg. The sample for the first in-solution digestion using 10 µg of acetone precipitated serum that was drawn from one healthy pilot sheep (Sheep ID 473) yielded only 25 protein IDs. This sample was prepared and analysed at the same time as the 10 µg sample of the first in-gel digestion discussed earlier. Protein detection was therefore likely to have been affected by unoptimised experimental processes at the time prior to running on the MS instrument. The second in-solution digestion using 20 µg of crude serum from the same sheep yielded 100 protein IDs. This result was considered substantial, as the number of protein IDs was comparable to those of other studies [10, 75–81]. Unexpectedly however, the third in-solution that utilised 100 µg of pooled crude serum from six sheep under the same experimental conditions yielded only 32 IDs. It is thought that this result was possibly due to the inhibition of trypsin by the presence of intravenous agents in the pooled sample from the anaesthetic cocktail used to anaesthetise the sheep, as this was not the case with the pilot sheep sample in which the sheep was not anaesthetised during sample collection.

BioVenn Software [51] was utilised for visualisation of the data presented in Fig. 4. This tool enabled the comparison of a protein identification list derived from in-gel digestion with that from in-solution digestion by displaying the data in an area-proportional Venn diagram. It showed protein IDs that were exclusive to in-solution and in-gel, and those common between the two digestions. The composite in-solution digestion workflow yielded 102 protein IDs. Of the 17 protein IDs that were exclusive to in-solution digestion workflow, five were mapped to the ox, two to the goat and the remaining 10 IDs were for sheep. Despite having known genes, the

vast majority of the identified proteins were either uncharacterised or unreviewed in UniProtKB. Another interesting observation was that the combined list of 284 protein IDs from in-gel and in-solution digestion displayed in BioVenn Software was marginally higher than the 274 IDs from a composite ProteinPilot™ search of the same datasets. It is likely that the subsequent composite ProteinPilot™ search helped to further group proteins, thereby improving the confidence of protein IDs by minimising false protein identifications – a known challenge when searching a multi-species protein database to identify proteins.

A combined search of the first, second and third in-gel and in-solution digestion datasets using a sheep-only database yielded 245 protein IDs in ProteinPilot™ (cf 274 protein IDs using the composite database of the ox, goat and sheep) and Mascot search yielded 379 IDs. The PeptideShaker validation search yielded 133 protein IDs. The comparatively low number of protein IDs made by PeptideShaker is because the protein entries were identified using validated unique peptides – a feature that is not obvious in either ProteinPilot™ or Mascot, whose protein ID entries were only based on at least two high-scoring peptides per protein, on the assumption that the peptides were unique to the protein.

The results from Mascot search were embraced and utilised for further analysis because this software platform is widely used by the proteomics community and it is considered the industry standard, as it implements a vast array of applications necessary for protein identification [82]. As of September, 2016, the 379 protein IDs complete with UniProtKB accessions was probably the highest number of sheep serum proteins to date using nanoLC-nanoESI-MS/MS. Of these protein IDs, only 74 were named in UniProtKB, whilst the vast majority (305) were yet to be characterised. This study can therefore be considered the first to provide a comprehensive MS/MS protein sequence data of serum proteins of normal sheep and by contributing to the efforts of annotating genes and charactering sheep proteins. Despite most of the proteins not being characterised in UniProtKB, their mapping to known genes and the available mass spectrometry-derived peptide sequence data alongside verification on more than one software platforms, constitute strong supportive evidence that the identified proteins do exist. The downside of the Mascot search is that it does not provide a user-friendly protein sequence output that can be readily tabulated as in the case of ProteinPilot™ IDs. For this reason, only protein names and UniProtKB entries were utilised mostly for the purposes the present study.

Regarding GO-term analysis, the significance of many of the enzymes that dominated catalytic activity in the molecular function domain (Fig. 5 a), remains to be documented in sheep, but the functions of some are known. For example, adenylyl cyclase-associated protein regulates cofilin function, actin cytoskeleton and cell adhesion [83]. Alpha-mannosidase participates in glycoprotein synthesis and endoplasmic reticulum quality control [84]. It has been reported to be down-regulated in locoweed *(Oxytropis sericea)* in sheep [85, 86], for example. The functions of other identified enzymes that were drawn from [24, 87–116] are provided in Additional file 7.

Serum samples of healthy adult female Merino sheep were utilised for this study. It is quite possible that a relatively low representation of the growth process domain in the biological process GO-term was because serum samples were derived from adult sheep. Also, the cellular component fractions could possibly vary depending on the physiological status of the sheep – which remains yet to be determined and documented. It can be argued that hormonal changes and the influence of age contribute to observations of serum proteome profiles and this should be accounted for. For instance, studies in sheep have shown that there is a diurnal variation metabolic and stress-responsive hormones [117].

In the present study, there were mechanisms in place to mitigate the effects of stress on the laboratory sheep. The sheep were reared together and acclimatised to their housing and handling by people as a standard management practice prior to blood sampling [33, 118]. Also, there was no variation in calorie intake because feed was supplemented as required [33, 118] in order to mitigate the well-established phenomenon of seasonal weight loss – a well-established major nutritional stress factor in sheep [119]. During agistment, there were wethers that belonged to other experiments of the research group, but there were no entire males to cause 'ram effect' that could have caused surges in reproductive hormones [120], for example. Nevertheless, gonadotropic activity would have occurred naturally in the ewes to cause hormonal changes [121], perhaps even with a synchronised hypothalamic-pituitary-ovarian axis in all the ewes, as this phenomenon is known to occur naturally [122]. All the sheep were approximately 2 years old and were therefore, practically in the same metabolic and physiological state during blood sampling. Also, the sheep belonged to an ovine model of blood transfusion [123], so most preventable adverse attributes had been catered for.

The fundamental 'method' for pulling proteins from the liquid fraction of blood using the explored approach is already well-developed in itself, but this study went beyond this to develop a tailored platform, comprising a series of refined methods, to give this practical application. The knowledge from this prototype study has illuminated a considerable number of bovine-aligned gene entries associated with protein pathways that can be valuably exploited by animal model studies using sheep serum as their analyte. A downside of the present study is that no males were represented in the dataset. Future studies should take into account hormonal changes, be gender and age inclusive in order to capture broad aspects of the proteome that could have been missed in this report.

Conclusion

This study has demonstrated for the first time that it is feasible to identify several hundred sheep serum proteins using nanoLC-nanoESI-MS/MS. By utilising the PANTHER tool, this serum-derived prototype of the ovine circulating acellular proteome revealed association of 349 genes with 127 protein pathway hits. When used with protein quantitative data, these findings have the potential to be applied as the foundation for establishing the baseline normal ovine serum proteome that could be used in comparison with samples from sick sheep. The peptide spectral data here also are a contribution towards a library that can be applied for targeted proteomics approaches, such as sequential acquisition of all theoretical fragrant mass spectra (SWATH)-MS to fulfil proteogenomics study efforts on sheep in future.

Additional files

Additional file 1: One-dimensional sodium dodecyl sulfate polyacrylamide gel electrophoresis (1D SDS-PAGE). In-gel fractionation (1D SDS-PAGE) of sheep serum protein samples. (DOCX 24 kb)

Additional file 2: Protein identification results from using ProteinPilot™ to search a composite *(Bos taurus, Ovis aries* and *Carpra hircus)* UniProtKB protein sequence database of serum samples derived from the first, second and third in-gel and in-solution digestion with a results quality of FDR ≤1%; ≥ 2 peptides for the highest scoring member of the protein group to be considered confidently identified. Each tab contains a list of protein IDs based on the quantity of protein loaded (µg), digestion workflow or sample conditions as follows: 1st_In-gel_digestion_2 µg = first in-gel (2 µg); 1st_In-gel_digestion_10 µg = first in-gel (10 µg); 1st_In-gel_digestion_22 µg = first in-gel (22 µg); All_1st_In-gel_digestion_IDs = all first in-gel samples; 2nd_In-gel_digestion_100 µgGelA = second in-gel digestion of Gel A (100 µg); 2nd_In-gel_digestion_50 µgGelA = second in-gel digestion of Gel A (50 µg); 2nd_In-gel_digestion_50 µgGelB = second in-gel digestion of Gel B (50 µg); 2nd_In-gel_digestion_100µgGelB = second in-gel digestion of Gel B (100 µg); 2nd_In-gel_digestion_50µgGelB = second in-gel digestion of Gel B (50 µg); All_2nd_In-gel_digestion_IDs = composite of all second in-gel digestion samples; 3rd_In-gel_digestion_200µg = crude serum protein of the third in-gel digestion (200 µg); 3rd_In-gel_digestion_100µg = crude serum protein of the third in-gel digestion (100 µg); 3rd_In-gel_digest_100µgx2CrudeI = crude serum protein with a protease inhibitor (Roche) of the third in-gel digestion (100 µg × 2); 3rd_In-gel_digest_100µgx2AcePPT = acetone precipitated serum protein without a protease inhibitor of the third in-gel digestion (100 µg × 2); 3rd_In-gel_digest_100µgx2No_I = crude serum protein without a protease inhibitor of the third in-gel digestion (100 µg × 2); All_3rd_In-gel_Digestion_IDs = composite of all third in-gel digestion samples; All_In-gel_digestion_IDs = all in-gel digestion workflow; 1st_In-solution_digestion_10 µg = acetone precipitated serum protein from the first in-solution digestion (10 µg); 2nd_In-solution_digestion_20 µg = crude serum protein from the second in-solution digestion (20 µg); 3rd_In-solution_digestion_100µg = crude serum protein from the third in-solution digestion (100 µg); All_In-solution_digestion_IDs = all in-solution workflow samples; All_Proteins + Peptide_Sequences = 274 protein IDs and peptide sequences of the entire in-gel and in-solution digestion experiments. (XLSX 423 kb)

Additional file 3: Details of known, novel and disease-associated sheep serum proteins identified by ProteinPilot™ by a searching a composite UniProtKB protein sequence database *of Bos taurus, Ovis aries and Capra hircus*. This 3-sheet Microsoft Excel file contains the details of 67 known (Known_Proteins_in_Literature), 207 novel (Novel_Proteins) and 83 disease-associated (Disease-Associated_Proteins) serum proteins identified using this composite database. The known proteins are those that have been cited in the literature and also have a confirmed status in UniProtKB. Novel proteins constitute those that previously appeared as predicted and proteins that had hitherto been inferred by homology. Disease-associated proteins refer to proteins that are expressed or alter during pathology in sheep and other species. (XLSX 161 kb)

Additional file 4: Protein identifications by ProteinPilot™, Mascot and PeptideShaker search engines from searching a composite of the first, second and third digestions sheep serum protein data using an *Ovis aries* only UniProtKB protein sequence database. This 3-sheet Microsoft® Excel™ file contains the details of protein identifications of a composite search of all the sample data from the three workflows (first, second and third in-gel and in-solution digestion) using an *Ovis aries* only UniProtKB database optimised for PeptideShaker Software. The tab details are as follows: ProteinPilot_IDs_ + _Sequences = 245 ProteinPilot™ protein IDs complete with peptide sequences; Mascot_IDs = 379 Mascot protein IDs and PeptideShaker_IDs = 133 PeptideShaker protein IDs. (XLSX 170 kb)

Additional file 5: Details of known, novel and disease-associated sheep serum proteins identified by Mascot from searching a composite of the first, second and third digestions sheep serum protein data using an *Ovis aries* UniProtKB protein sequence database. This 3-sheet Microsoft Excel file contains the details of 77 known (Known_Proteins_in_Literature), 301 novel (Novel_Proteins) and 83 disease-associated (Disease-Associated_-Proteins) serum proteins identified using an *Ovis aries* UniProtKB database. (XLSX 14 kb)

Additional file 6: List of 349 bovine aligned gene entries derived from inputting 379 *Ovis aries* protein data in the PANTHER classification tool. This 2-sheet Microsoft Excel file contains the list of 349 bovine aligned gene entries (349_Bovine_aligned_gene_entries) derived from analysing gene information of 379 serum proteins identified by Mascot (Input_of_379_Ovine_Mascot_IDs) using an *Ovis aries* UniProtKB database. (XLSX 49 kb)

Additional file 7: Functions of dominant enzymes identified in healthy sheep serum. The functions of enzymes identified in healthy sheep serum that dominated catalytic activity in the molecular function domain after gene ontology analysis. (DOCX 67 kb)

Abbreviations

1D SDS-PAGE: One-dimensional sodium dodecyl sulfate polyacrylamide gel electrophoresis; ACN: Acetonitrile; ARCBS: Australian Red Cross Blood Service; BCA: Bicinchoninic acid assay; BSA: Bovine serum albumin; CARF: Central Analytical Facility; DDA: Data dependent acquisition; DNA: Deoxyribonucleic acid; DTT: Dithiothreitol; EDTA: Ethylenediaminetetraacetic acid; ESI-QUAD-TOF: Electrospray ionisation quadrupole time-of-flight; FA: Formic acid; FDR: False discovery rate; GO: Gene ontology; IAM: Iodoacetamide; ID: Identification; LC: Liquid chromatography; MGRF: Molecular Genetics Research Facility; MHC: Major histocompatibility complex; MS: Mass spectrometry; MS/MS: tandem mass spectrometry; nanoLC-nanoESI-MS/MS: nano liquid chromatography nano electrospray ionisation tandem mass spectrometry; NCBI: National Center for Biotechnology Information; PANTHER: Protein analysis through evolutionary relationships; PRIDE: Proteomics identifications; QUT: Queensland University of Technology; RT: Room temperature; SP: Serum protein; SWATH: Sequential acquisition of all theoretical fragrant mass spectra; TEMED: Tetramethylethylenediamine; TFA: Trifluoroacetic acid; TOF: Time-of-flight; UQ: The University of Queensland

Acknowledgements

Thank you to Dr John-Paul Tung and the Australian Red Cross Blood Service in Brisbane for making available the primary samples used in this manuscript under Agreement No. 15-03QLD-19.

Funding

This work was undertaken as part of SC's PhD which was financially supported from personal means, an Australian Postgraduate Award scholarship and H. George Osborne Research Scholarship, both of which were through The University of Queensland. All the facilities for the experimental work were funded by a collaborative arrangement with Queensland University of Technology Central Analytical Research Facility (QUT-CARF).

Authors' contributions

SC originated the concept of the study and conducted the experiments with PS. SC, RG and PS developed the methods and analytical procedures of the samples. SC drafted the manuscript, and PCM and SRK made substantial revisions. All authors have read the final manuscript.

Competing interests

The authors declare that they have no competing interests.

Author details

[1]School of Veterinary Science, The University of Queensland, Gatton, Australia. [2]Proteomics and Small Molecule Mass Spectrometry, Central Analytical Research Facility, Queensland University of Technology, Brisbane, Australia.

References

1. Pieper R, Gatlin CL, Makusky AJ, Russo PS, Schatz CR, Miller SS, Su Q, McGrath AM, Estock MA, Parmar PP, et al. The human serum proteome: display of nearly 3700 chromatographically separated protein spots on two-dimensional electrophoresis gels and identification of 325 distinct proteins. Proteomics. 2003;3:1345–64.
2. Adkins JN, Varnum SM, Auberry KJ, Moore RJ, Angell NH, Smith RD, Springer DL, Pounds JG. Toward a human blood serum proteome: analysis by multidimensional separation coupled with mass spectrometry. Mol Cell Proteomics. 2002;1:947–55.
3. Kijas JW, Menzies M, Ingham A. Sequence diversity and rates of molecular evolution between sheep and cattle genes. Anim Genet. 2006;37:171–4.
4. Vanselow J, Furbass R, Rehbock F, Klautschek G, Schwerin M. Cattle and sheep use different promoters to direct the expression of the aromatase cytochrome P450 encoding gene, Cyp19, during pregnancy. Domest Anim Endocrinol. 2004;27:99–114.
5. Walters EM, Agca Y, Ganjam V, Evans T. Animal models got you puzzled?: think pig. Ann N Y Acad Sci. 2011;1245:63–4.
6. Wang S, Liu Y, Fang D, Shi S. The miniature pig: a useful large animal model for dental and orofacial research. Oral Dis. 2007;13:530–7.
7. Dehoux JP, Gianello P. The importance of large animal models in transplantation. Front Biosci. 2007;12:4864–80.
8. Casal M, Haskins M. Large animal models and gene therapy. Eur J Hum Genet. 2005;14:266–72.
9. Schaefer A, Schneeberger Y, Stenzig J, Biermann D, Jelinek M, Reichenspurner H, Eschenhagen T, Ehmke H, Schwoerer AP. A New Animal Model for Investigation of Mechanical Unloading in Hypertrophic and Failing Hearts: Combination of Transverse Aortic Constriction and Heterotopic Heart Transplantation. PLoS One. 2016;11, e0148259.
10. Chiaradia E, Avellini L, Tartaglia M, Gaiti A, Just I, Scoppetta F, Czentnar Z, Pich A. Proteomic evaluation of sheep serum proteins. BMC Vet Res. 2012;8:66.
11. Di Girolamo F, D'Amato A, Lante I, Signore F, Muraca M, Putignani L. Farm animal serum proteomics and impact on human health. Int J Mol Sci. 2014; 15:15396–411.
12. Meling S, Kvalheim OM, Arneberg R, Bardsen K, Hjelle A, Ulvund MJ. Investigation of serum protein profiles in scrapie infected sheep by means of SELDI-TOF-MS and multivariate data analysis. BMC Res Notes. 2013;6:466.
13. Batxelli-Molina I, Salvetat N, Andreoletti O, Guerrier L, Vicat G, Molina F, Mourton-Gilles C. Ovine serum biomarkers of early and late phase scrapie. BMC Vet Res. 2010;6:49.
14. Zhong L, Taylor D, Begg DJ, Whittington RJ. Biomarker discovery for ovine paratuberculosis (Johne's disease) by proteomic serum profiling. Comp Immunol Microbiol Infect Dis. 2011;34:315–26.
15. Sun D, Zhang H, Guo D, Sun A, Wang H. Shotgun Proteomic Analysis of Plasma from Dairy Cattle Suffering from Footrot: Characterization of Potential Disease-Associated Factors. PLoS One. 2013;8:e55973.

16. Bradshaw RA, Burlingame AL, Carr S, Aebersold R. Reporting protein identification data: the next generation of guidelines. Mol Cell Proteomics. 2006;5:787–8.

17. Seymour SL, Farrah T, Binz P-A, Chalkley RJ, Cottrell JS, Searle BC, Tabb DL, Vizcaíno JA, Prieto G, Uszkoreit J, et al. A standardized framing for reporting protein identifications in mzIdentML 1.2. Proteomics. 2014;14: 2389–99.

18. Anderson NL, Anderson NG. The human plasma proteome: history, character, and diagnostic prospects. Mol Cell Proteomics. 2002;1:845–67.

19. Omenn GS, States DJ, Adamski M, Blackwell TW, Menon R, Hermjakob H, Apweiler R, Haab BB, Simpson RJ, Eddes JS, et al. Overview of the HUPO Plasma Proteome Project: results from the pilot phase with 35 collaborating laboratories and multiple analytical groups, generating a core dataset of 3020 proteins and a publicly-available database. Proteomics. 2005;5:3226–45.

20. Omenn GS. Plasma Proteomics, The Human Proteome Project, and Cancer-Associated Alternative Splice Variant Proteins. Biochim Biophys Acta. 1844; 2014:866–73.

21. States DJ, Omenn GS, Blackwell TW, Fermin D, Eng J, Speicher DW, Hanash SM. Challenges in deriving high-confidence protein identifications from data gathered by a HUPO plasma proteome collaborative study. Nat Biotechnol. 2006;24:333–8.

22. Cai XW, Shedden KA, Yuan SH, Davis MA, Xu LY, Xie CY, Fu XL, Lawrence TS, Lubman DM, Kong FM. Baseline plasma proteomic analysis to identify biomarkers that predict radiation-induced lung toxicity in patients receiving radiation for non-small cell lung cancer. J Thorac Oncol. 2011;6:1073–8.

23. Clement CC, Aphkhazava D, Nieves E, Callaway M, Olszewski W, Rotzschke O, Santambrogio L. Protein expression profiles of human lymph and plasma mapped by 2D-DIGE and 1D SDS–PAGE coupled with nanoLC–ESI–MS/MS bottom-up proteomics. J Proteomics. 2013;78:172–87.

24. Magrane M, Consortium U. UniProt Knowledgebase: a hub of integrated protein data. Database. 2011;2011.

25. Consortium TU. UniProt: a hub for protein information. Nucleic Acids Res. 2015;43:D204–12.

26. Armengaud J, Trapp J, Pible O, Geffard O, Chaumot A, Hartmann EM. Non-model organisms, a species endangered by proteogenomics. J Proteomics. 2014;105:5.

27. Trapp J, Geffard O, Imbert G, Gaillard J-C, Davin A-H, Chaumot A, Armengaud J. Proteogenomics of Gammarus fossarum to Document the Reproductive System of Amphipods. Mol Cell Proteomics. 2014;13:3612–25.

28. Nesvizhskii AI. Proteogenomics: concepts, applications and computational strategies. Nat Methods. 2014;11:1114–25.

29. Renuse S, Chaerkady R, Pandey A. Proteogenomics. Proteomics. 2011;11:620–30.

30. Sigdel TK, Sarwal MM. The proteogenomic path towards biomarker discovery. Pediatr Transplant. 2008;12:737–47.

31. Perkins DN, Pappin DJ, Creasy DM, Cottrell JS. Probability-based protein identification by searching sequence databases using mass spectrometry data. Electrophoresis. 1999;20:3551–67.

32. Vaudel M, Burkhart JM, Zahedi RP, Oveland E, Berven FS, Sickmann A, Martens L, Barsnes H. PeptideShaker enables reanalysis of MS-derived proteomics data sets. Nat Biotechnol. 2015;33:22–4.

33. Chemonges S, Shekar K, Tung JP, Dunster KR, Diab S, Platts D, Watts RP, Gregory SD, Foley S, Simonova G, et al. Optimal Management of the Critically Ill: Anaesthesia, Monitoring, Data Capture, and Point-of-Care Technological Practices in Ovine Models of Critical Care. Biomed Res Int. 2014;2014:468309.

34. Smith PK, Krohn RI, Hermanson GT, Mallia AK, Gartner FH, Provenzano MD, Fujimoto EK, Goeke NM, Olson BJ, Klenk DC. Measurement of protein using bicinchoninic acid. Anal Biochem. 1985;150:76–85.

35. Laemmli UK. Cleavage of structural proteins during the assembly of the head of bacteriophage T4. Nature. 1970;227:680–5.

36. Schägger H. Electrophoretic isolation of membrane proteins from acrylamide gels. Appl Biochem Biotechnol. 1994;48:185–203.

37. Schägger H. Tricine-SDS-PAGE. Nat Protoc. 2006;1:16–22.

38. Schägger H, von Jagow G. Tricine-sodium dodecyl sulfate-polyacrylamide gel electrophoresis for the separation of proteins in the range from 1 to 100 kDa. Anal Biochem. 1987;166:368–79.

39. Brunelle JL, Green R. One-dimensional SDS-polyacrylamide gel electrophoresis (1D SDS-PAGE). Methods Enzymol. 2014;541:151–9.

40. Herbert B, Galvani M, Hamdan M, Olivieri E, MacCarthy J, Pedersen S, Righetti PG. Reduction of alkylation of proteins in preparation of two-dimensional map analysis: Why, when, and how? Electrophoresis. 2001;22: 2046–57.

41. Shevchenko A, Tomas H, Havlis J, Olsen JV, Mann M. In-gel digestion for mass spectrometric characterization of proteins and proteomes. Nat Protocols. 2007;1:2856–60.

42. Villén J, Gygi SP. The SCX/IMAC enrichment approach for global phosphorylation analysis by mass spectrometry. Nat Protoc. 2008;3:1630–8.

43. Yu Y, Smith M, Pieper R. A spinnable and automatable StageTip for high throughput peptide desalting and proteomics. 2014.

44. Rappsilber J, Mann M, Ishihama Y. Protocol for micro-purification, enrichment, pre-fractionation and storage of peptides for proteomics using StageTips. Nat Protocols. 2007;2:1896–906.

45. Kulak NA, Pichler G, Paron I, Nagaraj N, Mann M. Minimal, encapsulated proteomic-sample processing applied to copy-number estimation in eukaryotic cells. Nat Meth. 2014;11:319–24.

46. Shilov IV, Seymour SL, Patel AA, Loboda A, Tang WH, Keating SP, Hunter CL, Nuwaysir LM, Schaeffer DA. The Paragon Algorithm, a Next Generation Search Engine That Uses Sequence Temperature Values and Feature Probabilities to Identify Peptides from Tandem Mass Spectra. Mol Cell Proteomics. 2007;6:1638–55.

47. Elias JE, Gygi SP. Target-decoy search strategy for mass spectrometry-based proteomics. Methods Mol Biol. 2010;604:55–71.

48. Vaudel M, Breiter D, Beck F, Rahnenfuhrer J, Martens L, Zahedi RP. D-score: a search engine independent MD-score. Proteomics. 2013;13:1036–41.

49. Taus T, Kocher T, Pichler P, Paschke C, Schmidt A, Henrich C, Mechtler K. Universal and confident phosphorylation site localization using phosphoRS. J Proteome Res. 2011;10:5354–62.

50. Barsnes H, Vaudel M, Colaert N, Helsens K, Sickmann A, Berven FS, Martens L. compomics-utilities: an open-source Java library for computational proteomics. BMC Bioinformatics. 2011;12:70.

51. Hulsen T, de Vlieg J, Alkema W. BioVenn - a web application for the comparison and visualization of biological lists using area-proportional Venn diagrams. BMC Genomics. 2008;9:488.

52. Vizcaino JA, Deutsch EW, Wang R, Csordas A, Reisinger F, Rios D, Dianes JA, Sun Z, Farrah T, Bandeira N, et al. ProteomeXchange provides globally coordinated proteomics data submission and dissemination. Nat Biotechnol. 2014;32:223–6.

53. Martens L, Hermjakob H, Jones P, Adamski M, Taylor C, States D, Gevaert K, Vandekerckhove J, Apweiler R. PRIDE: the proteomics identifications database. Proteomics. 2005;5:3537–45.

54. Zougman A, Mann M, Nagaraj N, Winiewski JR. Universal sample preparation method for proteome analysis. Nat Methods. 2009;6:359–62.

55. Seymour SL, Hunter CJ. ProteinPilot™ Software Overview: High Quality, In-Depth Protein Identification and Protein Expression Analysis. USA: AB Sciex Pte. Ltd; 2015. p. 1–5.

56. Spivak M, Weston J, Tomazela D, MacCoss MJ, Noble WS. Direct maximization of protein identifications from tandem mass spectra. Mol Cell Proteomics. 2012;11:M111 012161.

57. States DJ, Omenn GS, Blackwell TW, Fermin D, Eng J, Speicher DW, Hanash SM. Challenges in deriving high-confidence protein identifications from data gathered by a HUPO plasma proteome collaborative study. Nat Biotech. 2006;24:333–8.

58. International Sheep Genomics C, Archibald AL, Cockett NE, Dalrymple BP, Faraut T, Kijas JW, Maddox JF, McEwan JC, Hutton Oddy V, Raadsma HW, et al. The sheep genome reference sequence: a work in progress. Anim Genet. 2010;41:449–53.

59. Bannach O, Birkmann E, Reinartz E, Jaeger K-E, Langeveld JPM, Rohwer RG, Gregori L, Terry LA, Willbold D, Riesner D. Detection of Prion Protein Particles in Blood Plasma of Scrapie Infected Sheep. PLoS One. 2012;7, e36620.

60. Katunguka-Rwakishaya E. Influence of Trypanosoma congolense infection on some blood inorganic and protein constituents in sheep. Rev Elev Med Vet Pays Trop. 1996;49:311–4.

61. Meling S, Bardsen K, Ulvund MJ. Presence of an acute phase response in sheep with clinical classical scrapie. BMC Vet Res. 2012;8.

62. Wells B, Innocent GT, Eckersall PD, McCulloch E, Nisbet AJ, Burgess ST. Two major ruminant acute phase proteins, haptoglobin and serum amyloid A, as serum biomarkers during active sheep scab infestation. Vet Res. 2013;44:103.

63. Ersdal C, Jørgensen HJ, Lie KI. Acute and Chronic Erysipelothrix rhusiopathiae Infection in Lambs. Vet Pathol. 2014;52:635–43.

64. Aytekin I, Aksit H, Sait A, Kaya F, Aksit D, Gokmen M, Baca AU. Evaluation of oxidative stress via total antioxidant status, sialic acid, malondialdehyde and

RT-PCR findings in sheep affected with bluetongue. Veterinary Record Open. 2015;2, e000054.

65. Burgess STG, Nunn F, Nath M, Frew D, Wells B, Marr EJ, Huntley JF, McNeilly TN, Nisbet AJ. A recombinant subunit vaccine for the control of ovine psoroptic mange (sheep scab). Vet Res. 2016;47:26.

66. Chemonges S, Tung JP, Fraser JF. Proteogenomics of selective susceptibility to endotoxin using circulating acute phase biomarkers and bioassay development in sheep: a review. Proteome Sci. 2014;12:12.

67. McDonald CI, Fung YL, Shekar K, Diab SD, Dunster KR, Passmore MR, Foley SR, Simonova G, Platts D, Fraser JF. The impact of acute lung injury, ECMO and transfusion on oxidative stress and plasma selenium levels in an ovine model. J Trace Elem Med Biol. 2015;30:4–10.

68. Mi H, Poudel S, Muruganujan A, Casagrande JT, Thomas PD. PANTHER version 10: expanded protein families and functions, and analysis tools. Nucleic Acids Res. 2016;44:D336–342.

69. Tirumalai RS, Chan KC, Prieto DA, Issaq HJ, Conrads TP, Veenstra TD. Characterization of the Low Molecular Weight Human Serum Proteome. Mol Cell Proteomics. 2003;2:1096–103.

70. Zheng X, Baker H, Hancock WS, Fawaz F, McCaman M, Pungor Jr E. Proteomic analysis for the assessment of different lots of fetal bovine serum as a raw material for cell culture. Part IV. Application of proteomics to the manufacture of biological drugs. Biotechnol Prog. 2006;22:1294–300.

71. Zhang G, Fenyö D, Neubert TA. Use of DNA ladders for reproducible protein fractionation by SDS-PAGE for quantitative proteomics. J Proteome Res. 2008;7:678–86.

72. Gillet LC, Navarro P, Tate S, Rost H, Selevsek N, Reiter L, Bonner R, Aebersold R. Targeted data extraction of the MS/MS spectra generated by data-independent acquisition: a new concept for consistent and accurate proteome analysis. Mol Cell Proteomics. 2012;11:O111 016717.

73. Gallagher SR. One-dimensional SDS gel electrophoresis of proteins. Curr Protoc Mol Biol 2006, Chapter 10:Unit 10 12A.

74. Wu X, Xiong E, Wang W, Scali M, Cresti M. Universal sample preparation method integrating trichloroacetic acid/acetone precipitation with phenol extraction for crop proteomic analysis. Nat Protocols. 2014;9:362–74.

75. Seth M, Lamont EA, Janagama HK, Widdel A, Vulchanova L, Stabel JR, Waters WR, Palmer MV, Sreevatsan S. Biomarker discovery in subclinical mycobacterial infections of cattle. PLoS One. 2009;4, e5478.

76. Guryca V, Lamerz J, Ducret A, Cutler P. Qualitative improvement and quantitative assessment of N-terminomics. Proteomics. 2012;12:1207–16.

77. Turk R, Piras C, Kovačić M, Samardžija M, Ahmed H, De Canio M, Urbani A, Meštrić ZF, Soggiu A, Bonizzi L, Roncada P. Proteomics of inflammatory and oxidative stress response in cows with subclinical and clinical mastitis. J Proteomics. 2012;75:4412–28.

78. Genini S, Paternoster T, Costa A, Botti S, Luini MV, Caprera A, Giuffra E. Identification of serum proteomic biomarkers for early porcine reproductive and respiratory syndrome (PRRS) infection. Proteome Sci. 2012;10:48.

79. Alonso-Fauste I, Andrés M, Iturralde M, Lampreave F, Gallart J, Álava MA. Proteomic characterization by 2-DE in bovine serum and whey from healthy and mastitis affected farm animals. J Proteomics. 2011;75:3015.

80. Barton C, Beck P, Kay R, Teale P, Roberts J. Multiplexed LC-MS/MS analysis of horse plasma proteins to study doping in sport. Proteomics. 2009;9:3058–65.

81. Faulkner S, Elia G, Mullen MP, O'Boyle P, Dunn MJ, Morris D. A comparison of the bovine uterine and plasma proteome using iTRAQ proteomics. Proteomics. 2012;12:2014–23.

82. Deutsch EW. File Formats Commonly Used in Mass Spectrometry Proteomics. Mol Cell Proteomics. 2012;11:1612–21.

83. Zhang H, Ghai P, Wu H, Wang C, Field J, Zhou GL. Mammalian adenylyl cyclase-associated protein 1 (CAP1) regulates cofilin function, the actin cytoskeleton, and cell adhesion. J Biol Chem. 2013;288:20966–77.

84. Herscovics A. Structure and function of Class I alpha 1,2-mannosidases involved in glycoprotein synthesis and endoplasmic reticulum quality control. Biochimie. 2001;83:757–62.

85. Stegelmeier BL, James LF, Panter KE, Gardner DR, Pfister JA, Ralphs MH, Molyneux RJ. Dose response of sheep poisoned with locoweed (Oxytropis sericea). J Vet Diagn Invest. 1999;11:448–56.

86. Stegelmeier BL, James LF, Panter KE, Molyneux RJ. Tissue and serum swainsonine concentrations in sheep ingesting Astragalus lentiginosus (locoweed). Vet Hum Toxicol. 1995;37:336–9.

87. Yu PH, Davis BA, Boulton AA, Zuo DM. Deamination of aliphatic amines by type B monoamine oxidase and semicarbazide-sensitive amine oxidase; pharmacological implications. J Neural Transm Suppl. 1994;41:397–406.

88. Sousse LE, Yamamoto Y, Enkhbaatar P, Rehberg SW, Wells SM, Leonard S, Traber MG, Yu YM, Cox RA, Hawkins HK, et al. Acute lung injury-induced collagen deposition is associated with elevated asymmetric dimethylarginine and arginase activity. Shock. 2011;35:282–8.

89. Sousse LE, Jonkam CC, Traber DL, Hawkins HK, Rehberg SW, Traber LD, Herndon DN, Enkhbaatar P. Pseudomonas aeruginosa is associated with increased lung cytokines and asymmetric dimethylarginine compared with methicillin-resistant Staphylococcus aureus. Shock. 2011;36:466–70.

90. Alemu P, Forsyth GW, Searcy GP. A comparison of parameters used to assess liver damage in sheep treated with carbon tetrachloride. Can J Comp Med. 1977;41:420–7.

91. Katsoulos PD, Christodoulopoulos G, Minas A, Karatzia MA, Pourliotis K, Kritas SK. The role of lactate dehydrogenase, alkaline phosphatase and aspartate aminotransferase in the diagnosis of subclinical intramammary infections in dairy sheep and goats. J Dairy Res. 2010;77:107–11.

92. Bodeker D, Oppelland G, Holler H. Involvement of carbonic anhydrase in ammonia flux across rumen mucosa in vitro. Exp Physiol. 1992;77:517–9.

93. Leonhard-Marek S, Gabel G, Martens H. Effects of short chain fatty acids and carbon dioxide on magnesium transport across sheep rumen epithelium. Exp Physiol. 1998;83:155–64.

94. Petrera A, Kern U, Linz D, Gomez-Auli A, Hohl M, Gassenhuber J, Sadowski T, Schilling O. Proteomic Profiling of Cardiomyocyte-Specific Cathepsin A Overexpression Links Cathepsin A to the Oxidative Stress Response. J Proteome Res. 2016;15:3188–95.

95. Kostadinov S, Shah BA, Alroy J, Phornphutkul C. A case of galactosialidosis with novel mutations of the protective protein/cathepsin a gene: diagnosis prompted by trophoblast vacuolization on placental examination. Pediatr Dev Pathol. 2014;17:474–7.

96. Hofmann F, Feil R, Kleppisch T, Schlossmann J. Function of cGMP-dependent protein kinases as revealed by gene deletion. Physiol Rev. 2006;86:1–23.

97. Bohr S, Patel SJ, Vasko R, Shen K, Golberg A, Berthiaume F, Yarmush ML. The Role of CHI3L1 (Chitinase-3-Like-1) in the Pathogenesis of Infections in Burns in a Mouse Model. PLoS One. 2015;10, e0140440.

98. Habib GM, Shi ZZ, Cuevas AA, Lieberman MW. Identification of two additional members of the membrane-bound dipeptidase family. FASEB J. 2003;17:1313–5.

99. Bebenek K, Kunkel TA. Functions of DNA polymerases. Adv Protein Chem. 2004;69:137–65.

100. Marcus F, Rittenhouse J, Gontero B, Harrsch PB. Function, structure and evolution of fructose-1,6-bisphosphatase. Arch Biol Med Exp (Santiago). 1987;20:371–8.

101. Visinoni S, Khalid NF, Joannides CN, Shulkes A, Yim M, Whitehead J, Tiganis T, Lamont BJ, Favaloro JM, Proietto J, et al. The role of liver fructose-1,6-bisphosphatase in regulating appetite and adiposity. Diabetes. 2012;61:1122–32.

102. Woolliams JA, Wiener G, Anderson PH, McMurray CH. Variation in the activities of glutathione peroxidase and superoxide dismutase and in the concentration of copper in the blood in various breed crosses of sheep. Res Vet Sci. 1983;34:253–6.

103. Carlile GW, Chalmers-Redman RME, Tatton NA, Pong A, Borden KLB, Tatton WG. Reduced Apoptosis after Nerve Growth Factor and Serum Withdrawal: Conversion of Tetrameric Glyceraldehyde-3-Phosphate Dehydrogenase to a Dimer. Mol Pharmacol. 2000;57:2–12.

104. Yugueros J, Temprano A, Berzal B, Sánchez M, Hernanz C, Luengo JM, Naharro G. Glyceraldehyde-3-Phosphate Dehydrogenase-Encoding Gene as a Useful Taxonomic Tool for Staphylococcus spp. J Clin Microbiol. 2000;38:4351–5.

105. Beavo JA. Cyclic nucleotide phosphodiesterases: functional implications of multiple isoforms. Physiol Rev. 1995;75:725–48.

106. Fritz TA, Hurley JH, Trinh LB, Shiloach J, Tabak LA. The beginnings of mucin biosynthesis: the crystal structure of UDP-GalNAc:polypeptide alpha-N-acetylgalactosaminyltransferase-T1. Proc Natl Acad Sci U S A. 2004;101:15307–12.

107. Tenno M, Kezdy FJ, Elhammer AP, Kurosaka A. Function of the lectin domain of polypeptide N-acetylgalactosaminyltransferase 1. Biochem Biophys Res Commun. 2002;298:755–9.

108. Ramadoss J, Liao WX, Morschauser TJ, Lopez GE, Patankar MS, Chen DB, Magness RR. Endothelial caveolar hub regulation of adenosine triphosphate-induced endothelial nitric oxide synthase subcellular partitioning and domain-specific phosphorylation. Hypertension. 2012;59:1052–9.

109. Filali H, Martín-Burriel I, Harders F, Varona L, Hedman C, Mediano DR, Monzón M, Bossers A, Badiola JJ, Bolea R. Gene expression profiling of mesenteric lymph nodes from sheep with natural scrapie. BMC Genomics. 2014;15:1–17.

110. Yamashita H, Kotani T, Park JH, Murata Y, Okazawa H, Ohnishi H, Ku Y, Matozaki T. Role of the Protein Tyrosine Phosphatase Shp2 in Homeostasis of the Intestinal Epithelium. PLoS One. 2014;9, e92904.

111. Gurung RB, Begg DJ, Purdie AC, Bach H, Whittington RJ. Immunoreactivity of protein tyrosine phosphatase A (PtpA) in sera from sheep infected with Mycobacterium avium subspecies paratuberculosis. Vet Immunol Immunopathol. 2014;160:129–32.

112. Steel EG, Witzel DA, Blanks A. Acquired coagulation factor X activity deficiency connected with Hymenoxys odorata DC (Compositae), bitterweed poisoning in sheep. Am J Vet Res. 1976;37:1383–6.

113. Tillman P, Carson SN, Talken L. Platelet function and coagulation parameters in sheep during experimental vascular surgery. Lab Anim Sci. 1981;31:263–7.

114. Filippovich I, Sorokina N, St Pierre L, Flight S, de Jersey J, Perry N, Masci PP, Lavin MF. Cloning and functional expression of venom prothrombin activator protease from Pseudonaja textilis with whole blood procoagulant activity. Br J Haematol. 2005;131:237–46.

115. Klee L, Zand R. Probable epitopes: Relationships between myelin basic protein antigenic determinants and viral and bacterial proteins. Neuroinformatics. 2004;2:59–70.

116. Bongaerts GP, Vogels GD. Mechanism of uricase action. Biochim Biophys Acta. 1979;567:295–308.

117. Rietema SE, Blackberry MA, Maloney SK, Martin GB, Hawken PA, Blache D. Twenty-four-hour profiles of metabolic and stress hormones in sheep selected for a calm or nervous temperament. Domest Anim Endocrinol. 2015;53:78–87.

118. Chemonges S. Suspected selective susceptibility to endotoxin in an ovine model. Online J Vet Res. 2014;18:941–63.

119. Almeida AM, Palhinhas RG, Kilminster T, Scanlon T, van Harten S, Milton J, Blache D, Greeff J, Oldham C, Coelho AV, Cardoso LA. The Effect of Weight Loss on the Muscle Proteome in the Damara, Dorper and Australian Merino Ovine Breeds. PLoS One. 2016;11, e0146367.

120. Knight TW, Lindsay DR, Oldham CM. Proceedings: the influence of rams on the fertility of the ewe. J Reprod Fertil. 1975;43:377–8.

121. Brown HM, Fabre Nys C, Cognie J, Scaramuzzi RJ. Short oestrous cycles in sheep during anoestrus involve defects in progesterone biosynthesis and luteal neovascularisation. Reproduction. 2014;147:357–67.

122. Mitchell LM, King ME, Aitken RP, Gebbie FE, Wallace JM. Ovulation, fertilization and lambing rates, and peripheral progesterone concentrations, in ewes inseminated at a natural oestrus during November or February. J Reprod Fertil. 1999;115:133–40.

123. Tung JP, Fung YL, Nataatmadja M, Colebourne KI, Esmaeel HM, Wilson K, Barnett AG, Wood P, Silliman CC, Fraser JF. A novel in vivo ovine model of transfusion-related acute lung injury (TRALI). Vox Sang. 2011;100:219–30.

124. Australian code of practice for the care and use of animals for scientific purposes [https://www.nhmrc.gov.au/guidelines-publications/ea28]. Accessed 4 Aug 2016.

13

Proteomic analysis of early salt stress responsive proteins in alfalfa roots and shoots

Junbo Xiong[1†], Yan Sun[2†], Qingchuan Yang[3], Hong Tian[1], Heshan Zhang[1], Yang Liu[1*] and Mingxin Chen[1]

Abstract

Background: Alfalfa (*Medicago sativa*) is the most extensively cultivated forage legume in the world, and salinity stress is the most problematic environmental factors limiting alfalfa production. To evaluate alfalfa tissue variations in response to salt stress, comparative physiological and proteomic analyses were made of salt responses in the roots and shoots of the alfalfa.

Method: A two-dimensional gel electrophoresis (2-DE)-based proteomic technique was employed to identify the differentially abundant proteins (DAPs) from salt-treated alfalfa roots and shoots of the salt tolerance cultivars Zhongmu No 1 cultivar, which was subjected to a range of salt stress concentrations for 9 days. In parallel, REL, MAD and H_2O_2 contents, and the activities of antioxidant enzymes of shoots and roots were determinand.

Result: Twenty-seven spots in the shoots and 36 spots in the roots that exhibited showed significant abundance variations were identified by MALDI-TOF-TOF MS. These DAPs are mainly involved in the biological processes of photosynthesis, stress and defense, carbohydrate and energy metabolism, second metabolism, protein metabolism, transcriptional regulation, cell wall and cytoskeleton metabolism, ion transpor, signal transduction. In parallel, physiological data were correlated well with our proteomic results. It is worth emphasizing that some novel salt-responsive proteins were identified, such as CP12, pathogenesis-related protein 2, harvest-induced protein, isoliquiritigenin 2′-O-methyltransferase. qRT-PCR was used to study the gene expression levels of the four above-mentioned proteins; four patterns are consistent with those of induced protein.

Conclusion: The primary mechanisms underlying the ability of alfalfa seedlings to tolerate salt stress were photosynthesis, detoxifying and antioxidant, secondary metabolism, and ion transport. And it also suggests that the different tissues responded to salt-stress in different ways.

Keywords: NaCl stress, *Medicago sativa* root and shoot, Two-dimensional electrophoresis, Differentially abundant proteins

Background

Soil salinity is a world-wide problem, but is most acute in North and Central Asia, South America, Australasia, and the Mediterranean area. The soil solution in saline soils is composed of a range of dissolved salts, such as NaCl, Na_2SO_4, $MgSO_4$, $CaSO_4$, $MgCl_2$, KCl, and Na_2CO_3, each of which contribute to salinity stress. However, NaCl is the most prevalent salt and has been the focus of much of the work on salinity to date [1, 2]. High NaCl concentrations affect plant physiology and metabolism at different levels. High concentrations can cause water deficits, ion toxicity, nutrient imbalance, and oxidative stress, leading to molecular damage, growth and yield reductions, and even plant death.

Alfalfa (*Medicago sativa* L.) is a perennial warm-season forage legume with a high yield and good nutrient contents (crude protein content can reach approximately 16% to 22%), and can be grown on more than 30 Mha worldwide. However, its yield is low in arid and semi-arid

* Correspondence: liuyang430209@126.com
†Equal contributors
[1]Hubei Key Laboratory of Animal Embryo and Molecular Breeding, Institute of Animal and Veterinary Science, Hubei Academy of Agricultural Science, Yaoyuan 1, Hongshan, Wuhan, Hubei 430017, China
Full list of author information is available at the end of the article

regions where salinity is the main problem. Alfalfa is moderately tolerant to salinity when the electrical conductivity (EC) is 2.0 dS/m (1280 ppm) and the soil osmotic potential threshold is 1.5 bars (1 bar = 0.987 atm) at field capacity. An additional 7% decrease in alfalfa yields can be expected with each dS/m increase in saturation extract salinity [3]. Excessive salinity in the crop root zone creates osmotic stress, which reduces root uptake of water and crop transpiration, leading to reduced forage yields [4].

Understanding the alfalfa tolerance mechanisms to high concentrations of NaCl in soils may ultimately help to improve yields on saline lands. Previous studies indicated that alfalfa salt tolerance is generally associated with modifications of morphological and physiological traits, such as changes in plant architecture and growth (shoots and roots), variations in leaf cuticle thickness, stomatal regulation, germination, and photosynthesis rate. These changes are linked to diverse cellular modifications, including, changes in membrane and protein stability, increased antioxidant capacity and activation of hormonal signaling pathways, notably those depending on the "stress hormone" abscissic acid [5]. The regulation of these changes at the cellular level are the main responses that cause alterations in gene expression and several attempts have been made to obtain a profile for gene expression in alfalfa under saline conditions [6, 7]. However, transcript profiles do not always provide a complete story due to limited correlations between the transcript and protein levels, and proteomics has become a critical complement to mRNA data and an improved biological view of plant biology. Currently, several studies have attempted to analyze alterations in protein expression in response to salt, and proteomics studies that focused on 34 plant species have identified 2171 salt-responsive protein identities, representing 561 unique proteins [8]. To date, few studies have investigated the effects of salt stress on alfalfa.

Salt stress induces many different proteomic changes in various plant tissues due to their distinct functions and growth environments. A comparative analysis of different plant tissue responses to salinity stress at the same time would improve understanding of different tissues protein compositions and their differential responses to salinity stress. Furthermore, it would provide further insights into the proteomic mechanisms controlling salt tolerance. A few previous studies examined protein change responses in different tissues to salinity stress, such as the report on soybean (*Glycine max* L.) leaves, hypocotyls, and roots [9, 10], creeping bentgrass (*Agrostis. stolonifera* L.) leaves and roots [11], and rice (*Oryza sativa* L.) leaves and roots [12]. They all suggested that protein responses to salt-stress in different tissues varied and some protein showed tissue specific abundance.

Alfalfa cultivar "Zhongmu No1", one salt tolerance cultivar commonly used in China agriculture, was released by the Chinese Academy of Agricultural Science in 2001. This germplasm represents the four cycle of recurrent mass selection for alfalfa genotypes that germinate at high levels of NaCl. In this study, we analyzed the "Zhongmu No1" cultivar shoot and root responses to different NaCl concentrations using physiological and biochemical methods, and comparative proteomics. Based on our findings, we produced a possible schematic representation of the mechanism associated with salt tolerance in alfalfa.

Methods

Plant materials and stress treatments

Alfalfa seeds (*Medicago sativa* L.cv. Zhongmu No 1) were germinated in the dark for 48 h at 28 °C, then transplanted into 1/2 Hoagland's nutrient solution and grown on for 7 days. Subsequently, the seedlings were subjected to 0 (control), 100, and 200 mM NaCl 1/2 Hoagland's nutrient solution for 9 d. The salt concentration was maintained by a daily input of 50 mM NaCl. The experiments were conducted in a glasshouse chamber that had an average temperature of 27 °C/18 °C day/night, and a light irradiance of 150 μmol m^{-2} s^{-1}.

H_2O_2, MDA, and relative electrolyte leakage analyses

For the H_2O_2 content analysis, 1 g each of root and shoot tissues were ground in liquid N_2 and then homogenized in 5 ml cold acetone. The supernatants were used for H_2O_2 content assays after centrifugation at 3000 *g* and 4 °C for 10 min. The H_2O_2 content was assayed by analyzing the production of titanium–hydroperoxide complex at 410 nm [13]. MDA was measured using a modified thiobarbituric acid (TBA) method as described previously [14]. Relative electrolyte leakage was determined by modifying a method described previously [15]. A total of 500 mg of tissues were rinsed with ddH$_2$O, placed in test tubes containing 10 ml of ddH$_2$O, and incubated at room temperature for 2 h. The electrical conductivity of the solution (C_1) was measured using a conductivity meter (DDS-307A; China). Then the tubes were boiled for 15 min, cooled to room temperature, and the electrical conductivity (C_2) measured again. The REL was calculated by the formula: $C_1 / C_2 \times 100\%$.

SOD, APX, POD, and CAT activity analyses

The enzyme extraction and enzyme activity assays were determined by methods modified from those previously described [16]. Root and shoot samples (200 mg each) were ground into fine powder with liquid nitrogen in a pre-chilled mortar and pestle. Further grinding was performed in a solution of 50 mM potassium phosphate buffer pH 7.0 containing 1 mM EDTA and 2% (w/v) polyvinylpolypyrrolidone (PVPP) for the APX and CAT assays, and in a solution of 50 mM potassium phosphate buffer at pH 7.0 containing 0.5 mM EDTA for the SOD and POD assays. The homogenates were centrifuged at 14000 g for 15 min at 4 °C. The resulting supernatants were centrifuged again and used immediately for enzyme activity assays or stored at −30 °C to be used later. Total SOD (EC 1.15.1.1) activity was determined by monitoring its ability to inhibit the photochemical reduction of nitro blue tetrazolium (NBT). APX activity (EC 1.11.1.11) was determined by following the decrease in ascorbate and measuring the change in absorbance at 290 nm over 2 min intervals.The POD (EC 1.11.1.7) and CAT (EC 1.11.1.6) activity were determined by following the decrease in H_2O_2, and measuring the change in absorbance at 240 nm over 2 min intervals.

Protein extraction and 2-DE

The total proteins were extracted by a modified TRIzol reagent method, which was recently developed to obtain high-quality proteins from *Medicago truncatula* tissues for 2-DE [17]. The whole roots and shoots were cut off the seedlings, frozen in liquid nitrogen and ground to a fine powder for protein extraction. Finally, the pellets were dried in a freeze-vacuum dryer for 10 min, resuspended in 1.5 mL lysis buffer (8 M urea, 4% v/v CHAPS, 2% w/v DTT), sonicated (10 min) at 4 °C and incubated at room temperature for 2 h. The supernatant was collected after centrifugation (40 min, 40,000 g, 4 °C). The protein concentration of the supernatant was determined using a 2-D Quant kit, following the manufacturer's protocol.

Samples containing 120 µg total protein in 450 µL rehydration buffer (8 M urea, 2% w/v CHAPS, 1% w/v DTT, 0.5% v/v IPG buffer pH 4–7, 0.002% w/v bromophenolblue) were loaded onto a 24 cm, pH 4 to 7 linear gradient IPGstrip (GE Healthcare, USA). IEF was carried out using an Ettan IPGphorII (GE Healthcare, Uppsala, Sweden). Focusing was performed at 20 °C as follows: active rehydration at 30 V for 12 h, 150 V for 1 h, 500 V for 1 h, 1000 V for 1 h, 8000 V for 2 h, and 8000 V up to 40,000 VH. After IEF, the proteins were equilibrated as described. First the IPG strips were incubated in 10 mL of equilibration buffer (6 M urea, 30% w/v glycerol, 2% w/v SDS, 50 mM Tris-HCl, pH 8.8) with 1% w/v DTT for 15 min, and then in the same solution containing 2.5% w/v iodoacetamide instead of DTT for 15 min. Following this, the strips were transferred to 12% SDS-PAGE gels for second dimension electrophoresis with the Ettan DALTsix gel system (GE Healthcare, Uppsala, Sweden), using SDS electrophoresis buffer (250 mM Tris-base, 1.92 M glycine, 1% w/v SDS) with a 0.2 W/strip for 1 h, and a 15 W/strip until the dye front reached the bottom of the gel. All 2-DE separations were repeated three times for each tissue extract.

Protein visualization, image analysis

Upon electrophoresis, Gels gels were stained with silver nitrate according to GE handbook (GE Healthcare, Uppsala, Sweden) with some modifications. Briefly, gels were fixed in 40% ethanol and 10% acetic acid for 60 min, and then sensitized with 30% ethanol, 0.2% sodium thiosulfate w/v, and 6.8% sodium acetate w/v for 30 min. Then gels were rinsed with distilled water three times, 5 min for each time, then incubated in silver nitrate (2.5 g/L) for 20 min. Incubated gels were rinsed with distilled water two times, and developed in a solution of sodium carbonate (25 g/L) with formaldehyde (37%, w/v) added (240 mL/L) for two times, first for 1 min, then stained for 4 min. Development was stopped with 1.46%w/v Ethylene Diamine Tetraacetic Acid for 10 min, then gels were rinsed with distilled water three times, 5 min for each time. Gels were stored in distilled water until they could be processed.Gels images were acquired using a PowerLook 2100XL color scanner (UMAX Technologies, CA, USA) and analyzed with Image master 2D Platinum Software Version 6.0 (GE Healthcare, Uppsala, Sweden).

Protein identification by MALDI-TOF-MS/MS

Proteins were identified by MALDI-TOF-MS/MS. Selected spots were excised from the gels and destained with a solution containing 20% w/v sodium thiosulphate and 1% w/v potassium ferricyanide for 5 min. The supernatant was removed and the gel spots were washed twice with 25 mM ammonium bicarbonate in 50% v/v acetonitrile for 20 min. The gel spots were then washed in acetonitrile, dried in a Speed-Vac and digested overnight with 20 µg/mL trypsin in 25 mM ammonium bicarbonate at 37 °C. Tryptic peptides were passed through C18 Zip-Tips and mixed with 5 mg/mL of R-cyano-4-hydroxycinnamic acid, as the matrix, and subject to MALDI-TOF/TOF analysis (4700 Proteomics Analyzer, Applied Biosystems). Data files obtained from MALDI-TOF/TOF mass spectra were submitted to the Mascot search engine using Daemon 2.1.0 (Matrix Science; http://www.matrixscience.com) on Mascot server version 2.2.1. The data were searched against the NCBInr database and the peptides were constrained to being tryptic with a maximum of one missed cleavage. Carbamidomethylation of cysteine was considered a fixed modification, and oxidation of methionine residues was considered as a variable modification. The identification was based on the combination of a high Mascot score and maximum peptide coverage.

qRT-PCR analysis

Total RNA was extracted from salt-treated and control alfalfa roots and shoot by Trizol reagent (TaKaRa), and cDNA was reverse transcribed from 1 µg of to total RNA using a First Strand cDNA Synthesis Kit (Invitrogen). Gene-specific primers (GSPs) used for qRT-PCR were designed using primer 5 according to cDNA sequences obtained from the alfalfa (Table 1). The alfalfa Actin gene was used as an endogenous control for normalization. The PCR reaction was carried out in a 20 uL volume containing 10 µL 2 × SYBR Green Master Mix reagent (TaKaRa), 1 µL template cDNA and 0.5 µL of each GSPs with the following reaction conditions: 95 °C for 30 s; followed by 40 cycles of 95 °C for 10 s; 55 °C for 10 s and 72 °C for 15 s. Relative gene expression was calculated using the ddCt alogorithm [18].

Immunoblot analysis

Protein samples (50 mg/lane) were separated using 12% one dimensional SDS-PAGE gel electrophoresis, transferred onto nitrocellulose membranes, and incubated at room temperature for 2 h with rabbit polyclonal antibodies raised against Rubisco activase, Heat shock

Table 1 The primers for qRT-PCR

Protein	Genes	Primers	Sequence
Actin	gi|378407816	Forward primer (5'-3')	GATACTCTTTCACCACAACAGCCG
		Reverse primer (5'-3')	ACTTCAGGACAACGGAAACGCT
CP12	gi|3,123,345	Forward primer (5'-3')	TGGCAACAATAGGTGGTCT
		Reverse primer (5'-3')	CTCGTCGGTTTCAGGGT
HI protein	gi|283,831,548	Forward primer (5'-3')	GCTGATGAAATCGTCCCA
		Reverse primer (5'-3')	ACCCTGTTCCTCCCACTAAGCTGTA
PR protein 2	gi|44,887,779	Forward primer (5'-3')	CTAAATTACCAGCATCAACGC
		Reverse primer (5'-3')	CCTCTACTTTCATCAGGGACAA
IOMT	gi|22,266,001	Forward primer (5'-3')	GCTGATGAAATCGTCCCA
		Reverse primer (5'-3')	AACCCTGTTCCTCCTACCA

protein 70 each (Agrisera, Sweden) at 1:5000 dilution. After washing three times with TBST buffer (0.01 M TBS, 0.1% Tween-20, pH 7.6), the membranes were exposed for 2 h at room temperature to horseradish peroxidase-conjugated goat anti-rabbit IgG at 1:300 dilution. Positive signals were visualized with 3, 3′-diaminobenzidine.

Statistical analysis

Data from repeated measurements are shown as mean. Comparison of differences among the groups was carried out using Student's test. Significant differences were determined relative to the P value [P-values <0.05 (*) and <0.01 (**)].

Results
Changes in REL and MAD contents

REL and MAD are indicators of membrane damage caused by NaCl stress. Stress-induced REL and MAD changes in the roots and shoots are shown in Fig. 1a, b. These data demonstrated a significant increase in the REL and MAD (P-values <0.05 and P-values <0.01) when alfalfa seedlings were treated with 100 mM and 200 mM NaCl. The roots had higher REL and MAD contents than the shoots.

Changes in H_2O_2 and antioxidant enzyme activities

As shown in Fig. 1c, a significant increase in the H_2O_2 when alfalfa seedlings were treated with 100 mM and 200 mM NaCl (P-values <0.05 and P-values <0.01). The shoots had higher H_2O_2 contents than the roots. Under normal conditions, the SOD activity was higher in the shoots than in the roots, and it was significant increase (P-values <0.01) in roots and shoots when alfalfa seedlings were treated with 100 mM and 200 mM NaCl (Fig. 1d). The SOD in the roots was 3.78 and 5.29 times higher in 100 and 200 mM NaCl, respectively, than in the control, and was 1.59 and 2.35 times higher than in the shoots. Similarly, the APX activity was significant increase (P-values <0.01) in the shoots and roots as the NaCl concentration increased. Furthermore, the rate of increase in APX activity in the shoots was slower than in the roots (Fig. 1f). Salinity effects on POD activity are shown in Fig. 1e. Under normal conditions, the POD activity in the roots was 5.48 times higher than in the shoots. Salt stress slightly increased the POD activity in the roots and shoots, but it was not significant (P-values >0.05). The salt stress treatments up-regulated CAT activity by 2.15 and 2.91 times respectively, in shoot. However, the CAT activity in the roots slightly decreased under salt stress, it was not significant (P-values >0.05) (Fig. 1g).

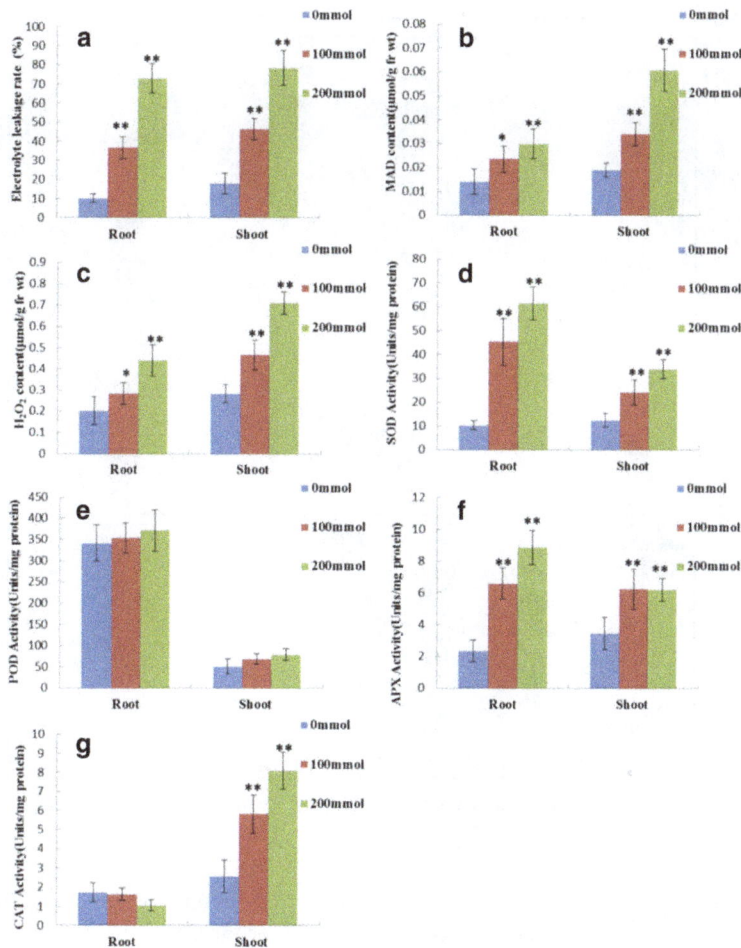

Fig. 1 Physiological responses induced by NaCl treatment (0,100, 200 Mm) for 9 days in *Medicago sativa* leaves and roots. Effects of salinity on the Relative electrolyte leakage (**a**), MAD content (**b**), H_2O_2 content (**c**), SOD activity (**d**), POD activity (**e**), APX activity (**f**), CTA activity (**g**) were presented. Significant differences were determined relative to each treatment using a student's t-test [P-values <0.05 (*) and < 0.01 (**)]. Bars: SD

Identification and functional classification of DAPs

More than 850 proteins were detected in each gel by ImageMaster software (Figs. 2 and 3). Comparison of control and salt-treated plants reference gels allowed the identification of differentially spots. Differentially spots were selected based on the following criteria: (i) relative vol% of the spot with fold change in a comparison >1.5 or <0.67; (ii) unadjusted significance level $p < 0.05$. Then the spots were analyzed by MALDI-TOF-TOF MS, and a total 61 DAPs were identified: 26 spots in the shoots and 35 spots in the roots (Table 2). Differentially expressed proteins were classified based on KEGG (http://www.kegg.jp/kegg/pathway.html) and previous literature (Fig. 4). In the shoots, the largest two groups were photosynthesis (31%), and stress and defense (20%) groups. In the roots, the largest three groups were stress and defense (26%), metabolism (17%), and protein translation, processing, and degradation (17%). It is

noteworthy that proteins involved in signaling and ion transport were only found in the roots.

Correlation of 2-DE data with qRT-PCR

Four mRNAs encoding novel salt-responsive proteins were selected for analysis. We compared the mRNA levels with the 2-DE data, and determined that all of the qRT-PCR results were in good agreement with the 2-DE data (Fig. 5).

Immunoblot analysis for RuBisCO activase and heat shock protein 70

In the current study, the accuracy of 2-DE analysis was further validated by immunoblot analysis. Proteins of alfalfa roots and shoots were separated by one-dimensional SDS-PAGE, and immunoblot analysis was performed for Heat shock protein 70 and RuBisCO activase (Fig. 6).

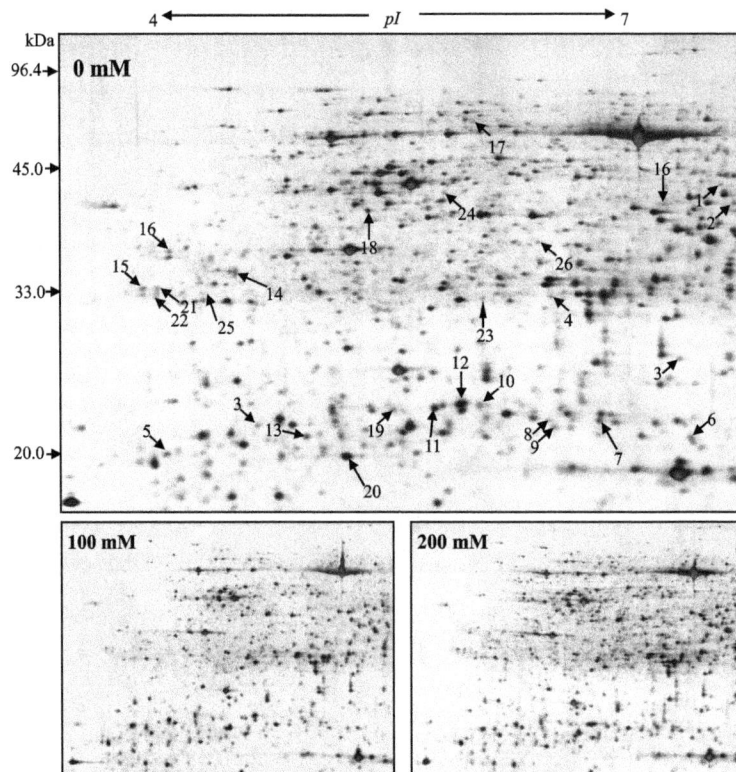

Fig. 2 2-DE analysis of proteins extracted from alfalfa shoot under different salinity. Arrows indicate protein changes induced by NaCl treatment

In agreement with the changes in protein abundance observed by 2-DE, Heat shock protein 70 showed an increased amount in response to 100 mM and 200 mM NaCl treatment. RuBisCO activase immunoblot analysis revealed an increase amount in response to 100 mM NaCl treatment, while the 200 mM value is not significantly different from control. This result is different in 2-DE analysis.

Discussion

Salt stress decreased the growth of both shoots and roots, and this is a well-known physiological change in alfalfa. However, the mechanisms that regulate salt adaptation in alfalfa are complicated and are not well understood. In this study, through a combination of biochemical and proteomic approaches, we were able to undertake a comprehensive analysis of salt stress responses and defense in alfalfa shoots and roots for the first time.

Proteins involved in photosynthesis

Photosynthesis is one of the most important processes to be affected by salinity. The effects of salt stresses on photosynthesis are either direct, such as diffusion limitations through the stomata and the mesophyll, and alterations in photosynthetic metabolism, or

secondary, such as the oxidative stress arising from the superimposition of multiple stresses [19]. Therefore, it was not surprising to observe that the abundance of eight proteins involved in photosynthesis were altered under NaCl treatment. Among these proteins, three thylakoid membrane proteins: cytochrome b6-f complex iron-sulfur (Cyt b6/f, spot 12), chlorophyll a/b binding protein (CAB, spot 16), and chloroplast oxygen-evolving enhancer protein 1 (OEE1, spot 26) were down-regulated by salt-stress. These proteins are involved in the light reactions, including electron transfer, light-harvesting, and light-induced oxidation of water. As previously pointed out, salt stress can limit CO_2 fixation, and the reducing power production rate is greater than the rate of its use by the Calvin cycle. The excess reducing power will induce the production of reactive oxygen species, thus the protection mechanisms against excess reducing power are an important strategy for combating salt stress [19]. In our study, the down-regulated proteins involved in the light reactions will help alfalfa to reduce reducing power production. However, in some salt-tolerant plants, such as *Thellungiella halophila*, *Agrostis stolonifera*, and *Kandelia candel*, salt stress induced the up-regulation of light reaction proteins [20–22].

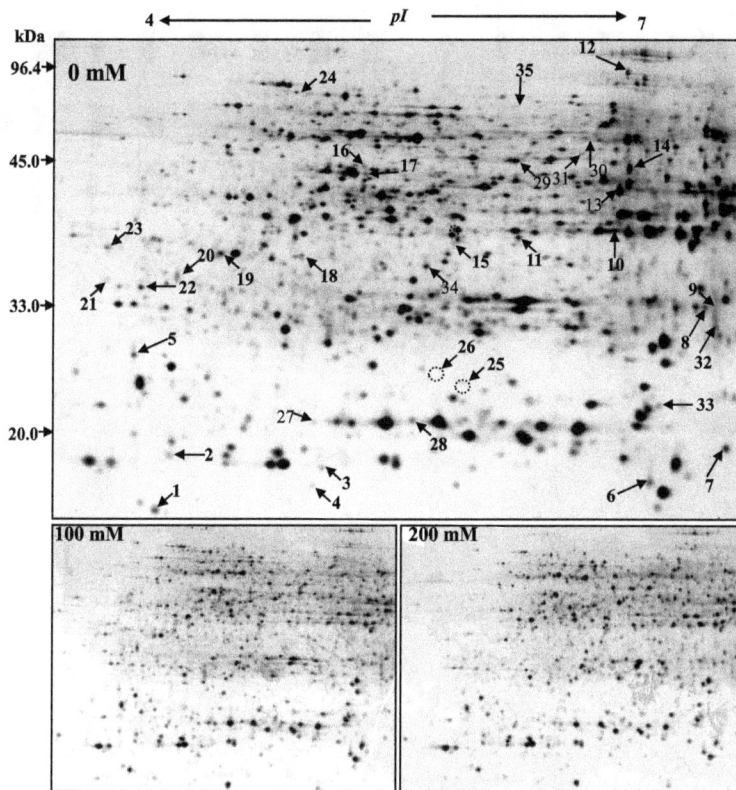

Fig. 3 2-DE analysis of proteins extracted from alfalfa roots under different salinity. Arrows indicate protein changes induced by NaCl treatment

Three spots (spots 2, 3, and 4) were identified as RuBisCO subunits. RuBisCO, created either through the carboxylation or oxygenation of ribulose-1,5-bisphosphate with carbon dioxide or oxygen, respectively, is composed of eight large subunits and eight small subunits. RuBisCOs are the most common enzymes in plants and salt stress induced altered abundance of RuBisCO subunits have been found in almost all green plant leaves. Previous studies showed that oxidative stress may lead to small-subunit degradation, which subsequently leads to translational arrest of the large subunit. Alternatively, oxidative stress could initially arrest large subunit translation, resulting in a rapid degradation of the unassembled small subunits [23]. It is noteworthy that a RuBisCO activase protein (spot 8) was up-regulated as the salt concentration rose. The principal role of RuBisCO activase is to release inhibitory sugar phosphates, such as ribulose-1,5-biphosphate, from the active RuBisCO sites so that CO_2 can activate the enzyme controlling carbamylation. Therefore it ultimately determines the proportion of available RuBisCO active sites that are catalytically competent [24, 25]. Salt-stress directed reduction in stomatal conductance and subsequent low CO_2 levels, together with the up-regulation of activase activity, may be required in order to induce salt stress tolerance. Previous studies have

shown that salt stress induced the up-regulation of RuBisCO activase in rice leaf lamina, barley, and wild halophytic rice [26–28].

A CP12 (spot 6) protein was down- regulated after the 200 mM NaCl treatment in the shoots. CP12 is a small nuclear encoded chloroplast protein, which, in chloroplasts, is oligomerized with phosphoribulokinase (PRK) and $NADP^+$-GAPDH in the presence of NAD(H) to generate a PRK/CP12/GAPDH complex. However, the complex dissociates in the presence of NADP(H). In Synechococcus, the oligomerization of CP12 with PRK and GAPDH regulates the activities of both enzymes and thus the carbon flow from the Calvin cycle to the oxidative pentose phosphate cycle [29]. In this manner, the down-regulated of CP12 seem to induce by the depression of photosynthesis.

Stress responsive proteins form the largest protein group
Salt stress causes the production of excessive reactive oxygen species (ROS), which oxidize cellular components and irreversibly damage plant cells. In the present study, a total of 8 identified proteins were found to obviously relate to anti-oxidative reactions in alfalfa seedling roots and shoots in response to salt stress. All these proteins were up-regulated under 100 mM and /or 200 mM NaCl stress in alfalfa shoots and/or roots. The 8 proteins

Table 2 Identification of salt-responsive proteins in alfalfa using MALDI-TOF-MS/MS

Spot No.[a]	Homologous protein (plant species)[b]	gi Number[c]	Theo. Mr/pI[d]	Exp. Mr/pI[e]	Scores[f]	M.P[g]	Relative Vol% ± SE[h] CK 100 mM 200 mM
Photosynthesis							
2	RuBisCO large subunit [Medicago sativa]	1,223,773	12.0/6.1	41.7/6.9	263	3	
3	RuBisCO small subunit [Medicago sativa]	2,342,980	52.3/6.1	27.5/6.7	132	3	
4	RuBisCO large subunit [Medicago sativa]	1,223,773	12.0/6.1	31.5/6.1	99	4	
8	RuBisCO activase [Oryza sativa Japonica Group]	1,778,414	24.6/8.6	21.7/6.1	126	5	
12	Cytochrome b6-f complex iron-sulfur [Pisum sativum]	136,707	23.1/5.9	23.5/5.7	138	5	
16	Chlorophyll a/b binding protein [Arabidopsis thaliana]	16,374	25.0/5.1	36.6/4.5	112	4	
26	Chloroplast oxygen-evolving enhancer protein 1 [Leymus chinensis]	147,945,622	34.5/6.0	35.5/6.1	95	4	
6	CP12 [Chlamydomonas reinhardtii]	3,123,345	12.5/5.7	15.3/6.6	92	3	
Stress and defense							
25	Glutathione peroxidase 1 [Lotus japonicus]	37,930,463	26.0/10.0	31.8/4.6	97	7	
23	Chloroplast thylakoid-bound ascorbate peroxidase [Vigna unguiculata]	45,268,437	45.1/8.2	31.2/5.8	115	4	
13	Harvest-induced protein [Medicago sativa]	283,831,548	16.6/5.1	21.6/5.1	133	5	
19	Quinone reductase family protein [Arabidopsis thaliana]	30,687,535	21.7/6.1	23.7/5.5	186	6	
20	Pathogenesis-related protein 2 [Medicago sativa]	22,266,001	16.5/5.8	20.1/5.3	344	9	
R4	Glutathione peroxidase [Medicago truncatula]	355,524,544	21.3/7.6	11.7/4.6	76	5	
R21	Ascorbate peroxidase [Medicago sativa]	16,304,410	20.1/5.3	34.7/4.3	99	6	
R23	Ascorbate peroxides [Medicago sativa]	16,304,410	20.1/5.3	40.1/4.2	81	5	
R16	Cytosolic ascorbate peroxidase [Cucumis sativus]	1,669,585	27.5/5.4	37.9/6.1	102	3	
R9	Pathogenesis-related protein 5 [Arabidopsis thaliana]	15,222,089	26.2/4.8	34.7/6.9	113	5	
R27	Pathogenesis-related protein 2 [Medicago sativa]	22,266,001	16.5/5.8	23.7/5.6	344	9	

Table 2 Identification of salt-resposive proteins in alfalfa using MALDI-TOF-MS/MS (Continued)

Spot No.[a]	Homologous protein (plant species)[b]	gi Number[c]	Theo. Mr/pI[d]	Exp. Mr/pI[e]	Scores[f]	M.P[g]	Relative Vol% ± SE[h] CK 100 mM 200 mM
R26	Alcohol dehydrogenase-1F [Phaseolus acutifolius]	113,361	41.8/6.1	21.3/5.1	101	4	
R10	Alcohol dehydrogenase [Dianthus caryophyllus]	33,149,683	41.9/6.6	38.1/6.4	79	3	
R18	Ferritin [Glycine max]	968,987	28.0/5.9	34.2/5.1	82	4	
Carbohydrate and energy metabolism							
18	Glyceraldehyde-3-phosphate dehydrogenases [Arabidopsis thaliana]	15,229,231	39.0/6.7	39.5/5.4	117	3	
11	ATP synthase bate subunit [Kerria japonica]	7,578,491	48.0/5.5	23.2/5.6	101	5	
R7	Malate dehydrogenase, mitochondrial [Medicago sativa]	32,328,905	48.6/6.5	18.7/6.9	97	3	
R34	Glyceraldehyde 3-phosphate dehydrogenases [Arabidopsis thaliana]	15,229,231	39.0/6.7	22.3/5.6	87	4	
R14	ATP synthase bate china 2 mitochondrial [Arabidopsis thaliana]	18,415,911	73.0/5.8	42.3/6.7	75	4	
R12	Nucleoside diphosphate kinase 1 [Pisum sativum]	134,667	16.4/5.9	95.0/6.5	69	3	
R24	Cytosolic phosphoglycerate kinase [Pisum sativum]	923,077	42.3/5.7	91.2/5.1	522	9	
Second metabolism							
1	Allene oxide cyclase 2 [Arabidopsis thaliana]	18,404,656	38.9/ 5.0	42.7/6.9	91	5	
17	Myo-inositol-3-phosphate synthase [Glycine max]	13,936,691	56.6/5.3	51.3/5.8	87	4	
R5	Isoliquiritigenin 2'-O-methyltransferase [Medicago sativa]	44,887,779	41.5/5.1	25.0/4.3	88	5	
R2	Glutamine synthetase [Arabidopsis thaliana]	28,393,681	44.0/5.15	18.5/4.5	121	5	
R11	3-Isopropylmalate dehydrogenase [Arabidopsis thaliana]	121,343	46.6/6.0	38.0/6.0	104	4	
R13	Chalcone reductase [Medicago sativa]	563,540	35.0/6.5	42.1/6.6	111	6	
R19	Isoflavone reductase [Medicago sativa]	19,620	35.5/5.3	38.5/4.7	129	5	
R33	Chalcone isomerase [Medicago sativa]	166,400	21.4/5.5	22.3/6.6	104	5	
Protein metabolism							

Table 2 Identification of salt-resposive proteins in alfalfa using MALDI-TOF-MS/MS *(Continued)*

Spot No.[a]	Homologous protein (plant species)[b]	gi Number[c]	Theo. Mr/pI[d]	Exp. Mr/pI[e]	Scores[f]	M.P[g]	Relative Vol% ± SE[h] CK 100 mM 200 mM
10	Small ribosomal protein 4 [Squamidium brasiliense]	37,992,679	21.9/10.0	23.3/5.8	79	3	
R25	Probable protein disulfide-isomerase A6 [Arabidopsis thaliana]	7,294,421	40.8/5.4	22.3/5.7	544	9	
R32	Ribosomal protein L32 [Medicago sativa]	71,534,997	15.7/10.9	30.2/6.8	69	5	
R29	Mitochondrial processing peptidase [Solanum tuberosum]	587,562	54.6/5.9	44.2/6.0	72	4	
R1	Heat shock protein 70 [Cucumis sativus]	1,143,427	75.3/5.1	12.0/4.4	75	4	
R30	Proteasome subunit alpha type-2-B [Arabidopsis thaliana]	15,219,317	25.7/5.5	46.2/6.4	78	4	
R28	Eukaryotic translation initiation factor 5A-2(eIF-5A-2) [Gossypium barbadense]	45,644,510	17.2/5.2	21.3/5.6	75	5	
Transcriptional regulation							
7	mRNA binding protein precursor [Solanum lycopersicum]	26,453,355	44.0/7.1	21.2/6.3	84	3	
15	Nucleic acid binding protein1 [Zea mays]	162,463,757	33.1/4.6	33.1/4.4	102	5	
5	Maturase [Kopsia fruticosa]	59,932,902	59.8/9.2	20.0/4.4	88	4	
24	Putative RNA binding protein [Mesembryanthemum crystallinum]	388,621	32.0/4.7	43.7/5.7	87	4	
R20	Putative polyprotein [Oryza sativa Japonica Group]	50,300,539	35.6/4.5	35.6/4.5	93	5	
R22	RNA-binding protein [Arabidopsis thaliana]	21,593,201	42.7/7.7	34.2/4.4	85	5	
R35	Nucleic acid binding protein1 [Zea mays]	162,463,757	33.1/4.6	48.8/6.0	89	4	
Cell wall and cytoskeleton metabolism							
21	Cell division cycle protein 48 [Arabidopsis thaliana]	1,841,493	90.6/5.0	33.0/4.5	81	5	
22	Cell division cycle protein 48 [Arabidopsis thaliana]	1,841,493	90.6/5.0	32.8/4.5	73	5	
R6	Actin-depolymerizing factor [Malus x domestica]	33,772,153	11.1/8.7	20.8/6.7	75	4	
R8	Profilin-2 [Lilium longiflorum]	14,423,862	14.3/4.6	31.2/6.8	226	8	
R31	Actin-depolymerizing factor [Malus domestica]	33,772,153	11.1/8.7	44.9/6.3	107	6	
Signaling							

Table 2 Identification of salt-resposive proteins in alfalfa using MALDI-TOF-MS/MS *(Continued)*

Spot No.[a]	Homologous protein (plant species)[b]	gi Number[c]	Theo. Mr/pI[d]	Exp. Mr/pI[e]	Scores[f]	M.P[g]	Relative Vol% ± SE[h] CK 100 mM 200 mM
R17	Annexin [*Lavatera thuringiaca*]	2,459,926	36.2/6.1	44.7/5.3	121	6	
R15	Vacuole-associated annexin VCaB42 [*Nicotiana tabacum*]	4,580,920	36.0/5.3	37.7/5.7	135	5	
Ion transport							
R3	Plasma membrane H$^+$- ATPase [*Hordeum vulgare subsp. vulgare*]	15,149,829	70.9/8.6	16.4/5.2	91	4	
Unknown							
9	Unknown protein [*Medicago truncatula*]	217,071,692	55.5/5.5	21.8/6.1	87	4	
14	Unknown protein [*Arabidopsis thaliana*]	22,329,503	54.9/5.5	34.2/4.8	145	5	

[a]The number of identification spot
[b]The homologous protein and plant species
[c]The number NCBInr databases
[d]Theoretical mass (Mr, kDa) and pI of identified proteins. Theoretical values were retrieved from the protein database
[e]Experimental mass (Mr, kDa) and pI of identified proteins. Experimental values were calculated with Image master software (GE Healthcare) and standard molecular mass markers
[f]The mascot scores
[g]Number of matched petides
[h]The Relative Vol% of spot

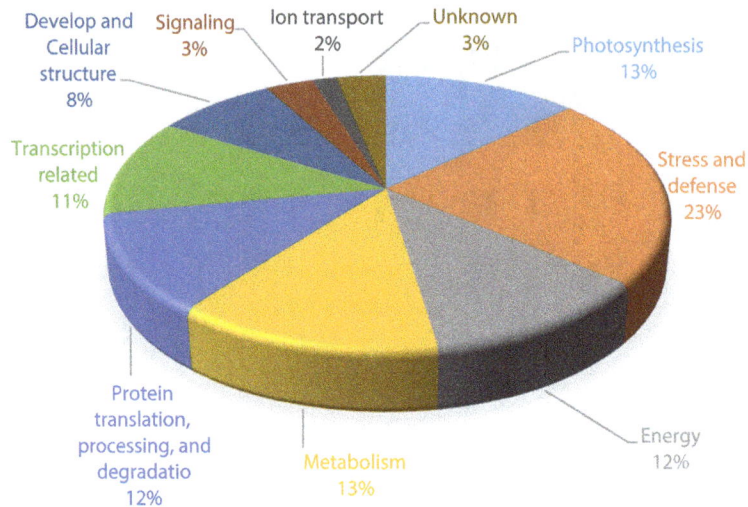

Fig. 4 Functional classification of differentially abundant proteins identified in the seedling shoot and root of alfalfa under salt stress. The pie chart shows the distribution of the salt-responsive proteins into their functional classes in percentage

included 4 ascorbate peroxidases (spots 23, R16, R21, R23), 2 glutathione peroxidase (spots 25, R4), 1 ferritin protein (spot R18), and 1 quinone reductase family protein (spot 19). These proteins are major ROS-scavenging proteins, providing plant cells with highly efficient machinery for detoxifying H_2O_2 and the other ROS. However, all of these identified proteins had more distribution in root than in shoots. Our proteomics results might indicate that alfalfa seedling would increase ROS-scavenging proteins in response to salt stress and root may have stronger ROS-scavenging capability than shoot. REL and MAD are important indicators of membrane damage caused by ROS stress. In this study, the H_2O_2, REL and MAD contents were similar for all the NaCl concentrations, and all three were higher in the shoots than in the roots. A possible reason is that roots

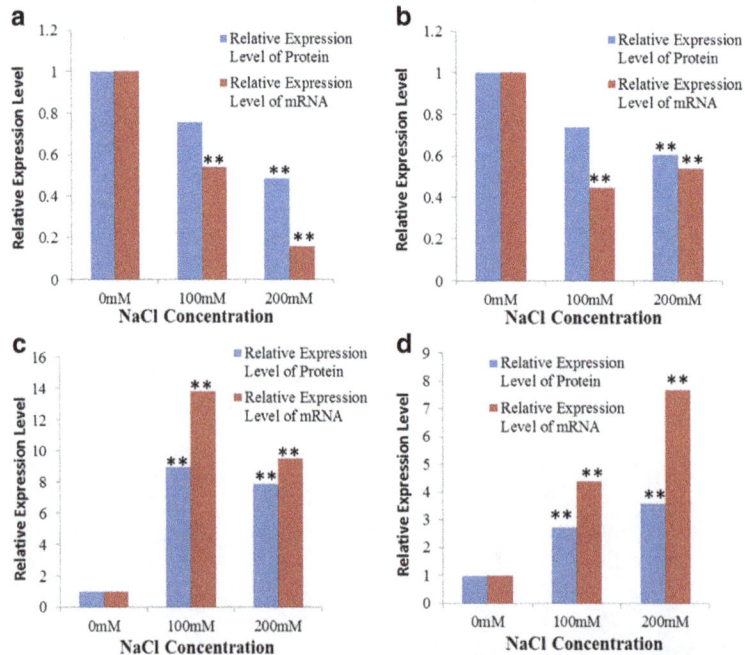

Fig. 5 Transcript abundances of mRNAs encoding four novel salt-responsive proteins were analyzed following salt stress treatment. The mRNA levels were compared with the 2-DE data. Significant differences were determined relative to each treatment using a student's t-test [*P*-values <0.05 (*) and <0.01 (**)]. **a** CP12 (gi|3,123,345). **b** Harvest-induced protein (HI,gi|283,831,548). **c** Pathogenesis-related protein 2 (PR2,gi|44,887,779). **d** Isoliquiritigenin 2′-O-methyltransferase(IOMT, gi|22,266,001)

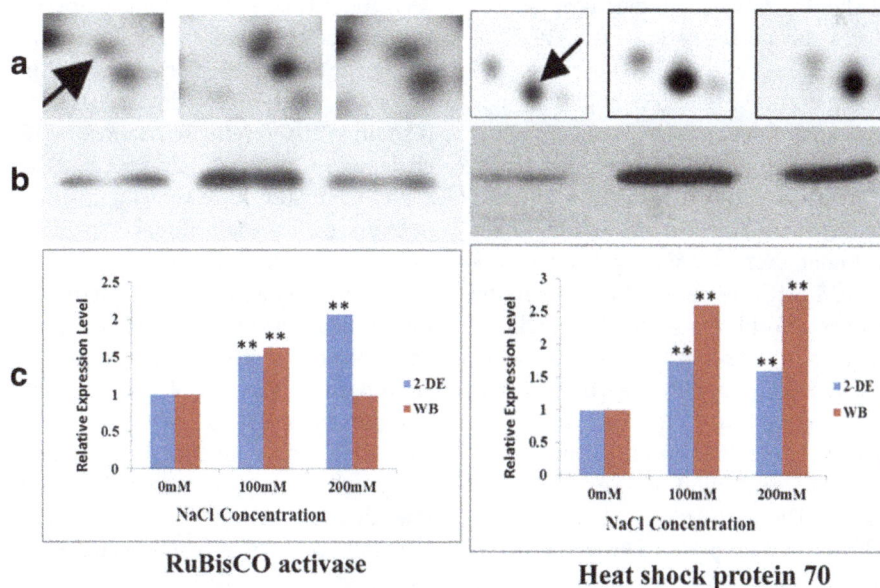

Fig. 6 Western blot and 2-DE analysis of RuBisCO activase and HSP70 abundant patterns and the relative adundance level in alfalfa. **a** Antibodies against RuBisCO activase and HSP70 were used to detect the change of protein levels in alfalfa in response to salt stress treatment of the plants; **b** Image of RuBisCO activase and HSP70 spots on 2-DE gels; **c** Gray analysis of the relative adundance of RuBisCO activase and HSP70 (spots 8 and R1) were compared with the 2-DE data. Significant differences were determined relative to each treatment using a student's t-test [P-values <0.05 (*) and <0.01 (**)]

may have a better antioxidative defense system than the shoots. To validate the hypothesis, four enzymes involved in ROS scavenging were selected for activity analysis. It is important to note that the roots had higher SOD, APX and POD activities, and the shoots had higher CAT activities. CAT is known to have a lower affinity to H_2O_2 than POD (mM and μM range, respectively), and lower CAT activities were correlated to salt tolerance simply because large increases in CAT activity was not essential as long as POD and APX imposed tight controls on the H_2O_2 concentration [30]. This suggests that different mechanisms control the response to ROS.

In addition to the redox related proteins, plants have developed cross-tolerance mechanisms to be able to cope with different stresses. Some biotic stress-related proteins were induced under salt stress conditions, such as Pathogenesis-related protein 5(PR5, spot R9), Pathogenesis-related protein 2(PR2, spots 20, R27). PR2 is encoded by β-1,3-glucanase gene, and plant β-1, 3-glucanases are induced not only by pathogen infection, but also by other factors. Stress factors like wounding, drought, exposure to heavy metals, air pollutant ozone, and ultraviolet radiation can stimulate synthesis of β-1,3-glucanases in some plants [31]. However, there was few reported about it induced by salt stress. Previously, PR5 was gradually increased in abundance with increasing concentrations of NaCl in *Arabidopsis*, but the change was the opposite in *Thellungiella* [21]. PR proteins have been found to be induced in several plant

species when they are infected by viruses, viroids, fungi or bacteria. In our study, the PR protein was induced by salt stress, which suggested that it had a special role in plant adaptation to salt stress, but whether it can be used as a potential salt stress marker in alfalfa needs further research. Moreover, some abiotic stress-related proteins, such as alcohol dehydrogenase (ADH, spots R10, R26) and harvest-induced protein (HI, spot 13) also respond to salt stress. ADH enzymes were traditionally of interest because of their activity during oxygen deprivation [32]. However, more recently, ADH gene expression and ADH activity have been shown to be affected by a number of other stresses [33, 34]. ADH1 has been found to be up-regulated in *Porteresia coarctata* under high salinity and this study also suggested that ADH1 was up-regulated when alfalfa was subjected to salt stress. HI proteins are involved in defense responses and the response to biotic stimulus, but their molecular details are poorly understood.

The main energy metabolism associated proteins were down regulated

Salt-stress led to a reduction in photosynthesis, and thus to decreased carbohydrate synthesis. It also inhibited energy production. Energy production declined as the NaCl concentration increased. There were three proteins involved in glycolysis and the citrate cycle decreased after NaCl treatment.

The abundance of two glyceraldehyde-3-phosphate dehydrogenases (GAPDH, spots 18, R34) were altered in both the roots and shoots. Previous studies have shown that salt stress induces the up-regulation of GAPDH in rice leaves, OSRK1 transgenic rice roots, sugarcane, and in *Arabidopsis thaliana* roots [15, 35, 36]. The up-regulation of GAPDH may increase soluble sugars accumulation and provide more energy for plants under stress. It is therefore an indicator of stress tolerance. In this study, GAPDH was down-regulated in the shoots under 200 mM NaCl treatment, but was up-regulated in the roots under the 100 mM NaCl treatment. In addition, the abundance of a cytosolic phosphoglycerate kinase (cPGK, spot R24) was also altered in the roots. PGK is the seventh enzyme in the cycle that catalyzes the reaction of 1,3-biphosphoglycerate and ADP to produce 3-phosphoglycerate and ATP. GAPDH and PGK are crucial enzymes in the glycolysis cycle and showed the same abundance trends as GAPDH in the roots, which presumably reflects altered carbon flux patterns in response to the increased need for osmotic adjustment in the roots. Furthermore, a malate dehydrogenase mitochondrial (miMD, spot R7) was down-regulated under salt-stress in the roots. Overexpression of malate dehydrogenase in transgenic alfalfa enhances organic acid synthesis and confers tolerance to aluminum [37]. Malate dehydrogenase (cytoplasmic) was up-regulated under NaCl stress in cucumber roots and young rice panicles [38, 39], whereas in our study, malate dehydrogenase was down-regulated under salt stress. A possible reason is that the MD cellular localization was different in each species.

Up abundance of the ATP β synthase subunit was observed in both the roots and shoots under salt stress (spots 11, R14). ATP synthase includes two regions: an F0 region and F1 region consisting of α, β, γ, δ, and ε subunits. ATP synthase β subunit induction by salt stress has been reported in plants [12, 40, 41] and those studies show a positive correlation between the abundance of ATP synthase and a plant's ability to resist salt stress. A nucleoside diphosphate kinase 1 (NDPK1, spot R12) was down-regulated under salt stress NDPKs are housekeeping enzymes, and their main function is to transfer a γ-phosphate from ATP to a cognate nucleoside diphosphate, thereby balancing the nucleoside pool. NDPK has been reported in response to drought [42, 43], cold [44], high temperature [45], and salt stresses [22, 38]. A recent study showed that NDPK 2 was involved in salt stress and H_2O_2 signaling in *Arabidopsis thaliana* [46].

Salt-stress induced some secondary metabolism proteins

Secondary metabolism is a unique plant characteristic, is critical growth and development, and also allows plants to adapt to changing environments. Plant cells produce a vast number of secondary products, and some compounds are restricted to single species.

Flavonoids are ubiquitous plant secondary products that are best known as the characteristic red, blue, and purple anthocyanin pigments seen in plant tissues. In our study, isoflavone reductase (IFR, spot R19), isoliquiritigenin 2′-O-methyltransferase (IOMT, spot R5), chalcone reductase (CHR, spot R13), and chalcone isomerase (CHI, spot R33) showed up-regulated by salt stress in the seedling roots under 100 mM. Previously, it has been reported that flavonoids act as attractants to pollinators and symbionts, as sunscreens to protect against UV irradiation, as allelochemicals, as antimicrobial and antiherbivory factors, and are involved in resistance to aluminum toxicity [47, 48]. It is noteworthy that the key enzyme involved in flavonoid metabolite production showed an up-regulation under moderate NaCl treatment, which suggested that flavonoids also respond to the salt stress.

It well known that plants accumulate compatible osmolytes and osmoprotectants that help them to resist salt and drought stress. A L-myo-inositol 1-phosphate synthase (MIPS, spot 17) was up-regulated by salt stress in the shoots. The structure of this protein has been well-studied and was found to be inherently salt-tolerant [49]. Previous studies have suggested that salt stress induced the accumulation of MIPS in *Mesembryanthemum crystallinum* and that it was slightly upregulated in *P. coarctata* [27, 50].

It has been reported that an number of amino acids increase in alfalfa following NaCl treatment [51]. Glutamine synthetase 58 (GS58, spot R2) was up-regulated by both the salt stress treatments. GS catalyzes the ATP-dependent condensation of ammonium with glutamate to yield glutamine, which then provides nitrogen groups for the biosynthesis of all nitrogenous compounds in the plant [52]. Because glutamate is a precursor of proline, GS activation may contribute to proline synthesis under salt stress [53]. Previous reports indicated that GS was up-regulated under salt stress in rice and *Arabidopsis* roots [15, 54, 55]. A 3-isopropylmalate dehydrogenase (IMD, spot R11), which is involved in Leu biosynthesis, and cobalamine-independent methionine synthase, decreased in abundance following 100 mM NaCl treatment, but were up-regulated following 200 mM NaCl treatment. In *Arabidopsis* roots, IMD decreased in abundance following NaCl treatment [15], and the abundances of two IMDs were also influenced by NaCl in *Oryza sativa* roots [12].

This study also revealed that many protein related hormones were synthesized in response to salt treatment in alfalfa. An allene oxide cyclase 2(AOC2, spot 1) protein was up-regulated under salt stress. AOC catalyzes the stereospecific cyclization of an unstable allene oxide to 9(S),13(S)-12-oxo-10,15(Z)-phytodienoic acid, the precursor of jasmonic acid (JA) [56]. JA is involved in a wide range of stress, defense, and developmental processes [57].Transgenic plants expressing a tomato allene oxide cyclase (AOC) also displayed enhanced salt tolerance [58]. Up-regulation of AOC2 protein has been previously reported in *Arabidopsis* under salt-stress [15, 21].Our results provide additional evidence that AOC improves plants survival under salt stress.

Salt stress induced protein metabolism

Several proteins, involved in protein translation, processing and degradation, were identified. In our study, ribosomal proteins S4(RP S4, spot 10) was down-regulated under 200 mM NaCl in the shoots, while, RP L32 (spot R32) was up-regulated under salt stress in root. Ribosomes are essential ribonucleoprotein complexes that are engaged in translation and thus play an important role in metabolism, cell division, and growth. The levels of some of the ribosomal proteins decreased while some specific ribosomal components increased under salt stress were also reported on *Arabidopsis* [59] and *Gossypium hirsutum* [60]. Moreover, our data showed a eukaryotic translation initiation factor, 5A-2(eIF 5A-2, spot R28), was up-regulated under 100 mM NaCl, but down-regulated under 200 mM NaCl. EIF 5A-2 is part of the start site selection for the eIF2-GTP-tRNAi ternary complex within the ribosomal-bound preinitiation complex, and also stabilizes the binding of GDP to eIF2. Alter abundance of eIF5A protein has also been reported in rice leaf lamina and *SnRK2* transgenic rice under salt stress [26, 55]. Other eukaryotic translation initiation factor, such as eIF3I, were also found down-regualted under salt-stress in *Arabidopsis* roots and *Gossypium hirsutum* roots [15, 60]. All of these studies suggest that complicated regulation mechanisms may govern protein synthesis in order to help plants cope with salt stress.

Several proteins that promote the proper folding of proteins and/or prevent the aggregation of nascent or damaged proteins were detected. A protein, disulfide-isomerase A6 (PDI A6, spot R25), was up-regulated under 100 mM NaCl treatment but down-regulated under 200 mM NaCl treatment. A major function of PDI is as a chaperone, where it helps wrongly folded proteins to reach a correctly folded state without the aid of enzymatic disulfide shuffling [61]. Moreover, increased abundance of PDI protein has also been reported in rice roots [62] and *Gossypium hirsutum* roots [60]. A heat shock protein, 70 (HSP70, spot R1), was up-regulated under NaCl stress. HSPs are grouped into five families: HSP100s, HSP90s, HSP70s, HSP60s, and sHSPs (small HSPs), and may prevent misfolding and promote the refolding and proper assembly of the unfolded polypeptides generated. Experiments in which

chrysanthemum HSP70 gene was overexpressed in *Arabidopsis thaliana* showed that an increase in HSP70 abundance led to a remarkable tolerance to heat, drought and salt [63]. In our study, HSP70 was up-regulated by exposure to high salinity, which suggested that the proteins play a crucial role in aiding the folding and assembly of proteins under salt stress in alfalfa seedlings. These results are similar to the results reported for *Kandelia candel*, *Saccharum spp.*, *Brachypodium distachyon*, and *Oryza sativa* under salt stress [12, 20, 40, 64]. A proteasome subunit alpha type-2-B (spot R30), which is involved in protein degradation, accumulated under salt stress in the roots. The proteasome is a very large protein complex (26S) containing a 20S core particle, and is a multicatalytic protease that degrades proteins using an ATP-dependent mechanism by which cells regulate the concentration of particular proteins and degrade misfolded proteins [65]. The degradation process yields peptides that are about seven to eight amino acids long, which can then be further degraded into shorter amino acid sequences that can be used to synthesize new proteins [66]. In *Brachypodium distachyon*, a proteasome subunit was down-regulated in the salt-treated group but up-regulated in the recovery group, which suggested that it was mainly involved in abnormal condition recovery rather than in the defense against salt stress [64]. In our study, the refold-associated proteins were up-regulated, which suggested that alfalfa handles misfolded proteins mainly through refolding. One possible reason is that energy production is depressed under salt stress and degradation is an energy-consuming process.

Transcriptional and translational control is a part alfalfa's response to salt stress

Under salt stress, many response and defense-related genes are stimulated by upstream transcription regulatory factors, but the genes involved in normal plant growth and development are inhibited [64]. Gene expression regulation is achieved at several levels, i.e. transcriptional, post-transcriptional, translational, and post-translational levels. For example, a maturase protein (spot 5) involved in post-transcription was down-regulated under 200 mM NaCl in the shoots. In vivo, most plant group II introns do not self-splice, but require the assistance of proteinaceous splicing factors, known as maturases [67, 68]. Maturase genes can be found in fungal and plant mitochondria, as well as in plant chloroplasts, and the down-regulation of this protein may be related to the translation of related genes. Two nucleic acid binding proteins (NABP1, spots 15, R35) showed altered abundance in both the shoots and roots under salt stress. NABP is a small and highly conserved protein with nucleic acid chaperone activity that binds single-stranded nucleic acids [69]. One group of

NABPs is the cold shock domain (CSD) containing proteins, and these CSDPs are involved in various cellular processes, including adaptation to low temperatures, cellular growth, nutrient stress, and the stationary phase [70, 71]. RNA-binding proteins (RBPs) have crucial roles in various cellular processes, such as cellular function, transport, and localization. They also play a major role in post- transcriptional control of RNAs, such as splicing, polyadenylation, mRNA stabilization, mRNA localization, and translation. In this study, three RNA binding proteins (spots 7, 24, R22) were identified.

Salt stress depressed the abundance of proteins involved in cellular processes

Salt stress decreased the growth of alfalfa, and several proteins associated with the dynamic changes of cellular processes were found in the current study. The actin cytoskeleton plays a critical role in plant development by regulating a number of fundamental cellular processes, including cell division, cell expansion, organelle motility, and vesicle trafficking [72, 73]. The dynamic reorganization of actin is modulated by the specific activity of a number of actin binding proteins (ABPs) that either promote or inhibit actin polymerization. Actin-depolymerizing factor (ADF) is one of the most highly expressed ABPs in plants that modulate the dynamic organization of the actin cytoskeleton by promoting filamentous actin disassembly [74]. ADFs were induced by salt stress, drought, and cold in cereal plants [43, 54, 75], which suggested that ADFs might be required for osmoregulation under osmotic stress. According to Yan [4], increased ADF levels under salt stress may result in depolymerization of actin filaments and enhanced K^+ influx through the inward rectification of potassium channels, which helps to restore ion homeostasis. In this study, two spots, down-regulated under 200 mM salt stress in the roots, were identified as ADF (spots R6, R31). Another important ABP is profilin(spot R8), which was also first up and then down-regulated as the salt concentrations rose in the roots. These differences may be due to the need for growth under salt stress.

Two cell division cycle proteins (CDC48, spots 21, 22) were up-regulated under 100 mM NaCl. CDC48 belongs to the ATPases associated with proteins and has many cellular activities. They are believed to function as chaperones and to regulate cell-cycle progression, membrane fusion, and the destruction misfolded secretory proteins [76, 77]. Recent studies have shown that virus movement is impaired by the overexpression of CDC48, suggesting that CDC48 controls virus movement by removal of movement proteins from the endoplasmic reticulum-transport pathway and by interfering with protein movement using microtubule dynamics [78].

Signal transduction and ion transport

Salt stress is first perceived by putative sensors in the root cell membranes and these signals are transmitted to the cellular machinery to regulate gene expression and changes in cellular metabolism designed to prevent or minimize the deleterious effects of abiotic stress. This signaling is mediated by different kinds of secondary messengers, such as Ca^{2+}. Salinity induced increases in cytoplasmic free calcium ($[Ca^{2+}]_{cyt}$) and fluctuations in $[Ca^{2+}]_{cyt}$ provide a means for relatively rapid responses and may lead to specific changes in gene expression programs [79]. Two annexin proteins (spots R15, R17) were identified in our study. Annexins are a multigene, multifunctional family of Ca^{2+}-dependent membrane-binding proteins found in both animal and plant cells, and certain annexins may be targets of $[Ca^{2+}]_{cyt}$ fluctuations [80]. In *Arabidopsis thaliana*, annexins have been shown to mediate osmotic stress and abscisic acid signal transduction [81]. In alfalfa, annexin is up-regulated in response to osmotic stress, abscisic acid (ABA), and drought [82].

A plasma membrane H^+-ATPase (PM H^+-ATPase, spot R3) was also up-regulated under salt stress. A response to the accumulation of Na^+ ions in the cytosol is their compartmentalization within the vacuole, while another response is to extrude them from the cell. In each case, the active Na^+ efflux requires a H^+ gradient across the vacuolar membrane enerated by stimulating protein expression of the vacuolar H^+-ATPase [83].The accumulation of PM H^+-ATPase gene mRNA was induced by NaCl and this has been found to occur in glycophytes and halophytes. In rice, a salt-tolerant mutant highly expressed PM H^+-ATPase in its roots, compared to the non-mutant variety [84]. Therefore, the increased abundanceof plant plasma membrane H^+-ATPase many play an important role in salt stress tolerance in alfalfa.

Conclusion

In summary, we found significant physiological and protein abundance differences during salt treatment in alfalfa. Quantitative analysis of more than 850 spots on 2D gels showed significant variations in of 36 protein spots from the roots and 27 protein spots from the shoots, which were confidently identified by MS/MS. These DAPs are mainly involved in the biological processes of photosynthesis, stress and defense, carbohydrate and energy metabolism, second metabolism, protein metabolism, transcriptional regulation, cell wall and cytoskeleton metabolism, membrane and transport, signal transduction. The diverse array of proteins affected by salt stress conditions indicates that there is a remarkable flexibility in alfalfa roots and shoots metabolism, which may contribute to its survival in salinity conditions. Further analysis demonstrated that the primary mechanisms underlying the ability of alfalfa seedlings to tolerate salt stress were photosynthesis, detoxifying and antioxidant, secondary metabolism, and ion transport. In parallel, physiological data, including REL, MAD and H_2O_2 contents, and the activities of antioxidant enzymes all correlated well with our proteomic results. It is worth emphasizing that some novel salt-responsive proteins were identified, such as CP12, pathogenesis-related family proteins, harvest-induced protein, isoflavone reductase, isoliquiritigenin 2′-O-methyltransferase, chalcone reductase, chalcone isomerase. qRT-PCR was used to study the gene expression levels of thefour above-mentioned proteins; four patterns are consistent with those of induced protein. These novel proteins provide a good starting point for further research into their functions using genetic or other approaches. These findings significantly improve the understanding of the molecular mechanisms involved in the tolerance of plants to salt stress.

Abbreviations

2-DE: Two-dimensional gel electrophoresis; ABP: Actin binding proteins; ADF: Actin-depolymerizing factor; ADH: Alcohol dehydrogenase; AOC: Allene oxide cyclase; APX: Ascorbic acid peroxidase; CAB: Chlorophyll a/b binding protein; cAPX: cytosolic ascorbate peroxidase; CAT: Catalase; CDC: Cell division cycle proteins; CHI: Chalcone isomerase; CHR: Chalcone reductase; cPGK: cytosolic phosphoglycerate kinase; ctbAPX: chloroplast thylakoid-bound ascorbate peroxidase; Cyt b6/f: Cytochrome b6-f complex iron-sulfur; DAPs: Differentially abundant proteins; elF: Eukaryotic translation initiation factor; GAPDH: Glyceraldehyde-3-phosphate dehydrogenases; GO: Gene ontology; GPX: Glutathione peroxidase; GS: Glutamine synthetase; HI: Harvest-induced protein; HSP: Heat shock protein; IFR: Isoflavone reductase; IMD: 3-Isopropylmalate dehydrogenase; IOMT: Isoliquiritigenin 2′-O-methyltransferase; MD: Malate dehydrogenase; MDA: Methane Dicarboxylic Aldehyde; MIPS: L-myo-inositol 1-phosphate synthase; NABP: Nucleic acid binding proteins; NDPK1: Nucleoside diphosphate kinase 1; OEE1: Chloroplast oxygen-evolving enhancer protein 1; PDI: Protein disulfide-isomerase; POD: Peroxidase; PR: Pathogenesis-related; RBP: RNA-binding proteins; REL: Electrolyte leakage analyses; ROS: Reactive oxygen species; RP: Ribosomal proteins; SOD: Superoxide dismutase

Acknowledgements

The authors thank Dr. Tiejun Zhang and Mrs. Junmei Kang of Chinese Academy of Agricultural Science for their suggestions and critical comments.

Funding

This work was supported by the National Natural Science Foundation of China (Grant no. 31402130, 31,401,903), Modern Agro-industry Technology Research System (Grant no. CARS-35-34), and Hubei Key Laboratory of Animal Embryo Engineering and Molecular Breeding (Grant no.2012ZD201).

Authors' contributions

JX carried out sample preparation, 2DE and bioinformatics analysis. YL carried out Na/K content, REL, MAD and H_2O_2 analysis. QY, HC, HT helped the research. YS conceived, designed, implemented and coordinated the study, and also carried out physiological experiments. All authors read and approved the final manuscript.

Competing interests

The authors declare that they have no competing interests.

Author details

[1]Hubei Key Laboratory of Animal Embryo and Molecular Breeding, Institute of Animal and Veterinary Science, Hubei Academy of Agricultural Science, Yaoyuan 1, Hongshan, Wuhan, Hubei 430017, China. [2]Institute of Grassland Science, China Agricultural University, 2 West Road, Yuan Ming Yuan, Beijing 100193, China. [3]Institute of Animal Science, Chinese Academy of Agricultural Science, West Road 2, Yuan Ming Yuan, Beijing 100193, China.

References

1. Tavakkoli E, Rengasamy P, GK MD. High concentrations of Na⁺ and Cl⁻ions in soil solution have simultaneous detrimental effects on growth of faba bean under salinity stress. J Exp Bot. 2010;61(15):4449–59.
2. Munns R, Tester M. Mechanisms of salinity tolerance. Annu Rev Plant Biol. 2008;59:651–81.
3. Maas E, Hoffman G. Crop salt tolerance: evaluation of existing data. In: Managing saline water for irrigation proceedings of the international salinity conference Ed HE Dregne; 1977. p. 187–98.
4. Sanden B, Sheesley B. Salinity tolerance and management for alfalfa; 2007. p. 2–4.
5. Limami AM, Ricoult C, Planchet E, González EM, Ladrera R, Larrainzar E, Arrese-Igor C, Merchan F, Crespi M, Frugier F. Response of Medicago truncatula to abiotic stress. The Medicago truncatula Handbook (http://www.noble.org/MedicagoHandbook). 2006.
6. Postnikova OA, Shao J, Nemchinov LG. Analysis of the alfalfa root transcriptome in response to salinity stress. Plant Cell Physiol. 2013; 54(7):1041–55.
7. Jin H, Sun Y, Yang Q, Chao Y, Kang J, Jin H, Li Y, Margaret G. Screening of genes induced by salt stress from alfalfa. Mol Biol Rep. 2010;37(2):745–53.
8. Zhang H, Han B, Wang T, Chen S, Li H, Zhang Y, Dai S. Mechanisms of plant salt response: insights from proteomics. J Proteome Res. 2012;11(1):49–67.
9. Sobhanian H, Razavizadeh R, Nanjo Y, Ehsanpour AA, Jazii FR, Motamed N, Komatsu S. Proteome analysis of soybean leaves, hypocotyls and roots under salt stress. Proteome Sci. 2010;8(1):19.
10. Aghaei K, Ehsanpour A, Shah A, Komatsu S. Proteome analysis of soybean hypocotyl and root under salt stress. Amino Acids. 2009;36(1):91–8.
11. Xu C, Sibicky T, Huang B. Protein profile analysis of salt-responsive proteins in leaves and roots in two cultivars of creeping bentgrass differing in salinity tolerance. Plant Cell Rep. 2010;29(6):595–615.
12. Liu CW, Chang TS, Hsu YK, Wang AZ, Yen HC, Wu YP, Wang CS, Lai CC. Comparative proteomic analysis of early salt stress-responsive proteins in roots and leaves of rice. Proteomics. 2014;14(15):1759–75.
13. Patterson BD, MacRae EA, Ferguson IB. Estimation of hydrogen peroxide in plant extracts using titanium (IV). Anal Biochem. 1984;139(2):487–92.
14. Hodges DM, DeLong JM, Forney CF, Prange RK. Improving the thiobarbituric acid-reactive-substances assay for estimating lipid peroxidation in plant tissues containing anthocyanin and other interfering compounds. Planta. 1999;207(4):604–11.
15. Jiang Y, Yang B, Harris NS, Deyholos MK. Comparative proteomic analysis of NaCl stress-responsive proteins in Arabidopsis roots. J Exp Bot. 2007;58(13):3591–607.
16. Babakhani B, Khavari-Nejad R A, Sajedi R H, et al. Biochemical responses of Alfalfa (Medicago sativa L.) cultivars subjected to NaCl salinity stress[J]. African J Biotechnol. 2011;10(55):11433–41.
17. Xiong J, Yang Q, Kang J, Sun Y, Zhang T, Margaret G, Ding W. Simultaneous isolation of DNA, RNA, and protein from Medicago Truncatula L. Electrophoresis. 2011;32(2):321–30.
18. Livak KJ, Schmittgen TD. Analysis of relative gene expression data using real-time quantitative PCR and the 2(−Delta Delta C(T)) method. Methods. 2001;25(4):402–8.
19. Chaves M, Flexas J, Pinheiro C. Photosynthesis under drought and salt stress: regulation mechanisms from whole plant to cell. Ann Bot. 2009;103(4):551–60.
20. Wang ZQ, XY X, Gong QQ, Xie C, Fan W, Yang JL, Lin QS, Zheng SJ. Root proteome of rice studied by iTRAQ provides integrated insight into aluminum stress tolerance mechanisms in plants. J Proteome. 2014;98:189–205.
21. Pang Q, Chen S, Dai S, Chen Y, Wang Y, Yan X. Comparative proteomics of salt tolerance in Arabidopsis Thaliana and Thellungiella Halophila. J Proteome Res. 2010;9(5):2584–99.
22. Xu C, Huang B. Differential proteomic responses to water stress induced by PEG in two creeping bentgrass cultivars differing in stress tolerance. J Plant Physiol. 2010;167(17):1477–85.

23. Cohen I, Knopf JA, Irihimovitch V, Shapira M. A proposed mechanism for the inhibitory effects of oxidative stress on Rubisco assembly and its subunit expression. Plant Physiol. 2005;137(2):738–46.

24. Jordan DB, Chollet R, Ogren WL. Binding of phosphorylated effectors by active and inactive forms of ribulose 1, 5-bisphosphate carboxylase. Biochemistry. 1983;22(14):3410–8.

25. Somerville C, Portis AR, Ogren WL. A mutant of Arabidopsis Thaliana which lacks activation of RuBP carboxylase in vivo. Plant Physiol. 1982;70(2):381–7.

26. Parker R, Flowers TJ, Moore AL, Harpham NV. An accurate and reproducible method for proteome profiling of the effects of salt stress in the rice leaf lamina. J Exp Bot. 2006;57(5):1109–18.

27. Sengupta S, Majumder AL. Insight into the salt tolerance factors of a wild halophytic rice, Porteresia coarctata: a physiological and proteomic approach. Planta. 2009;229(4):911–29.

28. Witzel K, Weidner A, Surabhi GK, Varshney RK, Kunze G, Buck-Sorlin GH, Börner A, Mock HP. Comparative analysis of the grain proteome fraction in barley genotypes with contrasting salinity tolerance during germination. Plant Cell Environ. 2010;33(2):211–22.

29. Tamoi M, Miyazaki T, Fukamizo T. The Calvin cycle in cyanobacteria is regulated by CP12 via the NAD(H)/NADP(H) ratio under light/dark conditions. Plant J. 2005;42(4):504–13.

30. Abogadallah GM. Antioxidative defense under salt stress. Plant Signal Behav. 2010;5(4):369–74.

31. Stintzi A, Heitz T, Prasad V, Wiedemann-Merdinoglu S, Kauffmann S, Geoffroy P, Legrand M, Fritig B. Plant 'pathogenesis-related' proteins and their role in defense against pathogens. Biochimie. 1993;75(8):687–706.

32. Hageman R, Flesher D. The effect of an anaerobic environment on the activity of alcohol dehydrogenase and other enzymes of corn seedlings. Arch Biochem Biophys. 1960;87(2):203–9.

33. Christie PJ, Hahn M, Walbot V. Low-temperature accumulation of alcohol dehydrogenase-1 mRNA and protein activity in maize and rice seedlings. Plant Physiol. 1991;95(3):699–706.

34. Matton DP, Constabel P, Brisson N. Alcohol dehydrogenase gene expression in potato following elicitor and stress treatment. Plant Mol Biol. 1990;14(5):775–83.

35. Nam MH, Huh SM, Kim KM, Park WJ, Seo JB, Cho K, Kim DY, Kim BG, Yoon IS. Comparative proteomic analysis of early salt stress-responsive proteins in roots of SnRK2 transgenic rice. Proteome Sci. 2012;10:25.

36. Ruan SL, Ma HS, Wang SH, YP F, Xin Y, Liu WZ, Wang F, Tong JX, Wang SZ, Chen HZ. Proteomic identification of OsCYP2, a rice cyclophilin that confers salt tolerance in rice (Oryza Sativa L.) seedlings when overexpressed. BMC Plant Biol. 2011;11:34.

37. Tesfaye M, Temple SJ, Allan DL, Vance CP, Samac DA. Overexpression of malate dehydrogenase in transgenic alfalfa enhances organic acid synthesis and confers tolerance to aluminum. Plant Physiol. 2001;127(4):1836–44.

38. Dooki AD, Mayer-Posner FJ, Askari H, Aa Z, Salekdeh GH. Proteomic responses of rice young panicles to salinity. Proteomics. 2006;6(24):6498–507.

39. C-X D, Fan H-F, Guo S-R, Tezuka T, Li J. Proteomic analysis of cucumber seedling roots subjected to salt stress. Phytochemistry. 2010;71(13):1450–9.

40. Murad AM, Molinari HB, Magalhaes BS, Franco AC, Takahashi FS, de Oliveira NG, Franco OL, Quirino BF. Physiological and proteomic analyses of Saccharum spp. grown under salt stress. PLoS One. 2014;9(6):e98463.

41. Li W, Zhang C, Lu Q, Wen X, Lu C. The combined effect of salt stress and heat shock on proteome profiling in Suaeda Salsa. J Plant Physiol. 2011;168(15):1743–52.

42. Hajheidari M, Abdollahian-Noghabi M, Askari H, Heidari M, Sadeghian SY, Ober ES, Hosseini Salekdeh G. Proteome analysis of sugar beet leaves under drought stress. Proteomics. 2005;5(4):950–60.

43. Salekdeh G, Siopongco J, Wade LJ, Ghareyazie B, Bennett J. Proteomic analysis of rice leaves during drought stress and recovery. Proteomics. 2002;2(9):1131–45.

44. Imin N, Kerim T, Rolfe BG, Weinman JJ. Effect of early cold stress on the maturation of rice anthers. Proteomics. 2004;4(7):1873–82.

45. Xu C, Huang B. Root proteomic responses to heat stress in two Agrostis grass species contrasting in heat tolerance. J Exp Bot. 2008;59(15):4183–94.

46. Verslues PE, Batelli G, Grillo S, Agius F, Kim Y-S, Zhu J, Agarwal M, Katiyar-Agarwal S, Zhu J-K. Interaction of SOS2 with nucleoside diphosphate kinase 2 and catalases reveals a point of connection between salt stress and H2O2 signaling in Arabidopsis Thaliana. Mol Cell Biol. 2007;27(22):7771–80.

47. Kidd PS, Llugany M, Poschenrieder C, Gunse B, Barcelo J. The role of root exudates in aluminium resistance and silicon-induced amelioration of aluminium toxicity in three varieties of maize (Zea Mays L.). J Exp Bot. 2001;52(359):1339–52.

48. Winkel-Shirley B. Biosynthesis of flavonoids and effects of stress. Curr Opin Plant Biol. 2002;5(3):218–23.

49. Majee M, Maitra S, Dastidar KG, Pattnaik S, Chatterjee A, Hait NC, Das KP, Majumder AL. A novel salt-tolerant L-myo-inositol-1-phosphate synthase from Porteresia coarctata (Roxb.) Tateoka, a halophytic wild rice molecular cloning, bacterial overexpression, characterization, and functional introgression into tobacco-conferring salt tolerance phenotype. J Biol Chem. 2004;279(27):28539–52.

50. Ishitani M, Majumder AL, Bornhouser A, Michalowski CB, Jensen RG, Bohnert HJ. Coordinate transcriptional induction of myo-inositol metabolism during environmental stress. Plant J. 1996;9(4):537–48.

51. Fougère F, Le Rudulier D, Streeter JG. Effects of salt stress on amino acid, organic acid, and carbohydrate composition of roots, bacteroids, and cytosol of alfalfa (Medicago Sativa L.). Plant Physiol. 1991;96(4):1228–36.

52. Skopelitis DS, Paranychianakis NV, Paschalidis KA, Pliakonis ED, Delis ID, Yakoumakis DI, Kouvarakis A, Papadakis AK, Stephanou EG, Roubelakis-Angelakis KA. Abiotic stress generates ROS that signal expression of anionic glutamate dehydrogenases to form glutamate for proline synthesis in tobacco and grapevine. Plant Cell. 2006;18(10):2767–81.

53. Wang Z-Q, Yuan Y-Z, Ou J-Q, Lin Q-H, Zhang C-F. Glutamine synthetase and glutamate dehydrogenase contribute differentially to proline accumulation in leaves of wheat (< i> Triticum aestivum</i>) seedlings exposed to different salinity. J Plant Physiol. 2007;164(6):695–701.

54. Yan S, Tang Z, Su W, Sun W. Proteomic analysis of salt stress-responsive proteins in rice root. Proteomics. 2005;5(1):235–44.

55. Nam MH, Huh SM, Kim KM, Park WJ, Seo JB, Cho K, Kim DY, Kim BG, Yoon IS. Comparative proteomic analysis of early salt stress-responsive proteins in roots of SnRK2 transgenic rice. Proteome Sci. 2012;10(25):2–19.

56. Ziegler J, Stenzel I, Hause B, Maucher H, Hamberg M, Grimm R, Ganal M, Wasternack C. Molecular cloning of Allene oxide Cyclase. The enzyme establishing the stereochemistry of octadecanoids and jasmonates. J Biol Chem. 2000;275(25):19132–8.

57. Devoto A, Turner JG. Jasmonate-regulated Arabidopsis stress signalling network. Physiol Plant. 2005;123(2):161–72.

58. Yamada A, Saitoh T, Mimura T, Ozeki Y. Expression of mangrove allene oxide cyclase enhances salt tolerance in Escherichia coli, yeast, and tobacco cells. Plant Cell Physiol. 2002;43(8):903–10.

59. Sutka M, Li G, Boudet J, Boursiac Y, Doumas P, Maurel C. Natural variation of root hydraulics in Arabidopsis grown in normal and salt-stressed conditions. Plant Physiol. 2011;155(3):1264.

60. Wu L, Zhao FA, Fang W, Xie D, Hou J, Yang X, Zhao Y, Tang Z, Nie L, Lv S. Identification of early salt stress responsive proteins in seedling roots of upland cotton (Gossypium Hirsutum L.) employing iTRAQ-based proteomic technique. Front Plant Sci. 2015;6:732.

61. Gruber CW, Čemažar M, Heras B, Martin JL, Craik DJ. Protein disulfide isomerase: the structure of oxidative folding. Trends Biochem Sci. 2006;31(8):455–64.

62. Nohzadeh MS, Habibi RM, Heidari M, Salekdeh GH. Proteomics reveals new salt responsive proteins associated with rice plasma membrane. Biosci Biotechnol Biochem. 2014;71(9):2144–54.

63. Song A, Zhu X, Chen F, Gao H, Jiang J, Chen S. A chrysanthemum heat shock protein confers tolerance to abiotic stress. Int J Mol Sci. 2014;15(3):5063–78.

64. Lv D-W, Subburaj S, Cao M, Yan X, Li X, Appels R, Sun D-F, Ma W, Yan Y-M. Proteome and phosphoproteome characterization reveals new response and defense mechanisms of Brachypodium Distachyon leaves under salt stress. Mol Cell Proteomics. 2014;13(2):632–52.

65. Babbitt SE, Kiss A, Deffenbaugh AE, Chang Y-H, Bailly E, Erdjument-Bromage H, Tempst P, Buranda T, Sklar LA, Baumler J. ATP hydrolysis-dependent disassembly of the 26S proteasome is part of the catalytic cycle. Cell. 2005;121(4):553–65.

66. Darnell J, Lodish H, Baltimore D. Molecular cell biology[J]. Yale J Biol Med. 2008;60(3):292–385.

67. Ahlert D, Piepenburg K, Kudla J, Bock R. Evolutionary origin of a plant mitochondrial group II intron from a reverse transcriptase/maturase-encoding ancestor. J Plant Res. 2006;119(4):363–71.

68. Matsuura M, Noah JW, Lambowitz AM. Mechanism of maturase-promoted group II intron splicing. EMBO J. 2001;20(24):7259–70.

69. Yang C, Wang L, Siva VS, Shi X, Jiang Q, Wang J, Zhang H, Song L. A novel cold-regulated cold shock domain containing protein from scallop Chlamys Farreri with nucleic acid-binding activity. PLoS One. 2012;7(2):e32012.

70. Wolffe AP. Structural and functional properties of the evolutionarily ancient Y-box family of nucleic acid binding proteins. BioEssays. 1994;16(4):245–51.

71. Graumann PL, Marahiel MA. A superfamily of proteins that contain the cold-shock domain. Trends Biochem Sci. 1998;23(8):286–90.

72. Hussey PJ, Ketelaar T, Deeks MJ. Control of the actin cytoskeleton in plant cell growth. Annu Rev Plant Biol. 2006;57:109–25.

73. Kandasamy MK, Burgos-Rivera B, McKinney EC, Ruzicka DR, Meagher RB. Class-specific interaction of profilin and ADF isovariants with actin in the regulation of plant development. Plant Cell. 2007;19(10):3111–26.

74. Pope BJ, Gonsior SM, Yeoh S, McGough A, Weeds AG. Uncoupling actin filament fragmentation by cofilin from increased subunit turnover. J Mol Biol. 2000;298(4):649–61.

75. Ouellet F, Carpentier É, Cope MJT, Monroy AF, Sarhan F. Regulation of a wheat actin-depolymerizing factor during cold acclimation. Plant Physiol. 2001;125(1):360–8.

76. Buchberger A. Control of ubiquitin conjugation by cdc48 and its cofactors. In: Conjugation and Deconjugation of Ubiquitin family modifiers. Springer. 2010;54(54):17–30.

77. Vembar SS, Brodsky JL. One step at a time: endoplasmic reticulum-associated degradation. Nat Rev Mol Cell Biol. 2008;9(12):944–57.

78. Niehl A, Amari K, Gereige D, Brandner K, Mély Y, Heinlein M. Control of tobacco mosaic virus movement protein fate by CELL-DIVISION-CYCLE protein48. Plant Physiol. 2012;160(4):2093–108.

79. Knight H, Trewavas AJ, Knight MR. Calcium signalling in Arabidopsis Thaliana responding to drought and salinity. Plant J. 1997;12(5):1067–78.

80. Mortimer JC, Laohavisit A, Macpherson N, Webb A, Brownlee C, Battey NH, Davies JM. Annexins: multifunctional components of growth and adaptation. J Exp Bot. 2008;59(3):533–44.

81. Lee S, Lee EJ, Yang EJ, Lee JE, Park AR, Song WH, Park OK. Proteomic identification of annexins, calcium-dependent membrane binding proteins that mediate osmotic stress and abscisic acid signal transduction in Arabidopsis. Plant Cell. 2004;16(6):1378–91.

82. Kovács I, Ayaydin F, Oberschall A, Ipacs I, Bottka S, Pongor S, Dudits D, Tóth ÉC. Immunolocalization of a novel annexin-like protein encoded by a stress and abscisic acid responsive gene in alfalfa. Plant J. 1998;15(2):185–97.

83. Zhang JL, Shi H. Physiological and molecular mechanisms of plant salt tolerance. Photosynth Res. 2013;115(1):1.

84. Zhang J-S, Xie C, Li Z-Y, Chen S-Y. Expression of the plasma membrane H+–ATPase gene in response to salt stress in a rice salt-tolerant mutant and its original variety. Theor Appl Genet. 1999;99(6):1006–11.

14

Comparative proteomic analysis between nitrogen supplemented and starved conditions in *Magnaporthe oryzae*

Yeonyee Oh[1], Suzanne L. Robertson[2], Jennifer Parker[2], David C. Muddiman[2] and Ralph A. Dean[1*]

Abstract

Background: Fungi are constantly exposed to nitrogen limiting environments, and thus the efficient regulation of nitrogen metabolism is essential for their survival, growth, development and pathogenicity. To understand how the rice blast pathogen *Magnaporthe oryzae* copes with limited nitrogen availability, a global proteome analysis under nitrogen supplemented and nitrogen starved conditions was completed.

Methods: *M. oryzae* strain 70–15 was cultivated in liquid minimal media and transferred to media with nitrate or without a nitrogen source. Proteins were isolated and subjected to unfractionated gel-free based liquid chromatography-tandem mass spectrometry (LC-MS/MS). The subcellular localization and function of the identified proteins were predicted using bioinformatics tools.

Results: A total of 5498 *M. oryzae* proteins were identified. Comparative analysis of protein expression showed 363 proteins and 266 proteins significantly induced or uniquely expressed under nitrogen starved or nitrogen supplemented conditions, respectively. A functional analysis of differentially expressed proteins revealed that during nitrogen starvation nitrogen catabolite repression, melanin biosynthesis, protein degradation and protein translation pathways underwent extensive alterations. In addition, nitrogen starvation induced accumulation of various extracellular proteins including small extracellular proteins consistent with observations of a link between nitrogen starvation and the development of pathogenicity in *M. oryzae*.

Conclusion: The results from this study provide a comprehensive understanding of fungal responses to nitrogen availability.

Keywords: Proteomics, *Magnaporthe oryzae*, Nitrogen starvation, Melanin biosynthesis, Extracellular protein

Background

In nature, fungi, including plant pathogens, are often exposed to environments where nutrients are insufficient. To deal with such conditions, several survival mechanisms are activated such as intercellular nutrient recycling, the scavenging of resources and/ or morphological changes to aid growth and proliferation in stressful surroundings. On the plant surface, potential pathogenic fungi are often limited for nutrients until the host has been successfully infected. Indeed, the expression of pathogenicity genes are frequently elevated during nitrogen limiting conditions

suggesting that nitrogen starvation is a driving force for successful fungal infection of their host organisms [1–3]. Unlike their plant hosts, fungi can utilize diverse compounds as nitrogen sources from primary preferred sources such as ammonium, glutamine and glutamate to other non-preferred secondary forms including nitrate, nitrite, other amino acids and proteins. Assimilation of nitrogen is tightly regulated by global regulators that ensure use of preferential nitrogen sources over less desirable sources [4]. AreA, Nit2 and NUT1, which encode GATA type transcription factors are well known as a key regulatory genes in *Aspergillus nidulans*, *Neurospora crassa* and *Magnaporthe oryzae* respectively [5–7]. A number of studies have linked nitrogen metabolism regulation with the ability to cause disease by plant pathogenic

* Correspondence: radean2@ncsu.edu
[1]Center for Integrated Fungal Research, Department of Entomology and Plant Pathology, North Carolina State University, Raleigh, NC 27695, USA
Full list of author information is available at the end of the article

fungi, including *Cladosporium fulvum*, *Colletotrichum lindemuthianum* and *M. oryzae* [7–9].

The rice blast pathogen, *M. oryzae*, is the most significant fungal pathogen of rice crops worldwide as it routinely destroys rice production by 10–30% [10]. *M. oryzae* also infects other agronomically important grass species including wheat, barley and millet [11]. As typical of many filamentous ascomycete fungal pathogens, *M. oryzae* develops a specialized infection structure, the appressorium, to attach and penetrate the host plant. After successful colonization of host tissues, necrotic lesions form within a few days, from which conidia are produced rapidly spreading the disease to neighboring plants under favorable conditions [12].

Over the past few years, transcriptome and proteome studies during nitrogen starvation are beginning to reveal how nitrogen availability affects phytopathogens [13, 14]. In transcriptome studies of *M. oryzae*, 520 genes showed increased gene expression during nitrogen starvation, the majority of which were involved in amino acid metabolism and uptake [13]. We found the important pathogenicity factor, SPM1, a putative subtilisin-like protease was significantly induced under nitrogen starvation as well as during appressorium formation and in response to exogenous cyclic AMP treatment [15]. Further study showed that SPM1 was localized in vacuoles and involved in autophagy during appressorium formation. Infectious growth of the spm1 deletion mutant in rice epidermal cells was very limited [16]. In other phytopathogenic fungi such as in *Verticillum dahliae*, the expression of 487 genes, which included genes involved in melanin biosynthesis, were significantly upregulated under nitrogen starvation [14]. Furthermore, analyses of upregulated genes under nitrogen stress has yielded numerous small secreted proteins, a feature typical of many fungal effectors [17, 18].

Proteomic analysis of liquid media from *V. dahliae* culture grown under nitrogen starvation showed enrichment of proteins for cell wall degradation, reactive oxygen species (ROS) scavenging and stress response as well as protein and carbohydrate metabolism [18]. Up to now, studies of the *M. oryzae* proteome under nitrogen starvation have been conducted through analysis of two-dimensional gels coupled with mass spectrometry analysis. An analysis of liquid culture media identified 85 putative secreted proteins upregulated under nitrogen starvation. The majority were cell wall hydrolase enzymes, protein and lipid hydrolases and proteins for ROS detoxification [17]. Another study reported 975 protein spots from complete media and 1169 spots under nitrogen limitation conditions. Forty three differentially accumulated proteins were identified, of which several were found to be involved in glycolysis, the tricarboxylic acid cycle and nitrogen metabolism [19].

In this comprehensive study, unfractionated protein samples coupled with advanced mass spectrometry technology was employed to identify and monitor more than 40% of the predicted *M. oryzae* entire proteome during nitrogen starvation, which revealed key biological information pertaining to its survival and pathogenicity.

Methods
Materials
All reagents were purchased from Sigma-Aldrich (St. Louis, MO) and all solvents were HPLC-grade from Honeywell Burdick & Jackson (Muskegon, MI), unless otherwise stated.

Fungal growth and protein extraction
M. oryzae strain 70–15 conidia were harvested from 1 week old V8 agar plates and inoculated into complete liquid medium (10 g sucrose; 1 ml of 1000X trace elements (2.2 g $ZnSO_4$, 1.1 g H_3BO_3, 0.5 g $MnCl_2$-$4H_2O$, 0.5 g $FeSO_4$-$7H_2O$, 0.17 g $CoCl_2$, 0.16 g $CuSO_4$-$5H_2O$, 0.15 g Na_2MoO_4-$2H_2O$ and 5 g disodium EDTA per 100 ml); 6 g casein acid hydrolysate; 6 g yeast extract in 1 L). The culture was grown at 28 °C on a 200 rpm shaker for 3 days. The mycelial mat was then collected on sterile filter paper, washed three times with sterile distilled water, and divided into six equal pieces. Biological replicates were grown by placing one piece of mycelial mat into each of the three flasks containing minimal media supplemented with nitrogen (N+), or three flasks containing nitrogen- limiting (N-) media. N+ media contained 10 g sucrose, 1 ml of 1000X trace element solution, 50 ml nitrate salts (60 g $NaNO_3$, 5.2 g KCl, 5.2 g $MgSO_4$·$7H_2O$, and 15.2 g KH_2PO_4 for 500 ml), 1 mg thiamine and 5 µg biotin per liter. N- media was the same as N+ except lacked $NaNO_3$. The pH of the media was adjusted to 6.5 with NaOH. These cultures were grown at 28 °C in a 200 rpm shaker. After 12 h, the mycelial mats were collected, washed with sterile distilled water and then ground into powder with liquid nitrogen. Proteins were extracted using lysis buffer (50 mM HEPES (pH 7.5), 0.5% Nonidet P-40, 250 mM NaCl, 10% (*v/v*) glycerol, 2 mM EDTA (pH 8.0), one cOmplete™ ULTRA tablet Protease Inhibitor cocktail (Roche, Germany) per 50 ml). Lysate was clarified by centrifugation at 16,000 g for 15 min. Protein concentration was estimated using the Pierce™ Coomassie Bradford assay kit (Thermo Fisher Scientific, Waltham, MA). Samples were stored at −80 C.

Sample preparation and digestion
Filter aided sample preparation was performed as in Loziuk et al. [20] with slight modifications. For each biological replicate, a volume of lysate containing 250µg of protein was reduced in 50 mM Dithiothreitol (DTT) in 8 M urea and 50 mM tris-HCl (denaturing/alkylating buffer) at 56 °C for 30 min. Cysteine residues were alkylated in the dark with 10 mM N-ethylmaleimide (NEM) for 30 min at room

temperature. Samples were transferred to a 10 kDa molecular weight cutoff centrifugation filter (EMD Millipore, Billerica, MA) and centrifuged at 14000 g for 15 min at 20 °C (all centrifugation steps performed with these settings). Buffer exchange with 2 M urea, 10 mM calcium chloride was performed three times by centrifugation. Samples were digested in the filter at 37 °C for 12 h using a 1:50 modified porcine trypsin to sample ratio. Peptides were eluted by centrifugation, and quenched with 1% formic acid, 0.001% Zwittergent 3–16 (Calbiochem, La Jolla, CA) for further analysis. The concentration of peptides was measured by the NanoDrop™ (Thermo Scientific).

LC-MS/MS

Reverse-phase nano-LC was performed with an EASY nLC 1000 (Thermo Fisher, Waltham, MA) using a 20 cm, 75 um I.D Picofrit column (New Objective, Woburn, MA) packed with Kinetix 2.6 μm (100 Å) stationary phase (Phenomenex, Torrance, CA). One microgram of peptides was loaded onto the column and eluted at a flow rate of 300 nL/min with a 240 min linear gradient (5–30%). Buffers consisted of mobile phase A (98% H_2O, 2% ACN, and 0.2% FA) and mobile phase B (2% H_2O, 98% ACN, and 0.2% FA). Three technical replicates were performed per sample, and data was collected on a Q-Exactive High Field mass spectrometer (Thermo Fisher Scientific, Waltham, MA). For MS^1 scans, resolving power was 120,000, the AGC target was 3e6, an injection time of 50 ms was applied, and the scan range was set to 300–1600 m/z. During top-20 data dependent MS^2 scans, resolution was 15,000, the AGC was 1e5, an injection time of 30 ms was applied, the scan range was 200–2000 m/z, a 2 m/z isolation window was used, normalized collision energy was set to 27, the underfill ratio was 2% with an intensity threshold of 6.7e4, and a dynamic exclusion time of 20 s was applied.

Database searching and data analysis

Proteins were searched against a concatenated target-reverse database MG8 (*Magnaporthe* comparative Sequencing Project, Broad Institute of Harvard and MIT), and identified using the Sequest HT algorithm in Proteome Discoverer 1.4 (Thermo Scientific, San Jose, CA). Search parameters used a 5 ppm precursor mass and 0.02 Da fragment mass tolerance and allowed up to 2 missed cleavage sites. False discovery rate calculations were generated using Percolator at a 1% protein false discovery rate (FDR). Peptide spectral matches (PSMs) were normalized across technical and biological replicates, and treatment conditions using the total spectral counts method as previously described [21]. Differential protein expression was calculated by dividing the normalized PSMs of the nitrogen-starved state by the nitrogen treated state. Statistical significance of unique protein

identification and protein fold-changes between the treatments was determined using a pairwise Student's t-test with a cutoff of $p \leq 0.05$.

Functional annotation was performed using David Algorithm (version 6.7) [22]. Biological pathways of identified proteins were predicted by searching the *M. oryzae* KEGG pathway database (release 79) [23]. Subcellular localizations to all predicted proteins of the *M. oryzae* version 8 genome were assigned by the WoLF PSORT program [24]. The 500 bp upstream of open reading frames were searched for HGATAR domains (H = A, C or T and R = A or G) using the POCO motif finding program [25] and the presence of HGATAR motifs were compared between gene of interests and background genes.

Results and discussion

Proteome identification in nitrogen supplemented and nitrogen depleted growth condition

In this study, the protein profiles of *M. oryzae* 70–15 mycelia grown under nitrate supplemented or nitrogen depleted conditions were compared. The fungus was initially cultivated in nutrient rich complete media for 3 days and then switched to minimal media with or without nitrogen sources for an additional 12 h. To generate a detailed interrogation of protein changes associated with nitrogen starvation, highly sensitive, global MS/MS technologies for protein detection coupled to advanced annotation tools were employed. A total of eighteen injections from samples consisting of two nitrogen treatments (nitrate supplemented (N+) and nitrogen depleted (N-)), three biological replicates and three technical replicates were investigated. On average, 153,000 tandem mass spectra were collected corresponding to an average of 23,648 MS^1 peptides per injection. Overall, the whole proteomic dataset mapped to a total of 5498 *M. oryzae* proteins (representing >42% of the 12,991 predicted proteins in *M. oryzae* V8 annotation proteome) with a 1% FDR (Fig. 1, Additional file 1: Table S1). There were 4098 proteins shared between the two treatments, 704 found only in N-, and 696 only in N +. In previous work, employing FASP and anion Stage-Tip fractionation analyzed on an Orbitrap XL, 3200 proteins were identified in *M. oryzae* [26]. The method used in this study did not include fractionation, which resulted in more complex samples. Nevertheless, the employ of newer MS technologies with significant higher MS/MS scan rates resulted in an increase in proteome coverage by more than 75%. Similar results have been recently reported in yeast where it is now possible to identity nearly 4000 proteins representing ~60% of the theoretical proteome without sample fractionation [27]. This important advancement enabling the use of non-

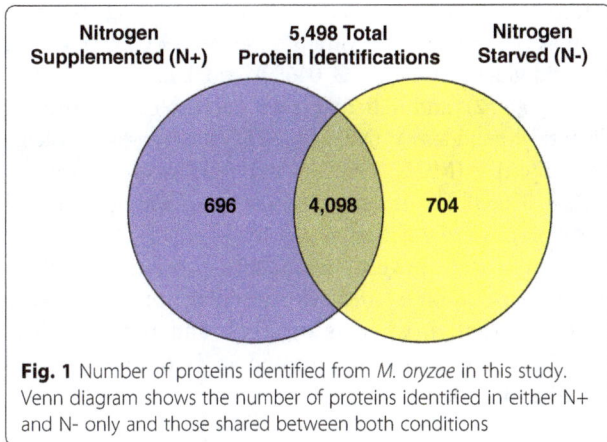

Fig. 1 Number of proteins identified from *M. oryzae* in this study. Venn diagram shows the number of proteins identified in either N+ and N- only and those shared between both conditions

fractionated samples now paves the way for more sophisticated and complex proteome studies in the future.

The cellular localization of *M. oryzae* proteins was analyzed using the prediction tool, WoLF PSORT [24]. The WoLF PSORT distribution of the theoretical *M. oryzae* database (12,991 proteins) localized 15%, 28%, 25%, and 16% of the proteins to cytosol, nucleus, mitochondria, and extracellular regions, respectively (Fig. 2a). By comparison, among the 4794 identified proteins in the N+ condition, 23% were cytosolic, 28% nuclear, 25% mitochondrial and 8% extracellular (Fig. 2b). Among the 4802 identified proteins in the N- condition, 23% were

cytosolic, 28% nuclear, 23% mitochondrial and 10% extracellular (Fig. 2d). The N+ and N- proteome both showed a clear enrichment of cytosolic proteins and reduction of extracellular proteins.

Protein detection is likely to be directly related to the amount of protein present in the tissue and thus may reflect the general level of protein expression in the N+ and N- conditions (high for cytosolic proteins and low for extracellular proteins). To test this hypothesis, the expression level according to cellular localization was evaluated. The relative cellular protein expression was measured by calculating the normalized spectral abundance factor (NSAF) and summed for each subcellular localization category [28, 29].

This analysis showed that in both nitrogen supplemented and starved conditions, the cytoplasmic and mitochondrial proteins represented not only the largest groups of identified proteins but also embodied proteins of high abundance. In contrast, extracellular proteins were underrepresented among the identified proteins and generally contained proteins of lower abundance. However, we cannot exclude the possibility that some extracellular protein were lost during sample preparation. Although, we identified the nuclear and plasma membrane proteins in the same proportion as the whole proteome, these proteins showed relatively lower protein abundance (Fig. 2c, e). These results

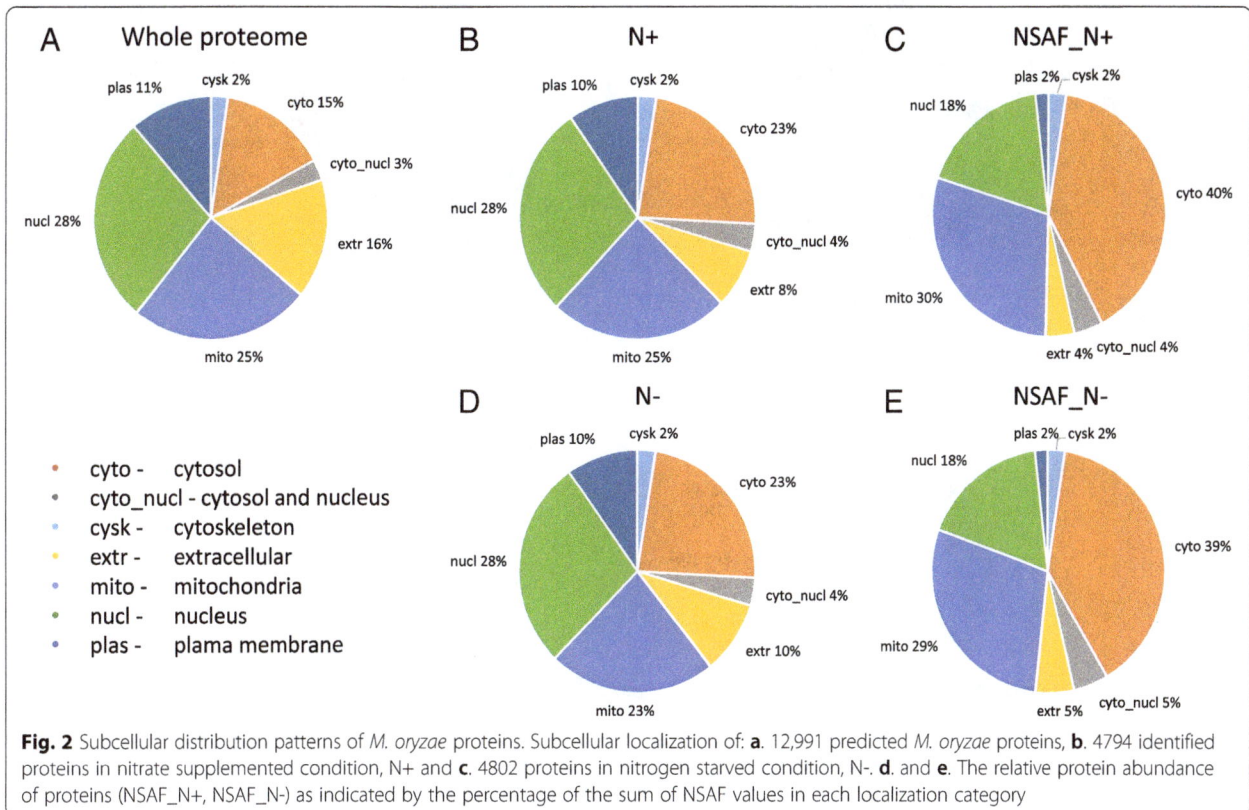

Fig. 2 Subcellular distribution patterns of *M. oryzae* proteins. Subcellular localization of: **a.** 12,991 predicted *M. oryzae* proteins, **b.** 4794 identified proteins in nitrate supplemented condition, N+ and **c.** 4802 proteins in nitrogen starved condition, N-. **d.** and **e.** The relative protein abundance of proteins (NSAF_N+, NSAF_N-) as indicated by the percentage of the sum of NSAF values in each localization category

were very similar to our previous analysis of *M. oryzae* conidial proteome [30] and suggests the general enrichment of cytosolic and mitochondrial proteins in quantity and quality and underrepresentation of extracellular proteins in *M. oryzae* tissues.

Follow-up studies were performed on the highest (≥ 90 percentile in NSAF value) and lowest abundant (≤ 10 percentile NSAF value) proteins. Among the most abundant proteins in the N+ group, the majority were cytosolic (43%) followed by mitochondrial (30%) proteins and 15% nuclear proteins (Fig. 3a). A similar distribution was observed for highly abundant proteins in the N- condition (Fig. 3b). The highly abundant cytosolic proteins in both N + and N- were annotated as subunits of ribosome and proteasome or involved in glycolysis, amino acid biosynthesis, aminoacyl-tRNA biosynthesis, starch and sucrose metabolism or fatty acid biosynthesis. Based on KEGG pathway analysis, among the highly expressed mitochondrial proteins, 12 proteins, including NADH dehydrogenases, F-type ATPases, cytochrome c oxidase and cytochrome c reductase, were implicated in the processes related to oxidative phosphorylation. TCA cycle and amino acid biosynthesis related proteins and components of the ribosome were also identified in the mitochondria proteins (Additional file 1: Table S1). The most abundant

extracellular proteins included cell wall modifying enzymes, such as cell wall glucanosyltransferase (MGG_00592), glucan 1,3-beta-glucosidase (MGG_04689) and beta-glucosidase 1 (MGG_09272) and different types of proteases, such as dipeptidyl-peptidase V (MGG_07877), subtilisin-like proteinase Spm1 (MGG_03670) and carboxypeptidase Y (MGG_05663). In sum, proteins associated with growth and metabolism were the most abundant.

Among the lowest abundant proteins, nuclear, plasma membrane and extracellular proteins were over-represented, whereas cytosolic and mitochondrial proteins were under-represented in both the N+ and N- (Fig. 3c, d).

Proteome changes during nitrogen starvation

Relative quantification of protein abundance between the N+ and N- conditions was determined using PSMs. One hundred fifteen and 70 proteins were found to be significantly unique (Student's T test, $p \leq 0.05$) in the N- and N + groups, respectively. Among the 4098 overlapping proteins, 444 proteins were identified as significantly regulated, with a 2-fold or greater change between the groups (p-value <0.05). 248 proteins were over-expressed and 196 proteins were repressed in the N- group compared with N + (Fig. 4). Combining the uniquely identified with the

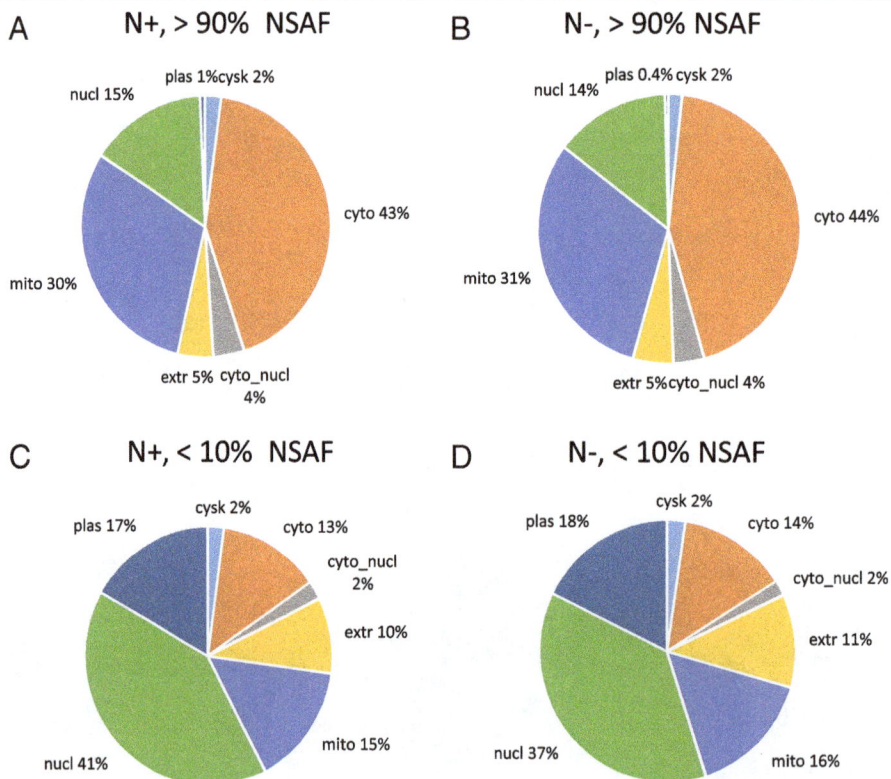

Fig. 3 Distribution of cellular localization of the most and the least abundant proteins identified in *M. oryzae* under nitrate nitrogen (N+) and nitrogen starved (N-) conditions. Proteins ranking in the top 90% (>90%, **a** and **b**) and bottom 10% (<10%, **c** and **d**) in NSAF value were grouped respectively according to the expected cellular location. Localization categories are shown in Fig. 2

differentially expressed proteins resulted in 363 induced and 266 repressed proteins in fungal mycelia undergoing nitrogen starvation (Additional file 2: Table S2), which represented 11% of the total proteins identified. To investigate the potential biological implications of these 629 differentially expressed proteins, David GO Functional analysis v6.7 [22] were applied (Table 1). Groups of proteins for melanin biosynthesis, amino acid transport, cell morphogenesis, carboxylic acid transport, ion transport and tyrosine metabolic process were enriched during nitrogen starvation. By contrast, proteins for protein synthesis, nitrate assimilation, carbohydrate biosynthetic process port and porphyrin biosynthesis (Table 1) were repressed. In-depth discussion of these proteins is provided below to gain further insight into the relationship between nitrogen starvation and fungal development, signaling and effector protein expression.

Redirection of nitrogen metabolism pathway

Fungi have evolved mechanisms to uptake and convert inorganic nitrogen compounds from the environment to organic nitrogen compounds, which are incorporated into cellular substances including amino acids [31]. By so-called nitrogen assimilation, extracellular nitrate is transported into the cell by nitrate transporters and reduced to nitrite by nitrate reductase. This is further reduced to ammonia by nitrite reductase [31]. Our data suggests that in *M. oryzae*, this nitrogen assimilation process is highly regulated by controlling the production of the key enzymes. During nitrogen starvation, the production of nitrate reductase (MGG_06062) and

nitrite reductase (MGG_00634) were significantly reduced, both by more than 100 fold. Further, levels of the nitrate transporter (MGG_13793) were significantly reduced in N- by more than 20 fold. Thus when nitrate nitrogen is limiting in the environment, fungal resources are likely directed elsewhere and enzymes required for nitrogen assimilation are greatly reduced.

Elevated levels of nitrogen regulators such as NUT1 (MGG_02755) were also observed under nitrogen starvation. NUT1 is a member of GATA family transcription factors, which includes AreA from *A. nidulans* and Nit2 from *N. crassa*, which are well known global nitrogen regulators in nitrogen catabolite repression (NCR) [5, 32]. Typically, these transcription factors activate expression of nitrogen metabolic genes under nitrogen starvation. They contain Cys2/Cys2 type zinc fingers which recognize the consensus DNA sequences HGATAR in promoter sequences of target genes. Interestingly, although NUT1 is highly upregulated by nitrogen starvation in our studies, proteins involved in assimilation of nitrate, which are typically regulated by NUT1, were not up-regulated as noted above. There are possible explanations for this apparent contradictory observation. In *Aspergillus nidulans*, for example, the activity of AreA is subject to co-repression by NmrA [33]. *M. oryzae* contains 3 homologs of NmrA, two of which Nmr1 and Nmr3 interact with NUT1 [34]. Also, other proteins such as the highly conserved AreB, have been shown to act as regulators of nitrogen metabolic genes [35]. Thus, it is possible that when nitrogen is lacking co-repressors or other

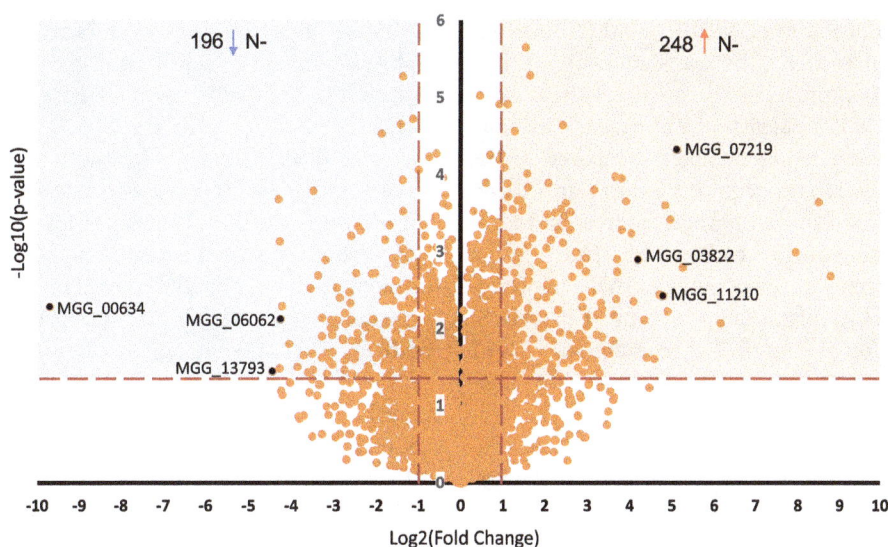

Fig. 4 Volcano plot of *M. oryzae* proteins from nitrate nitrogen (N+) and nitrogen starved (N-) conditions. Out of 5498 proteins identified, 444 were differentially expressed. Two hundred and forty eight were induced and 196 repressed (≥2 fold-change, *p*-value ≤0.05). MGG_00634 (nitrite reductase), MGG_06062 (nitrate reductase), MGG_13793 (nitrate transporter), MGG_07219 (melanin biosynthesis polyketide synthase), MGG_03822 (peptidase family T4) and MGG_11210 (beta-glucosidase 1) are highlighted

Table 1 Enriched functional groups among differentially expressed proteins during nitrogen starvation

Expression	GO ID	GO Description	Protein ID	pValue	Enrichment Score
ID	GO:0042438	melanin biosynthetic process	MGG_07216, MGG_07219, MGG_02252, MGG_05059	0.00	23.8
ID	GO:0046942	carboxylic acid transport	MGG_05128, MGG_00289, MGG_14115, MGG_01054, MGG_07606, MGG_11327	0.02	4.0
ID	GO:0000902	cell morphogenesis	MGG_04703, MGG_06033, MGG_03703, MGG_02781	0.00	13.2
ID	GO:0006570	tyrosine metabolic process	MGG_02252, MGG_06691, MGG_05059	0.02	12.7
ID	GO:0006811	ion transport	MGG_04159, MGG_03299, MGG_10027, MGG_06118, MGG_01054, MGG_02124, MGG_09119, MGG_04135, MGG_10634, MGG_05281, MGG_09063	0.04	2.0
ID	GO:0006865	amino acid transport	MGG_05128, MGG_00289, MGG_14115, MGG_07606, MGG_11327	0.05	3.5
RP	GO:0006412	translation	MGG_13783, MGG_01165, MGG_00161, MGG_07154, MGG_06935, MGG_08323, MGG_04042, MGG_02511, MGG_05031, MGG_04455, MGG_10825, MGG_06468, MGG_05275, MGG_09301, MGG_06744, MGG_14349, MGG_05647	0.00	2.3
RP	GO:0042128	nitrate assimilation	MGG_00634, MGG_06062, MGG_04144	0.01	18.0
RP	GO:0042401	biogenic amine biosynthetic process	MGG_10533, MGG_11574, MGG_07454	0.02	12.0
RP	GO:0008610	lipid biosynthetic process	MGG_06288, MGG_03343, MGG_08474, MGG_13185, MGG_09239, MGG_06935, MGG_06133, MGG_07543, MGG_00806	0.03	2.5
RP	GO:0006418	tRNA aminoacylation for protein translation	MGG_13783, MGG_00161, MGG_01165, MGG_05275, MGG_04042	0.04	3.9
RP	GO:0034637	cellular carbohydrate biosynthetic process	MGG_00865, MGG_13185, MGG_06935, MGG_00450	0.04	5.1
RP	GO:0006779	porphyrin biosynthetic process	MGG_04860, MGG_06446, MGG_10321	0.05	8.3

*Functional annotation clustering analysis was performed with 629 differentially expressed proteins using David GO v 6.7. Groups with enrichment score ≥ 2 and p-value ≤0.05 are presented for the induced (ID) and repressed (RP) proteins during nitrogen starvation

proteins regulate the action of NUT1. In addition, other unknown factors may affect the translation or stability of nitrate assimilation proteins.

NUT1 plays a role in virulence of *M. oryzae*. Nut1 null mutants cause reduced lesion numbers compared to the wild-type [7]. NUT1 is also required for the growth on several non-preferred secondary nitrogen sources and regulates gene expression of the hydrophobin like effector, MPG1 under nitrogen starvation [7, 36]. GATA transcription factors from other plant pathogenic fungi have been shown to be key for pathogenesis. For example, CLNR1 mutants of *Colletotrichum lindemuthianum* are non-pathogenic [8]. Like in *M. oryzae*, these mutants can produce appressoria, but invasive growth is hampered. Reduced virulence was also reported in the FNR1 disrupted mutant in *Fusarium oxysporum* f. sp. *lycopersici* [37].

Plants contain a number of nitrogen sources that may be used by pathogenic fungi. For example, γ-aminobutyric acid (GABA) is a major metabolite in apoplast of tomato and other plants. During infection, GABA levels have been shown to rise [38, 39]. Fungi uptake GABA via GABA permease. In *A. nidulans*, expression of GABA permease, GabA is under NCR and regulated by the GATA transcription factor AreA. Here, we observed protein expression of a GABA permease MGG_14115 in *M. oryzae* increased more than three-fold during nitrogen starvation. However, there is no direct evidence that GABA permease is regulated by NUT1. In our previous work, gene expression of both GABA permease and NUT1 were found to be increased by nitrogen starvation [13]. This strongly suggests that in *M. oryzae*, nitrogen limitation triggers both gene expression and accumulation of proteins of major players for nitrogen scavenging including uptake of available nitrogen sources, potentially including those encountered during infection.

Our data also showed co-occurrence between nitrate metabolism and sulfate metabolism during nitrogen starvation. Key enzymes for sulfate assimilation including phosphoadenosine phosphosulfate reductase (MGG_03662), sulfite reductase flavoprotein component (MGG_00929), sulfite reductase (MGG_04144), sulfate adenylyltransferase (MGG_15027) were significantly reduced during nitrogen starvation. It has been reported in other biological systems that sulfate assimilation is regulated by availability of nitrogen sources and nitrogen starvation represses gene expression and enzyme activity of proteins involved in sulfate assimilation [40–42]. This may be a result of general repression of pathways involved in protein synthesis preventing the accumulation of high levels of sulfur containing amino acids such as cysteine and methionine [43].

Melanin biosynthetic process
Availability of nitrogen regulates the synthesis of broad range of secondary metabolites in fungi [44]. Melanin is one of the most thoroughly studied secondary metabolites in fungi including pathogenic fungi. Most fungi produce DHN melanins using 1,8-dihydroxynaphthalene (DHN) as precursor. Highly conserved in pathogenic filamentous fungi, DHN melanin is synthesized via the polyketide pathway. Proteins involved in this process include polyketide synthase (PKS), 1,3,6,8-tetrahydroxy-naphthalene reductase (4HNR), trihydroxy-naphthalene reductase (3HNR) and scytalone dehydratase (SCD). In *M. orzyae*, melanin biosynthesis is known to be essential for infection of the host plant [45]. A highly melanized outer cell wall of the appressorium facilitates high turgor pressure that is essential for successful penetration into the host plant [46]. Genes involved in melanin synthesis including PKS (MGG_07219), 3HNR (MGG_07216), 4HNR (MGG_02252) and SCD (MGG_05059) have been functionally characterized and shown to be essential for fungal development and pathogenicity [45]. For example, gene deletion of PKS (MGG_07219) resulted in the production of non-functional appressoria, which failed to infect the host plant [15, 47].

In a previous study, we showed that genes involved in this pathway were highly induced during appressorium formation and were under cAMP signaling pathways in *M. orzyae* [15]. Here, we found that proteins involved in melanin biosynthesis increased in response to nitrogen starvation in *M. oryzae*. All the principal enzymes in this pathway, PKS (MGG_07219), 3HNR (MGG_07216), 4HNR (MGG_02252) and SCD (MGG_05059), were significantly induced during nitrogen starvation. In addition, we observed mycelia were more pigmented under nitrogen starvation compared to the non-starved condition, which is likely the result of increased melanin production at least in part (Additional file 3: Figure S1). PKS (MGG_07219), 4HNR (MGG_02252) and SCD (MGG_05059) genes contains the consensus DNA sequences HGATAR in their promoter region (Additional file 2: Table S2), which suggests that expression of the melanin biosynthesis genes may be controlled by a GATA transcription factor under nitrogen starvation, possibly by NUT1.

In other fungal pathosystems, increased melanin biosynthesis has been suggested to enable infection. For example, based on the transcriptomic profiling, increased melanin biosynthesis during nitrogen starvation was proposed to be important for pathogenesis by the wilt pathogen, *V. dahliae* [14]. Increased melanin may also confer other beneficial properties. Increased melanin biosynthesis may provide protection of fungal cells from ROS during infection. For example in *Colletotrichum acutatum*, ROS accumulation has been reported during nitrogen limiting conditions [48]. With a strong affinity for metals, melanin acts as very effective scavenger of those free radicals. DHN melanin also protects fungal cells against permanganate, hypochlorite and neutrophil oxidative burst [49, 50]. Melanin extracted from the medical fungus, *Auricularia auricular*, exhibited strong radical scavenging activities [51]. Thus, for pathogenic fungi, melanin biosynthesis is crucial for both the infection process and protection against the oxidative burst associated with host defense responses.

In other studies, expression of genes in a number of secondary metabolite gene clusters, which include polyketide synthase, have been shown to be dependent on the quantity and quality of the nitrogen source. For example, in *Fusarium fujikuroi* among 20 PKS gene clusters, the expression of 13 was influenced by nitrogen source [52]. In our experiments, with the exception of the PKS involved in melanin biosynthesis (MGG_07219), the expression of only one other PKS (MGG_15100) was affected by nitrogen source.

Increased activity of protein degradation
The largest group of proteins induced during nitrogen starvation was proteins involved in recycling of nitrogen sources. Exopeptidases (MGG_03822, MGG_07704, MGG_07981, MGG_09530) and metallopeptidases (MGG_01970, MGG_06643, MGG_07704, MGG_07981, MGG_09530) were induced as were 7 different endopeptidases (MGG_00922, MGG_02514, MGG_02849, MGG_02898, MGG_04031, MGG_06643, MGG_11021). It is noteworthy that certain metallopeptidases, such as AVR-PITA in *M. oryzae*, have roles associated with virulence [53]. Once proteins are degraded into amino acids, they must be translocated quickly through amino acid transporters. Five amino acid transporters (MGG_05128, MGG_00289, MGG_14115, MGG_07606 and MGG_11327) were enriched under nitrogen starvation. Increased synthesis of protein degrading enzymes and amino acid transporters suggests that during nitrogen starvation, fungi aggressively and efficiently recycle/scavenge available nitrogen sources from their own cells and/or surrounding resources.

To identify nitrogen metabolism proteins which are potentially regulated by GATA transcription factors, HGATAR motifs within the promoters of identified proteins were searched using the POCO motif finding program [25]. The occurrence and frequency of this motif, according to the protein expression pattern, was compared in regard to nitrogen starvation (Additional file 4: Table S3). Among 5498 *M. oryzae* proteins we identified, genes encoding 3831 proteins (69.7%) have HGATAR motifs in their promoter region. Most proteins had single (48.2%) or double (32.2%) motifs. A similar percentage of proteins containing the motif was found across the subset of induced and repressed proteins. However, promoters with high numbers of HGATAR motifs were enriched in induced proteins. For induced proteins, 7.6% contained 4 or more HGATAR motifs per promoter compared to 5.8% in repressed or not differentially expressed proteins. Surprising, these induced proteins with high numbers of HGATAR motifs had similar biological function; protein degradation and amino acid modification. These proteins contained an amino acid transferase (MGG_09919), an amidohydrolase (MGG_10507), an L-asparaginase 1 (MGG_04119), a dipeptidase (MGG_09530), a urease (MGG_01324) and a L-serine dehydratase (MGG_06950). These findings suggests that protein catabolism during nitrogen starvation is tightly controlled by GATA transcription factors, likely including NUT1 in *M. oryzae*.

Increased production of putative extracellular proteins during nitrogen starvation

Among the proteins enriched during nitrogen starvation, 27% of the proteins (99 proteins out of 363 proteins) were predicted to be extracellular proteins in contrast to only 8% of proteins (22 proteins out of 266 proteins) among repressed proteins. On the other hand, nuclear and mitochondrial proteins were more represented among repressed proteins (Fig. 5a). In plant pathogenic fungi, a number of small secreted proteins have been shown to act as effector proteins enabling pathogen virulence and inplanta growth [54, 55]. Interestingly, all

22 small secreted proteins (< 250 a.a) were identified exclusively among proteins induced by nitrogen starvation (Fig. 5b). Most of these proteins are unknown hypothetical proteins without any functional domains or orthologs in other organisms (Table 2).

Among the small proteins, the hydrophobin effector MPG1 (MGG_10315) was significantly (>6 fold) induced during nitrogen starvation. MPG1 is a small hydrophobic protein involved in the surface interaction during plant infection. The gene is highly expressed in fungi during plant infection and when starved for nitrogen and carbon sources. Mpg1 mutants are non-pathogenic [2, 56]. In addition, MPG1 is regulated by NUT1 under nitrogen starvation conditions [36, 57]. Other effector genes, such as Avr9 from *Cladosporium fulvum*, are highly expressed during infection and under nitrogen starvation conditions. Interestingly, the promoter of Avr9 contains mutiple GATA binding sites that are necessary for induction by the GATA transcription factor NRF1. NRF1 is also required for virulence by *C. fulvum* [1, 2].

We also found MGG_05344, Snodprot1, to be significantly induced by nitrogen starvation. Snodprot1 is associated with pathogenicity and was first identified to be expressed during plant infection by the wheat blotch pathogen *Parastagonospora nodorum* [58]. Application of the Snodprot1 protein from *M. oryzae* elicited host defense responses in rice and induced host cell death [59]. The protein was found to be essential for rice infection by *M. oryzae* [60].

M. oryzae has several lysin motif (LysM) containing proteins and one of them, MGG_07571, was expressed only in nitrogen starved condition in this study. Other LysM containing secreted proteins such as ECP6 in *Cladosporium fulvum* and SLP1 in *M. oryzae* are well characterized

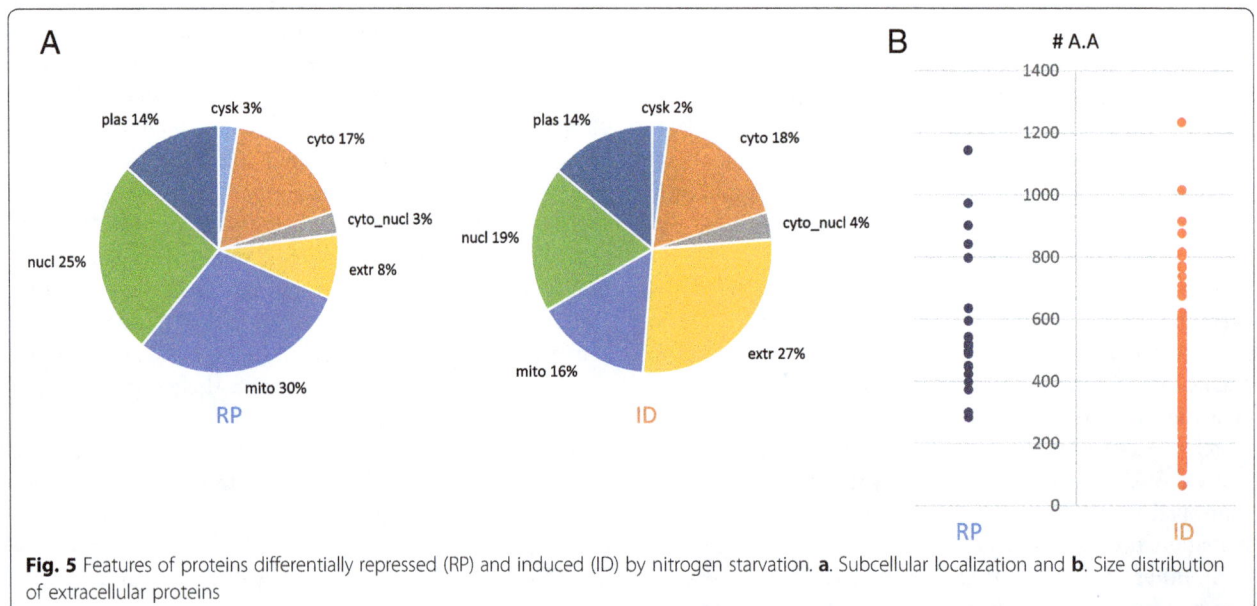

Fig. 5 Features of proteins differentially repressed (RP) and induced (ID) by nitrogen starvation. **a**. Subcellular localization and **b**. Size distribution of extracellular proteins

Table 2 List of small extracellular proteins enriched during nitrogen starvation

Protein ID	Annotation	# A.A	PSM_N+	PSM_N-	FC	p value	# HGATAR	Cys > 3%
MGG_09842	hypothetical protein	171	0	521.8	–	0.00	0	Y
MGG_13009	hypothetical protein	226	0	250.4	–	0.00	0	
MGG_00052	hypothetical protein	225	0	231.7	–	0.00	1	
MGG_05344	SnodProt1	138	56	120.7	2.15	0.00	0	
MGG_04323	hypothetical protein	244	46.2	97.1	2.10	0.04	1	
MGG_07571	LysM domain-containing protein	143	0	64.9	–	0.00	3	Y
MGG_00081	hypothetical protein	190	0	51.3	–	0.01	1	
MGG_07850	hypothetical protein	245	13.4	50.7	3.78	0.01	3	
MGG_07782	dehydroquinase class II	158	10.4	46.3	4.45	0.00	3	
MGG_03791	hypothetical protein	130	0	41.1	–	0.02	0	Y
MGG_13654	hypothetical protein	124	0	38.1	–	0.00	1	Y
MGG_06234	hypothetical protein	142	0	34	–	0.01	2	Y
MGG_06359	hypothetical protein	248	0	28.9	–	0.01	0	
MGG_15022	hypothetical protein	143	0	28.8	–	0.02	1	
MGG_12247	hypothetical protein	221	0	20.6	–	0.01	0	
MGG_10456	hypothetical protein	150	0	14.4	–	0.00	0	Y
MGG_10315	hydrophobin-like protein MPG1	113	2.1	13.4	6.47	0.02	0	Y
MGG_03442	hypothetical protein	246	4.2	12.4	2.96	0.02	2	
MGG_07246	hypothetical protein	200	0	9.3	–	0.00	1	Y
MGG_07791	surface protein 1	135	0	9.2	–	0.03	0	Y
MGG_02085	FAD-linked sulfhydryl oxidase ALR	218	0	8.3	–	0.02	2	
MGG_10604	hypothetical protein	66	0	5.2	–	0.04	1	

*Number of spectral counts in nitrogen starved (PSM_N-) and nitrate nitrogen supplemented (PSM_N+), fold change (FC) and p-value are presented. The number of HGATAR domains in the promoter region and cysteine content of the protein are also shown

effector proteins. LysM domains bind chitin and prevent chitin triggered immune responses in the host plant [61, 62].

MGG_07791, a secreted protein homologous to a major cell surface protein CLSP1 in the bean pathogen *C. lindemuthianum*, was also induced by nitrogen starvation. This protein may be involved in adhesion of the pathogen to the host. However, the gene is conserved in other filamentous fungi including *F. graminearum* and *A. nidulans* [63].

Nine of the 22 small secreted proteins, including MPG1, Surface protein1 (MGG_07791), and LysM protein (MGG_07571) have more than 3% cysteine content (Table 2). High cysteine content would likely confer structural stability to the secreted protein that in a hostile external environment such as at the host interface may enable them to work as effector proteins [64].

Reduction of de novo protein production

De novo protein synthesis is a very energy consuming process. The largest functional group (17 of 266) of repressed proteins was associated with translation. These included five tRNA synthetases (MGG_01165, isoleucyl-tRNA synthetase; MGG_00161, lysyl-tRNA synthetase; MGG_04042, leucyl-tRNA synthetase; MGG_05275, glutamyl-tRNA synthetase; and MGG_13783, aspartyl-tRNA synthetase) and MGG_01021, a tRNA (guanine-N(7)-)-methyltransferase. Protein synthesis is mediated by the ribosome complex, which is composed of dozens of proteins in both the large and small units. In our study, a number of structural constituents of the large subunit (MGG_02511, MGG_04455, MGG_05647, MGG_06468, MGG_10825) and small subunit (MGG_06744, MGG_06935) as well as mitochondrial ribosomal units (MGG_08323, MGG_09301) were repressed during nitrogen starvation. MGG_14349, a translation release factor was also down regulated.

Under nitrogen limiting conditions, active growth becomes severely challenging. Thus, an overall down-turn in the production of proteins involved in protein synthesis machinery would be not unexpected. Instead, efforts are directed to processes that can help fungi survive nitrogen limitation, including scavenging and in the case of pathogens exploiting host resources.

Conclusion

Technological improvements in tandem mass spectroscopy and data analysis has enabled a thorough investigation of proteome changes when the rice blast fungus encounters nitrogen starvation. Representing more than 40% of the entire proteome, 5498 proteins were identified during mycelial growth with/without nitrogen sources employing total unfractionated protein samples. In depth analysis of 629 differentially enriched proteins afforded new insight into fungal responses to nitrogen starvation. Proteins associated with melanin accumulation and nitrogen scavenging were observed to increase under nitrogen stress, whereas protein synthesis and proteins associated with nitrogen assimilation decreased. This study further uncovered that nitrogen limitation triggers accumulation of secreted proteins, which may enable host plant infection or function as effector proteins. We expect further functional characterization of those differentially expressed proteins will help broaden the scope of future studies.

Additional files

Additional file 1: Table S1. Protein Identification. (XLSX 316 kb)

Additional file 2: Table S2. List of differentially expressed proteins. (XLSX 66 kb)

Additional file 3: Figure S1. Increased pigmentation during nitrogen starvation. (PPTX 161 kb)

Additional file 4: Table S3. Distribution of HGATAR motifs in *M. oryzae* gene promoters. (XLSX 11 kb)

Abbreviations

3HNR: trihydroxy-naphthalene reductase; 4HNR: tetrahydroxy-naphthalene reductase; A.A.: Amino acid; ACN: Acetonitrile; AGC: Automatic gain control; Cyclic AMP: Cyclic adenosine monophosphate; DHN: Dihydroxynaphthalene; DTT: Dithiothreitol; EDTA: Ethylenediaminetetraacetic acid; FA: Formic acid; FASP: Filter aided sample preparation; FDR: False discovery rate; GABA: Gamma-aminobutyric acid; HPLC: High performance liquid chromatography; KEGG: Kyoto Encyclopedia of Genes and Genomes; LC: Liquid chromatography mass spectrometry; LC-MS/MS: Liquid chromatography-tandem mass spectrometry; LysM: Lysin motif; MS: Mass spectrometry; NCR: Nitrogen catabolic repression; NEM: N-ethylmaleimide; NSAF: Normalized spectral abundance factor; PSM: Peptide spectral match; ROS: Reactive oxygen species; SCD: Scytalone dehydratase; TCA cycle: Tricarboxylic acid cycle

Acknowledgements

Not applicable.

Funding

This material is based upon work supported by the U.S Department of Agriculture (National Institute of Food and Agriculture) under Grant Number 2014-67013-21722.

Authors' contributions

YO and RD designed the experiments. YO, SR and JP carried out the proteomic studies. JP and DM helped to draft the manuscript. YO and SR analyzed the data and wrote the manuscript. YO, RD and DM take full responsibility for the integrity of the data analysis. All authors read and approved the final manuscript.

Competing interests

The authors declare that they have no competing interests.

Author details

¹Center for Integrated Fungal Research, Department of Entomology and Plant Pathology, North Carolina State University, Raleigh, NC 27695, USA. ²W. M. Keck FT-ICR Mass Spectrometry Laboratory, Department of Chemistry, North Carolina State University, Raleigh, NC 27695, USA.

References

1. Van den Ackerveken GFJM, Dunn RM, Cozijnsen AJ, Vossen JPMJ, Van den Broek HWJ, De Wit PJGM. Nitrogen limitation induces expression of the avirulence gene avr9 in the tomato pathogen *Cladosporium fulvum*. MGG. Mol Gen Genet. 1994;243:277–85.
2. Talbot NJ, Ebbole DJ, Hamer JE. Identification and characterization of MPG1, a gene involved in pathogenicity from the rice blast fungus *Magnaporthe grisea*. Plant Cell. 1993;5:1575–90.
3. Snoeijers SS, Pérez-García A, Joosten MHAJ, De Wit PJGM. The effect of nitrogen on disease development and gene expression in bacterial and fungal plant pathogens. Eur J Plant Pathol. 2000;106:493–506.
4. Marzluf GA. Genetic regulation of nitrogen metabolism in the fungi. Microbiol Mol Biol Rev. 1997;61:17–32.
5. Caddick MX. Characterization of a major Aspergillus regulatory gene, Area. Mol. Biol. In: Filamentous Fungi. New York: V C H Publishers; 1992. p. 141–52.
6. Fu YH, Marzluf GA. Characterization of nit-2, the major nitrogen regulatory gene of *Neurospora crassa*. Mol Cell Biol. 1987;7:1691–6.
7. Froeliger EH, Carpenter BE. NUT1, a major nitrogen regulatory gene in Magnaporthe grisea, is dispensable for pathogenicity. Mol Gen Genet. 1996; 251:647–56.
8. Pellier A-L, Lauge R, Veneault-Fourrey C, Langin T. CLNR1, the AREA/NIT2-like global nitrogen regulator of the plant fungal pathogen *Colletotrichum lindemuthianum* is required for the infection cycle. Mol Microbiol. 2003;48: 639–55.
9. Pérez-García A, Snoeijers SS, Joosten MHAJ, Goosen T, De Wit PJGM. Expression of the avirulence gene Avr9 of the fungal tomato pathogen Cladosporium Fulvum is regulated by the global nitrogen response factor NRF1. Mol Plant-Microbe Interact. 2001;14:316–25.
10. Greer CA, Webster RK. Occurrence, distribution, epidemiology, cultivar reaction, and management of rice blast disease in California. Plant Dis. 2001; 85:1096–102.
11. Couch BC, Kohn LM. A multilocus gene genealogy concordant with host preference indicates segregation of a new species, *Magnaporthe oryzae*, from *M. grisea*. Mycologia. 2002;94:683–93.
12. Dean RA, Talbot NJ, Ebbole DJ, Farman ML, Mitchell TK, Orbach MJ, et al. The genome sequence of the rice blast fungus *Magnaporthe grisea*. Nature. 2005;434:980–6.
13. Donofrio NM, Oh Y, Lundy R, Pan H, Brown DE, Jeong JS, et al. Global gene expression during nitrogen starvation in the rice blast fungus, *Magnaporthe grisea*. Fungal Genet Biol. 2006;43:605–17.
14. Xiong D, Wang Y, Tian C. Transcriptomic profiles of the smoke tree wilt fungus *Verticillium dahliae* under nutrient starvation stresses. Mol Gen Genomics. 2015;290:1963–77.
15. Oh Y, Donofrio N, Pan H, Coughlan S, Brown DE, Meng S, et al. Transcriptome analysis reveals new insight into appressorium formation and function in the rice blast fungus *Magnaporthe oryzae*. Genome Biol. 2008;9:R85.
16. Saitoh H, Fujisawa S, Ito A, Mitsuoka C, Berberich T, Tosa Y, et al. SPM1 encoding a vacuole localized protease is required for infection related autophagy of the rice blast fungus *Magnaporthe oryzae*. FEMS Microbiol Lett. 2009;300:115.
17. Wang Y, Wu J, Park ZY, Kim SG, Rakwal R, Agrawal GK, et al. Comparative secretome investigation of *Magnaporthe oryzae* proteins responsive to nitrogen starvation. J Proteome Res. 2011;10:3136–48.
18. Chu J, Li W-F, Cheng W, Lu M, Zhou K-H, Zhu H-Q, et al. Comparative analyses of secreted proteins from the phytopathogenic fungus *Verticillium dahliae* in response to nitrogen starvation. Biochim Biophys Acta Proteins Proteomics. 2015;1854:437–48.
19. Zhou X-G, Yu P, Yao C-X, Ding Y-M, Tao N, Zhao Z-W. Proteomic analysis of mycelial proteins from *Magnaporthe oryzae* under nitrogen starvation. Genet Mol Res. 2016;15(2). doi:10.4238/gmr.15028637.

20. Loziuk PL, Parker J, Li W, Lin C-Y, Wang JP, Li Q, et al. Elucidation of xylem specific transcription factors and absolute quantification of enzymes regulating cellulose biosynthesis in Populus trichocarpa. J Proteome Res. 2015;14:4158–68.

21. Gokce E, Shuford CM, Franck WL, Dean RA, Muddiman DC. Evaluation of normalization methods on GeLC-MS/MSlabel-free spectral counting data to correct for variation during proteomic workflows. J Am Soc Mass Spectrom. 2011;22:2199–208.

22. Huang DW, Lempicki RA, Sherman BT. Systematic and integrative analysis of large gene lists using DAVID bioinformatics resources. Nat Protoc. 2009;4:44–57.

23. Kanehisa M, Sato Y, Kawashima M, Furumichi M, Tanabe M. KEGG as a reference resource for gene and protein annotation. Nucleic Acids Res. 2016;44:D457–62.

24. Horton P, Park K-J, Obayashi T, Fujita N, Harada H, Adams-Collier CJ, et al. WoLF PSORT: protein localization predictor. Nucleic Acids Res. 2007;35: W585–7.

25. Kankainen M, Holm L. POCO: discovery of regulatory patterns from promoters of oppositely expressed gene sets. Nucleic Acids Res. 2005;33:427–31.

26. Franck WL, Gokce E, Oh Y, Muddiman DC, Dean RA. Temporal analysis of the Magnaporthe oryzae proteome during conidial germination and cyclic AMP (cAMP)-mediated appressorium formation. Mol Cell Proteomics. 2013; 12:2249–65.

27. Hebert AS, Richards AL, Bailey DJ, Ulbrich A, Coughlin EE, Westphall MS, et al. The one hour yeast proteome. Mol Cell Proteomics. 2014;13:339–47.

28. Paoletti AC, Parmely TJ, Tomomori-Sato C, Sato S, Zhu D, Conaway RC, et al. Quantitative proteomic analysis of distinct mammalian mediator complexes using normalized spectral abundance factors. Proc Natl Acad Sci. 2006;103: 18928–33.

29. Zybailov B, Mosley AL, Sardiu ME, Coleman MK, Florens L, Washburn MP. Statistical analysis of membrane proteome expression changes in Saccharomyces cerevisiae. J Proteome Res. 2006;5:2339–47.

30. Gokce E, Franck WL, Oh Y, Dean RA, Muddiman DC. In-depth analysis of the Magnaporthe oryzae conidial proteome. J Proteome Res. 2012;11:5827–35.

31. Crawford NM, Arst HNJ. The molecular genetics of nitrate assimilation in fungi and plants. Annu. Rev. Genet. 1993;27:115–46.

32. Stewart V, Vollmer SJ. Molecular cloning of nit-2, a regulatory gene required for nitrogen metabolite repression in Neurospora crassa. Gene. 1986;46:291–5.

33. Andrianopoulos A, Kourambas S, Sharp JA, Davis MA, Hynes MJ. Characterization of the Aspergillus nidulans nmrA gene involved in nitrogen metabolite repression. J Bacteriol. 1998;180:1973–7.

34. Wilson RA, Gibson RP, Quispe CF, Littlechild JA, Talbot NJ. An NADPH-dependent genetic switch regulates plant infection by the rice blast fungus. Proc Natl Acad Sci. 2010;107:21902–7.

35. Todd RB. 11 regulation of fungal nitrogen metabolism. In: Hoffmeister D, editor. Biochemistry and molecular biology. The Mycota (a comprehensive treatise on fungi as experimental systems for basic and applied research); 2016. p. 281–303.

36. Lau G, Hamer JE. Regulatory genes controlling MPG1 expression and pathogenicity in the rice blast fungus Magnaporthe grisea. Plant Cell. 1996;8:771–81.

37. Divon HH, Ziv C, Davydov O, Yarden O, Fluhr R. The global nitrogen regulator, FNR1, regulates fungal nutrition-genes and fitness during Fusarium oxysporum pathogenesis. Mol Plant Pathol. 2006;7:485–97.

38. Solomon PS, Oliver RP. The nitrogen content of the tomato leaf apoplast increases during infection by Cladosporium fulvum. Planta. 2001;213:241–9.

39. Solomon PS, Oliver RP. Evidence that gamma-aminobutyric acid is a major nitrogen source during Cladosporium fulvum infection of tomato. Planta. 2002;214:414–20.

40. Kopriva S, Suter M, von Ballmoos P, Hesse H, Krähenbühl U, Rennenberg H, et al. Interaction of sulfate assimilation with carbon andntrogen metabolism in lemna minor. Plant Physiol. 2002;130:1406–13.

41. Davidian J-C, Kopriva S. Regulation of sulfate uptake and assimilation–the same or not the same? Mol Plant. 2010;3:314–25.

42. Koprivova A, Suter M, den Camp RO, Brunold C, Kopriva S. Regulation of sulfate assimilation by nitrogen in arabidopsis. Plant Physiol. 2000;122:737–46.

43. Lee S, Kang BS. Interaction of sulfate assimilation with nitrate assimilation as a function of nutrient status and enzymatic co-regulation inbrassica juncea roots. J Plant Biol. 2005;48:270–5.

44. Tudzynski B. Nitrogen regulation of fungal secondary metabolism in fungi. Front Microbiol. 2014;5:656.

45. Chumley FG, Valent B. Genetic analysis of melanin deficient, nonpathogenic mutants of Magnaporthe grisea. Mol Plant-Microbe Interact. 1990;3:135–43.

46. Howard RJ, Valent B. Breaking and entering: host penetration by the fungal rice blast pathogen Magnaporthe grisea. Annu Rev. Microbiol. 1996;50:491–512.

47. Takano Y, Kubo Y, Kuroda I, Furusawa I. Temporal transcriptional pattern of three melanin biosynthesis genes, PKS1, SCD1, and THR1, in appressorium-differentiating and nondifferentiating conidia of Colletotrichum lagenarium. Appl Environ Microbiol. 1997;63:351–4.

48. Brown SH, Yarden O, Gollop N, Chen S, Zveibil A, Belausov E, et al. Differential protein expression in Colletotrichum acutatum: changes associated with reactive oxygen species and nitrogen starvation implicated in pathogenicity on strawberry. Mol Plant Pathol. 2008;9:171–90.

49. Schnitzler N, Peltroche-Llacsahuanga H, Bestier N, Zundorf J, Lutticken R, Haase G. Effect of melanin and carotenoids of Exophiala (Wangiella) dermatitidis on phagocytosis, oxidative burst, and killing by human neutrophils. Infect Immun. 1999;67:94–101.

50. Jacobson ES, Hove E, Emery HS. Antioxidant function of melanin in black fungi. Infect Immun. 1995;63:4944–5.

51. Zou Y, Zhao Y, Hu W. Chemical composition and radical scavenging activity of melanin from Auricularia auricula fruiting bodies. Food Sci Technol. 2015;35:253–8.

52. Wiemann P, Sieber CMK, von Bargen KW, Studt L, Niehaus EM, Espino JJ, et al. Deciphering the cryptic genome: genome-wide analyses of the rice pathogen Fusarium fujikuroi reveal complex regulation of secondary metabolism and novel metabolites. PLoS Pathog. 2013;9:e1003475.

53. Orbach MJ. A telomeric avirulence gene determines efficacy for the rice blast resistance gene pi-ta. Plant Cell. 2000;12:2019–32.

54. Rep M. Small proteins of plant-pathogenic fungi secreted during host colonization. FEMS Microbiol Lett. 2005;253:19–27.

55. Kim K-T, Jeon J, Choi J, Cheong K, Song H, Choi G, et al. Kingdom-wide analysis of fungal small secreted proteins (SSPs) reveals their potential role in host association. Front Plant Sci. 2016;7:186.

56. Talbot NJ, Kershaw MJ, Wakley GE, De Vries O, Wessels J, Hamer JE. MPG1 encodes a fungal hydrophobin involved in surface interactions during infection-related development of Magnaporthe grisea. Plant Cell. 1996;8: 985–99.

57. Soanes DM, Kershaw MJ, Cooley RN, Talbot NJ. Regulation of the MPG1 hydrophobin gene in the rice blast fungus Magnaporthe grisea. Mol Plant-Microbe Interact. 2002;15:1253–67.

58. Hall N, Keon JPR, Hargreaves JA. A homologue of a gene implicated in the virulence of human fungal diseases is present in a plant fungal pathogen and is expressed during infection. Physiol Mol Plant Pathol. 1999;55:69–73.

59. Wang Y, Wu J, Gon Kim S, Tsuda K, Gupta R, Park S-Y, et al. Magnaporthe oryzae-secreted protein MSP1 induces cell death and elicits defense responses in rice. Plant-Microbe Interact. 2016;29:299–312.

60. Jeong JS, Mitchell TK, Dean RA. The Magnaporthe grisea snodprot1 homolog, MSP1, is required for virulence. FEMS Microbiol Lett. 2007;273:157–65.

61. de Jonge R, van Esse HP, Kombrink A, Shinya T, Desaki Y, Bours R, et al. Conserved fungal LysM effector Ecp6 prevents chitin-triggered immunity in plants. Science. 2010;329:953–5.

62. Mentlak TA, Kombrink A, Shinya T, Ryder LS, Otomo I, Saitoh H, et al. Effector-mediated suppression of chitin-triggered immunity by Magnaporthe oryzae is necessary for rice blast disease. Plant Cell. 2012;24:322–35.

63. Hoi JWS, Herbert C, Bacha N, O'Connell R, Lafitte C, Borderies G, et al. Regulation and role of a STE12-like transcription factor from the plant pathogen Colletotrichum lindemuthianum. Mol Microbiol. 2007;64:68–82.

64. Trivedi MV, Laurence JS, Siahaan TJ. The role of thiols and disulfides on protein stability. Curr Protein Pept Sci. 2009;10:614–25.

15

Gender-specific effects of intrauterine growth restriction on the adipose tissue of adult rats: a proteomic approach

Adriana Pereira de Souza[1], Amanda Paula Pedroso[1], Regina Lúcia Harumi Watanabe[1], Ana Paula Segantine Dornellas[1], Valter Tadeu Boldarine[1], Helen Julie Laure[2], Claudia Maria Oller do Nascimento[1], Lila Missae Oyama[1], José Cesar Rosa[2] and Eliane Beraldi Ribeiro[1*]

Abstract

Background: Intrauterine growth restriction (IUGR) may program metabolic alterations affecting physiological functions and lead to diseases in later life. The adipose tissue is an important organ influencing energy homeostasis. The present study was aimed at exploring the consequences of IUGR on the retroperitoneal adipose tissue of adult male and female rats, using a proteomic approach.

Methods and Results: Pregnant Wistar rats were fed with balanced chow, either *ad libitum* (control group) or restricted to 50 % of control intake (restricted group) during the whole gestation. The offspring were weaned to ad libitum chow and studied at 4 months of age. Retroperitoneal fat was analyzed by two-dimensional gel electrophoresis followed by mass spectrometry.

Both male and female restricted groups had low body weight at birth and at weaning but normal body weight at adulthood. The restricted males had normal fat pads weight and serum glucose levels, with a trend to hyperinsulinemia. The restricted females had increased fat pads weight with normal glucose and insulin levels.

The restricted males showed up-regulated levels of proteasome subunit α type 3, branched-chain-amino-acid aminotransferase, elongation 1- alpha 1, fatty acid synthase levels, cytosolic malate dehydrogenase and ATP synthase subunit alpha. These alterations point to increased proteolysis and lipogenesis rates and favoring of ATP generation. The restricted females showed down-regulated levels of L-lactate dehydrogenase perilipin-1, mitochondrial branched-chain alpha-keto acid dehydrogenase E1, and transketolase. These findings suggest impairment of glycemic control, stimulation of lipolysis and inhibition of proteolysis, pentose phosphate pathway and lipogenesis rates.

In both genders, several proteins involved in oxidative stress and inflammation were affected, in a pattern compatible with impairment of these responses.

Conclusions: The proteomic analysis of adipose tissue showed that, although IUGR affected pathways of substrate and energy metabolism in both males and females, important gender differences were evident. While IUGR males displayed alterations pointing to a predisposition to later development of obesity, the alterations observed in IUGR females pointed to a metabolic status of established obesity, in agreement with their increased fat pads mass.

Keywords: Intrauterine growth restriction, Maternal undernutrition, Adipose tissue, Proteome

* Correspondence: eliane.beraldi@gmail.com
[1]Departamento de Fisiologia, Universidade Federal de São Paulo, Rua Botucatu, 862 - 2 andar, Vila Clementino, São Paulo, SP 04023-062, Brazil
Full list of author information is available at the end of the article

Background

The concept of fetal programming describes that the exposition to adverse stimuli or insults, during critical phases of intrauterine development, may induce permanent changes in physiological functions and lead to adulthood diseases [1]. Increased risks of type 2 diabetes, insulin resistance, cardiovascular diseases and obesity have been associated with intrauterine growth restriction (IUGR) induced by undernutrition [2, 3]. These consequences have been shown to depend on the severity, duration and gestational period of the insult and also to be gender-dependent [4, 5]. Importantly, a mismatch between intrauterine and post-natal nutritional environment has been shown to be relevant for the expression of the programmed metabolic dysfunctions [6].

Previous reports have found that the adult offspring of IUGR rats displayed hyperphagia, obesity, hypertension, high serum leptin and insulin levels, increased hypothalamic density of leptin and serotonin receptors, and impairment of serotonin and insulin hypothalamic signaling [7–13]. Additionally, IUGR led to increased circulating levels of catecholamines in rats [14] and decreased levels of branched-chain amino acids in mice [15].

Lipogenesis and lipolysis are important physiologic pathways in the adipose tissue. Fatty acids for triacylglycerols synthesis may be taken up from the circulation or derive from *de novo* synthesis from glucose. Glucose degradation also yields glycerol 3-phosphate for fatty acids esterification and storage [16, 17]. Conversely, fatty acids and glycerol derived from triacylglycerols lipolysis may be released into the circulation. Those are hormone-controlled pathways. Lipolysis is inhibited by insulin and stimulated by cathecolamines and growth hormone. On the other hand, lipogenesis is stimulated by insulin while growth hormone is inhibitory [16–18]. Rat studies have indicated that IUGR due to maternal food restriction decreased adipose tissue lipolysis while it increased lipogenesis and/or adipogenesis, due to impairment of sympathetic activity [2].

The adipose tissue is also an endocrine organ whose secretions influence the onset of metabolic disorders [19]. Increased production of pro-inflammatory adipokynes in obesity plays a relevant role in the linking of adiposity, metabolic syndrome and cardiovascular diseases [20, 21]. Recent reports have shown that fetal leptin and adiponectin levels closely related to birth weight and IUGR has been shown to increase leptin but not adiponectin levels. Moreover, TNF-α levels have been found to be either normal or increased while IL-6 levels were either increased or decreases in IUGR [19, 22, 23].

Proteomic analysis allows the exam of hundreds of proteins in a sample and the identification of modification on their expression pattern in response to physiologic, pathologic and nutritional alterations, possibly leading to the identification and characterization of biological markers [24, 25]. Two-dimensional gel electrophoresis (2DE) followed by mass spectrometry remains an effective methodology in proteomics, especially as an initial approach [26, 27]. Recent proteomic studies have focused on the consequences of IUGR in tissues of animals and humans. Down-regulation of proteins related to oxidative phosphorylation has been found in the liver of both male and female IUGR rats [28]. In piglets, IUGR up-regulated subcutaneous adipose tissue levels of proteins related to glucose and fatty acid metabolism, lipid transport and apoptosis [29]. In humans, IUGR has been shown to increase serum levels of proteins related to signal transduction, blood coagulation and antioxidant response, while immune response proteins were down-regulated [30].

The above data indicate that IUGR may injure multiple aspects pertinent to adipocytes physiology that are relevant to the development of metabolic impairment and obesity in adulthood. Considering the above, the objective of this study was to further explore the consequences of IUGR in the adipose tissue of adult rats through proteomic approach.

Results

Body and white adipose tissue weight and blood and tissue parameters

Restricted male and female rats had low body weight at birth and at weaning but this difference was no longer observed at four months of age (Table 1). Food intake was similar between control and restricted animals, from weaning to 4 months of age (data not shown).

White adipose tissues weight were similar between control and restricted males. The female restricted rats showed increased weight of mesenteric and gonadal white adipose tissue and the sum of the three fat pads was higher than that of the control females (Table 1).

Serum glucose, adiponectin, corticosterone and triglycerides were similar between the groups of male rats. Serum insulin levels showed a tendency to increase in the restricted males (p = 0.083). Serum and tissue cytokines levels were similar between the male groups (Table 2).

For female rats, no differences were found in serum glucose, insulin, corticosterone, adiponectin and triglycerides between the control and restricted groups. The restricted females showed low serum IL-1β (Table 2).

Proteomic analysis

The 2DE gels of retroperitoneal adipose tissue showed 425 ± 2.9 spots in control males ($N = 8$) and 417 ± 4.5 spots in restricted males ($N = 8$). Of these, 37 spots showed significant density changes, with 15 spots under- and 22 over-expressed. Spots optic densities are shown in Additional file 1: Table S1. The significantly affected

Table 1 Body and white adipose tissue weight of male and female control and restricted offspring

	Male		Female	
	Control (16)	Restricted (17)	Control (14)	Restricted (14)
BW at birth (g)	6.08 ± 0.11	4.84 ± 0.12***	5.69 ± 0.14	4.80 ± 0.13***
BW at weaning (g)	85.44 ± 2.32	76.05 ± 2.52**	79.83 ± 2.15	70.24 ± 2.68**
BW at 4-months (g)	390.5 ± 7.2	385.4 ± 10.6	234.9 ± 3.3	227.2 ± 4.2
BW gain (g)	263.6 ± 8.4	271.2 ± 10.8	121.2 ± 4.8	127.2 ± 5.5
Retroperitoneal (g/100 g bw)	1.15 ± 0.11	1.24 ± 0.11	0.80 ± 0.06	0.95 ± 0.06
Mesenteric (g/100 g bw)	0.91 ± 0.07	0.98 ± 0.07	1.01 ± 0.05	1.19 ± 0.04**
Gonadal	1.31 ± 0.11	1.50 ± 0.08	2.46 ± 0.19	3.0 ± 0.17*
Total weight (g/100 g bw)	3.37 ± 0.28	3.72 ± 0.23	4.27 ± 0.28	5.22 ± 0.22*

Data are means ± SEM; (number of animals)
BW body weight, *g/100 g bw* grams/100 g of body weight, *Total weight* sum of retroperitoneal, mesenteric and gonaldal fat pads
*$p < 0.01$ vs. control
**$p < 0.01$ vs. control
***$p < 0.001$ vs. control

spots were analyzed by mass spectrometry for proteins identification. Figure 1 shows a representative image of a 2DE gel of a control male with indication of the spots significantly affected by IUGR.

The MS analysis identified 11 of the 15 under-expressed proteins and 17 of the 22 over-expressed proteins. One down-regulated protein (Murinoglobulin-1) and 2 up-regulated proteins (Actin, cytoplasmic I and Voltage-dependent anion-selective channel protein 1) were identified in 2 adjacent spots. Table 3 shows the Swiss-Prot Accession Numbers (available at http://www.expasy.ch/sprot), full protein names, theoretical molecular weight (MW) and isoelectric point as well as the mass spectrometry data of the identified proteins having statistically significant Mascot score results ($p < 0.05$) in males. Additional file 2: Table S2 shows gene names and biological processes of the proteins significantly up-regulated and down-regulated proteins, as assessed by Panther software. Metabolic

process was the most common biological process class for both the down-regulated (7 out of 10) and the up-regulated (9 out of 15) proteins. The metabolic processes included lipidic, amino acid and carbohydrate metabolism (Table 4).

The 2DE gels of females had 404 ± 3.6 spots in the controls ($N = 6$) and 397 ± 3.9 spots in the restricted ones ($N = 6$). Of these, 27 spots showed significant density changes, with 20 spots under- and 7 over-expressed. Spots optic densities are shown in Additional file 1: Table S1. The significantly affected spots were analyzed by mass spectrometry for proteins identification. Figure 2 shows a representative image of a 2DE gel of a control female with indication of the spots significantly affected by IUGR.

The MS analysis identified 18 of the 20 down-regulated proteins and 4 of the 7 up-regulated proteins. Two down-regulated proteins (Serotransferrin and Ig gamma 2-A chain C) were identified in 2 adjacent spots.

Table 2 Blood and retroperitoneal adipose tissue parameters of adult male and female control and restricted offspring

	Male		Female	
	Control	Restricted	Control	Restricted
Serum glucose (mg/dl)	100.2 ± 5.2 (18)	92.4 ± 3.1 (9)	95.2 ± 3.9 (13)	100.0 ± 5.2 (9)
Serum Insulin (ng/ml)	0.50 ± 0.07 (8)	0.79 ± 0.13 (9)	0.35 ± 0.06 (9)	0.34 ± 0.12 (9)
Serum adiponectin (μg/ml)	12.3 ± 1.4 (9)	11.3 ± 1.1 (9)	17.45 ± 1.81 (9)	18.47 ± 2.27 (9)
Serum corticosterone (ng/ml)	94.2 ± 7.9 (9)	97.0 ± 8.8 (8)	63.58 ± 9.54 (9)	71.06 ± 5.58 (9)
Serum triglycerides (mg/dl)	55.1 ± 4.3 (18)	68.6 ± 9.3 (9)	37.2 ± 3.1 (14)	38.7 ± 6.0 (8)
Serum TNF-α (pg/ml)	7.6 ± 1.7 (8)	18.2 ± 6.4 (8)	20.07 ± 1.8 (8)	14.78 ± 4.3 (9)
Serum IL-1β (ng/ml)	0.10 ± 0.04 (6)	0.06 ± 0.01 (6)	0.14 ± 0.04 (9)	0.04 ± 0.01*(8)
Tissue TNF- α (pg/ml)	49.3 ± 4.5 (8)	55.0 ± 3.2 (8)	63.93 ± 4.9 (8)	71.55 ± 4.3 (8)
Tissue IL-6 (pg/ml)	107.3 ± 7.0 (8)	106.6 ± 6.9 (8)	126.24 ± 3.8 (8)	134.13 ± 6.9 (8)
Tissue IL-10 (pg/ml)	187.6 ± 19.2 (8)	225.0 ± 28.7 (8)	275.89 ± 19.5 (8)	311.09 ± 19.5 (8)

Data are means ± SEM. (number of animals)
*$p < 0.05$ vs control

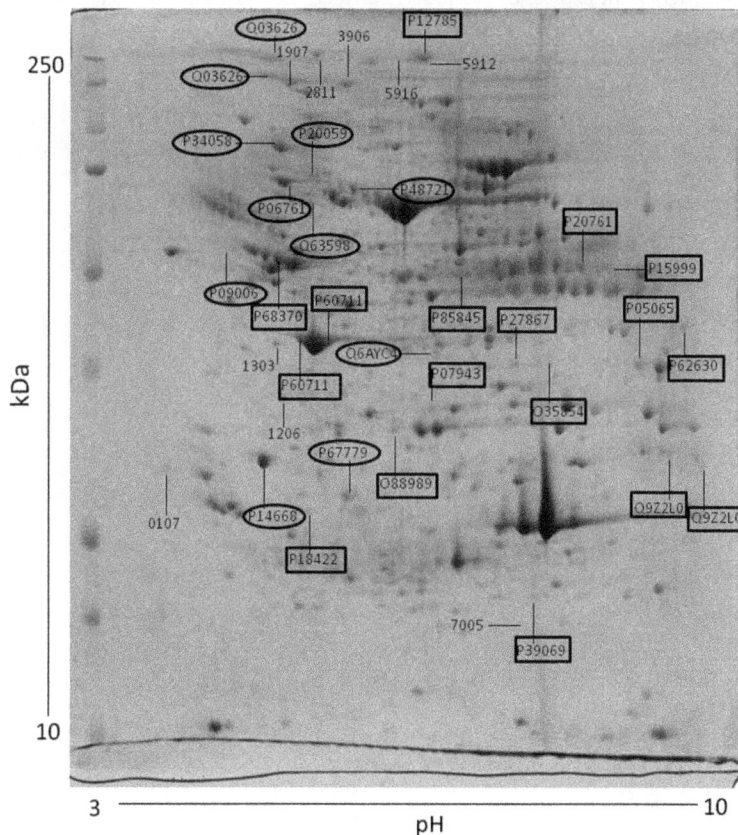

Fig. 1 Representative image of a 2DE gel of a control male retroperitoneal adipose tissue depicting the proteins significantly affected by IUGR. The numbers indicate the protein acession number. Numbers in squares indicate the over-expressed identified proteins. Numbers in circles indicate the under-expressed identified proteins

Female proteins data of significant Mascot score results ($p < 0.05$) are shown in Table 5. Additional file 3: Table S3 shows gene names and biological processes of the significantly up-regulated and down-regulated proteins in females. Metabolic process was the most common biological process class for the down-regulated (13 out of 16) and the up-regulated (4 out of 4) proteins. The metabolic processes included lipidic, amino acid and carbohydrate metabolism (Table 6).

In both males and females, some proteins were identified in 2 adjacent spots, what may possibly be attributed to the existence of either different isoforms or post-translational modifications of the protein.

Western blot analysis

A sub-set of selected proteins was analyzed by Western blotting to confirm the proteome results. Corroborating the male proteome result of a 40 % decrease in expression of 78 kDa glucose-regulated protein in the restricted males, the western blot analysis showed a 33 % decrease. The mitochondrial stress-70 protein showed a 48 % decrease in the proteome experiment and a 36 % decrease in the western blot experiment (Fig. 3).

In the females, the proteome results showed a 68 % decrease of Glutathione S-transferase theta-2 in the restricted group while the western blot analysis showed a 25 % decrease (Fig. 4).

Discussion

Male and female rats submitted to intrauterine growth restriction had normal food intake and body weight as adults, indicating catch-up growth, an adaptive mechanism against obesity in adult life [31, 32].

The restricted females, unlike the restricted males, showed increased fat pads weight, without overt peripheral insulin resistance, in accordance with a previous work from our laboratory [11]. This observation agrees with other reports showing the gender-dependency of the late consequences of rat maternal nutritional restriction [4, 5, 7–9]. In humans, early intrauterine undernutrition increased body mass index in 50 year-old women but not men [33].

The proteomic analysis of the adipose tissue of the males showed that IUGR caused alterations in the protein levels of 28 identified proteins. Levels of fatty acid synthase, enzyme of the *de novo* lipogenesis pathway

Table 3 Identified proteins with significant expression alteration between control and restricted males

Accession Number	Protein Name	Matched Peptides	Score	Coverage (%)	Fold Change (R/C)	MW (Da)/ pI
Down-regulated Proteins						
P06761	78 kDa glucose-regulated protein	6	334	11	0.60	71476/5.07
P14668	Annexin A5	6	338	18	0.39	35780/4.93
P34058	Heat shock protein HSP 90 β	9	245	12	0.68	83577/4.97
P20059	Hemopexin	4	66	6	0.55	52072/7.58
Q6AYC4	Macrophage-capping protein	1	50	3	0.63	39065/6.11
Q03626	Murinoglobulin-1	3	66	3	0.47	166614/5.68
Q03626	Murinoglobulin-1	2	81	1	0.24	166614/5.68
Q63598	Plastin-3	1	46	1	0.22	71157/5.32
P67779	Prohibitin	4	196	17	0.58	28860/5.57
P09006	Serine protease inhibitor A3N	3	109	6	0.37	46796/5.33
P48721	Stress-70 protein, mitochondrial	3	66	6	0.52	74102/5.97
Up-regulated Proteins						
P60711	Actin, cytoplamic I	6	191	21	1.77	42058/5.29
P60711	Actin, cytoplamic I	4	244	15	1.83	42058/5.29
P39069	Adenylate kinase izoenzyme 1	2	30	12	1.96	21686/7.66
P07943	Aldose reductase	3	94	10	2.05	36238/6.26
P15999	ATP synthase subunit alpha, mitochondrial	1	57	1	1.94	59833/9.22
O35854	Branched-chain-amino-acid aminotransferase. mitochondrial	4	52	9	1.67	44827/8.46
P62630	Elongation factor 1-alpha 1	2	52	4	1.79	50430/9.10
P85845	Fascin	3	39	7	2.27	55211/5.96
P12785	Fatty acid synthase	5	151	2	2.40	275146/5.96
P05065	Fructose-biphosphate aldolase A	7	245	17	1.49	39791/8.31
P20761	Ig gamma-2B chain C region	2	38	4	2.07	37112/7.70
O88989	Malate dehydrogenase, cytoplasmic	3	81	11	2.44	36634/6.16
P18422	Proteasome subunit α-type 3	4	86	15	1.93	28633/5.29
P27867	Sorbitol dehydrogenase	2	94	7	2.05	38790/7.14
P68370	Tubulin α-1A chain	3	121	9	2.37	50800/4.94
Q9Z2L0	Voltage-dependent anion-selective channel protein 1	2	60	8	2.16	30853/8.62
Q9Z2L0	Voltage-dependent anion-selective channel protein 1	1	42	4	4.94	30853/8.62

Accession number, protein name, number of matched peptides, proein score, percentage coverage, fold change (restricted/control) and theoretical molecular mass (Da) and pI of identified proteins

[16], were increased by IUGR, in agreement with other reports [34, 35]. A study comparing lean and obese subjects found that increased fatty acid synthase gene expression was linked to visceral fat accumulation [36].

Prohibitin levels were down-regulated in the restricted males. This protein has been shown to attenuate insulin-stimulated oxidation of glucose and fatty acids in adipose tissue [37]. Over-expression of prohibitin in mice adipose tissue increased fat pads [38]. In contrast, knockdown of prohibitin in 3 T3-L1 pre-adipocytes increased oxidative stress due to impairment of mitochondrial function [39].

These protein expression alterations found in the restricted males, one favoring and the other counteracting lipid accumulation, may represent a pre-obese condition. Although the restricted males did not have augmented fat pads mass, they did show a tendency to hyperinsulinemia, suggesting that an increase in lipid synthesis could lead to obesity later in life.

The increased levels of proteasome subunit α type-3 suggest that IUGR caused stimulation of proteolysis in males. The adipose tissue has been shown to be an important site of proteolysis and to contribute to the circulating amino acids pool [40, 41]. Obese women showed

Table 4 Biological process classification of identified proteins of male rats

Protein name	Metabolic process
Down-regulated in restricted males	
Serine protease inhibitor A3N	Proteolysis
Murinoglubulin-1	
Hemopexin	
78 kDa glucose-regulated protein	Protein folding
Heat shock protein HSP 90-beta	
Prohibitin	DNA replication
Annexin A5	Lipidic
Up-regulated in restricted males	
Branched-chain-amino-acid aminotransferase, mitochondrial	Amino acid
Proteasome subunit α type-3	Proteolysis
Fatty acid synthase	Lipidic
Sorbitol dehydrogenase	Carbohydrate
Malate dehydrogenase, cytoplasmic	Carbohydrate, tricarboxilic acid cycle (TCA)
ATP synthase subunit alpha, mitochondrial	Respiratory chain
Adenylate kinase isoenzyme 1	Nucleotide
Elongation factor 1-alpha 1	Translation
Aldose reductase	Transport

a decreased rate of amino acids release from the tissue, in response to fasting [42].

In the adipose tissue of obese humans, levels of mitochondrial branched-chain-amino-acid aminotransferase were reportedly decreased from lean levels in the metabolically unhealthy but not in the healthy subset of obese subjects [43]. Here, tissue levels of mitochondrial branched-chain-amino-acid aminotransferase were increased in the restricted males, indicating that their metabolism was not affected at the same extent as that seen in unhealthy obesity.

IUGR up-regulated the levels of elongation factor 1-alpha 1, a GTPase that delivers aminoacyl–tRNAs to ribosomes during protein translation [44, 45]. This protein has been shown to interact with nascent proteins ubiquitinated during translation, facilitating their delivery to proteasome [46] and to be associated with stimulation of cell proliferation in cancer cells [47]. In kidneys of streptozotocin diabetic rats, increased expression of elongation factor-1A has been related to hypertrophy of the adipose organ and to diabetes-associated oxidative stress [48].

Obesity has recently been associated with increased levels of several amino acids in the visceral adipose tissue of humans [49]. Moreover, metabolomic analysis showed increased levels of phenylalanine, tryptophan and glutamate in the umbilical vein blood of IUGR

neonates [50]. The increased levels of proteins related to proteolysis stimulation, as observed in the present study, may increase adipose tissue levels of amino acids. These may be converted to intermediates of the tricarboxylic acid (TCA) cycle. It is important to point out that, once entering the TCA cycle, these amino acids could be directed to either complete oxidation or generation of citrate [51], an important precursor for *de novo* lipogenesis. It is thus reasonable to suggest that proteolysis stimulation in the restricted males may provide amino acids for metabolic reactions in the tissue, rather than for release. A recent review has indicated that impairment of TCA cycle metabolites by IUGR could be an important biomarker of this condition [52].

Cytosolic malate dehydrogenase levels were up-regulated in the restricted males. This enzyme is active in the malate/aspartate shuttle, where it catalyzes the reduction of oxaloacetate to malate, using NADH. Malate enters mitochondria and is oxidized to oxaloacetate by mitochondrial malate dehydrogenase, with production of NADH. This shuttle not only channels the NADH produced during glycolysis to ATP production but also maintains the cytosolic NAD+/NADH ratio, essential for the oxidative metabolism of glucose [53–55]. Increased levels of mitochondrial malate dehydrogenase have been reported in pancreatic islets of adult rats with IUGR. However, ATP levels were not altered, which was attributed to the concomitant decrease of ATP synthase subunit 6 levels [55]. In the present study, ATP synthase subunit alpha was up-regulated, indicating that ATP production could be increased.

Some proteins down-regulated by IUGR in males are related to inflammation and cellular stress. Murinoglobulin-1 is a serino-protease inhibitor [56] that plays a protective role in the inflammatory response. Hemopexin is a positive acute-phase reactant that plays a protective role in lipid peroxidation through its heme binding effect [57], its levels being negatively associated with the severity of chronic sepsis [58]. In diet-induced obese mice, up-regulation of serum hemopexin levels has been suggested to represent a dysfunctional response in this chronic inflammatory condition [59].

The 78 kDa glucose-regulated protein is related to proper protein folding, protecting the cell from endoplasmic reticulum stress [60, 61], which has been described to link obesity to insulin resistance [61]. Obese mice overexpressing 78 kDa glucose-regulated protein in pancreas were protected against endoplasmic reticulum stress and had improvement of insulin sensitivity [62]. Heat shock protein HSP 90-beta is an important chaperone whose levels reportedly increase in obese humans, playing a role in mitigating the inflammatory stress present in obesity [63, 64]. Taken together, these protein alterations indicate that the restricted males presented

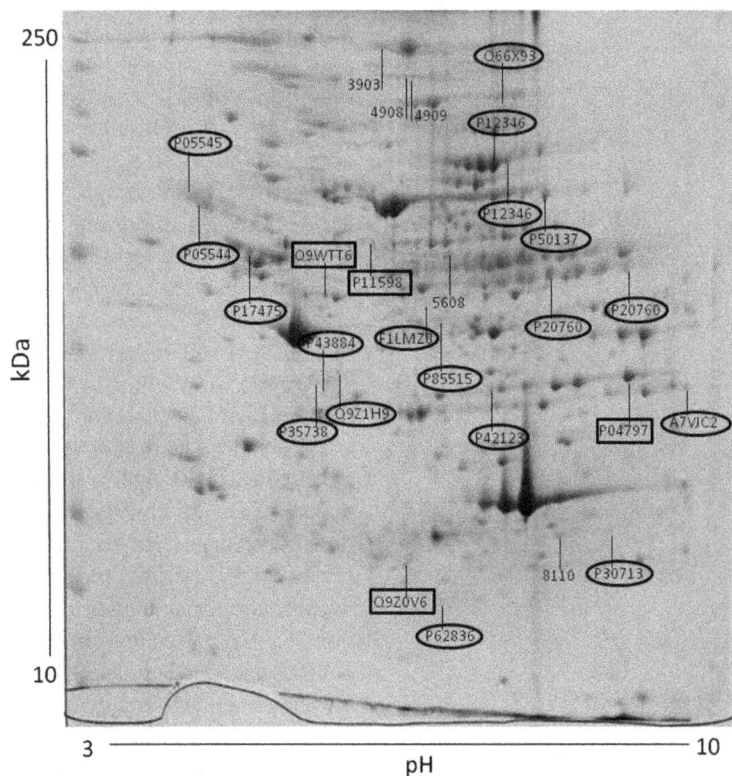

Fig. 2 Representative image of a 2DE gel of a control female retroperitoneal adipose tissue depicting the proteins significantly affected by IUGR. The numbers indicate the protein acession number. Numbers in squares indicate the over-expressed identified proteins. Numbers in circles indicate the under-expressed identified proteins

impairment of anti-inflammatory reactions in the adipose tissue.

Overall, the results found in the male rats indicate that, even though the restricted males did not have augmented fat pads or glucose intolerance, the alterations in adipose tissue metabolism point to a tendency to develop obesity.

In the females, IUGR affected the glycolysis/gluconeogenesis pathway. L-lactate dehydrogenase B was down-regulated in the restricted females, indicating low production of lactate from pyruvate. Due to its low blood supply, the adipose tissue produces considerable amounts of lactate, which can serve either as precursor to energy production or fatty acid synthesis [65, 66] or be released to the systemic circulation [67], even in normoxia conditions [68]. Adipose tissue lactate production has been shown to correlate with lactate dehydrogenase activity, both under normal and cafeteria diet feeding, and suggested to contribute to glycemic control, through consumption of excess circulating glucose [69].

Glyceraldehyde-3-phosphate dehydrogenase, the enzyme catalyzing the reversible conversion of glyceraldehyde-3-phosphate to 1,3 bisphosphoglycerate and NADH, was up-regulated in the restricted females. Stimulation of

glyceraldehyde-3-phosphate dehydrogenase has been reported in pre-obese, normoinsulinemic, Zucker rats [70]. Maternal peri-conceptional overnutrition, but not food restriction, increased fat mass of postnatal female lambs and glyceraldehyde-3-phosphate dehydrogenase gene expression correlated positively with perirenal fat amount [71].

Transketolase was down-regulated in the restricted females. This enzyme catalyzes the formation of glyceraldehyde-3-phospate in the non-oxidative branch of the pentose phosphate pathway, in which ribose is re-converted to glucose. Moreover, the pentose phosphate pathway generates NADPH for lipid synthesis. In obese individuals, decreased activity of the lipogenic pathway, with down-regulation of transketolase, has been interpreted as a mechanism aimed at reducing the growth of adipose tissue [72].

Reduced levels of mitochondrial branched-chain-amino-acid aminotransferase and mitochondrial 2-oxoisovalerate dehydrogenase (also known as branched-chain alpha-keto acid dehydrogenase E1), enzymes participating in the pathway of degradation of branched-chain amino acids, were found in the subcutaneous adipose tissue of unhealthy obese humans but not in the healthy obese subset [43]. Here, the latter enzyme was down-regulated in the

Table 5 Identified proteins with significant expression alteration between control and restricted females

Accession number	Protein name	Matched peptides	Score	Coverage (%)	Fold change (R/C)	MW (Da)/ pI
Down-regulated proteins						
F1LMZ8	26S proteasome non-ATPase regulatory subunit 11	2	89	4	0.44	47724/6.08
P35738	2-oxoisovalerate dehydrogenase subunit β. mitochondrial	1	45	4	0.46	43550/6.41
P30713	Glutathione S-transferase theta-2	1	47	5	0.38	27596/7.75
A7VJC2	Heterogeneous nuclear ribonucleoproteins A2/B1	4	96	13	0.35	37513/8.97
P20760	Ig gamma 2-A chain C	3	79	9	0.53	35685/7.72
	Ig gamma 2-A chain C	4	109	13	0.23	
P42123	L-lactate dehydrogenase B	1	30	2	0.57	36879/5.70
P43884	Perilipin 1	2	74	5	0.41	55986/6.37
Q9Z1H9	Protein kinase C delta binding protein	5	119	18	0.33	27894/5.79
P62836	Ras-related protein Rap-1A	1	39	5	0.32	21322/6.38
P05545	Serine protease inhibitor A3K	3	119	10	0.45	46764/5.31
P05544	Serine protease inhibitor A3L	1	46	2	0.29	46442/5.48
P12346	Serotransferrin	6	133	8	0.38	78550/7.14
P12346	Serotransferrin	2	110	3	0.31	78550/7.14
Q66X93	Staphylococcal nuclease domain-containing protein 1	3	38	4	0.39	103585/6.76
P50137	Transketolase	3	115	9	0.45	68355/7.23
P17475	α-1 antiproteinase	6	255	14	0.50	46281/5.70
P85515	α-centractin	2	57	8	0.63	42703/6.19
Up-regulated proteins						
P04797	Glyceraldehyde-3-phosphate dehydrogenase	1	34	4	2.49	36095/8.14
Q9WTT6	Guanine deaminase	6	312	16	1.54	51564/5.56
P11598	Protein disulfide-isomerase A3	5	117	119	3.20	57052/5.88
Q9Z0V6	Thioredoxin-dependent peroxide reductase. mitochondrial	2	78	9	2.09	28567/7.14

Accession number, protein name, number of matched peptides, proein score, percentage coverage, fold change (restricted/control) and theoretical molecular mass (Da) and pI of identified proteins

restricted females. It is possible to suggest that the metabolism of branched-chain-amino-acids in the adipose tissue of the restricted females resembled that found in obesity associated with metabolic derangements. This contrasts with the result in the restricted males, in which mitochondrial branched-chain-amino-acid aminotransferase was increased.

The restricted females also showed down-regulation of perilipin-1, an enzyme active in lipid droplet formation [73] and inversely correlated with adipocyte size and basal lipolysis [74]. Perilipin gene suppression increased basal lipolysis and prevented high-fat diet obesity in mice [75].

Glutathione S-transferase theta-2 was down-regulated in the restricted females. This protein is part of the antioxidant enzymes family, which catalyzes the conjugation of glutathione to a wide variety of compounds. Decreased glutathione or glutathione-S transferase levels have been linked to diabetes, due to its role in antioxidant pathways [76, 77]. On the other hand, high levels of glutathione-S transferase P in obese subjects activated inflammatory pathways and endoplasmic reticulum stress [78].

Other proteins related to antioxidant pathways were up-regulated by IUGR in the females. Protein disulfide-isomerase A3 is a thiol-disulfide oxidoreductase present in the endoplasmic reticulum and it catalyzes the formation, breakdown and rearrangement of disulfide bonds [79]. Increased levels of protein disulfide-isomerase A3 in the adipose tissue of obese subjects have been suggested to activate inflammatory pathway and endoplasmic reticulum stress [79]. Mitochondrial thioredoxin-dependent peroxide reductase regulates H_2O_2 levels, protecting the cell from the toxicity resulting from its accumulation [80], and depletion of this protein accelerated apoptosis [81].

Table 6 Biological process classification of identified proteins of female rats

Protein name	Metabolic process
Down-regulated in restricted females	
Serine protease inhibitor A3K	Proteolysis
Serine protease inhibitor A3L	
26S proteasome non-ATPase regulatory subunit 11	
Alpha-1-antiproteinase	
Transketolase	Carbohydrate, Amino acid, Lipidic
2-oxoisovalerate dehydrogenase subunit beta, mitochondrial	
Perilipin-1	Lipidic
Glutathione S-transferase theta-2	Protein
L-lactate dehydrogenase B chain	Glycolysis, TCA
Heterogeneous nuclear ribonucleoproteins A2/B1	Nucleotide
Protein kinase C delta-binding protein	Transcription
Staphylococcal nuclease domain-containing protein 1	
Ras-related protein Rap-1A	
Up-regulated in restricted females	
Guanine deaminase	Purine
Protein disulfide-isomerase A3	Protein folding
Glyceraldehyde-3-phosphate dehydrogenase	Glycolysis
Thioredoxin-dependent peroxide reductase, mitochondrial	

Classification of proteins in metabolic process

A protein down-regulated in the restricted females, α-1-antiproteinase, also known as serpin A1 and α1-antitrypsin, is a serine protease inhibitor with anti-inflammatory effects. It caused inhibition of lipopolysaccharide-mediated activation of in vitro human monocytes [82] and inhibited lung neutrophil chemotaxis [83–85]. Inhalation of α-1-antiproteinase decreased protein levels of IL-1β and IL-8 [86] while addition of purified plasma α-1-antiproteinase to pancreatic β-cells in vitro inhibited cytokine-induced apoptosis [87]. Alpha-1-antiproteinase gene therapy prevented the development of type 1 diabetes in non-obese mice [88]. Decreased levels of α-1-antiproteinase was reported by proteomic analysis of adipose tissue of women with gestational diabetes mellitus [89] impair the protection against inflammation and oxidative stress, compensatory mechanisms were recruited in the restricted females.

Conclusions
In the restricted males, the high levels of proteasome subunit α type 3, branched-chain-amino-acid aminotransferase and elongation 1- alpha 1 indicate increased proteolysis rate in the adipose tissue. High tissue levels of amino acids could generate lipogenesis precursors, a suggestion supported by the high levels of fatty acid

synthase. The increased levels of cytosolic malate dehydrogenase and ATP synthase subunit alpha may favor ATP production. These results indicate that, in the restricted males, the alterations in protein expression induced by IUGR pointed to a metabolic status favoring the development of obesity.

In the restricted females, the decreased levels of perilipin-1 are indicative of increased lipolysis while the low levels of mitochondrial branched-chain alpha-keto acid dehydrogenase E1 indicate low proteolysis rate. The low levels of transketolase could represent low activity of the pentose phosphate pathway and, consequently, decreased lipogenesis rate. Down-regulation of L-lactate dehydrogenase may lead to impairment of glycemic control. These alterations point to a metabolic status of established obesity in the restricted females. In both genders, the protein variations indicated impairment of pathways involved in the responses to oxidative stress and inflammation (Fig. 5).

Methods
Rats
Wistar rats were cared for in accordance with the guidelines of the committee on animal research ethics of the Federal University of São Paulo (approval 486691). Three months-old rats were mated and the first day of pregnancy was determined by examination of vaginal smears for the presence of sperm. From day 1 of pregnancy, the dams were randomly assigned to be a control or a restricted dam. The control dams were fed ad libitum throughout pregnancy and lactation. The restricted dams received only 50 % of control intake during the whole pregnancy and were fed ad libitum during lactation. On the day of delivery, the pups were adjusted to eight per dam.

After weaning, the male and female offspring from control and restricted dams were housed four/five per cage and fed ad libitum until 4 months of age. The food provided to dams and offspring consisted of standard rat chow (Nuvital Nutrients, Columbo, PR, Brazil) containing (w/w) 4.5 % fat, 23 % protein, and 33 % carbohydrate, with 2.7 kcal/g, as determined at the Bromatology Division of the Federal University of São Paulo. All animals were maintained in controlled conditions of lighting (12-h light/12-h dark cycle, lights off at 18:00 h) and temperature (24 ± 1 °C) and had free access to water throughout the experimental period.

The numbers of animals used in the study were 16 male and 14 female controls and 17 male and 14 female restricted rats.

Weight gain, weight of white adipose tissue and blood and tissue measurements
Food intake and body weight were measured once a week since weaning. At 4 months of age, the animals

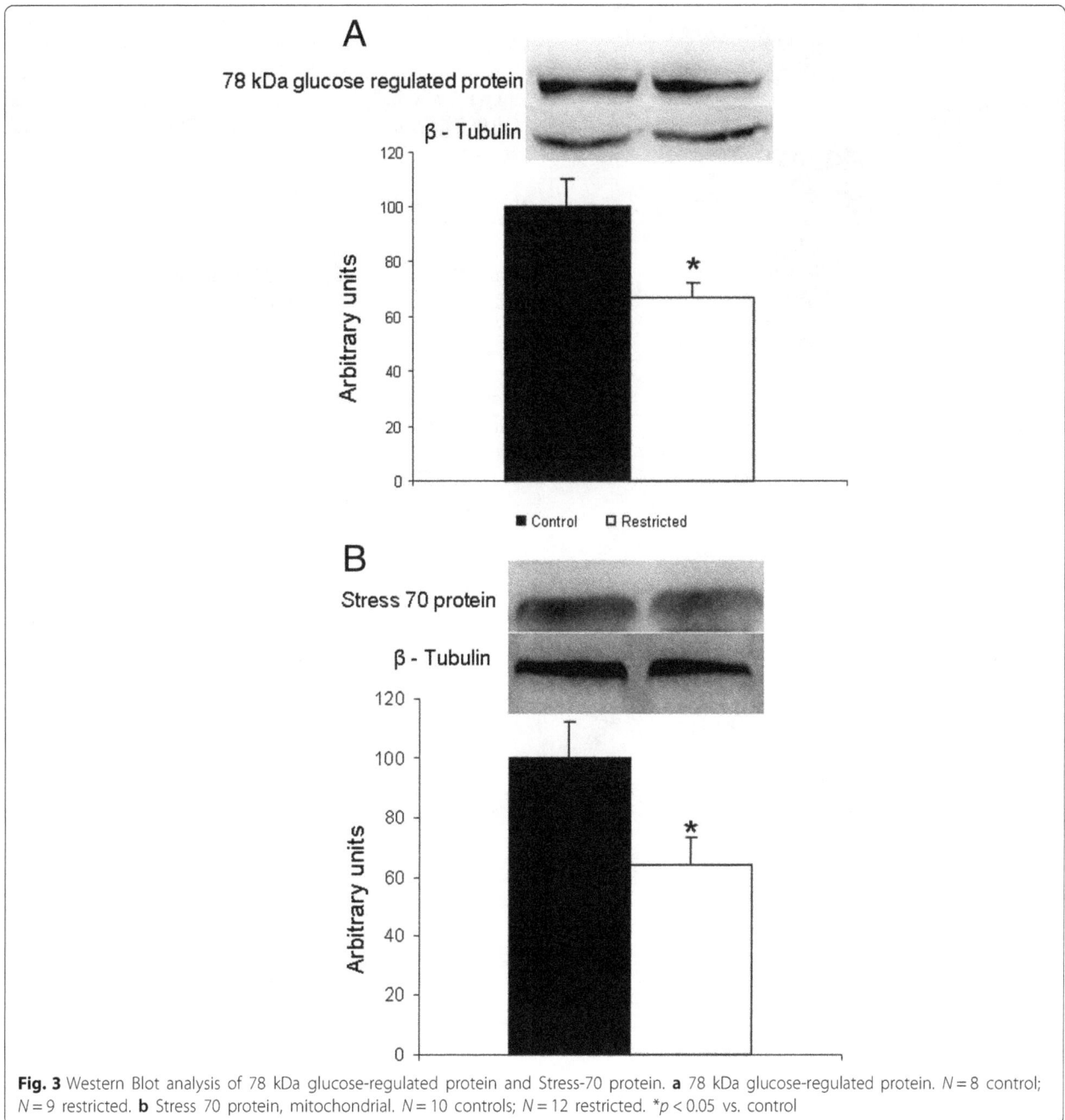

Fig. 3 Western Blot analysis of 78 kDa glucose-regulated protein and Stress-70 protein. **a** 78 kDa glucose-regulated protein. $N = 8$ control; $N = 9$ restricted. **b** Stress 70 protein, mitochondrial. $N = 10$ controls; $N = 12$ restricted. $*p < 0.05$ vs. control

were killed by decapitation. The retroperitoneal adipose tissue was rapidly removed, weighed and frozen in liquid nitrogen. The tissue was stored at - 80 °C until analysis. The gonadal and mesenteric white adipose tissues were dissected and weighed.

Trunk blood was centrifuged and the serum stored at - 80 °C. Glucose analysis was performed by the glucose oxidase method, using a commercially available kit with detection limit of 0.32 mg/dL (Glucose Pap Liquiform, Labtest Diagnostica, São Paulo, Brazil). Triglycerides levels were determined using a commercially available kit

with detection limit of 0.82 mg/dL (Labtest Diagnostica, São Paulo, Brazil). Insulin, corticosterone, adiponectin, TNF-α and IL-1β levels were measured by multiplex kit (Millipore, Bedford, MA, USA). The measurements of TNF-α, IL-10 and IL-6 in tissue were performed by Elisa (Millipore, Bedford, MA, USA).

Proteome analysis
Sample preparation

An aliquot of 700 mg of retroperitoneal adipose tissue was homogenized in 1 ml of extraction buffer (7 M urea,

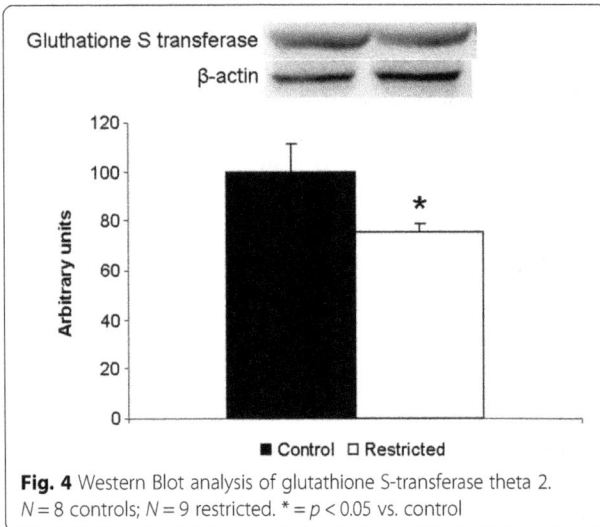

Fig. 4 Western Blot analysis of glutathione S-transferase theta 2. $N = 8$ controls; $N = 9$ restricted. $* = p < 0.05$ vs. control

2 M thiourea, 4 % (w/v) CHAPS, 0.5 % (v/v) Triton X-100) containing complete Mini Protease Inhibitor Cocktail Tablets (Roche Diagnostics, Germany), added immediately before use. Sample lysates were centrifuged (19,000 g/30 min.) and supernatants stored at - 80 °C until analysis.

Protein assay

Protein concentration of supernatants was determined using 2-D Quant Kit (GE Healthcare, Pittsburgh, USA) and bovine albumin as standard, according to manufacturer's recommendations.

Protein precipitation

Aliquots of 900 µg of protein were precipitated with a solution of 35 % KCl, 44 % chloroform, and 21 % methanol (v/v). The mixture was homogenized and centrifuged at 19,000 g and 4 °C for 15 min. The pellet was air-dried at room temperature.

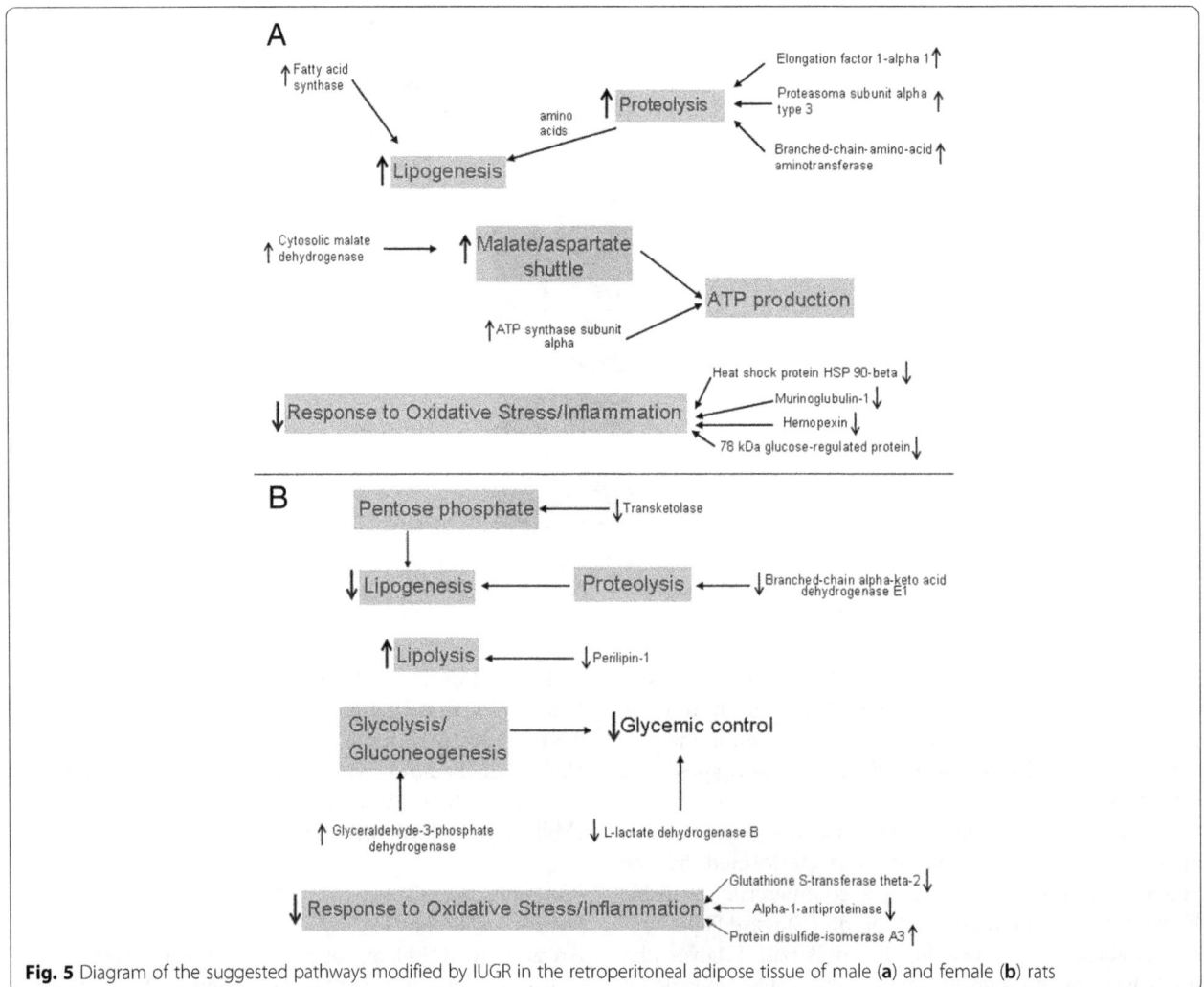

Fig. 5 Diagram of the suggested pathways modified by IUGR in the retroperitoneal adipose tissue of male (**a**) and female (**b**) rats

Two-dimensional gel electrophoresis and image analysis

For isoelectric focusing (IEF), the pellet was dissolved in 500 μL of rehydration buffer (7 M urea, 2 M thiourea, 4 % (w/v) CHAPS, 0.5 % (v/v) Triton X-100, 100 mM DTT, 0.2 % (v/v) IPG Buffer pH 3–10, and traces of Bromofenol blue). IEF was carried out on a Protean IEF cell (Bio-Rad, CA, USA) using immobiline dry strips (18 cm linear gradient, pH 3–10) previously rehydrated for 12–14 h. IEF was performed with the current limit set at 50 mA per IPG strip with the following conditions at 18 °C: 100 V for 30 min, 250 V for 2 h, 500 V for 30 min, 1000 V for 30 min, 2000 V for 30 min, 4000 V for 1 h, 8000 V for 1 h followed by 8000 V until 30000 Vh.

After focusing, strips were equilibrated for 25 min in buffer containing 6 M urea, 50 mM Trisma base pH 8.8, 34 % (v/v) glycerol, 2 % (w/v) SDS, and 1 % (w/v) DDT, followed by an additional 25 min in the same buffer containing 2.5 % (w/v) iodoacetamide instead of DTT. Strips were then loaded onto 12 % SDS- polyacrylamide gels. After running in Protean II Multi-Cell (Bio-Rad, CA, USA), at 50 mA per gel for 6 h, the gels were stained for 48 h with Coomassie Blue G-250 (Bio-Rad, CA, USA). Stained gels were scanned (GS-710 Calibrated Imaging Densitometer) and analyzed using PDQuest Image Analysis Software version 7.2 (Bio-Rad, CA, USA).

Matrix-assisted laser desorption ionization time-of-flight mass spectrometry

The selected spots were manually excised, distained and digested. The spots were excised and distained in 50 % methanol and 5 % acetic acid overnight. The excised spots were treated with 25 mM ammonium bicarbonate and 50 % acetonitrile (1:1) and dried in SpeedVac. To the dried spots, 10 mM DTT was added and incubated for 1 h at 56 °C, followed by 55 mM IAA for 45 min on the dark. The spots were dehydrated with 25 mM ammonium bicarbonate followed by 25 mM ammonium bicarbonate with 50 % acetonitrile and dried in SpeedVac. Digestion was performed overnight with 15 ng of trypsin (Promega, WI, USA) in 25 mM ammonium bicarbonate, at 37 ° C. Digested samples were desalted using C18 Zip Tips (Millipore, Bedford, MA, USA). Two microliters of sample were applied on the spectrometer plate and air-dried at room temperature. The matrix solution (10 mg/mL α-cyano-4 hydroxycinnamic acid in 70 % acetonitrile/0.1 % trifluoroacetic acid) was applied on the spectrometer plate and air-dried at room temperature.

MALDI-TOF/TOF MS was performed using an Axima Performance ToF-ToF, (Kratos-Shimadzu Biotech, Manchester, UK) mass spectrometer. The instrument was externally calibrated with $[M + H]^+$ ions of bradykinin (1–7 fragment, 757.4 Da), human angiotensin II (1046.54 Da), P14R synthetic peptide (1533.86 Da), and human ACTH (18–39 fragment, 2465.20 Da). Following MALDI MS analysis, MALDI MS/MS was performed on the 7 most abundant ions from each spot.

MASCOT (Matrix Science, UK) server was used to search Swiss-Prot protein database (http://www.matrix science.com). The following parameters were used in this search: no restrictions on protein molecular weight, trypsin digest with one missing cleavage, monoisotopic mass, taxonomy limited to Rattus, carbamidomethylation of cysteine as fixed modification, possible oxidation of methionine and tryptophan, peptide mass tolerance of 0.5 Da, fragment mass tolerance of 0.8 Da, and peptide charge +1. False discovery rate (FDR) assessment was estimated using Mascot decoy database approach and only proteins identified with 0 % FDR were included in the results. Protein matching probabilities were determined using MASCOT protein scores, with identification confidence indicated by the number of matching and the coverage of protein sequence by the matching peptides. The presence of at least one peptide with significant ion score was required for positive protein identification. Only statistically significant MASCOT score results ($p < 0.05$) were included in the analysis.

The identified proteins were classified in Panther (http://www.pantherdb.org/) according to biological process.

Western Blot analysis

A sub-set of adipose tissue samples was used in western blot experiments. A 700 mg aliquot was homogenized in 1.0 ml of solubilization buffer (10 mM EDTA, 100 mM Tris pH 7.5, 10 mM sodium pyrophosphate, 100 mM sodium fluoride, 10 mM sodium orthovanadate, 2 mM PMSF, aprotinin 2 μg/mL, and 1 % Triton X-100). Insoluble material was removed by centrifugation (19,000 g at 4 ° C for 40 min.). The supernatant was collected and one aliquot was separated for protein concentration determination. Tissue extracts were denatured by boiling for 5 min in Laemmli buffer [90] containing 100 mM DTT. The protein concentration was determined by colorimetric method (BCA Protein Assay, Bioagency Biotecnologia, Brazil).

Subsequently, protein extracts (100 μg) were resolved in 12 % SDS polyacrylamide gels and transferred to nitrocellulose membranes using a semi-dry transfer system (Bio-Rad, CA, USA). Non-specific binding sites were blocked for 2 h in 1 % bovine serum albumin. The nitrocellulose membranes were then incubated overnight with primary antibody and for 1 h with the appropriate secondary antibody conjugated with horseradish peroxidase. The quantitative analysis was performed by densitometry using Scion Image software (Scion Corporation, Frederick, MD, USA).

The results were expressed in arbitrary units, as percentage changes in relation to the control group. For evaluation of protein loading, all membranes were stripped and reblotted with anti-β-tubulin (for male) and anti-β-actin (for female) primary antibody. The antibodies against 78 kDa glucose regulated protein (1:1000; ab53068), mitochondrial stress 70 protein (1:1000; ab106654), and glutathione S-transferase theta-2 (1:2500; ab102045) were obtained from ABCAM (Cambridge, UK). The antibody against β-tubulin (1:5000; #2146S) was purchased from Cell Signaling (Danvers, MA, USA). The antibody against β-actin (1:1000; sc-130657) was purchased from Santa Cruz (Dallas, TX, USA).

Statistical analysis

The data are expressed as mean ± SEM. Comparisons between groups (control and restricted) were performed by Student t test. Statistical significance was set at $p < 0.05$.

Additional files

> **Additional file 1: Table S1.** Optic density of the spots with significant differences between the control and the restricted groups. (XLSX 12 kb)
>
> **Additional file 2: Table S2.** Protein name, gene name and biological process of the proteins identified in the adipose tissue of males. (XLSX 11 kb)
>
> **Additional file 3: Table S3.** Protein name, gene name and biological process of the proteins identified in the adipose tissue of females. (XLSX 10 kb)

Abbreviation

2DE: two-dimensional gel electrophoresis; ATP: adenosine triphosphate; BW: body weight; FC: Fold change; IL: interleukin; IUGR: intrauterine growth restriction; MW: Molecular weight; NADH: Nicotinamide adenine dinucleotide; TCA: trycarboxilic acid cycle; TNF-α: tumor nuclear factor alpha.

Competing interests

The authors declare that they have no competing interests.

Authors' contribution

APS contributed to the overall conception and design of the project and carried out the experiments, analysis and interpretation of the data, and preparation of the manuscript. APP and RLHW contributed to the overall conception and design of the project and carried out the experiments. APSD and VTB contributed with technical support and carried out the experiments. CMON and LMO contributed with critical discussions. HJL and JCR performed the mass spectrometry measurements and data analysis. EBR conceived the study, and participated in its design and coordination, data analysis and interpretation and preparation of the manuscript. All authors have read and approved the final manuscript.

Acknowledgements

This research was supported by grants from the Brazilian agencies: State of São Paulo Research Foundation (FAPESP), National Council for Scientific and Technological Development (CNPq), and Coordination for the Enhancement of Higher Education Personnel (CAPES).

Author details

[1]Departamento de Fisiologia, Universidade Federal de São Paulo, Rua Botucatu, 862 - 2 andar, Vila Clementino, São Paulo, SP 04023-062, Brazil. [2]Centro de Química de Proteínas – Hemocentro, Universidade de São Paulo, Ribeirão Preto, Brazil.

References

1. Langley-Evans SC. Nutritional programming of disease: unravelling the mechanism. J Anat. 2009;215:36–51.
2. Breton C. The hypothalamus-adipose axis is a key target of developmental programming by maternal nutrition manipulation. J Endocrinol. 2013;216(2): R19–31.
3. Hajj NE, Schneider E, Lehnen H, Haaf T. Epigenetics and life-long consequences of an adverse nutritional and diabetic intrauterine environment. Reproduction. 2014;148:R111–20.
4. Aiken CE, Ozanne SE. Sex differences in developmental programming models. Reproduction. 2013;145:R1–13.
5. Picó C, Palou M, Priego T, Sánchez J, Palou A. Metabolic programming of obesity by energy restriction during the perinatal period: different outcomes depending on gender and period, type and severity of restriction. Front Physiol. 2012;3:1–14.
6. Fisher RE, Steele M, Karrow NA. Fetal programming of the neuroendocrine-immune system and metabolic disease. J Preg. 2012. doi:10.1155/2012/792934.
7. Howie GJ, Sloboda DM, Vickers MH. Maternal undernutrition during critical windows of development results in differential and sex-specific effects on postnatal adiposity and related metabolic profiles in adult rat offspring. Br J Nutr. 2012;108:298–307.
8. Manuel-Apolinar L, Rocha L, Damasio L, Tesoro-Cruz E, Zarate A. Role of prenatal undernutrition in the expression of serotonin, dopamine and leptin receptor in adult mice: implication of food intake. Mol Med Reports. 2014;9:407–12.
9. Anguita RM, Sigulem DM, Sawaya AL. Intrauterine food restriction is associated with obesity in young rats. J Nutr. 1993;123:1421–8.
10. Porto LCJ, Sardinha FLC, Telles MM, Guimarães RB, Albuquerque KT, Andrade IS, et al. Impairment of the serotonergic control of feeding in adults female rats expose to intra-uterine malnutrition. Br J Nutr. 2009;101: 1255–61.
11. Sardinha FLC, Telles MM, Albuquerque KT, Oyama LM, Guimarães PAMP, Santos OFP, et al. Gender difference in the effect of intrauterine malnutrition on the central anorexigenic action of insulin in adult rats. Nutrition. 2006;22:1152–61.
12. Vickers MH, Breier BH, Cutfield WS, Hofman PL, Gluckman PD. Fetal origins of hyperphagia, obesity, and hypertension and postnatal amplification by hypercaloric nutrition. Am J Physiol. 2000;279:E83–7.
13. Bieswal F, Ahn M, Reusens B, Holvoet P, Raes M, Rees WD, et al. The importance of catch-up growth after early malnutrition for the programming of obesity in male rat. Obesity. 2006;14:1330–43.
14. Delahaye F, Lukaszewski M-A, Wattez J-S, Cisse O, Dutriez-Casteloot I, Fajardy I, et al. Maternal perinatal undernutrition programs a "brown-like" phenotype of gonadal white fat in male rat at weaning. Am J Physiol Regul Integr Comp Physiol. 2010;299:R101–10.
15. Alexandre-Goubau M-CF, Courant F, Le-Gall G, Moyon T, Darmaun D, Parnet P, et al. Offspring metabolomic response to maternal protein restriction in a rat model of intrauterine growth restriction (IUGR). J Proteome Res. 2011;10: 3292–302.
16. Proença ARG, Sertié RAL, Oliveira AC, Campaña AB, Caminhotto RO, Chimin P, et al. New concepts in White adipose tissue physiology. Braz J Med Biol Res. 2014;47(3):192–205.
17. Belfiore F, Rabuazzo AM, Napoli E, Borzi V, Vecchio LL. Enzymes of glucose metabolism and of citrate cleavage pathway in adipose tissue of normal and diabetes subjects. Diabetes. 1975;24:865–73.
18. Langin D. Adipose tissue lipolysis as a metabolic pathway to define pharmacological strategies against obesity and metabolic syndrome. Pharmacol Res. 2006;53:482–91.
19. Dessì A, Pravettoni C, Marincola FC, Schirru A, Fanos V. The biomarkers of fetal growth in intrauterine growth retardation and large for gestational age cases: from adipocytokines to a metabolomic all-in-one tool. Expert Rev Proteomics. 2015;12(3):309–16.
20. Kershaw EE, Flier JS. Adipose tissue as an endocrine organ. J Clin Endocrinol Metab. 2004;89(6):2548–56.
21. Nascimento CMO, Ribeiro EB, Oyama LM. Metabolism and secretory function of white adipose tissue: effect of dietary fat. Anais Acad Bras Cienc. 2009;81(3):453–66.

22. Briana DD, Malamitsi-Puchner A. Intrauterine growth restriction and adult disease: the role of adipocytokines. Eur J Endocrinol. 2009;160:337–47.

23. Ibáñez L, Sebastiani G, Lopez-Bermejo A, Díaz M, Gómez-Roig MD, de Zegher F. Gender specificity of body adiposity and circulating adiponectin, visfatin, insulin, and insulin growth factor-I at term birth: relation to prenatal growth. J Clin Endocrinol Metab. 2008;93(7):2774–8.

24. Fuchs D, Winkelmann I, Johnson IT, Mariman E, Wenzel U, Daniel H. Proteomics in nutrition research: principles, technologies and applications. Br J Nutr. 2005;94:302–14.

25. Wang J, Li D, Dangott LJ, Wu G. Proteomics and its role in nutrition research. J Nutr. 2006;136:1759–62.

26. Roepstorff P. Mass spectrometry-based proteomics, background, status and future needs. Protein Cell. 2012;3(9):641–7.

27. Silva TS, Richard N, Dias JP, Rodrigues PM. Data visualization and futures selection methods in gel-based proteomics. Cur Protein Pep Sci. 2014;15:4–22.

28. You Y-A, Lee JH, Kwon EJ, Yoo JY, Kwon W-S, Pang M-G, Kim YJ. Proteomic analysis of one-carbon metabolism-related marker in liver of rat offspring. Mol Cel Proteomics. 2015. Paper in press.

29. Sarr O, Louveau I, Kalbe C, Metges CC, Rehfeldt C, Gondret F. Prenatal exposure to maternal low or high protein diets induces modest changes in the adipose tissue proteome of newborn piglets. J Anim Sci. 2010;88:1626–41.

30. Ruis-Gonzáles MD, Cañete MD, Gómez-Chaparro JL, Abril N, Cañete R, López-Barea J. Alteration of protein expression in serum of infants with intrauterine growth restriction and different gestational age. J Proteomics. 2015;119:169–82.

31. Alexandre-Goubau M-CF, Bailly E, Moyon TL, Grit IC, Coupé B, Drean GL, et al. Postnatal growth velocity modulates alterations of proteins involved in metabolism and neuronal plasticity in neonatal hypothalamus in rats born with intrauterine growth restriction. J Nutr Biochem. 2012;23:140–52.

32. Fabricius-Bjerre S, Jensen RB, Faerch K, Larsen T, Molgaard C, Michaelsen KF, et al. Impact of birth weight and early infant weight gain on insulin resistance and associated cardiovascular risk factors in adolescence. Plos One. 2011;6(6):e20595. doi:10.1371/journal.pone.0020595.

33. Ravelli ACJ, van der Meuelen JHP, Osmond C, Barker DJP, Bleker OP. Obesity at the age 50 y in men and women exposed to famine prenatally. Am J Clin Nutr. 1999;70:811–6.

34. Desai M, Han G, Ferelli M, Kallichanda N, Lane RH. Programmed upregulation of adipogenic transcriptions factors in intrauterine growth-restricted offspring. Reprod Sci. 2008;15(8):785–96.

35. Lukaszewski M-A, Mayer S, Fajardy I, Delahaye F, Dutriez-Casteloot I, Montel V, et al. Maternal prenatal undernutrition programs adipose tissue gene expression in adult male rat offspring under high-fat diet. Am J Physiol Endocrinol Metab. 2011;301:E548–59.

36. Berndt J, Kovacs P, Ruschke K, Klöting N, Fasshauer M, Schön MR, et al. Fatty acid synthase gene expression in humana adipose tissue: assossiation with obesity and type 2 diabetes. Diabetologia. 2007;50:1472–80.

37. Vessal M, Mishra S, Moulik S, Murphy LJ. Prohibitin attenuates insulin-stimulated glucose and fatty acid oxidation in adipose tissue by inhibition of pyruvate carboxylase. FEBS J. 2006;273:568–76.

38. Ande SR, Nguyen KH, Padilla-Meier GP, Wahida W, Nyomba BLG, Mishra S. Prohibitin overexpression in adipocytes induces mitochondrial biogenesis, leads to obesity development, and affects glucose homeostasis in a sex-specific manner. Diabetes. 2014;63:3734–41.

39. Liu D, Lin Y, Kang T, Huang B, Xu W, Garcia-Barro M, et al. Miochondrial dysfunction and adipogenic reduction by prohibitin silencing in 3 T3-L1 cells. PLoS One. 2012;7(3):e34315. doi:10.1371/journal.pone.0034315.

40. Herman MA, She P, Peroni OD, Lynch CJ, Kahn BB. Adipose tissue branched chain amino acid (BCAA) metabolism modulates circulating BCAA levels. J Biol Chem. 2010;285(15):11348–56.

41. Kowalski TJ, Wu G, Watford M. Rat adipose tissue amino acid metabolism in vivo as assessed by microdialysis and arteriovenous techniques. Am J Physiol. 1997;273(3):E613–22.

42. Patterson BW, Horowitz JF, Wu G, Watford M, Coppack SW, Klein S. Regional muscle and adipose tissue amino acid metabolism in lean and obese women. Am J Physiol Endocrinol Metab. 2011;282:E931–6.

43. Badoud F, Lam KP, DiBattista A, Perreault M, Zulyniak MA, Cattrysse B, et al. Serum and adipose tissue amino acid homeostasis in the metabolically healthy obese. J Proteome Res. 2014;13:3455–66.

44. Hershey JW. Translational control in mammalian cell. Annu Rev Biochem. 1991;60:717–55.

45. Thornton S, Anand N, Purcell D, Lee J. Not just for housekeeping: protein initiation and elongation factors in cell growth and tumorigenesis. J Mol Med. 2003;81:536–48.

46. Chuang S-M, Chen L, Lambertson D, Anand M, Kinzy TG, Madura K. Proteasome-mediated degradation of cotranslationally damage proteins involves translation elongation factor 1A. Mol Cell Biol. 2005;25(1):403–13.

47. Al-Maghrebi M, Anin JT, Olalu AA. Up-regulation of eukaryotic elongation factor 1 subunits in breast carcinoma. Anticancer Res. 2005;25:2573–8.

48. Al-Maghrebi M, Cojocel C, Thompson MS. Regulation of elengation factor 1 expression by vitamin E in diabetic rat kidney. Mol Cell Biochem. 2005;273:177–83.

49. Hanzu FA, Vinaixa M, Papageourgiou A, Párrizas M, Correig X, Delgado S, et al. Obesity rather than regïonal fat depots marks the metabolomic pattern of adipose tissue: an untargeted metabolomic approach. Obesity. 2014;22:698–704.

50. Favretto D, Cosmi E, Ragazzi E, Visentin S, Tucci M, Fais P, et al. Cord blood metabolomic profiling in intrauterine growth restriction. Anal Bioanal Chem. 2012;402:1109–21.

51. Lee S-M, Dho SH, Ju S-K, Maeng J-S, Kim J-Y, Kwon K-S. Cytosolic malate dehydrogenate regulates senescence in human fybroblasts. Biogereontology. 2012;13:525–36.

52. Dessì A, Puddu M, Ottonello G, Fanos V. Metabolomics and fetal-neonatal nutrition: between "not enough" and "too much". Molecules. 2013;18:11724–32.

53. Mali Y, Zisapels N. Gain of interaction of ALS-linked G93A superoxide dismutase with cytosolic malate dehydrogenase. Neurobiol Dis. 2008;32:133–41.

54. Minárik P, Tomásková N, Kollárová M, Antalík M. Malate dehydrogenase – structure and function. Gen Physiol Biophys. 2002;21:257–65.

55. Theys N, Ahn M-T, Bouckenooghe T, Reusens B, Remacle C. Maternal malnutrition programs pancreatic islet mitochondrial dysfunction in the adult offspring. J Nutr Biochem. 2011;22:985–94.

56. Saito A, Shinohara H. Rat plasma murinoglobulin: isolation, characterization, and comparison with rat α-1- and α-2-macroglobulins. J Biochem. 1985;98:501–16.

57. Tolosano E, Altruda F. Hemopexin: structure, function, and regulation. DNA Cell Biol. 2002;21(4):297–306.

58. Jung JY, Kwak YH, Kim KS, Kwon WY, Suh GJ. Change of hemopexin level is associated with the severity of sepsis in endotoxemic rat model and the outcome of septic patients. J Crit Care. 2015. doi:10.1016/j.jcrc.2014.12.009.

59. Gianazza E, Sensi C, Eberini I, Gilardi F, Giudici M, Crestani M. Inflammatory serum proteome pattern in mice fed a high-fat diet. Amino Acids. 2013;44:1001–8.

60. Walter P, Ron D. The unfolded protein response: to stress pathway to homeostatic regulation. Science. 2011;334:1081–6.

61. Özcan U, Cao Q, Yilmaz E, Lee A-H, Iwakoshi NN, Özdelen E, et al. Endoplasmic reticulum stress links obesity, insulin action, and type 2 diabetes. Science. 2004;306:457–61.

62. Teodoro-Morrison T, Schuiki I, Zhang L, Belsham DD, Volchuk A. GRP78 overproduction in pancreatic beta cells protects against high-fat-diet-induced diabetes in mice. Diabetologia. 2013;56:1057–67.

63. Lanneau D, Brunet M, Frisan E, Solary E, Fontenay M, Garrido C. Heat shock proteins : essential proteins for apoptosis regulation. J Cell Mol Med. 2008;12(3):743–61.

64. Tiss A, Khadir A, Abubaker J, Abu-Farha M, Al-Khairi I, Cherian P, et al. Immunohistochemical profiling of the heat shock response in obese non-diabetic subjects revealed impairment expression of heat shock proteins in the adipose tissue. Lipids Health Dis. 2014. doi:10.1186/1476-511X-13-106.

65. Saggerson ED, McAllister TWJ, Bath HS. Lipogenesis in rat brown adipocytes – effects of insulin and noradrenaline, contributions from glucose and lactate as precursors and comparisons with white adipose tissue. Biocem J. 1988;251:701–9.

66. O'Hea EK, Leveille G. Significance of adipose tissue and liver as sites of fatty acid synthesis in the pig and the efficiency of utilization of various substrates for lipogenesis. J Nutr. 1969;99:338–44.

67. van Hall G. Lactate kinects in human tissues at rest and during exercise. Acta Physiol. 2010;199:499–508.

68. Sabbater D, Arriarán S, Romero MM, Agnelli S, Remesar X, Fernández-López JA, et al. Cultured 3 T3-L1 adipocytes dispose of excess medium glucose as lactate under abundant oxygen availability. Sci Rep. 2014;4:3663. doi:10.1038/srep03663.

69. Arriaran S, Agnelli S, Sabater D, Remesar X, Fernádez-López JA, Alemany M. Evidences of basal lactate production in the main white adipose tissue sites of rats. Effect of sex and a cafeteria diet. PLoS One. 2015;10(3):e0119572. doi:10.1371/journal.pone.0119572.

70. Dugail I, Quignard-Boulange A, Bazin R, Le Liepvre X. Adipo-tissue-specific increase in glyceraldehyde-3-phosphate dehydrogenase activity and mRNA amounts in suckling pre-obese Zucker rats. Biochem J. 1988;254:483–7.

71. Rattanatray L, MacLaughlin SM, Kleemann DO, Walker SK, Muhlausler BS, McMillen IC. Impact of maternal periconceptional overnutrion on fat mass and expression of adipogenic and lipogenic genes in visceral and subcutaneous fat depots in the postnatal lamb. Endocrinology. 2010;151(11):5195–205.

72. Pérez-Pérez R, García-Santos E, Ortega-Delgado FJ, López JA, Camafeita E, Ricart W, et al. Attenuated metabolism is a hallmark of obesity as revealed by comparative proteomic analysis of human omental adipose tissue. J Proteomics. 2012;75:783–95.

73. Brasaemle DL. The perilipin family of structural lipid droplet proteins: stabilization of lipid droplets and control of lipolysis. J Lipid Res. 2007;48:2547–59.

74. Ray H, Pinteur C, Frering V, Beylot M, Large V. Depot-specific differences in perilipin and hormone-sensitive lipase expression in lean and obese. Lipid Health Dis. 2009;8(58). doi:10.1186/1476-511X-8-58.

75. Tansey J, Sztalryd C, Gruia-Gray J, Roush DL, Zee JV, Gavrilova O, et al. Perilipin ablation results in a lean mouse with aberrant adipocyte lipolysis, enhanced leptin production, and esistance to diet-induced obesity. Proc Natl Acad Sci. 2001;98:6494–9.

76. Ballatory N, Krance SM, Notenboom S, Shi S, Tieu K, Hammond CL. Glutathione dysregulation and the etiology and progression of human disease. Biol Chem. 2009;390(3):191–214.

77. Kharb S. Low whole blood glutathione levels in pregnancies complicated by preeclampsia and diabetes. Clin Chim Acta. 2000;294:179–83.

78. Boden G, Duan X, Homko C, Molina EJ, Song W, Perez O, et al. Increase in endoplasmic reticulum stress-related proteins and gene in adipose tissue of obese, insulin-resistant individuals. Diabetes. 2008;57:2438–44.

79. Fuentes-Almagro CA, Prieto-Álamo M-J, Pueyo C, Jurado J. Identification of proteins containing redox-sensitive thiols after PRDX1, PRDX3 and GCLC silencing and/or glucose oxidase treatment in Hepa 1-6 cells. J Prot. 2012;77:262–79.

80. Nonn L, Berggren M, Powis G. Increased expression of mitochondrial peroxiredoxin-3 (thioredoxin peroxidase-2) protects cancer cells against hypoxia and drug-induced hydrogen peroxide-dependent apoptosis. Mol Cancer Res. 2003;1:682–9.

81. Chang T-S, Cho C-S, Park S, Yu S, Kang SW, Rhee SG. Peroxiredoxin III, a mitochondrion-specific peroxidase, regulates apoptotic signaling by mitochondria. J Biol Chemis. 2004;279(40):41975–84.

82. Janciauskiene S, Larsson S, Larsson P, Virtala R, Jansson L, Stevens T. Inhibition of lipopolysaccharide-mediated human monocyte activation, in vitro, by α1-antitrypsin. Biochem Biophys Res Comm. 2004;321:592–600.

83. Stockley RA, Shaw J, Afford SC, Morrison HM, Burnett D. Effect of alpha-1-proteinase inhibitor on neutrophil chemotaxis. Am J Respir Cell Mol Biol. 1990;2(2):163–70.

84. Bergin DA, Reeves EP, Meleady P, Henry M, McElvaney OJ, Carroll TP, et al. α-1 Antitrypsin regulates human neutrophil chemotaxis induced by soluble immune complexes and IL-8. J Clin Invest. 2010;120(12):4236–50.

85. Al-Omari M, Korenbaum E, Ballmaier M, Lehmann U, Jonigk D, Manstein DJ, et al. Acute-phase protein α1-antitrypsin inhibits neutrophil calpain I and induces random migration. Mol Med. 2011;17:865–74.

86. Griese M, Latzin P, Kappler M, Weckerle K, Heinzlmaier T, Bernhardt T, et al. α1-antitripsin inhalation reduces airway inflammation in cystic fibrosis patients. Eur Respir J. 2007;29:240–50.

87. Kalis M, Kumar R, Janciauskiene S, Salehi A, Cílio CM. α1-antitripsinenhances insulin secretion and prevents cytokine-mediated apoptosis in pancreatic β-cells. Islet. 2010;2(3):185–9.

88. Lu Y, Tang M, Wasserfall C, Kou Z, Campbell-Thompson M, Gardemann T, et al. α1-antitrypsin gene therapy modulates cellular immunity and efficiently prevents type 1 diabetes in nonobese diabetic mice. Human Gene Terapy. 2006;17:625–34.

89. Oliva K, Barker G, Rice GE, Bailey MJ, Lappas M. 2D-DIGE to identify proteins associated with gestational diabetes in omental adipose tissue. J Endocrinol. 2013;218:165–78.

90. Laemmli UK. Cleavage of structural protein during the assembly of the head of bacteriophage T4. Nature. 1970;227(5259):680–5.

16

Proteome analysis of shell matrix proteins in the brachiopod *Laqueus rubellus*

Yukinobu Isowa[1], Isao Sarashina[2], Kenshiro Oshima[3], Keiji Kito[4], Masahira Hattori[3] and Kazuyoshi Endo[1*]

Abstract

Background: The calcitic brachipod shells contain proteins that play pivotal roles in shell formation and are important in understanding the evolution of biomineralization. Here, we performed a large-scale exploration of shell matrix proteins in the brachiopod *Laqueus rubellus*.

Results: A total of 40 proteins from the shell were identified. Apart from five proteins, i.e., ICP-1, MSP130, a cysteine protease, a superoxide dismutase, and actin, all other proteins identified had no homologues in public databases. Among these unknown proteins, one shell matrix protein was identified with a domain architecture that includes a NAD(P) binding domain, an ABC-type transport system, a transmembrane region, and an aspartic acid rich region, which has not been detected in other biominerals. We also identified pectin lyase-like, trypsin inhibitor, and saposin B functional domains in the amino acid sequences of the shell matrix proteins. The repertoire of brachiopod shell matrix proteins also contains two basic amino acid-rich proteins and proteins that have a variety of repeat sequences.

Conclusions: Our study suggests an independent origin and unique mechanisms for brachiopod shell formation.

Keywords: Brachiopoda, Biomineralization, Shell matrix protein, Proteome, Transcriptome

Background

Many organisms have a diversity of biominerals that have a large number of ecologically important functions, including body support and defense against predators. Most metazoan biominerals are likely acquired independently in different phyla in the early Cambrian [1]. On the other hand, presence of some biomineralization genes shared by several phyla suggests that at least a part of the mechanisms was derived from a common ancestor [2]. A crucial factor in understanding the mechanisms and evolution of biomineralization is an in-depth analysis of the organic matrix present in hard tissues. A large number of skeletal proteins have been identified and their roles in the biomineralization process have been analyzed [3]. Recently, it became possible to identify skeletal proteins comprehensively using transcriptome analysis combined with proteome analysis across a wide range of phyla [4–9]. Brachiopods are marine invertebrates that appeared in the Cambrian and have two valves composed of calcium carbonate or calcium phosphate.

Organic shell matrices in brachiopods have been studied in some detail [10, 11]. Previous SDS-PAGE analyses showed several major bands stained by Coomassie Brilliant Blue (CBB) [10, 12], and an amino acid composition analysis revealed that glycine and alanine account for a large proportion of bulk shell extracts [13, 14]. In addition to these studies, shell matrix proteins of brachiopods have been studied for the characterization of peptides and amino acids preserved in fossil brachiopod shells and immunological assays showed that some peptide sequences in the shells are preserved for about 1 million years [15, 16]. However, the amino acid sequences of the shell matrix proteins in brachiopods have not been identified, except for the partial amino acid sequencing of a chromoprotein, named ICP-1 [17–19]. In this study, we performed a large-scale analysis of shell matrix proteins of the brachiopod *Laqueus rubellus* using proteomics combined with transcriptomics. The analysis identified 40 shell matrix proteins. These new proteome data contained 35 protein sequences that have no database-related homologues and indicate that the mechanism and evolution of brachiopod shell formation are unique.

* Correspondence: endo@eps.s.u-tokyo.ac.jp
[1]Department of Earth and Planetary Science, Graduate School of Science, The University of Tokyo, 7-3-1 Hongo, Bunkyo-ku, Tokyo 113-0033, Japan
Full list of author information is available at the end of the article

Results

SDS-PAGE of the soluble organic matrix extracted from the secondary layer showed two major bands (< 6.5 and 12 kDa) and a minor band of 62 kDa when stained with CBB (Fig. 1a). Silver staining showed a major band at 35 kDa and many minor bands in addition to the three bands stained with CBB (Fig. 1b). Transcriptome sequencing in mantle tissue generated a total of 125,437 reads. To identify shell matrix proteins from *Laqueus rubellus*, we performed a proteome analysis of the material from these different sources, namely (1) the soluble organic matrix from the whole shell, (2) insoluble organic matrix from the whole shell and (3) soluble organic matrix from the shell secondary layer. The generated MS/MS data were subjected to SEQUEST searches against the protein sequence database obtained from the transcriptome analysis. As a result, we identified 40 shell matrix proteins (Tables 1, 2, 3, 4 and Fig. 2). Among these proteins, 18 proteins were identified from the soluble organic matrix of the shell secondary layer, and the soluble and insoluble organic matrix from the whole shell (Table 1). While five proteins were identified from both the soluble and insoluble organic matrix of the whole shell, 17 proteins were only identified from either the soluble or insoluble organic matrices of the whole shell (Tables 2, 3, 4). A total of 22 proteins were deduced to be possibly complete sequences, because these sequences have a stop codon and either an in-frame start codon just after an in-frame stop codon in the 5' region of the sequence or a potential signal peptide (Tables 1, 2, 3, 4). A blast search against the GenBank non-redundant database showed that 36 out of the 40 shell matrix proteins have no annotated homologous

sequences. Among the proteins that have sequence homology, isotig 00281 showed relatively high sequence similarity with MSP130, which is a skeletal protein identified from sea urchins (Fig. 3) [20, 21]. Isotig 01587, isotig 00949, and isotig 00959 showed high sequence similarity with extracellular copper/zinc superoxide dismutase, actin I, and cathepsin L cysteine proteinase, respectively (Tables 2 and 4). Predicted molecular mass and isoelectric point are shown in Tables 1, 2, 3, 4, and domains predicted by the SMART program and a National Center for Biotechnology Information (NCBI) conserved domain search are schematically shown in Fig. 2. Using the known amino acid sequences of ICP-1 [19], we performed local blast searches against sequence data obtained from this study. As a result, isotig 00046 was found to have a high sequence similarity with ICP-1 (Fig. 4). Gene expression analysis showed that ICP-1 gene is expressed in lophophore as well as in mantle tissues (Additional file 1). To estimate the abundance of each protein, we calculated the relative copy number based on the identified spectral counting [22–25]. The number of identified spectra of each protein was divided by the number of theoretically observable tryptic peptide ions, which have a mass-to-charge ratio of 400 to 1500 at two or three charge states, to generate an abundance index as the relative copy number. The result showed that isotig 00046 (ICP-1) is the most abundant protein in the shell extracts among the proteins identified in this study (Fig. 5). The percentage of spectra that were matched to peptide contained in protein sequences translated from transcriptome data was about 10 % of the total MS/MS spectra acquired. The proportion in this case was lower than those of other mass spectrometric analyses for organisms of which genome has been completely sequenced (*S. cerevisiae*, 40-50 %), suggesting presence of many more proteins in shell matrix samples than proteins identified in this study. Messenger RNA sequences in the transcriptome data may not include complete list for proteins in shell matrix or unknown chemical modifications that were not considered in our database search may occur in a large proportion of shell matrix proteins. The depth of isotigs is shown in Additional file 2.

Discussion

Repertoire of matrix proteins found in brachiopod shells
ICP-1

ICP-1 (IntraCrystalline Protein-1) is a shell matrix protein extract from the calcitic shell of three brachiopod species: *Neothyris lenticularis*, *Calloria inconspicua*, and *Terebratella sanguinea*. The partial N-terminal amino acid sequences of ICP-1 from these species have been determined by Edman degradation [17–19]. The local blast searches showed that the partial sequences of ICP-1 have a sequence similarity with a part of isotig 00046 (Fig. 4). Isotig 00046 showed the highest abundance in the shell among the shell

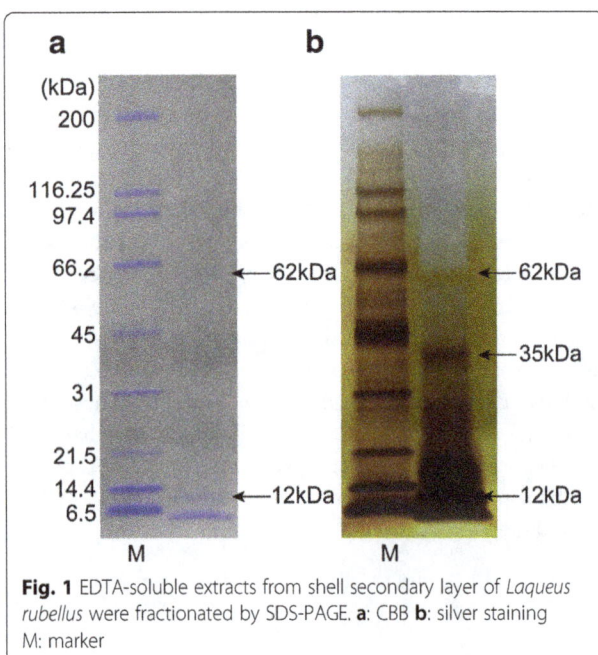

Fig. 1 EDTA-soluble extracts from shell secondary layer of *Laqueus rubellus* were fractionated by SDS-PAGE. **a**: CBB **b**: silver staining M: marker

Table 1 Shell matrix proteins identified from the whole shell and shell secondary layer

Isotig no.	Accession no.	Matching peptides	Complete sequence	Blast hit (e-value <1e⁻¹⁰)	Molecular mass (kDa)	pI
Isotig 00046	FX982984	11	–	–	–	–
Isotig 00149a	FX982985	4	○	–	16.9	10.1
Isotig 00227	FX982987	4	○	–	27.0	11.6
Isotig 00281	FX982988	7	–	MSP130	–	–
Isotig 00337	FX982989	6	○	–	58.0	10.30
Isotig 00341	FX982990	5	–	–	–	–
Isotig 00515	FX982992	4	○	–	9.2	9.38
Isotig 00543	FX982993	4	○	–	13.4	5.24
Isotig 00776	FX982995	4	–	–	–	–
Isotig 01016	FX983001	8	○	–	27.3	8.63
Isotig 01158	FX983004	6	–	–	–	–
Isotig 01176	FX983005	6	○	–	29.4	4.16
Isotig 01202	FX983006	11	○	–	25.0	4.68
Isotig 01252	FX983007	7	○	–	28.9	9.68
Isotig 01382	FX983009	5	–	–	–	–
Isotig 01556	FX983013	8	○	–	19.9	8.96
Isotig 01886	FX983016	6	○	–	10.0	8.47
Isotig 02671	FX983022	2	○	–	5.0	7.14

'-'means not applicable

matrix proteins identified in this study (Fig. 5). This is consistent with the results of a previous study [19]. ICP-1 was originally identified as a 6.5-kDa major band revealed by SDS-PAGE analysis of shell extracts from *Neothyris lenticularis*, *Calloria inconspicua*, and *Terebratella sanguinea* [19]. The predicted molecular mass of isotig 00046 is just over 11.9 kDa, while the thickest band in the SDS-PAGE analysis of *Laqueus rubellus* is < 6.5 kDa (Fig. 1). One possibility to explain this size discrepancy is that ICP-1 undergoes proteolytic cleavage after translation. We have identified cathepsin L cysteine proteinase from the shell extracts. This protein could be involved in post-translational modification. HPLC analysis in a previous study indicated that carotenoids are bound to ICP-1 [19]. The observation that ICP-1 shows the highest abundance in the shell suggests that ICP-1 plays key roles in the biomineralization processes in brachiopods. RT-PCR

analysis showed that ICP-1 gene is also expressed in the lophophore tissues, suggesting that ICP-1 is also involved in the formation of the calcareous loop structure embraced by the lophophore.

MSP130

Isotig 00281 showed relatively high sequence similarity with MSP130 (The e-value against MSP130 from *Saccoglossus kowalevskii* was 1e⁻¹⁰). MSP130 was originally identified from primary mesenchyme cells in the sea urchin [20, 21] and was subsequently detected in the hard tissues of sea urchins [26, 27]. Homologues and closely related proteins of MSP130 were also reported to be present in molluscan shells [28]. In addition, MSP130 have been found in genomes of hemichordate, cephalochordate, bacteria, and green algae [29]. MSP130 is predicted to have been acquired by independent horizontal gene transfer in

Table 2 Shell matrix proteins identified from the whole shell (the soluble and insoluble organic matrix)

Isotig no.	Accession no.	Matching peptides	Complete sequence	Blast hit (e-value <1e⁻¹⁰)	Molecular mass (kDa)	pI
Isotig 00149b	FX983023	6	–	–	–	–
Isotig 00435	FX982991	2	○	–	9.1	5.95
Isotig 01587	FX983014	2	○	Extracellular copper/zinc Superoxide dismutase	22.4	8.26
Isotig 02447	FX983019	3	–	–	–	–
Isotig 02555	FX983020	2	–	–	–	–

'-'means not applicable

Table 3 Shell matrix proteins identified from the whole shell (the soluble organic matrix)

Isotig no.	Accession no.	Matching peptides	Complete sequence	Blast hit (e-value <1e^{-10})	Molecular mass (kDa)	pI
Isotig 00213	FX982986	3	–	–	–	–
Isotig 01414	FX983010	3	o	–	26.2	8.53
Isotig 01423	FX983011	3	o	–	18.1	10.58
Isotig 01670	FX983015	3	–	–	–	–
Isotig 01967	FX983017	2	o	–	8.7	6.09
Isotig 02613	FX983021	2	–	–	–	–

'-'means not applicable

Cambrian, because this gene exists in bacteria and has an extremely wide phylogenetic distribution [29]. However, it appears possible that MSP130 gene was already present in the metazoan or bilaterian last common ancestor because the phylogenetic tree of MSP130 constructed in the previous study was divided into two clusters of the bacterial/green algae clade and the metazoan (bilaterian) clade [29]. If the MSP130 gene was transferred to animals horizontally many times independently, the animal MSP130 genes would not form a monophyletic cluster. The functions of MSP130 have not been determined, but this protein is predicted to function at the cell surface [30]. Identification of MSP130 from the brachiopod shells in addition to the skeletons of sea urchin and molluscs suggests that this protein plays an important role in biomineralization processes.

Digestive enzymes and inhibitors

Many digestive enzymes have been identified from the shell of *Laqueus rubellus*. Isotig 00227 has a pectin lyase-like domain, which enzymatically breaks down pectin. Pectin is usually found in plant cells, but some neutral sugars have been detected in terebratulide brachiopod shells [12]. Thus, isotig 00227 could be involved in the breakdown of these neutral sugars. Two lysosomal

proteins were identified from the shell: (i) isotig 00213 is a saposin protein that is involved in lipid degradation and (ii) isotig 00959 is cathepsin L, which is a cysteine protease that plays a major role in intracellular protein catabolism. However, a trypsin inhibitor domain was found in the amino acid sequence of isotig 01016. A number of protease inhibitors have been identified from molluscan shells and a possible function of these proteins could be to protect the shell matrix proteins against particular proteases [31, 32]. In brachiopod shell formation, a protease inhibitor could play a similar role, but since secreted proteases and protease inhibitors are involved in modification of extracellular and membrane bound proteins in many systems, their roles may not be specific to shell matrix modifications.

Membrane protein

Isotig 01176 has a transmembrane region, indicating that this protein binds to the cell membranes of the mantle tissue. Interestingly, several skeletal proteins of a coral also have a transmembrane region [8]. If this protein is involved in shell formation, the mantle epithelium would be closely attached to the inner shell surface when mineralization occurs. Isotig 01176 also has a NAD(P) binding domain. NADP is a hydrogen carrier that is used in metabolic

Table 4 Shell matrix proteins identified from the whole shell (the insoluble organic matrix)

Isotig no.	Accession no.	Matching peptides	Complete sequence	Blast hit (e-value <1e^{-10})	Molecular mass (kDa)	pI
Isotig 00601	FX982994	2	–	Hypothetical protein	19.3	8.48
Isotig 00914	FX982996	2	o	Predicted protein	57.6	9.69
Isotig 00916	FX982997	2	–	–	–	–
Isotig 00949	FX982998	2	o	Actin I	41.7	5.18
Isotig 00959	FX982999	3	o	Cathepsin L cysteine proteinase	39.8	6.87
Isotig 00996	FX983000	2	–	–	–	–
Isotig 01095	FX983002	4	o	–	41.3	9.76
Isotig 01124	FX983003	2	–	Hypothetical protein	–	–
Isotig 01312	FX983008	2	–	Hypothetical protein	–	–
Isotig 01521	FX983012	2	o	–	25.1	11.85
Isotig 02158	FX983018	2	–	–	–	–

'-'means not applicable

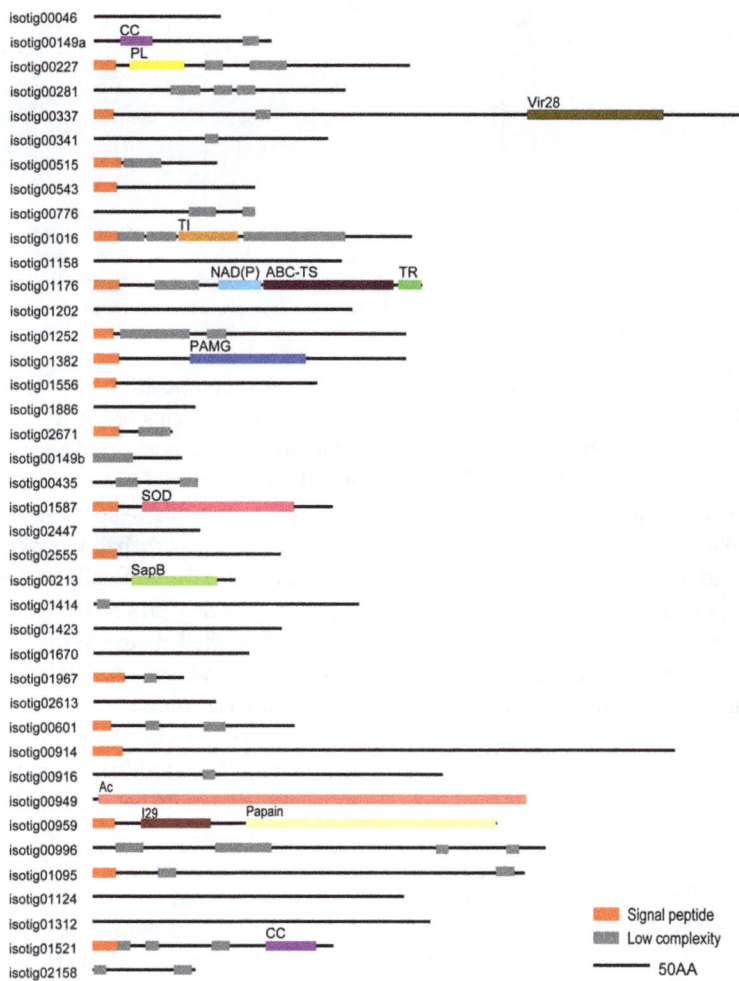

Fig. 2 Schematic of the domains in shell matrix proteins identified in this study. CC: coiled coil; PL: pectin lyase-like; Vir28: variable surface protein Vir28; TI: trypsin inhibitor-like cysteine-rich domain; NAD(P): NAD(P)-binding Rossmann-fold domains; ABC-TS: ABC-type transport system; TR: transmembrane region; PAMG: Pneumovirinae attachment membrane glycoprotein G; SOD: copper/zinc superoxide dismutase; SapB: saposin B domains; Ac: Actin; I29: cathepsin propeptide inhibitor domain; and Papain: papain family cysteine protease

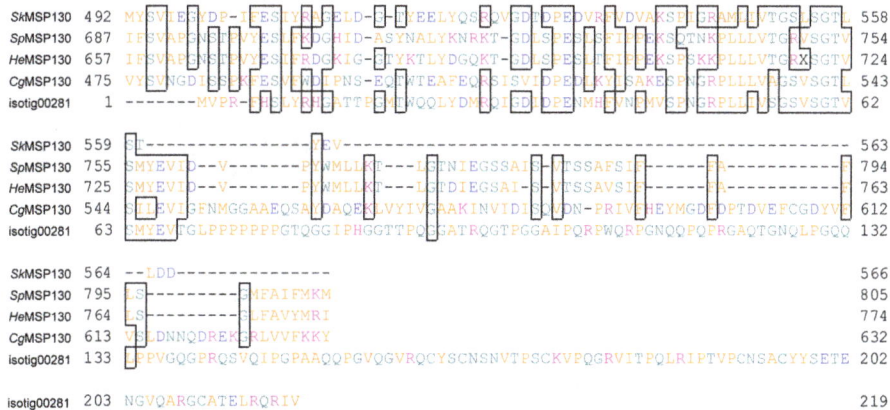

Fig. 3 Alignment of the amino acid sequences of MSP-130 and isotig 00281. Sk: *Saccoglossus kowalevskii* (NCBI Acc. No. XP_002739468.1); Sp: *Strongylocentrotus purpuratus* (NCBI Acc. No. NP_001116986.1); He: *Heliocidaris erythrogramma* (NCBI Acc. No. CAC20358.1); and Cg: *Crassostrea gigas* (NCBI Acc. No. EKC20477.1)

Fig. 4 Alignment of the amino acid sequences of ICP-1 and isotig 00046. Nl: *Neothyris lenticularis*; Ts: *Terebratella sanguinea*; and Ci: *Calloria inconspicua*

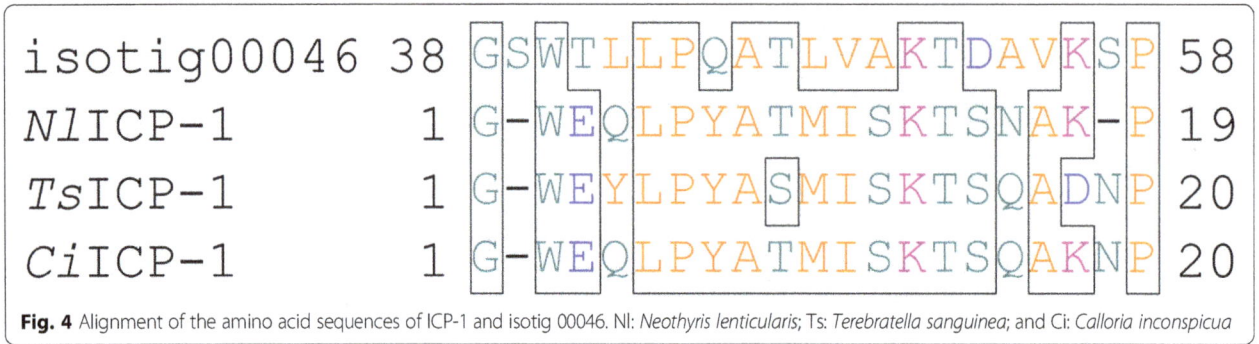

pathways such as the photosynthetic pathway and glycolysis as a reducing agent. Although the function of this domain in isotig 01176 is unknown, the high importance of H^+ in $CaCO_3$ synthesis suggests that the NAD(P) binding domain of isotig 01176 could function to sequester H^+ from the crystallization milieu. Isotig 01176 may also have an ABC-type transport system, which is involved in the transportation of many types of substrates. This suggests that isotig 01176 could also control ion concentrations in the space where crystallization occurs. As isotig 01176 has all these functional domains, this protein is likely to have key roles in brachiopod shell formation.

Secreted protein

Among the 40 shell matrix proteins, 18 have a signal peptide, indicating that many shell matrix proteins are secreted from cells. Generating the calcium carbonate in the extracellular space is consistent with brachiopod shells being an exoskeleton. However, seven proteins do not have a signal peptide even though they have complete sequences. One possibility is that these proteins bind with other proteins that have a signal peptide. Actin, which is one of the major intracellular proteins, was identified. However, there is a possibility that this protein is occluded from trapping of proteins involved in the secretory processes.

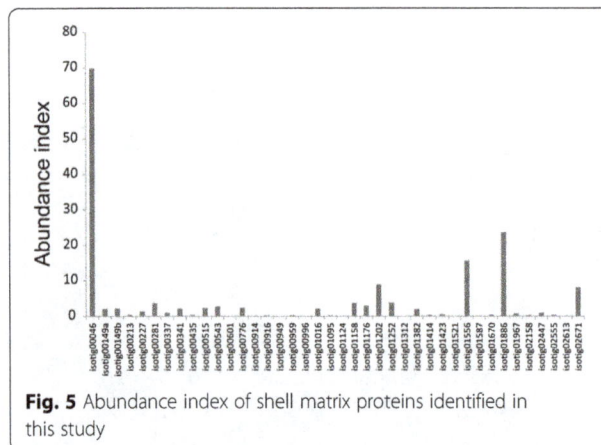

Amino acid composition and isoelectric points

Shell matrix proteins of molluscs contain aspartic acid-rich proteins that are postulated to interact with Ca^{2+} [33]. Isotig 01176 has a repeat sequence comprising aspartic acid, suggesting that this region also binds Ca^{2+} (Fig. 6). However, the brachiopod shell matrix proteins identified in this study do not have unusually low isoelectric points seen for aspartic acid-rich proteins in the molluscan shell (Fig. 7). Among the shell matrix proteins identified in this study, isotig 01521 and isotig 02158 have a relatively high concentration of basic amino acids (Fig. 6). Although it is possible that unusually acidic shell matrix proteins have not been identified in this study, there may exist a general difference in isoelectric points between mollusc and brachiopod shell matrix proteins.

Repeat sequences

Many of the shell matrix proteins identified in this study have repeat sequences, which are generally represented as a low complexity domain in the SMART prediction (Fig. 2). Repeat sequences exist in many skeletal proteins identified from other phyla and are thought to have important roles in biomineralization processes. The GXN (Glycine-X-Asparagine) repeat that is present in the amino acid sequence of isotig 02671 also exists in Nacrein, the

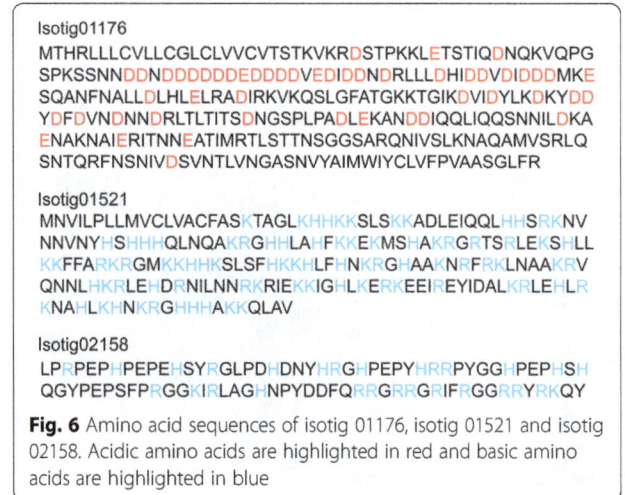

Fig. 5 Abundance index of shell matrix proteins identified in this study

Isotig01176

MTHRLLLCVLLCGLCLVVCVTSTKVKRDSTPKKLETSTIQDNQKVQPG
SPKSSNNDDNDDDDDDEDDDDVEDIDDNDRLLLDHIDDVDIDDDMKE
SQANFNALLDLHLELRADIRKVKQSLGFATGKKTGIKDVIDYLKDKYDD
YDFDVNDNNDRLTLTITSDNGSPLPADLEKANDDIQQLIQQSNNILDKA
ENAKNAIERITNNEATIMRTLSTTNSGGSARQNIVSLKNAQAMVSRLQ
SNTQRFNSNIVDSVNTLVNGASNVYAIMWIYCLVFPVAASGLFR

Isotig01521

MNVILPLLMVCLVACFASKTAGLKHHKKSLSKKADLEIQQLHHSRKNV
NNVNYHSHHHQLNQAKRGHHLAHFKKEKMSHAKRGTSRLEKSHLL
KKFFARKRGMKKHHKSLSFHKKHLFHNKRGHAAKNRFRKLNAAKRV
QNNLHKRLEHDRNILNNRKRIEKKIGHLKERKEEIREYIDALKRLEHLR
KNAHLKHNKRGHHHAKKQLAV

Isotig02158

LPRPEPHPEPEHSYRGLPDHDNYHRGHPEPYHRRPYGGHPEPHSH
QGYPEPSFPRGGKIRLAGHNPYDDFQRRGRRGRIFRGGRRYRKQY

Fig. 6 Amino acid sequences of isotig 01176, isotig 01521 and isotig 02158. Acidic amino acids are highlighted in red and basic amino acids are highlighted in blue

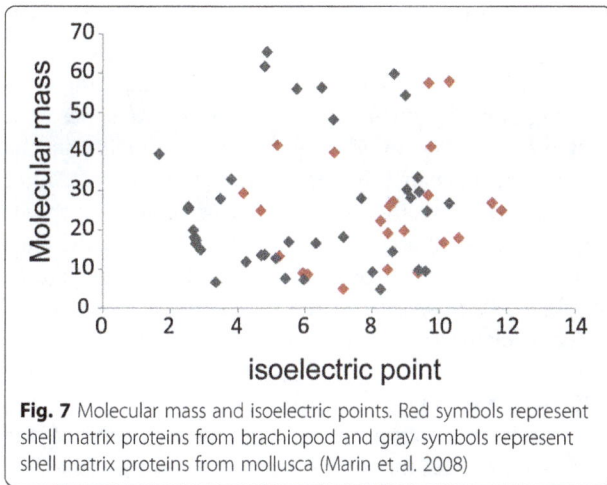

Fig. 7 Molecular mass and isoelectric points. Red symbols represent shell matrix proteins from brachiopod and gray symbols represent shell matrix proteins from mollusca (Marin et al. 2008)

carbonic anhydrase in the molluscan shell (Fig. 8). This repeat sequence has been proposed to function by inhibiting the precipitation of calcium carbonate [34]. Therefore, the GXN repeat in isotig 02671 could also be involved in the control of CaCO₃ growth. Besides this repeat sequence, the repeat motifs PPRG, GGX, and GGQNTGX are also present in sequences of the shell matrix proteins (Fig. 8). Although the exact function of these repeat sequences is not clear, the existence of a variety of repeat sequences in skeletal proteins suggests that this sequence structure has fundamental roles in biomineralization.

Possible mechanisms of brachiopod biomineralization

Many shell matrix proteins identified in this study have a signal peptide, suggesting that crystallization occurs in the outer cell space. This is consistent with shell matrix proteins in mollusca. Isotig 01176, which has an asparatic acid-rich region, NAD(P) binding domain, and ABC-type transport system, is a possible candidate that controls Ca^{2+} and H^+ levels, whereas isotig 01521, which contains many basic amino acids, may interact with HCO_3^-. These proteins do not exist in skeletal proteins of other phylum, suggesting that the brachiopod has a unique ion control system for shell formation. In addition to these proteins, MSP130 and ICP-1 are thought to have important roles in shell formation.

Evolution of brachiopod biomineralization

Skeletal proteins from brachiopods and other phyla share common structural features. The existence of MSP130 in brachiopod shell suggests that this protein have significant functions for biomineralization, and the several phyla share a part of biomineralization mechanisms. This protein could have been acquired by horizontal gene transfer from bacteria, but as discussed above, the data shown by Ettensohn (2014) are not conclusive, and MSP130 could also have been present in the last common ancestor of bilaterians or metazoans. On the other hand, proteinase and proteinase inhibitors did not show significant sequence similarities to the skeletal proteins in other phyla, suggesting that these proteins have been recruited as shell matrix proteins independently. In addition, almost all the brachiopod shell matrix proteins identified in this study had no homologous sequences in the public data banks, suggesting unique origins for those proteins. These results support the hypothesis that brachiopod biominerals were acquired independently from other phyla in the Cambrian explosion. On the other hand, there is also a possibility that brachiopod biominerals share the same ancestral biomineralization system with some other phyla, but after acquisition of biominerals, the matrix proteins evolved rapidly and/or novel lineage-specific proteins were added to the ancestral biomineralization system. The shell proteome of the mollusc *Lottia gigantea* showed that 25 out of 39 shell matrix proteins had no homologues, and only a few proteins showed high sequence similarity to shell matrix proteins in bivalves and skeletal proteins in other phyla [7]. This observation may well be reflecting the rapid nature of the evolution of shell matrix proteins, but can also be considered as reflecting the independent origin of the skeletal matrix proteins of molluscs from those of other phyla. Although more data are needed to address this problem, extremely low numbers of homologous shell matrix proteins between brachiopods and molluscs, combined with the presence of a possible unique ion control system involving basic shell matrix proteins, tend to support an independent origin for the brachiopod shells as expected from the phylogenetic relationships and the fossil record [1, 35].

Conclusions

Our results identified two interesting shell matrix proteins, ICP-1 and MSP130. ICP-1 is a brachiopod shell

isotig00227	PPRGPPRGPPRGPPRGPPRGPPRGPPRG	138–165
isotig00281	GGIPHGGTTPQGGATRQGTPGGA	81–103
isotig01016	GQGGGQNTGTVLQNTGTGGQNTGAGGQNTGTGGQNTGTGGQNTGT	152–195
isotig02671	GNNGDNGDNGDNGYNGNNGNNGNNGNNEK	40–69

Fig. 8 Repeat motifs found in shell matrix proteins identified in this study

matrix protein sequenced partially in previous studies, and MSP130 is a skeletal protein identified originally in sea urchins and oysters. Our data also showed novel shell proteins containing unique structures, including NAD(P) binding domains, suggesting the involvement of a hitherto unknown ion control system for shell formation. In addition, most other shell matrix proteins of *Laqueus rubellus* do not have a homologue in skeletal proteins of other phyla, suggesting an independent origin of the brachiopod shell. To further address the mechanisms and evolution of shell formation in brachiopods, additional studies, such as gene expression analysis and functional analysis, using this large-scale sequence information are necessary.

Materials and methods
RNA extraction and amplification
Live individuals of *Laqueus rubellu*s were collected by dredging operations in Sagami Bay, Japan (2 km off Jogashima, 90 m water depth). The mantle tissues from a single individual were separated from the shell using tweezers and homogenized in 500 μL of Isogen (Nippon Gene, Tokyo, Japan) in a 1.5-mL microcentrifuge tube. Total RNA was extracted following the manufacturer's protocol and purified using the RNeasy Mini Kit (Qiagen, Hilden, Germany). Amplification of mRNA was performed using MessageAmp II aRNA Amplification Kit (Life Technologies, Carlsbad, CA, USA) to obtain sufficient quantities of mRNA (200 ng) for transcriptome analysis. The amount of the initial total RNA subjected to the amplification was 1 μg and the incubation time for in vitro transcription was 8 h.

Transcriptome
The nucleotide sequences of the cDNA expressed in the mantle tissue were determined using 454 GS Junior (Roche, Basel, Switzerland). Template DNA was prepared according to the supplier's protocol. A total of 125,437 reads of the GS Junior sequences were generated. The obtained reads were assembled using Newbler v.2.8. software with default settings, resulting in 2,342 isotigs.

Extraction of shell matrix proteins
The shells were incubated overnight in a 5 % bleach solution with gentle shaking at room temperature to remove surface contaminants. After thorough washing with ultra-pure water, the shells were crushed in water and dried. Organic materials were extracted by dissolution of the calcium carbonate using 0.5 M EDTA (pH 8.0) at a ratio of 23 mL to 1 g shell with shaking at 4 °C. After the solution was centrifuged at 20,000 g for 1 h, the insoluble organic materials separated by the centrifugation step were dissolved in an aqueous solution containing 9 M urea and 2 % (v/v) Triton X-100. The supernatant was concentrated

and the EDTA removed using an Amicon Ultra-15 centrifugal filter unit with an Ultracel-3 membrane (Millipore, Billerica, CA, USA). The solution was then centrifuged at 15,000 g for 15 min and the supernatant was used for proteome analysis to identify the shell proteins in the soluble organic matrix. In addition to the matrix proteins of the whole shell, we also extracted proteins contained specifically in the shell secondary layer. The fibers of the shell secondary layer were collected by decantation after crushing the shells in water, and the soluble organic matrix was collected by the same method described above.

SDS-PAGE analysis
To check the concentration and heterogeneity of the extracted proteins, the soluble organic matrix from the shell secondary layer was subjected to SDS-PAGE analysis using 4–20 % Mini-PROTEAN TGX Precast Gels (Bio-Rad, Hercules, CA, USA) with the standard method [36], and stained with CBB and silver staining.

Peptide sample preparation by trypsin digestion
The soluble matrix proteins extracted with 0.5 M EDTA and the proteins that was insoluble in the first extraction step with EDTA and re-dissolved in solution containing 9 M urea and 2 % (v/v) triton X-100 were diluted in 0.1 M Tris–HCl (pH 8.5) to a final volume of 200 μL. Methanol (600 μL), chloroform (150 μL), and distilled water (450 μL) were added one-by-one and mixed thoroughly. After centrifugation at 12,000 rpm for 5 min at 4 °C, the upper aqueous layer was removed and 500 μL of methanol was added. After centrifugation at 15,000 rpm for 5 min at 4 °C, the supernatant was removed and the resultant precipitated proteins were dried using a Speed Vac for 2 min. Protein samples were re-dissolved in 10 μL of 8 M urea, 0.1 M Tris–HCl (pH 8.5), and mixed for 1 h. Subsequently, 0.5 μL of 0.1 M DTT was added and incubated at 37 °C for 1 h. Then, 0.5 μL of 208 mM iodoacetamide was added and incubated for 1 h in the dark. After adding 30 μL of 0.1 M Tris–HCl and 60 μL of ultra-pure water, sequencing grade modified trypsin (Promega, Fitchburg, WI, USA) was added, and the solution was incubated at 37 °C for over 15 h.

LC-MS/MS analysis
The tryptic peptides were analyzed with a LTQ Orbitrap mass spectrometer (Thermo Fisher Scientific, Waltham, MA, USA) coupled with a DiNa nanoLC system (KYA Technologies, Tokyo, Japan). Precursor ions were detected over a range of 400–1,500 m/z, and the top four high-intensity ions were selected for MS/MS analyses in a data-dependent mode. Acquired MS/MS spectra were subjected to a database search against the protein sequence database translated from the transcriptome data from the mantle tissues of *Laqueus rubellus* with the SEQUEST program

using Proteome Discoverer software version 1.2 (Thermo Fisher Scientific). The parameters was set as below: the charge state of the precursor ions: automatically recognized, the mass range of tryptic peptides: 800 to 4,500, mass tolerances for precursor ions: 10 ppm mass tolerances for fragment ions: 1 Da. Up to two missed cleavages and modifications of carbamidomethylation (+57.021) of cysteine and oxidation (+15.995) of methionine were considered for calculation of the theoretical masses. False discovery rates (FDRs) was calculated based on a decoy database and using the Proteome Discoverer software. A list of the identified peptides that include a <1 % false discovery rate was obtained after filtering low confidence identification.

Sequence analyses

We performed BlastP similarity searches using the non-redundant protein sequence data stored in GenBank (http://blast.ncbi.nlm.nih.gov) using the default settings. The e-value cutoff was set at 1e-10. The domains in the protein sequences were predicted using the SMART program [37, 38] (http://smart.embl-heidelberg.de), including the optional searches for outlier homologues and homologues of known structure, Pfam domains and signal peptides, and the NCBI conserved domain search with the default settings [39–41] (http://www.ncbi.nlm.nih.gov/Structure/cdd/wrpsb.cgi). Isoelectric points and molecular masses of the predicted proteins were calculated using Genetyx version 6 (Genetyx, Tokyo, Japan). Sequence alignments were performed using Genetyx version 6.

RT-PCR for gene expression analysis

The mantle and lophophore tissues from a single individual were homogenized in 500 μL of Isogen (Nippon Gene, Tokyo, Japan) in a 1.5-mL microcentrifuge tube. Total RNA was extracted following the manufacturer's protocol, and treated with RQ1 RNase-Free DNase (Promega, Fitchburg, WI, USA). Reverse transcription was catalyzed by ReverTra Ace (Toyobo, Osaka, Japan), primed with a random primer. The amount of resultant cDNA was quantified by Qubit 2.0 Fluorometer (Life Technologies, Carlsbad, CA, USA), and an amount of 1 ng each of cDNA was used as template for polymerase chain reaction (PCR). PCR was catalyzed by Ex Taq Hot Start Version (Takara, Otsu, Japan). Partial sequence of ICP-1was amplified using the primer pair of Lr46S-1 (5'-GGC CAC ACC TCT GAT GGA TCA T) and Lr46A-1 (5'-TAC ACA CTT AAT GGA GAC CAG GC), and the annealing temperature was set at 58 °C. The following primer pairs were used for amplification of EF-1α; EF-B (5'-CCN CCD ATY TTR TAN ACR TCY TG) and EF-3 (5'-GGN CAY MGN GAY TTY RTN AAR AAY ATG AT), and the annealing temperature was set at 50 °C. Size and amount of RT-PCR products were verified by 1.5 % agarose gel electrophoresis.

Competing interests
The authors declared that they have no competing interests.

Authors' contributions
YI collected Laqueus rubellus samples, and curried out RNA extraction, shell matrix protein extraction, SDS-PAGE analysis, sequence analysis, and drafted the manuscript. IS participated in extraction of shell matrix proteins and SDS-PAGE analysis. KK carried out LC-MS/MS analysis. KO and MH carried out transcriptome analysis. KE conceived of the study, participated in its design and coordination, and helped to draft the manuscript. All authors read and approved the final manuscript.

Acknowledgments
We thank the staff of the Misaki Marine Biological Station (The University of Tokyo) for collecting brachiopod samples. We also thank Elika Iioka and Misa Kiuchi (The University of Tokyo) for DNA sequencing. This study was supported by the Japan Society for the Promotion of Science Grant-in-Aid for Scientific Research No. 15104009.

Author details
[1]Department of Earth and Planetary Science, Graduate School of Science, The University of Tokyo, 7-3-1 Hongo, Bunkyo-ku, Tokyo 113-0033, Japan. [2]The University Museum, The University of Tokyo, 7-3-1 Hongo, Bunkyo-ku, Tokyo 113-0033, Japan. [3]Center for Omics and Bioinformatics, Department of Computational Biology, Graduate School of Frontier Sciences, The University of Tokyo, 5-1-5 Kashiwanoha, Kashiwa, Chiba 277-8561, Japan. [4]Department of Life Science, School of Agriculture, Meiji University, 1-1-1 Higashimita, Tama, Kawasaki, Kanagawa 214-8571, Japan.

References
1. Murdock DJ, Donoghue PC. Evolutionary origins of animal skeletal biomineralization. Cells Tissues Organs. 2011;194:98–102.
2. Jackson DJ, Macis L, Reitner J, Degnan BM, Wörheide G. Sponge paleogenomics reveals an ancient role for carbonic anhydrase inskeletogenesis. Science. 2007;316:1893–5.
3. Marin F, Luquet G, Marie B, Medakovic D. Molluscan shell proteins: primary structure, origin, and evolution. Curr Top Dev Biol. 2008;80:209–76.
4. Mann K, Wilt FH, Poustka AJ. Proteomic analysis of sea urchin (Strongylocentrotus purpuratus) spicule matrix. Proteome Sci. 2010;8:33.
5. Joubert C, Piquemal D, Marie B, Manchon L, Pierrat F, Zanella-Cleon I, et al. Transcriptome and proteome analysis of Pinctada margaritifera calcifying mantle and shell: focus on biomineralization. BMC Genomics. 2010;11:613.
6. Marie B, Joubert C, Tayale A, Zanella-Cleon I, Belliard C, Piquemal D, et al. Different secretory repertoires control the biomineralization processes of prism and nacre deposition of the pearl oyster shell. Proc Natl Acad Sci U S A. 2012;109:20986–91.
7. Marie B, Jackson DJ, Ramos-Silva P, Zanella-Cleon I, Guichard N, Marin F. The shell-forming proteome of Lottia gigantea reveals both deep conservations and lineage-specific novelties. Febs j. 2013;280:214–32.
8. Ramos-Silva P, Kaandorp J, Huisman L, Marie B, Zanella-Cleon I, Guichard N, et al. The skeletal proteome of the coral Acropora millepora: the evolution of calcification by co-option and domain shuffling. Mol Biol Evol. 2013;30:2099–112.
9. Drake JL, Mass T, Haramaty L, Zelzion E, Bhattacharya D, Falkowski PG. Proteomic analysis of skeletal organic matrix from the stony coral Stylophora pistillata. Proc Natl Acad Sci U S A. 2013;110:3788–93.
10. Cusack M, Walton D, Curry G. Shell biochemistry. In: Kaesler R, editor. Treatise on Invertebrate Paleontology, Part H Brachiopoda (Revised) 1. Boulder: Geological Society of America Inc., and The University of Kansas; 1997. p. 243–66.
11. Cusack M, Williams A. Biochemistry and Diversity of Brachiopod Shells. In: Selden P, editor. Treatise on Invertebrate Paleontology, Part H Brachiopoda (Revised) 6. Boulder: Geological Society of America Inc., and The University of Kansas; 2007. p. 2373–95.
12. Collins MJ, Muyzer G, Curry GB, Sandberg P, Westbroek P. Macromolecules in brachiopod shells - characterization and diagenesis. Lethaia. 1991;24:387–97.

13. Jope M. The protein of brachiopod shell—I. Amino acid composition and implied protein taxonomy. Comp Biochem Physiol. 1967;20:593–600.

14. Walton D, Cusack M, Curry GB. Implications of the amino-acid-composition of recent new-zealand brachiopods. Palaeontology. 1993;36:883–96.

15. Collins MJ, Muyzer G, Westbroek P, Curry GB, Sandberg PA, Xu SJ, et al. Preservation of fossil biopolymeric structures - conclusive immunological evidence. Geochim Cosmochim Acta. 1991;55:2253–7.

16. Endo K, Walton D, Reyment RA, Curry GB. Fossil intra-crystalline biomolecules of brachiopod shells - diagenesis and preserved geo-biological information. Org Geochem. 1995;23:661–73.

17. Curry G, Cusack M, Endo K, Walton D, Quinn R. Intracrystalline molecules from brachiopod shells. In: Suga S, Nakahara H, editors. Mechanisms and phylogeny of mineralization in biological systems. Tokyo: Springer; 1991. p. 35–9.

18. Curry GB, Cusack M, Walton D, Endo K, Clegg H, Abbott G, et al. Biogeochemistry of brachiopod intracrystalline molecules. Philos Trans R Soc Lond Ser B Biol Sci. 1991;333:359–66.

19. Cusack M, Curry G, Clegg H, Abbott G. An intracrystalline chromoprotein from red brachiopod shells: implications for the process of biomineralization. Comp Biochem Physiol B. 1992;102:93–5.

20. Anstrom JA, Chin JE, Leaf DS, Parks AL, Raff RA. Localization and expression of msp130, a primary mesenchyme lineage-specific cell surface protein in the sea urchin embryo. Development. 1987;101:255–65.

21. Leaf DS, Anstrom JA, Chin JE, Harkey MA, Showman RM, Raff RA. Antibodies to a fusion protein identify a cDNA clone encoding msp130, a primary mesenchyme-specific cell surface protein of the sea urchin embryo. Dev Biol. 1987;121:29–40.

22. Kito K, Ito T. Mass spectrometry-based approaches toward absolute quantitative proteomics. Curr Genomics. 2008;9:263–74.

23. Ishihama Y, Oda Y, Tabata T, Sato T, Nagasu T, Rappsilber J, et al. Exponentially modified protein abundance index (emPAI) for estimation of absolute protein amount in proteomics by the number of sequenced peptides per protein. Mol Cell Proteomics. 2005;4:1265–72.

24. Zybailov B, Mosley AL, Sardiu ME, Coleman MK, Florens L, Washburn MP. Statistical analysis of membrane proteome expression changes in Saccharomyces cerevisiae. J Proteome Res. 2006;5:2339–47.

25. Lu P, Vogel C, Wang R, Yao X, Marcotte EM. Absolute protein expression profiling estimates the relative contributions of transcriptional and translational regulation. Nat Biotechnol. 2007;25:117–24.

26. Mann K, Poustka AJ, Mann M. In-depth, high-accuracy proteomics of sea urchin tooth organic matrix. Proteome Sci. 2008;6:33.

27. Mann K, Poustka AJ, Mann M. The sea urchin (Strongylocentrotus purpuratus) test and spine proteomes. Proteome Sci. 2008;6:22.

28. Marie B, Zanella-Cleon I, Guichard N, Becchi M, Marin F. Novel proteins from the calcifying shell matrix of the Pacific oyster Crassostrea gigas. Mar Biotechnol (N Y). 2011;13:1159–68.

29. Ettensohn CA. Horizontal transfer of the msp130 gene supported the evolution of metazoan biomineralization. Evol Dev. 2014;16:139–48.

30. Parr BA, Parks AL, Raff RA. Promoter structure and protein sequence of msp130, a lipid-anchored sea urchin glycoprotein. J Biol Chem. 1990;265:1408–13.

31. Bedouet L, Duplat D, Marie A, Dubost L, Berland S, Rousseau M, et al. Heterogeneity of proteinase inhibitors in the water-soluble organic matrix from the oyster nacre. Mar Biotechnol (N Y). 2007;9:437–49.

32. Marie B, Zanella-Cleon I, Le Roy N, Becchi M, Luquet G, Marin F. Proteomic analysis of the acid-soluble nacre matrix of the bivalve Unio pictorum: detection of novel carbonic anhydrase and putative protease inhibitor proteins. Chembiochem. 2010;11:2138–47.

33. Weiner S, Hood L. Soluble protein of the organic matrix of mollusk shells: a potential template for shell formation. Science. 1975;190:987–9.

34. Miyamoto H, Miyoshi F, Kohno J. The carbonic anhydrase domain protein nacrein is expressed in the epithelial cells of the mantle and acts as a negative regulator in calcification in the mollusc Pinctada fucata. Zoolog Sci. 2005;22:311–5.

35. Skovsted CB, Balthasar U, Brock GA, Paterson JR. The tommotiid Camenella reticulosa from the early Cambrian of South Australia: morphology, scleritome reconstruction, and phylogeny. Acta Palaeontol Pol. 2009;54:525–40.

36. Laemmli UK. Cleavage of structural proteins during the assembly of the head of bacteriophage T4. Nature. 1970;227:680–5.

37. Letunic I, Doerks T, Bork P. SMART 7: recent updates to the protein domain annotation resource. Nucleic Acids Res. 2012;40:D302–5.

38. Schultz J, Milpetz F, Bork P, Ponting CP. SMART, a simple modular architecture research tool: identification of signaling domains. Proc Natl Acad Sci U S A. 1998;95:5857–64.

39. Marchler-Bauer A, Bryant SH. CD-Search: protein domain annotations on the fly. Nucleic Acids Res. 2004;32:W327–31.

40. Marchler-Bauer A, Anderson JB, Chitsaz F, Derbyshire MK, DeWeese-Scott C, Fong JH, et al. CDD: specific functional annotation with the Conserved Domain Database. Nucleic Acids Res. 2009;37:D205–10.

41. Marchler-Bauer A, Lu S, Anderson JB, Chitsaz F, Derbyshire MK, DeWeese-Scott C, et al. CDD: a Conserved Domain Database for the functional annotation of proteins. Nucleic Acids Res. 2011;39:D225–9.

17

Comparative proteomics of Bt-transgenic and non-transgenic cotton leaves

Limin Wang[1,2], Xuchu Wang[1*], Xiang Jin[1], Ruizong Jia[1], Qixing Huang[1], Yanhua Tan[1] and Anping Guo[1*]

Abstract

Background: As the rapid growth of the commercialized acreage in genetically modified (GM) crops, the unintended effects of GM crops' biosafety assessment have been given much attention. To investigate whether transgenic events cause unintended effects, comparative proteomics of cotton leaves between the commercial transgenic Bt + CpTI cotton SGK321 (BT) clone and its non-transgenic parental counterpart SY321 wild type (WT) was performed.

Results: Using enzyme linked immunosorbent assay (ELISA), Cry1Ac toxin protein was detected in the BT leaves, while its content was only 0.31 pg/g. By 2-DE, 58 differentially expressed proteins (DEPs) were detected. Among them 35 were identified by MS. These identified DEPs were mainly involved in carbohydrate transport and metabolism, chaperones related to post-translational modification and energy production. Pathway analysis revealed that most of the DEPs were implicated in carbon fixation and photosynthesis, glyoxylate and dicarboxylate metabolism, and oxidative pentose phosphate pathway. Thirteen identified proteins were involved in protein-protein interaction. The protein interactions were mainly involved in photosynthesis and energy metabolite pathway.

Conclusions: Our study demonstrated that exogenous DNA in a host cotton genome can affect the plant growth and photosynthesis. Although some unintended variations of proteins were found between BT and WT cotton, no toxic proteins or allergens were detected. This study verified genetically modified operation did not sharply alter cotton leaf proteome, and the target proteins were hardly checked by traditional proteomic analysis.

Keywords: Bacillus thuringiensis, Comparative proteomics, Cotton leaf, Cry1Ac gene, Toxin protein, Transgenic plant

Background

Since the first genetically modified (GM) crops were commercialized in 1996, the global GM crops have increased more than 100-fold from 1.7 million hectares in 1996 to over 175 million hectares in 2013 [1]. GM crops offer farmers opportunities to improve their products by planting disease resistance, drought resistance or nutrient components which incorporates new genes into crop plants [2,3]. Despite the many benefits of GM crops, the biggest problem is controversial on the safety of food that derived from GM crops. An important issue is whether the existence of unintended effects which are caused by random insertion of exogenous specific genes into plant genomes that may result in disruption, modification or rearrangement of the genome [4,5]. These unintended processes may further result in the formation of new biochemical processes or new proteins (especially new allergens or toxins), which have been an important matter of concerns [6,7]. So, evaluation of whether transgenic events have caused unintended changes is essential to guarantee the food safety and solve the controversial issue on the GM crops.

The concept of substantial equivalence was proposed as a major principle and guiding tool of biological safety assessment according to the Organization for Economic Cooperation and Development [8,9]. Also, more and more approaches involving in targeted and non-targeted genes were applied to assess the safety of GM crops. Traditional methods to detect the safety of GM crops mainly focused on the analysis of key nutritional and non-nutritional components, including the enzyme linked immunosorbent assay (ELISA) and PCR detection of some specific genes, which are considered as targeted approaches [6,10]. At present, non-targeted approaches including the profiling techniques (such as genomics, transcriptomics, proteomics,

* Correspondence: xchwanghainan@163.com; xchwanghainanlab@163.com
[1]Chinese Academy of Tropical Agricultural Sciences, The Institute of Tropical Biosciences and Biotechnology, Haikou, Hainan 571101, China
Full list of author information is available at the end of the article

and metabolomics) allow for simultaneously measuring and comparing the entire sets of transcripts, proteins, and metabolites in organisms [9,11-13]. These non-targeted approaches have been considered to provide unbiased results and more complete insights into any unpredicted changes.

Many studies have been conducted using profiling techniques to evaluate GM crops. Among the profiling techniques, proteomics is a direct method of investigation unpredicted alteration [14,15]. It has a broad application prospects in the safety assessment of genetically modified crops [16]. Proteins are not only the key players in gene function and directly involved in metabolism and cellular development, but also have roles as toxin, antinutrients, or allergens, which have great impact on human health [5,17]. Comparative proteomics by 2-DE combined with mass spectrometry (MS) technologies have been widely used to assess the safety of GM crops, such as soybean [18,19], rice [10,20], maize [8,21-23], potato [24,25], tomato [26,27], and wheat [28,29]. These studies mainly focused on detecting the unintended effects and researching the functional characterization of GM crops. However, no comparative proteomics on GM cotton was reported till now.

Transgenic insect-resistant cotton is the fastest one of global commercialization GM crops because of its economic advantages and environmental impacts, increasing income and reducing environmental pollution by reducing usage of pesticides [30,31]. The global cultivated area of GM cotton was reaching 23.9 million hectares in 2013. Previous studies mainly focused on detecting the biochemical compounds differences between transgenic and non-transgenic cotton, including amino acids fatty acids, carbohydrate content [32]. Fourier transform infrared spectroscopy (FTIR) was also used to detect the chemical and conformational changes between transgenic cotton seeds and their non-transgenic counterparts, and found both the indigenous and exogenous proteins structural changes in genetically modified organism (GMO) [33]. However, it didn't mention that the transgenic cotton might result in some protein changes and the formation of new metabolites or altered levels of existing metabolites.

Leaves are key organs for plant biomass and seed production because of their roles in energy capture and carbon conversion [34]. In the present study, we carried out comparative proteomics between transgenic cotton line with a toxin CrylAc gene from *Bacillus thuringiensis* (BT) and non-transgenic cotton (WT) leaves combined with 2-DE and MS to study the protein changed level for evaluating the unintended effects in the transgenic cotton. The transgenic cotton lines contain the inserted Cry1Ac and CpTI gene. Hypothetically, the only expected difference between BT and WT should be the presence of BT and CpTI proteins. However, none of these proteins were detected by 2-DE and MS. In addition, none of the DEPs was a toxic protein but

related to central carbon metabolism, starch synthesis, protein folding and modification.

Result
PCR and ELISA detection of target protein
A 119 bp DNA band only detected in BT leaves by PCR using gene specific primers, confirmed the exist of exogenous CrylAc gene in BT cotton (Additional file 1A). Envirologix's plate kits for Cry1Ac were used to study expression of Cry1Ac gene in transgenic and the non-transgenic cotton leaves. Cry1Ac expressed protein was not detected in non-transgenic cotton, but was detected at expressed level of 0.31 pg/g in the transgenic cotton leaves (Additional file 1B). The result suggested that Bt toxin protein was really existed in transgenic cotton, but the protein abundance was extremely low. RT-PCR revealed BT line had one detectable DNA fragment with a size of 282 bp. The DNA fragments were undetected in their nontransgenic controls (Additional file 1C).

Physiological parameters were compared between WT and BT lines (Figure 1). In BT lines, the plant heights (Figure 1A) and water content (Figure 1B) were significantly increased. In contrast, the net photosynthetic rate (Figure 1C) and chlorophyll content (Figure 1D) decreased in BT lines. The result suggested that the inserted Cry1Ac and CpTI gene directly or indirectly effect the plant growth and photosynthesis.

Analysis of protein profiles of non-Transgenic and Bt-Transgenic cotton leaves
2-DE and image analysis of the protein profiles were carried out to detect the DEPs between the WT (Figure 2A) and BT (Figure 2B) lines. Total proteins of 2-DE reference maps were obtained using IPG strips with pH 4–7 and 12% SDS-PAGE (Figure 2A-C). Protein spots were detected and quantified using Image Master 2D Platinum Software (Version 5.0, GE Healthcare). Our results showed that more than 600 protein spots were detected in each 2-DE image with good reproducibility, respectively. Only the DEPs with abundance change more than 1.5 fold (confidence above 95%, $p < 0.05$) were selected for MS analysis. Compared to the WT line, a total of 58 DEPs (Figure 2C) were selected, including 34 up-regulated and 24 down-regulated protein spots (Table 1).

Protein identification by MALDI TOF/TOF MS
Among the 58 DEPs, 35 (60.3%) proteins were positively identified *via* MALDI TOF/TOF MS (Figure 2), with 23 up-regulated protein spots and 12 down-regulated ones compared to WT. Among these identified proteins, 30 protein species were assigned to potential functions, and the other 5 protein species were identified as hypothetical proteins or unknown proteins (Table 1; Additional files 2 and 3).

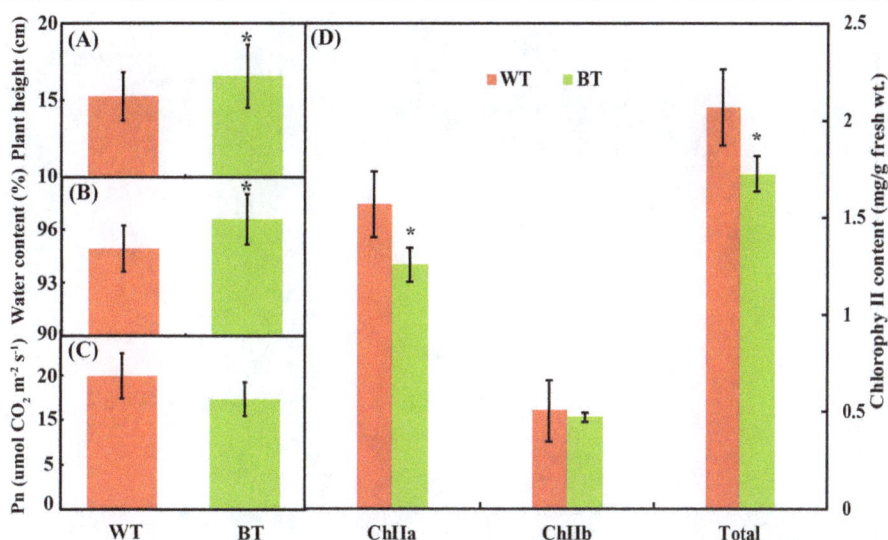

Figure 1 Growth patterns and physiological changes of the Bt-trangenic and non-transgenic cottons. The plant height **(A)**, leaf water content **(B)**, plant net photosynthetic rate **(C)**, and chlorophyll content **(D)** were highlighted. Statistically significant differences relative to the control plants were calculated by independent Student T-test. *indicated $p < 0.05$.

To evaluate the quality of the proteins identification by MALDI TOF/TOF MS, the theoretical and experimental ratios of molecular weight (Mr) and isoelectric point (pI) were determined, respectively (Table 1). These ratios were presented as radar axis labels (the Mr ratio for the radial value and the pI ratio for the annular value) in radial chart (Figure 3A). When the theoretical and experimental values of the identified proteins are the same, both the radial values and the annular values will be 1.0 and all these identified proteins will be located on the cyclical line 1.0 in radial chart. The closer a spot is to line 1.0, the greater the certainty that the identification made by means of MS/database searching will be the MS identification obtained. More than 80% of the identified protein spots were closely located on the cyclical line 1.0, indicating the high quality of the MS data (Figure 3A).

Protein function analysis

The identified proteins were obtained from 15 plant species (Figure 3B). The sequence homologies of these identified proteins to those of proteins from other plant species were also determined. Among the identified proteins, 22% showed strong sequence homology to *Ricinus* proteins, followed by 19% of *Gossypium* proteins, and 16% of *Vitis* proteins.

The 35 identified proteins were classified into 10 groups based on their main cellular functions as defined by COG functional catalogue (Table 1; Figure 3C), including: 35% proteins in carbohydrate transport and metabolism, 15% proteins in chaperones related to post-translational modification, 12% proteins in energy production and

conversion, 3% proteins in cell division and chromosome partitioning, 3% proteins in amino acid transport and metabolism, 3% proteins in coenzyme transport and metabolism, 3% proteins in inorganic ion transport and metabolism, 3% proteins in cell envelope biogenesis, outer membrane, 3% proteins in nucleotide transport and metabolism, 20% proteins with no-related or could not be classified by COG classification (Table 1; Figure 3C).

The subcellular locations of the identified 35 proteins were also predicted. Among them, the largest portion including 16 proteins were located in chloroplast. Followed by the 14 proteins which were in cytoplasmic. Then, several proteins were located on the periplasmic, mitochondrial, outermembrane or extracellular (Figure 3D; Additional file 2). These results suggested large number of DEPs related to carbohydrate transport and metabolism mainly located on chloroplast and cytoplasm.

Pathway analysis of all identified proteins using GO and KEGG

To reveal the functions of DEPs between WT and BT, GO analysis was performed using WEGO software to confirm the cellular component, biological process and molecular function (Figure 4; Additional file 2). Twenty-eight out of the 35 identified proteins were classified into 3 large groups containing 23 subgroups based on their functional annotation. At GO-cellular level, the largest part including 18 proteins were in the cell (GO: 0005623), another 18 proteins occur in the cell part (GO: 0043226), and 16 proteins occur in the organelle (GO: 0043226), with the remainder occurring in the extracellular region (GO: 0005576), membrane-enclosed

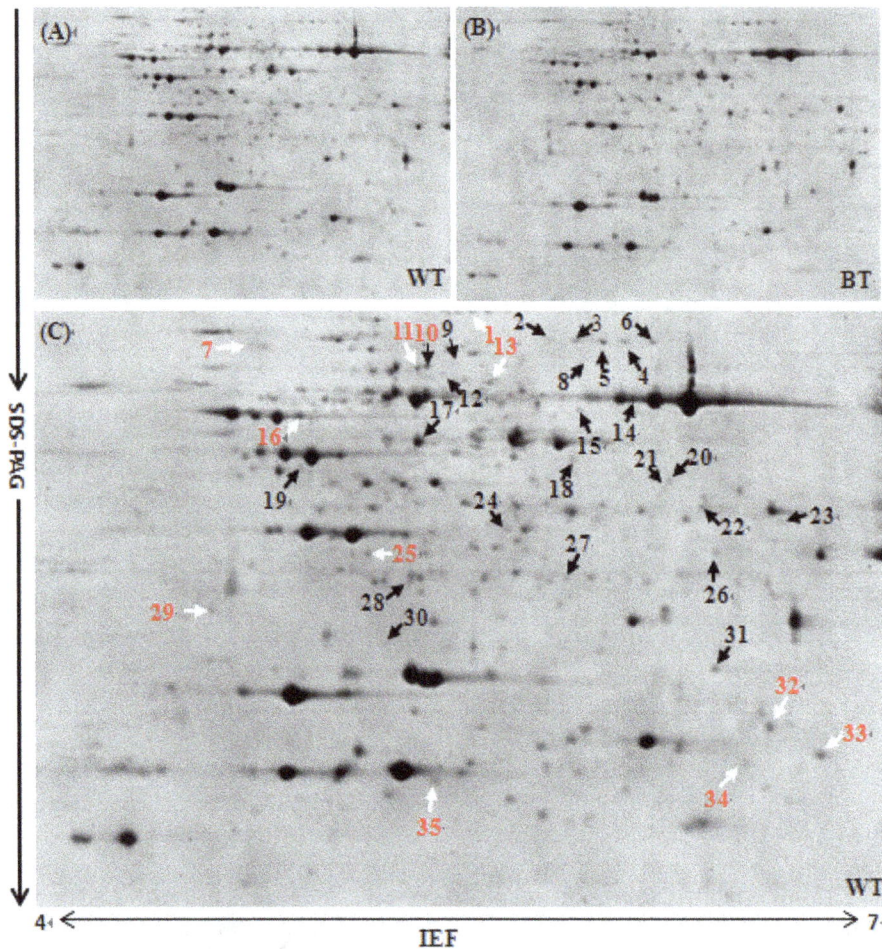

Figure 2 Typical 2-DE profile of leaf proteins from the transgenic cotton line and its control. 2-DE protein profiles of the WT **(A)** and BT **(B)** were presented, and the identified DEPs were marked with the number on the 2-DE gels **(C)**. Arrows indicated the 35 positively identified protein spots by MALDI TOF/TOF MS. Their identities were listed in Table 1 and Additional file 2.

lumen (GO: 0031974), envelope (GO: 0031975), macromolecular complex (GO: 0032991), and organelle part (GO: 0044422) (Figure 4; Additional file 2). For the molecular function ontology, 4 subcategories were assigned. The largest portion was catalytic activity (GO: 0003824) including 20 proteins, followed 19 proteins with binding function (GO: 0005488). Then, several proteins had transporter activity (GO: 0005215) and electron carrier activity (GO: 0009055) (Figure 4; Additional file 2). In the biological process category, 11 subgroups were overexpressed. The largest part including 25 proteins was related to metabolic process (GO: 0008152), followed by cellular process (GO: 0009987) involving in 20 proteins, with the other important biological processes including cellular component organization (GO: 0016043), multicellular organismal process (GO: 0032501), developmental process (GO: 0032502), pigmentation (GO: 0043473), response to stimulus (GO: 0050896), localization (GO: 0051234), multi-organism process (GO: 0051704) and

biological regulation (GO: 0065007) (Figure 4; Additional file 2).

KEGG pathway analysis was performed using Blast2GO program to determine their molecular interaction and reaction networks and which pathways were significant. The 35 identified proteins were involved in 13 kinds of KEGG pathways (Figure 5; Additional files 4 and 5), including carbon fixation in photosynthetic organisms, glyoxylate and dicarboxylate metabolism, purine metabolism, pentose phosphate pathway, nitrogen metabolism, photosynthesis, oxidative phosphorylation, amino acid metabolism, etc. The most important pathway is carbon fixation in photosynthetic organisms and photosynthesis, which contains 4 identified enzymes named ribulose-bisphosphate carboxylase (EC 4.1.1.39, spots 8, 14, 22, 26 and 32), transketolase (EC 2.2.1.1, spots 2, 3, 4, 5 and 6), ferredoxin–NADP$^+$ reductase (EC 1.18.1.2, spot 23), ion-sulfur reductase (EC 1.10.9.1, spot 33). Two enzymes belonging to the porphyrin and

Table 1 Proteins identified by MALDI TOF/TOF MS from transgenic cotton leaves

Spot no[a].	GI no.[b]	Function category protein name	Plant species	Exper.[c] pI/Mr	Thero.[d] pI/Mr	M P[e]	SC%[f]	M. S.[g]	Relative change (WT/BT)[h]
Posttranslational modification, protein turnover, chaperones (O)									
1	255556934	ATPase	R. communis	5.49/90	5.13/90.1	11	21	324	
7	147819511	Hypothetical protein	V. vinifera	4.86/72	5.20/61.4	5	10	256	
9	225433375	Chaperonin	V. vinifera	5.40/63	5.85/61.7	6	14	204	
16	12620883	Rubisco activase	G. hirsutum	4.99/49	5.06/48.6	13	41	822	
27	14594915	Proteasome subunit	N. tabacum	5.77/27	6.12/18.2	3	32	113	
Carbohydrate transport and metabolism (G)									
2	110224784	Transketolase	P. acerifolia	5.73/76	6.25/26.0	2	11	82	
3	255541252	Transketolase	R. communis	5.80/76	6.52/81.6	5	8	277	
4	255541252	Transketolase	R. communis	5.95/76	6.52/81.6	5	8	262	
5	255541252	Transketolase	R. communis	5.89/76	6.52/81.6	6	10	404	
6	255541252	Transketolase	R. communis	6.05/76	6.52/81.6	6	11	337	
8	3560664	Rubisco	C. ensifolium	5.73/66	6.4/49.7	5	13	211	
14	11230404	Rubisco, large subunit	C. Pettersson	5.96/53	5.96/52.9	12	40	998	
15	33415263	Enolase	G. hirsutum	5.69/49	6.16/47.9	11	36	502	
22	329317332	Rubisco, large subunit	G. barbadense	6.20/35	6.00/53.7	5	16	326	
24	449442663	Phosphoglycolate phosphatase	C. sativus	5.58/32	6.47/41.7	8	26	442	
26	1881499	Rubisco, large subunit	P. pendula	6.24/30	6.61/52.6	9	22	366	
32	548699	Rubisco, large chain	(–)	6.40/19	6.12/52.6	4	9	104	
Energy production and conversion (C)									
10	40850676	Betaine-aldehyde dehydrogenase	G. hirsutum	5.43/63	5.6/55.4	5	12	124	
11	91208909	ATP synthase, beta subunit	G. hirsutum	5.33/59	5.22/53.6	13	42	939	
20	225451308	Auxin-induced protein	V. vinifera	6.09/39	5.96/37.9	1	3	55	
33	315364830	Ion-sulfur protein	C. lanatus	6.54/19	8.45/24.6	3	18	231	
Cytoskeleton (Z)									
17	281485191	actin	P. americana	5.34/44	5.31/41.9	12	46	984	

Table 1 Proteins identified by MALDI TOF/TOF MS from transgenic cotton leaves *(Continued)*

Spot	GI number	Protein	Species	c	d	MP	SC	M.S.	Ratio
Amino acid transport and metabolism (E)									
18	211906462	Glutamine synthase	*G. hirsutum*	5.80/43	5.77/39.4	5	16	160	
Coenzyme transport and metabolism (H)									
19	44943772	Magnesium-chelatase subunit	*C. sativus*	4.98/40	5.72/46.0	5	15	150	
21	255558669	Porphobilinogen deaminase	*R. communis*	6.07/38	6.55/40.3	3	10	50	
Inorganic ion transport and metabolism (P)									
23	225431122	Ferredoxin–NADP reductase	*V. vinifera*	6.46/34	8.91/40.8	7	22	270	
Cell envelope biogenesis, outer membrane (M)									
25	292668595	Sanguinarine reductase	*E. californica*	5.18/29	4.97/29.6	2	6	104	
Nucleotide transport and metabolism (F)									
34	225457446	Nucleoside diphosphate kinase,	*V. vinifera*	6.41/18	9.28/26.0	3	13	147	
No related to COG (NO)									
12	255558986	Hypothetical protein	*R. communis*	5.39/58	8.23/54.9	2	2	69	
13	255558986	Hypothetical protein	*R. communis*	5.55/59	8.23/54.9	2	2	53	
28	226358407	Chlorophyll binding protein	*G. hirsutum*	5.30/27	5.53/25.6	3	20	166	
29	118489712	Unknown	*P. trichocarpa*	4.73/26	4.77/24.4	3	9	156	
30	147767601	Hypothetical protein	*V. vinifera*	5.23/42	8.46/25.5	2	11	59	
31	302595736	Oxygen-evolving enhancer protein	(–)	6.26/22	8.67/28.2	3	9	215	
35	211906510	Major latex-like protein	*G. hirsutum*	5.38/17	5.46/17.2	3	27	116	

[a] Assigned spot number as indicated in Figure 2.
[b] Database GI numbers according to NCBInr.
[c,d] The experimental (c) and theoretical (d) mass (kDa) and pI of the identified proteins.
[e] Number of the matched peptides (MP).
[f] The amino acid sequence coverage (SC) for the identified proteins.
[g] The Mascot searched score (M. S.) against the database NCBInr.
[h] The relatively changed ratios of protein amount on different 2-D gels.

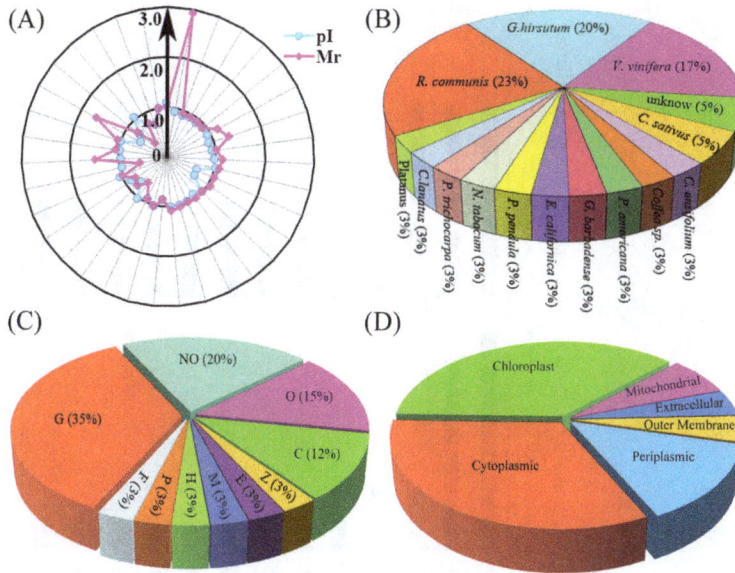

Figure 3 Classification and functional analysis of the identified 35 DEPs. To evaluate the quality of identified proteins, the theoretical and experimental ratios of molecular mass (Mr) and isoelectric point (pI) were determined and presented in radial chart as radial and annular radar axis labels respectively **(A)**. Then, the distributions of the identified proteins in different plant species were also presented **(B)**. Each protein was functionally classified by COG **(C)**. The proportion of each functional category was the sum of the proportion of all identities. The subcellular locations of the identified 35 proteins were presented **(D)**. The abbreviations were: G, carbohydrate transport and metabolism; O, Posttranslational modification, protein turnover, chaperones; C, energy production and conversion; D, Cell division and chromosome partitioning; E, Amino acid transport and metabolism; H, Coenzyme transport and metabolism; P, Inorganic ion transport and metabolism; M, Cell envelope biogenesis, outer membrane; F, Nucleotide transport and metabolism; NO, No related COG.

chlorophyll metabolism pathway, which are important for photosynthesis in green plants, were also identified. They were hydroxymethylbilane synthase (EC 2.5.1.61, spot 21) and chelatase (EC 6.6.1.1, spot 19). Another important pathway was glyoxylate and dicarboxylate metabolism, for which 3 enzymes were identified from 7 differentially sized protein spots. These enzymes were glutamine synthase (EC 6.3.1.2, spot 18), Rubisco carboxylase (EC 4.1.1.39, spot 8, 14, 22, 26, 32), and phosphatase (EC 3.1.3.18, spot 24). They are key enzymes for

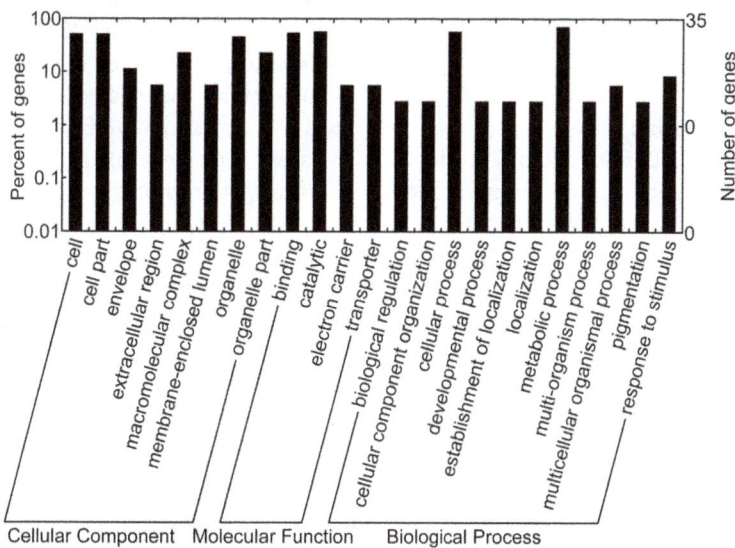

Figure 4 GO classification of the identified DEPs. To reveal the functions of the identified 35 DEPs between WT and BT, GO analysis was performed using WEGO software. The 28 proteins among the identified 35 DEPs were available and then classified into 3 main categories including cellular component, biological process, and molecular function with 23 subgroups. The number of genes denotes that of proteins with GO annotations.

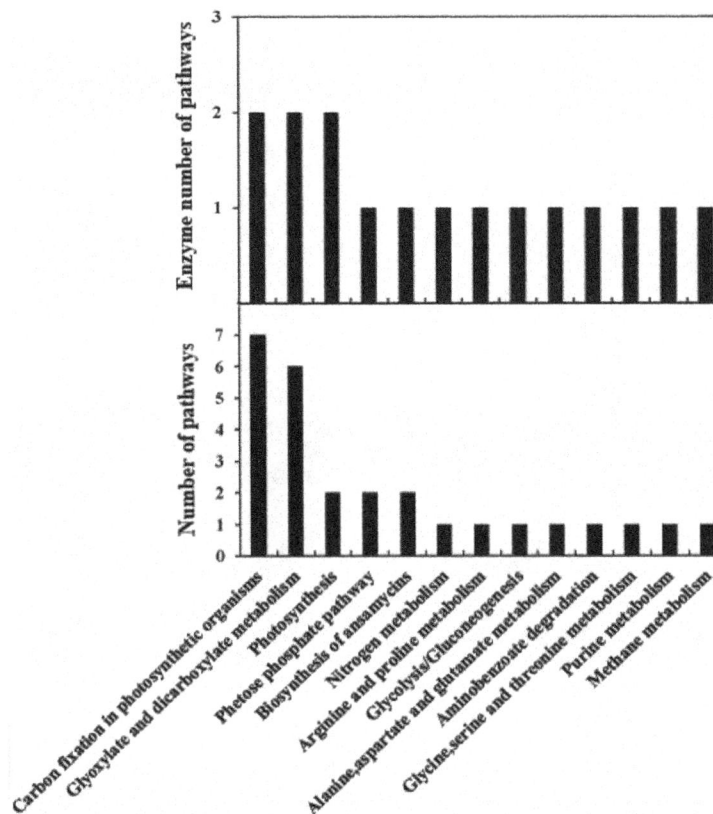

Figure 5 KEGG pathway analysis of the identified 35 DEPs. To determine the molecular interaction and reaction networks of the identified proteins, KEGG pathway analysis was performed. The related pathways were classified into 13 main categories. The number of sequences and enzymes corresponding to each pathway were illustrated.

carbon fixation and photosynthesis. It is noteworthy that most of enzymes involved in carbon fixation and photosynthesis pathways were considerably up-regulated after target gene over-expression.

Protein-protein interaction analysis

The DEPs were subjected to STRING database to identify the interaction of these proteins. Protein interaction network was constructed and visualized with Cytoscape software. Among the 35 identified proteins, 13 were involved in protein-protein interaction, and three major clusters of interacting proteins were constructed (Figure 6). The proteins interactions mainly participated in photosynthesis pathway (Figure 6A) and energy metabolism (Figure 6B). Rubisco activase (spot 16) and chlorophyll binding protein (spot 28) are the central core protein of the interacting network, due to their interactions with many other proteins.

Immunoblot and qRT-PCR analysis

Among the DEPs, several proteins with the different molecular weight and pI value were identified as Rubisco

(spot 8, 14, 22). We used 1-D western blot analysis to determine the expression abundance (Figure 7A). The expression profile showed that a higher level of protein abundance was observed in BT lines.

To explore the changes of DEPs at transcriptional level, 20 representative DEPs were chosen for qRT-PCR to assess their gene expression. The transcriptional expression patterns of these genes were divided into three groups as show in Figure 7B and D. The first group was up-regulated including three genes encoding Rubisco with similar changed pattern both at protein and gene level (Figure 7B). In the second group, DEPs except spot 16 related to photosynthesis were up-regulated with gene encoding Magnesium-chelatase subunit (spot 19), porphobilinogen deaminase (spot 21), Ferredoxin–NADP reductase (spot 23), and chlorophyll binding protein (spot 28) (Figure 7C). The last group displayed the other 12 representative transcripts expression patterns at transcriptional level (Figure 7D). Compared with the expression patterns at transcriptional and translational levels of the 20 coding genes, the transcriptional expression trends of 4 genes named ATPase (spot 1), chaperonin (spot 9),

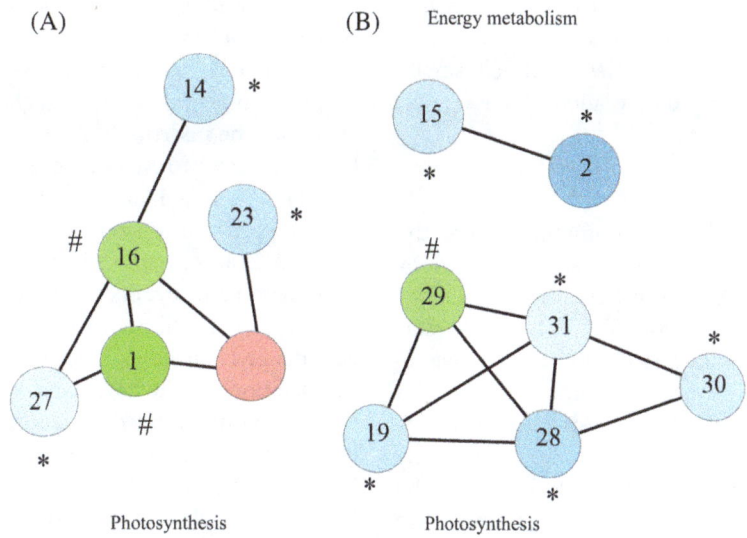

Figure 6 Protein-protein interaction network analysis by STRING. Protein interaction network was generated with STRING and visualized with Cytoscape software. Highly interacting proteins are divided into three clusters which mainly involved in photosynthesis **(A)** and energy metabolism **(B)**. Among them, the 10 up-regulated proteins were marked with * and 3 down-regulated proteins were marked with #.

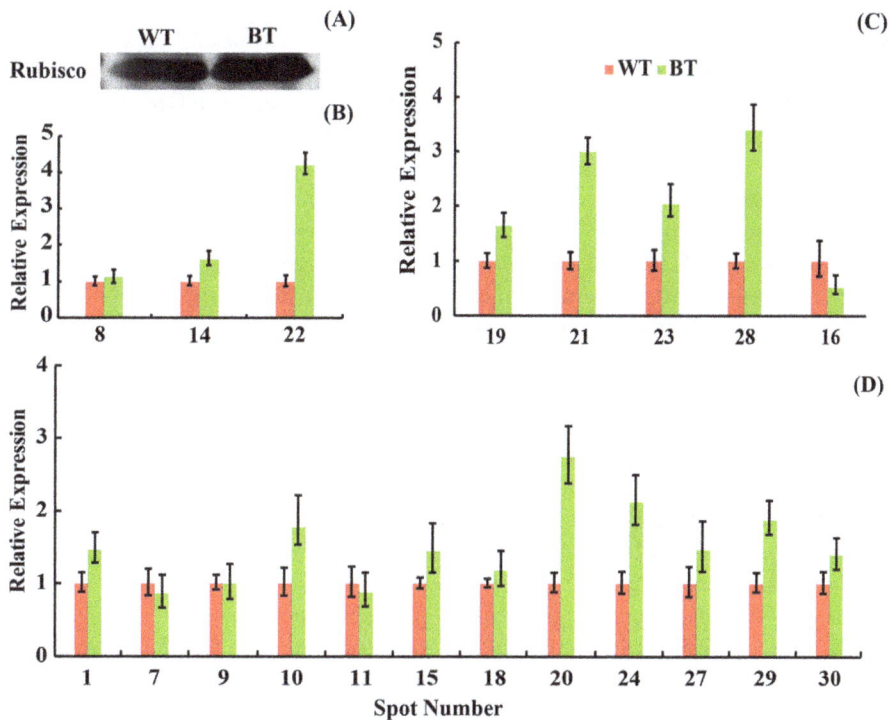

Figure 7 Immunoblot and quantitative RT-PCR analysis of the 20 representative DEPs. The expression profile for Rubisco was detected using 1-D Western blot analysis **(A)**. The identified three different members of Rubisco large subunit genes were up-regulated at transcriptional level in the Bt-transgenic lines **(B)**. The changed patterns of genes for the five DEPs related to photosynthesis were determined **(C)**. The gene expression patterns of the other 12 representative DEPs in both the Bt-transgenic (right) and non-transgenic (left) plants were highlighted **(D)**. The primers used for qRT-PCR were provided in Additional file 6.

betaine-aldehyde dehydrogenase (spot 10), and a function unknown protein (spot 29) were different with their translational expression. The other 16 genes displayed similar trends at both transcriptional and translational levels.

Discussion

Since genetically modified crops commercialized, the biosafety assessment of GM crops was concerned by more and more people [35]. To provide more evidence for the biosafety assessment of GM cottons, in this study, we applied proteomics-based approach to investigate the differentially expressed proteins between transgenic cotton leaves and their non-transgenic counterparts. To perform the proteomic analysis, not only the homozygous GM material SGK321, but also the exact non-GM counterpart SY321 was used to minimize the background differences in this study. Also, to ensure that the DEPs mainly come from the transgenic insertion event rather than the genetic background or others, only the protein spots with good reproducibility and which the fold-change in intensity was > 1.5 were further selected to identification *via* MS. Of course, we still cannot exclude the possibility that a few DEPs may come from the genetic background or others, though there was very little possibility. Our results suggested the changes among them were not obviously. The study is consistent with the other GM crops lines finding that no new or toxin proteins were detected in transgenic plants by comparative proteomics [3,8,10,16].

GM didn't dramatically alter proteomes of cotton leaves

Some reports referred that random insertion of exogenous genes in plant genomes could lead to disruption of endogenous genes and rearrangement of genome, which could produce new proteins especially new allergens or toxins proteins [10,16]. To evaluate the effected caused Cry1Ac + CpTI genes insertion, 2-DE combining with mass spectrometric techniques was conducted. Approximately 35 DEPs were identified in the transgenic cotton leaves in comparison with their non-transgenic lines. Nevertheless, neither allergens nor BT toxics were detected in transgenic cotton leaves in 2-DE gels. It was possibly due to the low abundance of Cry1Ac protein, which was detected as only 0.31 pg/g in transgenic cotton leaves (Figure 1B), so that it was undetectable in 2-DE gels. Similar result has been noted in other studies. This is expected because proteomics is a useful method for comprehensive analyses but not if the level of a target protein is extremely low. The result implying that GM did not sharply alter the proteome of cotton leaves, and also did not lead to the unintended effects, if it exists, was slight or not easy to detect.

Carbon fixation in photosynthesis is a major biological process in DEPs

The metabolic variations between the transgenic plant and its non-transgenic line might be due to the position effect of gene insertion [32]. According to the KEGG analysis, the present results revealed that DEPs between WT and BT lines mainly involved in photosynthetic organisms to take part into carbon fixation, photosynthesis, glyoxylate and dicarboxylate, oxidative phosphorylation, pentose phosphate pathway, and so on (Additional files 4 and 5). The largest portion of metabolic-related DEPs whose abundance changed significantly was connected with carbon fixation in photosynthetic organisms and photosynthesis. The unintended variations and effects could have effects on plant growth and developments. Photosynthesis is the process that plant converts light energy into chemical energy including light reaction and carbon reaction (dark reaction). It is not only the basis of biological survival, but also an important to meet energy and food needs. The recent in basic and applied research on photosynthesis more and more focused on the carbon fixation efficiencies improvements, due to the crops yield and energy requirement [36]. Our research revealed that 1 ribulose-bisphosphate carboxylase (Rubisco) (spots 8), 4 Rubisco large subunits (spots 14, 22, 26 and 32) and 5 transketolases (spots 2, 3, 4, 5 and 6) participated in the carbon fixation, with more expression in transgenic cotton line except for spot 32 (Table 1; Additional files 4 and 5). Rubisco has a pivotal role in photosynthetic organisms [37]. This enzyme catalyzes the carboxylation step in the Calvin cycle of carbon fixation, accompanying the process that stores the energy trapped by photosynthesis and also catalyzes the oxygenation step in photorespiration, during which a considerable amount of the stored energy is converted to heat thereby limiting crop yield [38]. In this study, most large subunits of Rubisco showed to be increased at both protein expression abundance and transcriptional expression patterns in the transgenic cotton lines (Table 1; Figure 7A and B), suggesting the efficiency of CO_2 fixation is increased in transgenic cotton. Additionally, 5 ribulose-bisphosphate carboxylases (spots 8, 14, 22, 26 and 32) also took part in the glyoxylate and dicarboxylate metabolism. In plants, transketolase related to energy metabolism can catalyze reactions in the Calvin cycle of photosynthesis and oxidative pentose phosphate pathway (OPPP). Related researches showed reduction of transketolase expression had a marked inhibited on photosynthesis, secondary metabolism, and plant growth but OPPP activity was not strongly inhibited by decreased transketolase activity [39]. In the present study, expression abundance of 5 transketolase isoforms (spots 2, 3, 4, 5 and 6) was up-regulated, implying the transgenic cotton could enhance photosynthesis ability.

In addition, the other related to photosynthesis and energy metabolism proteins also were identified and showed higher abundance in the transgenic cotton. Chlorophyll A-B binding protein is an important component in the light harvesting complex, and is considered as one of the most abundant proteins in chloroplast of plants [40,41]. Its key function is to collect and transfer light energy to photosynthetic reaction center [42]. In our experiment, the abundance of chlorophyll A-B binding protein increased in transgenic cotton line, but the chlorophyll content and Pn decreased in the transgenic cotton. These results demonstrate that photosynthesis changed in the Bt-transgenic line. The unintended effect could be caused by random insertion of exogenous Cry1Ac and CpTI genes in plant genomes. Enolase is a glycolytic enzyme that is responsible for the ATP-generated conversion of 2-phosphoglycerate to phosphoenolpyruvate [43]. In transgenic cotton leaves, the increased enolase helped to the need of cells for extra energy to deal with insertion of exogenous genes. These data revealed that the DEPs related to carbon fixation in photosynthesis organisms and photosynthesis, glyoxylate and dicarboxylate metabolism pathway, oxidative pentose phosphate pathway and energy metabolism were up-regulated, thus resulting in the higher photosynthesis ability in transgenic cotton line, which need further evidence to confirm. In contrast, the net photosynthesis rate decreased in BT lines as shown in Figure 1C. The results suggested the inserted Cry1Ac and CpTI genes can directly or indirectly affect the plant growth and photosynthesis.

Conclusions

In conclusion, our comparative proteomic data suggested the GM operation did not sharply alter cotton leaf proteome. Less than 10% of 2-DE detectable protein spots were DEPs, which mainly involving in carbon fixation and photosynthesis, glyoxylate and dicarboxylate metabolism pathway, oxidative pentose phosphate pathway. Our data demonstrated that exogenous DNA into a host cotton genome effected the plant growth and photosynthesis.

Materials and methods
Plant materials

The transgenic Bt + CpTI cotton SGK321 (BT) and their non-transgenic parental counterparts SY321 (WT) were used as the host plants in all experiments. The SGK321 plant species was bred by introducing the synthetic Cry1Ac gene and modified CpTI (cowpea trypsin inhibitor) gene into the cotton cultivar SY321 by way of the pollen tube pathway technique [44]. Then, SGK321 were self-pollinated to obtain homozygous BT plants. Also, the cotton cultivar SGK321 has been developed into a homozygous cotton species science 1999 and were planted commercialized with a new crop species number 2001ED782014 in china since

2002 [45]. Seeds of transgenic Cry1Ac and CpTI cotton cultivar SGK321 and their non-transgenic parental counterparts SY321 were obtained from Biotechnology Research Center of Chinese Academy of Agriculture Sciences. The seeds were germinated in the plastic pots containing 1:1 (v/v) mixture of vermiculite and nutrient soil moistened with distilled water in a growth chamber maintained at a thermo period of 30/22°C of day/night temperature, under long-day conditions (16 h of light and 8 h of dark) and a relative humidity $65 \pm 5\%$. After germination, seedlings were irrigated weekly with Hoagland's nutrient solution. One month after germination, the cotton leaves were harvested for physiological and proteomic analyses.

PCR, ELISA and RT-PCR detection

Genomic DNA from transgenic cotton leaves and their non-transgenic controls were extracted using cetyl trimethyl ammonium bromide (CTAB) method as described [46]. PCR analysis was performed to confirm the presence of the exogenous gene Cry1Ac in the transgenic cotton leaves. PCR reactions were carried out in 25 μl volume containing 12.5 μl 2X Taq PCR Master Mix (Trans Gene), 0.5 μl 10 pm/μl of each primer, 2.5 μl 10 ng/μl of template DNA, and 9 μl sterilized H₂O. The cry1Ac gene-specific primers used were Cry1Ac F (5'-GTTCC AGCTA CAGCTA CCTCC-3') and Cry1Ac R (5'-CCACT AAAGT TTCTA ACACC CAC-3') with expected PCR products size 119 bp. The amplification program was performed as follows: initial denaturation at 94°C for 5 min followed by 40 cycles of 45 s at 94°C for denaturation, 45 s at 56°C for primer annealing, 60 s at 72°C for elongation, final elongation at 72°C for 10 min. PCR amplification products were separated using agarose gel electrophoresis in 1X TAE buffer.

The Bt toxin protein content in cotton leaves was measured by ELISA using the Quantiplate Kit for Cry1Ab/Cry1Ac (Envirologix, Inc., USA), which was precoated with Cry1Ac antibody containing 96 well solid microplates. The ELISA experiment was performed according to the protocols provided by manufacturers. Absorbance was measured at 450 nm using a Varioskan Flash Spectral Scan Multimode Plate Reader (Thermo Fisher Scientific, Waltham, MA). A standard curve was established using Cry1Ac standard protein at concentration ranged from 0.1 to 0.5 pg/ml.

Total RNA was isolated to generate cDNA using Reverse Transcriptase kit reagents (TaKaRa, Tokyo, Japan). RT-PCR was used to detect the CPTI gene. The CPTI gene-specific primers were CPTI F (5'-GATTTGAAC CACCTCGGAGG-3') and CPTI R (5'-CTCATCATCTT CATCCCTGG-3').

Determination of plant height, water content, photosynthetic rate and chlorophyll content

The plant height was determined immediately after harvesting. The cotton leaves were collected and immediately weighed (fresh weight (FW)). Dry weight (DW) was determined by oven drying at 60°C for 72 h. The total water content (TWC) was calculated as follows: TWC = [(FW-DW)/FW]*100. The collected cotton leaves were washed, cut in small pieces, and ground in 80% chilled acetone. The supernatant was taken for determination of photosynthetic pigments: chlorophylla (mg/g) = (12.7*A663-2.69*A645) V/W, chlorophyllb (mg/g) = (22.9*A645-4.68*A663) V/W, chlorophyll Total (mg/g) = (8.02*A663 + 20.21*A645) V/W. The net photosynthetic rate (Pn) was measured using a LI-6400 Portable Photosynthesis System (Li-Cor, Lincoln, NE, USA) with chamber setting of 400 ppm. And, photosynthetic photon flux density (PPFD) was set at 1000 umol m^{-2} s^{-1}.

Protein preparation

Total leaf protein was extracted using TCA-acetone precipitation method as described [47]. Approximately 1 g of lyophilized powders was precipitated by 10 ml acetone solution containing 10% (w/v) TCA and 0.07% (w/v) β-mercaptoethanol. The mixture was stored at −20°C for 10 h and centrifuged at 15,000 g at 4°C for 30 min to collect precipitates. The precipitates were resuspended by acetone solution containing 0.07% (w/v) β-mercaptoethanol. The mixture was stored at −20°C for 1 h and centrifuged at 15,000 g at 4°C for 30 min to collect the precipitates. The proteins were collected from precipitates by centrifugation at 15,000 g at 4°C for 30 min, washed with 100% ice-cold methanol twice and 100% ice-cold acetone twice, and then air-dried. Resulting proteins were dissolved in lysis buffer (7 M urea, 2 M thiourea, 2% CHAPS, 13 mM DTT) for 2 hours at room temperature. Protein concentration was determined by the Bradford assay using a UV-160 spectrophotometer (Shimadzu, Kyoto, Japan) and bovine serum albumin as the protein standard [48]. The proteins underwent 2-DE immediately or were stored at −80°C.

2-DE and image analyses

2-DE was performed according to the manufacturer's instruction (2-DE Manual, GE Healthcare). A total of 1,200 μg proteins mixed with lysis buffer (7 M urea, 2 M thiourea, 2% CHAPS, 13 mM DTT) were loaded onto an IPG (immobilized pH gradient) strips with linear pH gradient 4–7 and 24 cm length (GE Healthcare, Uppsala, Sweden). The strips were hydrated for 18 h at room temperature. Then isoelectric focusing was performed on an Ettan IPGphor isoelectric focusing system (GE Healthcare, Uppsala, Sweden) under the following conditions: 250 V for 3 h, 500 V for 2 h, 1000 V for 1 h, a gradient to 8000 V for 4 h, and 8000 V up to 140000 Vhr. Subsequently, these strips were equilibrated with equilibration solution (50 mM Tris–HCl, pH 8.8, 6 M urea, 30% glycerol, 2% SDS, 0.002% bromophenol blue) containing 1% DTT for 15 min, followed with equilibration for

another 15 min in alkylation buffer containing 50 mM Tris–HCl, pH 8.8, 6 M urea, 30% glycerol, 2% SDS, 0.002% bromophenol blue, and 4% iodoacetamide. Then, IPG strips were transferred to SDS-PAGE gels for separating proteins with an Ettan Dalt system (GE Healthcare). Program was set up as follows: 4 W/gel for 1 h and then 8 W/gel for 6 h [49]. After electrophoresis, the gels were visualized by GAP staining methods [50]. Image analysis was performed using Image Master 2D Platinum Software (Version 5.0, GE Healthcare). The apparent molecular weight (Mr) of each visible protein was determined through comparison with protein markers with known Mr values. Biological variation analysis module was employed to identify spots differentially expressed (more than 1.5 fold) in different salt treated samples with statistically significant differences (confidence above 95%, p < 0.05). Three biological repeats for each sample were examined, and the results were shown in average ± SD (n = 3). Then, spots of interests were manually excised from the GAP stained 2-DE gels.

In-Gel trypsin digestion

The collected protein spots were washed with MilliQ water three times, for 30 min each until removing impurities on the surface of gels. Then, protein spots were destained three times with destaining solution containing 50 mM NH$_4$HCO$_3$ and 50% ACN for 30 min each at 37°C, and then incubated in 100 μL of 100% ACN until gel pieces became white and shrunken. They were air dried at room temperature for 1 h. Proteins were digested in-gel with bovine trypsin (Roche, Cat. 11418025001) as described [51]. After digestion, the remaining trypsin buffer were discarded, and then centrifuged at 10,000 g for 30 min to collect peptides extracts. 1 μL of peptides extracts was mixed with 1 μL of α-cyano-4-hydroxycinnamic acid (CHCA) and spotted on the target plate.

Protein Identification *via* MALDI TOF/TOF MS

Proteins were identified by using AB SCIEX MALDI TOF-TOF 5800 system (AB SCIEX, Foster City, CA, USA) equipped with a neodymium with laser wavelength 349 nm as described [51,52]. The laser can shot at a rate of up to 1000 Hz. CHCA was used as the matrix with TFA for an ionization auxiliary reagent. The spectrum was calibrated using the TOF/TOF calibration mixtures (AB SCIEX). All peptide mass fingerprint spectra were internally calibrated with trypsin autolysis peaks, and all known contaminants were excluded during this process. Peptide mass was used to database search.

Database searching

The raw MS and MS/MS data were combined to search against the taxonomy of Viridiplantae (Green Plants, including 1,196,615 sequences) in NCBI (NCBInr) database with 23,290,086 sequences using an in-house MASCOT server.

The searched parameters were set as followings: one missed cleavage, P < 0.05 significance threshold, 100 ppm mass tolerance for precursor ions, MS/MS ion tolerance of 0.1 Da, carbamidomethylation of cysteine as fixed modification, and oxidation of methionine as variable modification. When individual ions scores were higher than threshold score (scores higher than 45), proteins were considered as confident identifications or extensive homology (p < 0.05). For protein scores confidence intervals above 95%, In-house BLAST search against NCBI (http://www.ncbi.nlm) was performed to confirm the protein identifications. The identified proteins were categorized to specific processes or functions by searching Gene Ontology (http://www.geneontology.org) [52].

Bioinformatic analysis

The cluster of orthologous groups of proteins (COG) analysis was carried out for the identified proteins. Following subcellular localization was predicted using CELLO V.2.5 (http://cello.life.nctu.edu.tw), which made predictions based on a two-level support vector machine system [53,54]. The sequences of the identified proteins were searched against the UniProt database in order to extract corresponding GO information [55]. Then, GO classification of these proteins was conducted with WEGO web service (http:// wego. genomics. org.cn), by which GO terms assigned to query sequences and catalogued groups were produced based on biological process, molecular functions, and cellular components [56-59]. In addition, identified proteins were further analyzed using the STRING V.9.1 database for protein-protein interactions, to statistically determine the functions and pathways most strongly associated with the protein list [60]. Finally, KEGG (http://www.genome.jp/kegg/pathway) pathway analysis was performed to determine their molecular interaction and reaction networks.

Western blotting analysis and quantitative Real-time PCR

Western blotting was performed as described [61]. About 10 ug proteins were subjected to SDS-PAGE, transferred to a membrane. The 5% nonfat milk was used for blocking protein. The blocked membranes were incubated with specific antibodies against Rubisco at the dilution of 1:8000 at 37°C for 1.5 h. Antibody-bound proteins were detected using appropriate HRP-conjugated secondary antibodies (Sigma, USA) and clarity western ECL substrate (Bio-Rad, CA, USA). The target proteins were then visualized and quantitated using a LAS- 4000 luminescent image analyzer.

Total RNA was isolated to generate cDNA using Reverse Transcriptase kit reagents (TaKaRa, Tokyo, Japan). The primer pairs used for quantitative Real-time PCR (qRT-PCR) are provided in additional file 6. qRT-PCR was performed in a 20ul volume containing 10 ul 2*GoTaq q PCR Master Mix, 2 ul of cDNA, 0.4 ul of each gene-specific primer, 7 ul of Nuclease-Free Water, and 0.2 ul of 100* CXR (Promega, Madison, WI). Reaction was conducted on an Mx3500P

Real-Time PCR Detection System according to the manufacturer's instructions. All data were analyzed using MxPro software.

Additional files

Additional file 1: PCR, ELISA and RT-PCR analysis of Cry1Ac in different cotton leaves.

Additional file 2: MS identification and bioinformatic analysis of the differentially expressed proteins.

Additional file 3: Supplemental spectra and MALDI TOF/TOF MS identification of the differentially expressed proteins.

Additional file 4: Blast2go analysis results of the identified proteins.

Additional file 5: The main pathways involved in transgenic cotton.

Additional file 6: Primers used in qRT-PCR.

Abbreviations
ELISA: Enzyme linked immunosorbent assay; DEPs: Differentially expressed proteins; GM: Genetically modified; FTIR: Fourier transform infrared spectroscopy; GMO: Genetically modified organism; CTAB: Cetyl trimethyl ammonium bromide.

Competing interests
The authors declare that they have no competing interests.

Authors' contributions
LMW conceived and designed the study, carried out the experiments, analyzed data, performed bioinformatics analyses and drafted the manuscript; XCW participated in the study design and helped to draft the manuscript; XJ helped to draft and revise the manuscript; RZJ, QXH and YHT helped to revised the manuscript; APG participated in the study design and coordination and helped to revise the manuscript. All authors have read and approved the final manuscript.

Acknowledgements
This research was supported by the Special Fund for Agro-scientific Research in the Public Interest of the People's Republic of China (Grant No. 201403075), the Major Technology Project of Hainan (NO. ZDZX2013010-3), and the Program for the Top Young Talents in Chinese Academy of Tropical Agricultural Sciences (No. ITBB130102). The authors thank Dr. Zheng Tong in Institute of Tropical Biosciences and Biotechnology for the helpful discussion and suggestions.

Author details
[1]Chinese Academy of Tropical Agricultural Sciences, The Institute of Tropical Biosciences and Biotechnology, Haikou, Hainan 571101, China. [2]Chinese Academy of Agricultural Sciences, The Oilcrops Research Institute, Wuhan 430062, China.

References
1. James C. Global status of commercialized biotech/GM crops. ISAAA Brief NO. 46. Ithaca, NY: ISAAA; 2013.
2. Conner AJ, Jacobs JME. Food risks from transgenic crops in perspective. Nutrition. 2000;16:709–11.
3. Ren Y, Lv J, Wang H, Li L, Peng Y, Qu LJ. A comparative proteomics approach to detect unintended effects in transgenic Arabidopsis. J Genet Genomics. 2009;36:629–39.
4. Conner AJ, Jacobs JME. Genetic engineering of crops as potential source of genetic hazard in the human diet. Mutat Res-Gen Tox En. 1999;443:223–34.
5. Gong C, Wang T. Proteomic evaluation of genetically modified crops: current status and challenges. Front Plant Sci. 2013;4:1–8.
6. Cellini F, Chesson A, Colquhoun I, Constable A, Davies HV, Engel KH, et al. Unintended effects and their detection in genetically modified crops. Food Chem Toxicol. 2004;42:1089–125.

7.   Ruebelt MC, Leimgruber NK, Lipp M, Reynolds TL, Nemeth MA, Astwood JD. Application of two-dimensional gel electrophoresis to interrogate alterations in the proteome of genetically modified crops. 1. Assessing analytical validation. J Agric Food Chem. 2006;54:2154–61.

8.   Albo AG, Mila S, Digilio G, Motto M, Aime S, Corpillo D. Proteomic analysis of a genetically modified maize flour carrying CRY1AB gene and comparison to the corresponding wild-type. Maydica. 2007;52:443–55.

9.   Ricroch AE, Bergé JB, Kuntz M. Evaluation of genetically engineered crops using transcriptomic, proteomic, and metabolomic profiling techniques. Plant Physio. 2011;155:1752–61.

10. Gong C, Li Q, Yu H, Wang Z, Wang T. Proteomics insight into the biological safety of transgenic modification of rice as compared with conventional genetic breeding and spontaneous genotypic variation. J Proteome Res. 2012;11:3019–29.

11. Kuiper HA, Kok EJ, Engel KH. Exploitation of molecular profiling techniques for GM food safety assessment. Curr Opin Biotech. 2003;14:238–43.

12. Baudo MM, Lyons R, Powers S, Pastori GM, Edwards KJ, Holdsworth MJ, et al. Transgenesis has less impact on the transcriptome of wheat grain than conventional breeding. Plant Biotechnol J. 2006;4:369–80.

13. Barros E, Lezar S, Anttonen MJ, Van Dijk JP, Röhlig RM, Kok EJ, et al. Comparison of two GM maize varieties with a near-isogenic non-GM variety using transcriptomics, proteomics and metabolomics. Plant Biotechnol J. 2010;8:436–51.

14. Kok EJ, Kuiper HA. Comparative safety assessment for biotech crops. Trends Biotechnol. 2003;21:439–44.

15. Jelenie S. Food safety evaluation of crops produced through genetic engineering-how to reduce unintended effects. Arh Hig Rada Toksikol. 2005;56:185–93.

16. Brandao AR, Barbosa HS, Arruda MAZ. Image analysis of two-dimensional gel electrophoresis for comparative proteomics of transgenic and non-transgenic soybean seeds. J Proteomics. 2010;73:1433–40.

17. Salekdeh GH, Komatsu S. Crop proteomics: aim at sustainable agriculture of tomorrow. Proteomics. 2007;7:2976–96.

18. Batista R, Martins I, Jenö P, Ricardo CP, Oliveira MM. A proteomic study to identify soya allergens-the human response to transgenic versus non-transgenic soya samples. Int Arch Allergy Imm. 2007;144:29–38.

19. Barbosa HS, Arruda SCC, Azevedo RA, Arruda MA. New insights on proteomics of transgenic soybean seeds: evaluation of differential expressions of enzymes and proteins. Anal Bioanal Chem. 2012;402:99–314.

20. Luo J, Ning T, Sun Y, Zhu J, Zhu Y, Lin Q, et al. Proteomic analysis of rice endosperm cells in response to expression of hGM-CSF. J Proteome Res. 2008;8:829–37.

21. Balsamo GM, Cangahuala-Inocente GC, Bertoldo JB, Terenzi H, Arisi AC. Proteomic analysis of four Brazilian MON810 maize varieties and their four non-genetically-modified isogenic varieties. J Agr Food Chem. 2011;59:11553–9.

22. Coll A, Nadal A, Rossignol M, Puigdomènech P, Pla M. Proteomic analysis of MON810 and comparable non-GM maize varieties grown in agricultural fields. Transgenic Res. 2011;20:939–49.

23. Zolla L, Rinalducci S, Antonioli P, Righetti PG. Proteomics as a complementary tool for identifying unintended side effects occurring in transgenic maize seeds as a result of genetic modifications. J Proteome Res. 2008;7:1850–61.

24. Careri M, Elviri L, Mangia A, Zagnoni I, Agrimonti C, Visioli G, et al. Analysis of protein profiles of genetically modified potato tubers by matrix-assisted laser desorption/ionization time-of-flight mass spectrometry. Rapid Commun Mass Sp. 2003;17:479–83.

25. Goulet C, Benchabane M, Anguenot R, Brunelle F, Khalf M, Michaud D. A companion protease inhibitor for the protection of cytosol-targeted recombinant proteins in plants. Plant Biotechnol J. 2010;8:142–54.

26. Corpillo D, Gardini G, Vaira AM, Basso M, Aime S, Accotto GP, et al. Proteomics as a tool to improve investigation of substantial equivalence in genetically modified organisms: the case of a virus-resistant tomato. Proteomics. 2004;4:193–200.

27. Dicarli M, Villani ME, Bianco L, Lombardi R, Perrotta G, Benvenuto E, et al. Proteomic analysis of the plant- virus interaction in cucumber mosaic virus (CMV) resistant transgenic tomato. J Proteome Res. 2010;9:5684–97.

28. Horvath-Szanics E, Szabo Z, Janaky T, Pauk J, Hajós G. Proteomics as an emergent tool for identification of stress-induced proteins in control and genetically modified wheat lines. Chromatographia. 2006;63:S143–7.

29. Scossa F, Laudencia-Chingcuanco D, Anderson OD, Vensel WH, Lafiandra D, D'Ovidio R, et al. Comparative proteomic and transcriptional profiling of a

bread wheat cultivar and its derived transgenic line overexpressing a low molecular weight glutenin subunit gene in the endosperm. Proteomics. 2008;8:2948–66.

30. Gao W, Long L, Zhu LF, Xu L, Gao WH, Sun LQ, et al. Proteomic and virus-induced gene silencing (VIGS) analyses reveal that gossypol, brassinosteroids, and jasmonic acid contribute to the resistance of cotton to verticillium dahliae. Mol Cell Proteomics. 2013;12:3690–703.

31. An J, Gao Y, Lei C, Gould F, Wu K. Monitoring cotton bollworm resistance to Cry1Ac in two counties of northern China during 2009–2013. Pest Manag Sci 2014, Publish online. doi:10.1002/ps.3807.

32. Modirroosta BH, Tohidfar M, Saba J, Moradi F. The substantive equivalence of transgenic (Bt and Chi) and non-transgenic cotton based on metabolite profiles. Funct Integr Genomic. 2014;14:237–44.

33. Sun C, Wu X, Wang L, Wang Y, Zhang Y, Chen L, et al. Comparison of chemical composition of different transgenic insect-resistant cotton seeds using Fourier transform infrared spectroscopy (FTIR). Afr J Agr Res. 2012;7:2918–25.

34. Baerenfaller K, Massonnet C, Walsh S, Baginsky S, Bühlmann P, Hennig L, et al. Systems-based analysis of Arabidopsis leaf growth reveals adaptation to water deficit. Mol Syst Biol. 2012;8:1–18.

35. Kuiper HA, Kleter GA, Noteborn HP, Kok EJ. Assessment of the food safety issues related to genetically modified foods. Plant J. 2001;27:503–28.

36. Ducat DC, Silver PA. Improving carbon fixation pathways. Curr Opin Chem Biol. 2012;16:337–44.

37. Gatenby AA, van der Vies SM, Bradley D. Assembly in E coli of a functional multi-subunit ribulose bisphosphate carboxylase from a blue-green alga. Nature. 1985;314:617–20.

38. Andersson I, Knight S, Schneider G, Lindqvist Y, Lundqvist T, Brändén CI, et al. Crystal structure of the active site of ribulose-bisphosphate carboxylase. Nature. 1989;337:229–34.

39. Henkes S, Sonnewald U, Badur R, Stitt M. A small decrease of plastid transketolase activity in antisense tobacco transformants has dramatic effects on photosynthesis and phenylpropanoid metabolism. Plant Cell. 2001;13:535–51.

40. Xu YH, Liu R, Yan L, Liu ZQ, Jiang SC, Shen YY, et al. Light-harvesting chlorophyll a/b-binding proteins are required for stomatal response to abscisic acid in Arabidopsis. J Exp Bot. 2012;63:1095–106.

41. Xia Y, Ning Z, Bai G, Li R, Yan G, Siddique KH, et al. Allelic variations of a light harvesting chlorophyll a/b-binding protein gene (Lhcb1) associated with agronomic traits in barley. PLoS One. 2012;7:1–9.

42. Paulsen H, Dockter C, Volkov A, Jeschke G. Folding and pigment binding of light-harvesting chlorophyll a/b protein (LHCIIb). The Chloroplast: Springer Netherlands. 2010;31:231–44.

43. Song Y, Luo Q, Long H, Hu Z, Que T, Zhang XA, et al. Alpha-enolase as a potential cancer prognostic marker promotes cell growth, migration, and invasion in glioma. Mol Cancer. 2014;13:1–12.

44. Guo SD, Cui HZ, Xia LQ, Wu DL, Ni WC, Zhang ZL, et al. The research of Transgenic bivalent insect-resistant cotton. Sci Agricult Sinica. 1999;32:1–7.

45. Sui SX, Zhao GZ, Li AG, Zhu QZ, Zhao LF, Li ZS. Analysis on the features of Transgenic (Bt + CpTI) insect-resistant cotton variety SGK321 and their application in germplasm. Sci Technol Rev. 2008;26:42–5.

46. Huang QX, Wang XC, Kong H, Guo YL, Guo AP. An efficient DNA isolation method for tropical plants. Afr J Biotechnol. 2013;12:2727–32.

47. Han JC, Cui H, Shi P, Zhu SJ, Ye ZH, Yu XP. Development of two-dimensional electrophoresis protocol suitable for proteomic analysis of cotton leaves. Cotton Science. 2012;24:27–34.

48. Wang XC, Chang LL, Wang BC, Wang D, Li P, Wang L, et al. Comparative proteomics of Thellungiella halophila leaves from plants subjected to salinity reveals the importance of chloroplastic starch and soluble sugars in halophyte salt tolerance. Mol Cell Proteomics. 2013;12:2174–95.

49. Wang XC, Fan PX, Song HM, Chen XY, Li XF, Li YX. Comparative proteomic analysis of differentially expressed proteins in shoots of Salicornia europaea under different salinity. J Proteome Res. 2009;8:3331–45.

50. Wang XC, Wang D, Wang DY, Wang HY, Chang LL, Yi XP, et al. Systematic comparison of technical details in CBB methods and development of a sensitive GAP stain for comparative proteomic analysis. Electrophoresis. 2012;33:296–306.

51. Wang DZ, Li C, Xie ZX, Dong HP, Lin L, Hong HS. Homology-driven proteomics of dinoflagellates with unsequenced genomes using MALDI-TOF/TOF and automated de novo sequencing. Evid Based Complement Alternat Med. 2011;2011:1–16.

52. Yi XP, Sun Y, Yang Q, Guo AP, Chang LL, Wang D, et al. Quantitative proteomics of *Sesuvium portulacastrum* leaves revealed that ion transportation by V-ATPase and sugar accumulation in chloroplast played crucial roles in halophyte salt tolerance. J Proteomics. 2014;99:84–100.

53. Ye J, Fang L, Zheng H, Zhang Y, Chen J, Zhang Z, et al. WEGO: a web tool for plotting GO annotations. Nucleic Acids Res. 2006;34:W293–7.

54. Ang CS, Binos S, Knight MI, Moate PJ, Cocks BG, McDonagh MB. Global survey of the bovine salivary proteome: integrating multidimensional prefractionation, targeted, and glycocapture strategies. J Proteome Res. 2011;10:5059–69.

55. Neilson KA, Mariani M, Haynes PA. Quantitative proteomic analysis of cold-responsive proteins in rice. Proteomics. 2011;11:1696–706.

56. Yu M, Ren C, Qiu J, Luo P, Zhu R, Zhao Z, et al. Draft genome sequence of the opportunistic marine pathogen Vibrio harveyi strain E385. Genome Announcements. 2013;1:1–2.

57. Gotz S, García-Gómez JM, Terol J, Williams TD, Nagaraj SH, Nueda MJ, et al. High-throughput functional annotation and data mining with the Blast2GO suite. Nucleic Acids Res. 2008;36:3420–35.

58. Conesa A, Gotz S, García-Gómez JM, Terol J, Talón M, Robles M. Blast2GO: a universal tool for annotation, visualization and analysis in functional genomics research. Bioinformatics. 2005;21:3674–6.

59. Zhou Z, Yang H, Chen M, Lou CF, Zhang YZ, Chen KP. Comparative proteomic analysis between the domesticated silkworm (*Bombyx mori*) reared on fresh mulberry leaves and on artificial diet. J Proteome Res. 2008;7:5103–11.

60. Zhang A, Sun H, Wu G, Sun W, Yuan Y, Wang X. Proteomics analysis of hepatoprotective effects for scoparone using MALDI-TOF/TOF mass spectrometry with bioinformatics. Omics. 2013;17:224–9.

61. Guo BJ, Chen YH, Li C, Wang TY, Wang R, Wang B, et al. Maize (Zea mays L.) seeding leaf nuclear proteome and differentially expressed proteins between a hybrid and its parental lines. Proteomics. 2014;14:1071–87.

PERMISSIONS

All chapters in this book were first published in PS, by BioMed Central; hereby published with permission under the Creative Commons Attribution License or equivalent. Every chapter published in this book has been scrutinized by our experts. Their significance has been extensively debated. The topics covered herein carry significant findings which will fuel the growth of the discipline. They may even be implemented as practical applications or may be referred to as a beginning point for another development.

The contributors of this book come from diverse backgrounds, making this book a truly international effort. This book will bring forth new frontiers with its revolutionizing research information and detailed analysis of the nascent developments around the world.

We would like to thank all the contributing authors for lending their expertise to make the book truly unique. They have played a crucial role in the development of this book. Without their invaluable contributions this book wouldn't have been possible. They have made vital efforts to compile up to date information on the varied aspects of this subject to make this book a valuable addition to the collection of many professionals and students.

This book was conceptualized with the vision of imparting up-to-date information and advanced data in this field. To ensure the same, a matchless editorial board was set up. Every individual on the board went through rigorous rounds of assessment to prove their worth. After which they invested a large part of their time researching and compiling the most relevant data for our readers.

The editorial board has been involved in producing this book since its inception. They have spent rigorous hours researching and exploring the diverse topics which have resulted in the successful publishing of this book. They have passed on their knowledge of decades through this book. To expedite this challenging task, the publisher supported the team at every step. A small team of assistant editors was also appointed to further simplify the editing procedure and attain best results for the readers.

Apart from the editorial board, the designing team has also invested a significant amount of their time in understanding the subject and creating the most relevant covers. They scrutinized every image to scout for the most suitable representation of the subject and create an appropriate cover for the book.

The publishing team has been an ardent support to the editorial, designing and production team. Their endless efforts to recruit the best for this project, has resulted in the accomplishment of this book. They are a veteran in the field of academics and their pool of knowledge is as vast as their experience in printing. Their expertise and guidance has proved useful at every step. Their uncompromising quality standards have made this book an exceptional effort. Their encouragement from time to time has been an inspiration for everyone.

The publisher and the editorial board hope that this book will prove to be a valuable piece of knowledge for researchers, students, practitioners and scholars across the globe.

LIST OF CONTRIBUTORS

Jize Zhang, Cong Li, Xiangfang Tang, Qingping Lu, Renna Sa and Hongfu Zhang
State Key Laboratory of Animal Nutrition, Institute of Animal Sciences, Chinese Academy of Agricultural Sciences, Beijing 100193, People's Republic of China

Emanuela Monari, Aurora Cuoghi, Elisa Bellei, Stefania Bergamini and Aldo Tomasi
Department of Diagnostic, Clinical and Public Health Medicine, University of Modena and Reggio Emilia, Largo del Pozzo, 71-41124 Modena, Italy

Andrea Lucchi
Private Practice, Modena, Italy

Pierpaolo Cortellini
European Research Group on Periodontology (ERGOPERIO), Berne, Switzerland

Davide Zaffe
Department of Biomedical, Metabolic and Neural Sciences, University of Modena and Reggio Emilia, Modena, Italy

Carlo Bertoldi
Department of Surgery, Medicine, Dentistry and Morphological Sciences with Transplant Surgery, Oncology and Regenerative Medicine Relevance, University of Modena and Reggio Emilia, Modena, Italy

Yang Cheng and Jianjie Chen
Department of liver disease, Hospital for Infectious Diseases of Pudong New Area, Shanghai 201299, P. R. China
Shuguang Hospital affiliated to Shanghai University of Traditional Chinese Medicine, Shanghai 201203, P. R. China

Tianlu Hou, Jian Ping and Gaofeng Chen
Shuguang Hospital affiliated to Shanghai University of Traditional Chinese Medicine, Shanghai 201203, P. R. China

Chiu-Ping Kuo
Division of Chest Medicine, Department of Internal Medicine, Mackay Memorial Hospital, 92, Sec 2, Chungshan North Road, Taipei, Taiwan

Chien-Liang Wu
Division of Chest Medicine, Department of Internal Medicine, Mackay Memorial Hospital, 92, Sec 2, Chungshan North Road, Taipei, Taiwan
Mackay Junior College of Medicine, Nursing, and Management, Taipei, Taiwan

Yen-Ta Lu
Division of Chest Medicine, Department of Internal Medicine, Mackay Memorial Hospital, 92, Sec 2, Chungshan North Road, Taipei, Taiwan
Department of Medicine, Mackay Medical College, New Taipei City, Taiwan

Kuo-Song Chang
Department of Emergency Medicine, Mackay Memorial Hospital, Taipei, Taiwan
Mackay Junior College of Medicine, Nursing, and Management, Taipei, Taiwan

Tsai-Yin Wei and I-Fang Tsai
Department of Medical Research, Mackay Memorial Hospital, Taipei, Taiwan

Andrew Boyd Lin
Biology Department, Case Western Reserve University, Cleveland,OH, USA.

Jue-Liang Hsu
Graduate Institute of Biotechnology, National Pingtung University of Science and Technology, Pingtung 91201, Taiwan

Qi Xiong, Lihai Zhang and Peifu Tang
Department of Orthopedics, General Hospital of Chinese PLA, Fuxing Road 28#, Haidian District, Beijing 100853, China

Shaohua Zhan and Wei Ge
National Key Laboratory of Medical Molecular Biology & Department of Immunology, Institute of Basic Medical Sciences, Chinese Academy of Medical Sciences, DongdanSantiao 5#, Dongcheng District, Beijing 100005, China

Soundharrajan Ilavenil, Srisesharam Srigopalram and Ki Choon Choi
Grassland and Forage Division, National Institute of Animal Science, RDA, Seonghwan-Eup, Cheonan-Si, Chungnam 330801, Korea

Naif Abdullah Al-Dhabi and Mariadhas Valan Arasu
Department of Botany and Microbiology, Addiriyah Chair for Environmental Studies, College of Science, King Saud University, Riyadh 11451, Saudi Arabia

Chun Geon Park, Kyung Hun Park and Young Ock Kim
Department of Medicinal Crop Research, Rural Development Administration, Eumseong, Chungbuk 369-873, Republic of Korea

Paul Agastian
Research Department of Plant Biology and Biotechnology, Loyola College, Nungambakkam, Chennai-34, Tamil Nadu, India

Rajasekhar Baaru
Labmate (Asia) Pvt. Ltd, Chennai, Tamil Nadu 600015, India

Idanya Serafín-Higuera, Eugenia Flores-Alfaro and Pavel Sierra-Martínez
Laboratorio de Citopatología e Histoquímica, Unidad Académica de Ciencias Químico Biológicas, Universidad Autónoma de Guerrero, Chilpancingo, Guerrero, México

Luz del Carmen Alarcón-Romero
Laboratorio de Citopatología e Histoquímica, Unidad Académica de Ciencias Químico Biológicas, Universidad Autónoma de Guerrero, Chilpancingo, Guerrero, México
Laboratorio de Investigación en Citopatología e Histoquímica, Unidad Académica de Ciencias Químico
Biológicas Universidad Autónoma de Guerrero Avenida Lázaro Cárdenas, Ciudad Universitaria, Chilpancingo, Guerrero C.P. 39090, México

Olga Lilia Garibay-Cerdenares and Berenice Illades-Aguiar
Laboratorio de Biomedicina Molecular, Unidad Académica de Ciencias Químico Biológicas, Universidad Autónoma de Guerrero, Chilpancingo, Guerrero, México

Marco Antonio Jiménez-López
Instituto Estatal de Cancerología "Dr. Arturo Beltrán Ortega", Acapulco, Guerrero, México

Dijun Zhang, Weina He, Qianqian Tong, Jun Zhou and Xiurong Su
School of Marine Science, Ningbo University, 818 Fenghua Road, Ningbo, Zhejiang Province 315211, People's Republic of China

Fang Chen, Carl Spiessens, Thomas D'Hooghe and Karen Peeraer
Leuven University Fertility Centre, UZ Leuven Campus Gasthuisberg, Herestraat 49, Leuven, Belgium

Sebastien Carpentier
Facility for Systems Biology based Mass Spectrometry (SYBIOMA), KU Leuven, Leuven, Belgium

Elisabet Wieslander
Centre of Excellence in Biological and Medical Mass Spectrometry, Biomedical Centre D13, Lund University, 221 84 Lund, Sweden

Johan Malm
Centre of Excellence in Biological and Medical Mass Spectrometry, Biomedical Centre D13, Lund University, 221 84 Lund, Sweden
Section for Clinical Chemistry, Department of Translational Medicine, Lund University, Skåne University Hospital Malmö, 205 02 Malmö, Sweden

Magnus Dahlbäck, Thomas E. Fehniger, Roger Appelqvist and Jonatan Eriksson
Centre of Excellence in Biological and Medical Mass Spectrometry, Biomedical Centre D13, Lund University, 221 84 Lund, Sweden
Clinical Protein Science & Imaging, Biomedical Centre, Department of Biomedical Engineering, Lund University, BMC D13, 221 84 Lund, Sweden

György Marko-Varga
Centre of Excellence in Biological and Medical Mass Spectrometry, Biomedical Centre D13, Lund University, 221 84 Lund, Sweden
Clinical Protein Science & Imaging, Biomedical Centre, Department of Biomedical Engineering, Lund University, BMC D13, 221 84 Lund, Sweden
First Department of Surgery, Tokyo Medical University, 6-7-1 Nishishinjiku Shinjiku-ku, Tokyo 160-0023, Japan

Simone Andersson
Encap Security, Øvre Slottsgate 7, 0157 Oslo, Norway

Bo Andersson
Clinical Protein Science & Imaging, Biomedical Centre, Department of Biomedical Engineering, Lund University, BMC D13, 221 84 Lund, Sweden

Mikael Truedsson and May Bugge
Örestadskliniken, 217 67, Eddagatan 4, 217 67 Malmö, Sweden

Lijuan Zhou, Lina Sun and Jiewan Wang
Department of Genetics, College of Life Science, Northeast Forestry University, Harbin 150040, People's Republic of China

Xingshun Song
Department of Genetics, College of Life Science, Northeast Forestry University, Harbin 150040, People's Republic of China
State Key Laboratory of Tree Genetic sand Breeding, Northeast Forestry University, Harbin 150040, People's Republic of China

Zepeng Yin
Department of Genetics, College of Life Science, Northeast Forestry University, Harbin 150040, People's Republic of China
State Key Laboratory of Tree Genetic sand Breeding, Northeast Forestry University, Harbin 150040, People's Republic of China
Horticulture Department, College of Horticulture, Shenyang Agricultural University, No. 120 Dongling Road, Shenhe District, Shenyang 110866, People's Republic of China

Jing Ren
College of Food Science; Key Laboratory of Dairy Science, Ministry of Education, Synergetic Innovation
Center of Food Safety and Nutrition, Northeast Agricultural University, Harbin, Heilongjiang 150030, People's Republic of China

Yulong Liu
Forest Engineering and Environment Research Institute of Heilongjiang Province, No. 134 Haping Road, Nangang District, Harbin, Heilongjiang 150081, People's Republic of China

Saul Chemonges, Paul C. Mills and Steven R. Kopp
School of Veterinary Science, The University of Queensland, Gatton, Australia

Pawel Sadowski and Rajesh Gupta
Proteomics and Small Molecule Mass Spectrometry, Central Analytical Research Facility, Queensland University of Technology, Brisbane, Australia

Junbo Xiong, Hong Tian, Heshan Zhang, Yang Liu and Mingxin Chen
Hubei Key Laboratory of Animal Embryo and Molecular Breeding, Institute of Animal and Veterinary Science, Hubei Academy of Agricultural Science, Yaoyuan 1, Hongshan, Wuhan, Hubei 430017, China

Yan Sun
Institute of Grassl and Science, China Agricultural University, 2 West Road, Yuan Ming Yuan, Beijing 100193, China

Qingchuan Yang
Institute of Animal Science, Chinese Academy of Agricultural Science, West Road 2, Yuan Ming Yuan, Beijing 100193, China

Yeonyee Oh and Ralph A. Dean
Center for Integrated Fungal Research, Department of Entomology and Plant Pathology, North Carolina State University, Raleigh, NC 27695, USA

Suzanne L. Robertson, Jennifer Parker and David C. Muddiman
W.M. Keck FT-ICR Mass Spectrometry Laboratory, Department of Chemistry, North Carolina State University, Raleigh, NC 27695, USA

Adriana Pereira de Souza, Amanda Paula Pedroso, Regina Lúcia Harumi Watanabe, Ana Paula Segantine Dornellas, Valter Tadeu Boldarine, Claudia Maria Oller do Nascimento, Lila Missae Oyama and Eliane Beraldi Ribeiro
Departamento de Fisiologia, Universidade Federal de São Paulo, Rua Botucatu, 862 - 2 andar, Vila Clementino, São Paulo, SP 04023-062, Brazil

José Cesar Rosa and Helen Julie Laure
Centro de Química de Proteínas – Hemocentro, Universidade de São Paulo, Ribeirão Preto, Brazil.

Yukinobu Isowa and Kazuyoshi Endo
Department of Earth and Planetary Science, Graduate School of Science, The University of Tokyo, 7-3-1 Hongo, Bunkyo-ku, Tokyo 113-0033, Japan.

Isao Sarashina
The University Museum, The University of Tokyo, 7-3-1 Hongo, Bunkyo-ku, Tokyo 113-0033, Japan

Masahira Hattori and Kenshiro Oshima
Center for Omics and Bioinformatics, Department of Computational Biology, Graduate School of Frontier Sciences, The University of Tokyo, 5-1-5 Kashiwanoha, Kashiwa, Chiba 277-8561, Japan

Keiji Kito
Department of Life Science, School of Agriculture, Meiji University, 1-1-1 Higashimita, Tama, Kawasaki, Kanagawa 214-8571, Japan

Xuchu Wang, Xiang Jin, Ruizong Jia, Qixing Huang, Yanhua Tan and Anping Guo
Chinese Academy of Tropical Agricultural Sciences, The Institute of Tropical Biosciences and Biotechnology, Haikou, Hainan 571101, China

Limin Wang
Chinese Academy of Tropical Agricultural Sciences, The Institute of Tropical Biosciences and Biotechnology, Haikou, Hainan 571101, China.
Chinese Academy of Agricultural Sciences, The Oilcrops Research Institute, Wuhan 430062, China

Index

www.ingramcontent.com/pod-product-compliance
Lightning Source LLC
Chambersburg PA
CBHW082036190326
41458CB00010B/3382